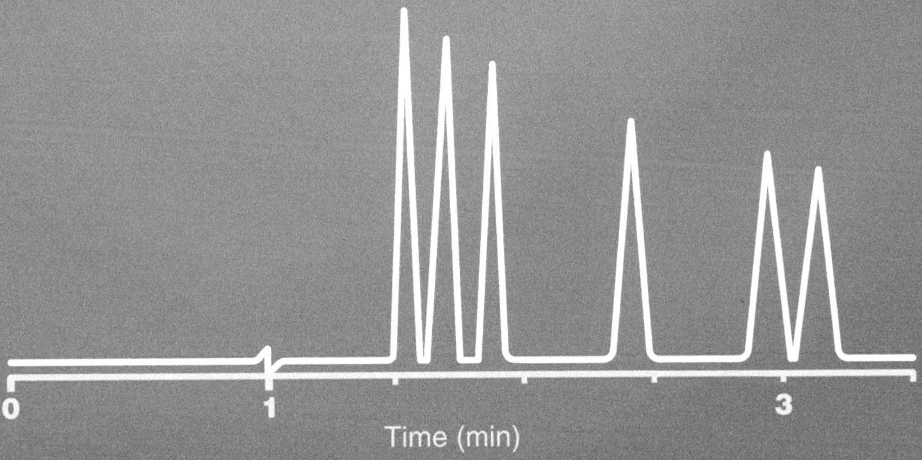

Chromatographic Method Development

Jenny Stanford Series on Pharmaceutical Analytics

Series Editors

Gregory K. Webster, J. Derek Jackson, and Robert G. Bell

Titles in the Series

Vol. 1
Supercritical Fluid Chromatography: Advances and Applications in Pharmaceutical Analysis
Gregory K. Webster, ed.
2014
978-981-4463-00-3 (Hardcover)
978-981-4463-01-0 (eBook)

Vol. 2
Poorly Soluble Drugs: Dissolution and Drug Release
Gregory K. Webster, J. Derek Jackson, and Robert G. Bell, eds.
2017
978-981-4745-45-1 (Hardcover)
978-981-4745-46-8 (eBook)

Vol. 3
Chromatographic Method Development
Gregory K. Webster and Laila Kott, eds.
2020
978-981-4800-53-2 (Hardcover)
978-0-429-20172-1 (eBook)

Jenny Stanford Series on Pharmaceutical Analytics — Volume 3

Chromatographic Method Development

edited by

Gregory K. Webster
Laila Kott

JENNY STANFORD
PUBLISHING

Published by

Jenny Stanford Publishing Pte. Ltd.
Level 34, Centennial Tower
3 Temasek Avenue
Singapore 039190

Email: editorial@jennystanford.com
Web: www.jennystanford.com

British Library Cataloguing-in-Publication Data
A catalogue record for this book is available from the British Library.

Chromatographic Method Development

ISBN 978-981-4800-53-2 (Hardcover)
ISBN 978-0-429-20172-1 (eBook)

Contents

Preface

From Greg Webster

Years ago, when starting my industrial analytical chemist career, the company I started working for directly out of graduate school sent me to the HPLC Methods Development course by Lloyd Snyder and Joseph Kirkland. Having spent most of my graduate years on spectroscopy-related research, I was excited to "fill in the gaps" from the several chromatography courses I had taken. I knew my practical experience was lacking.

Snyder and Kirkland's love of chromatography was evident in their teaching. It was contagious and set the foundation for an analytical career that has spanned several decades and many chromatography applications. After moving back to Chicago, I was asked to teach a graduate chromatographic methods development class at Governors State University. The previous instructors focused the class on regulatory applications and requirements of chromatography. My view was that students need more method development theory to be successful in industry—regulations are not much help without suitable selectivity for the analytical method. The standard text for liquid chromatographic development is and remains Snyder and Kirkland's *Practical HPLC Method Development* and I used it for several classes.[1]

I joked after finishing my second book with Pan Stanford Publishing (now Jenny Stanford Publishing), *Poorly Soluble Drugs: Dissolution and Drug Release*, that never again would I take on such an undertaking.[2] I also realized an update in text was needed after driving home from a lecture at Governors State in which I was getting tired of telling students that they needed to focus on the principles of Snyder and Kirkland's text, mostly because the specific columns and instruments used in 1997 were no longer

[1]*Practical HPLC Method Development*, 2nd edition, L. R. Snyder, J. L. Kirkland, J. L. Glajich, ©1997, Wiley-Interscience, Print ISBN 978-0471007036.

[2]*Poorly Soluble Drugs: Dissolution and Drug Release*, G. K. Webster, J. D. Jackson, R. G. Bell (editors). ©2017, Pan Stanford Publishing, Print ISBN 978-981-4745-45-1.

in use in current industrial laboratories. For example, few of us will ever use a 250 × 4.6 mm column packed with 10 μm particles again. I told them my true life experiences of the battles I personally had with sister laboratories trying to get them to move from 5 to 3 μm particles and then ultimately having to do it all over again with UHPLC packings. After further reflection, I called my dear, very energetic and motivated friend and former co-worker Laila to see if she was interested in undertaking what at this time is truly to be my last chemistry book venture. I told Laila that no one rewrites a "bible." If we were going to do something, we had to be different. We decided on a text that covers not just liquid chromatographic methods development but all the methods that an industrial chromatographer may see. Our idea was born. Laila and I approached Jenny Stanford Publishing with our idea and outline of the proposed chapters. With their interest assured, we then contacted those we knew who were subject matter experts in their assigned chapter.

Of course, life and work get in the way. A text we hoped to complete in 2017 was finished in 2019. I thank Laila for the periods where she drove the bus. We both thank our authors for their expertise, patience, and efforts. My hope is that this text may help an aspiring chromatographer, as Snyder and Kirkland's text has helped me.

From Laila Kott

My path to this point has some similarities to Greg's and some differences. As an undergrad at the University of Toronto, I took my time to declare my major and after almost two years at the university, chemistry it was and still is. The difference is that there were always core functions of chemistry that I considered just part of the chemists' toolbox as I followed one research project after another. I made the materials I needed to study; I analyzed the materials I needed to study. I never considered myself a certain type of chemist, just a chemist.

Several degrees later and having taken a sabbatical from the scholarly life between my master's and doctoral degree to do chemistry in the former Soviet Union, I ended up in the pharmaceutical industry, where I found my journey through many different projects had afforded me a very strong background in a variety of analysis techniques. Since, as Greg mentioned above, people often focus in one area or one technique, I found that looking at a problem using several different techniques often gave one a more complete picture of what you were looking at.

This was the exact time when Greg and I started collaborating and we have been tackling interesting problems as well as pharmaceutical industry standard problems ever since. We have always pushed the newest columns, the newest technology. This has its drawbacks as not all things new are readily accepted by the regulatory bodies during new drug filings. Nevertheless, we noted that there was a decided lack of method development information for the modern chromatographer out there, and this is our attempt to rectify the situation. We have had many challenges during the compilation of this book, but our drive never wavered as the need is out there. Happy separations!

Chapter 1

A Lesson Learned from UHPLC

Gregory K. Webster[a] and Laila Kott[b]

[a]*Abbvie, 1 N. Waukegan Rd,*
North Chicago, Illinois 60064, USA
[b]*Xenon, 200-3650 Gilmore Way, Burnaby, British Columbia V5G 4W8, Canada*

gregory.webster@abbvie.com, lkott@lu.lv

1.1 Introduction

In the late 1990s, it was not uncommon for laboratories to argue over whether it was robust to use the new 3 μm packed particle columns. Colleagues would argue over the merits of using this "new" particle over the trusted industry standard 5 μm particle. These memories bring a smile to those of us trying to bring Ultra High-Performance Liquid Chromatography (UHPLC) and its 1.2 to 2 μm columns into routine use.

The differences between systems designed for the 3 μm and UHPLC columns are significant. A change to 3 μm particle did not require new instrumentation. Not only does a change to UHPLC

Chromatographic Method Development
Edited by Gregory K. Webster and Laila Kott

ISBN 978-981-4800-53-2 (Hardcover), 978-0-429-20172-1 (eBook)
www.jennystanford.com

require a change in instrumentation, but it required analytical chemists to go back and learn their chromatography theory again. In order to use these smaller particle packings, mobile phase pumps had to be designed that are jokingly able to "pump liquid through a brick." New autosamplers that could reproducibly inject nanoliter volumes had to be devised and built. Detectors capable of tracking much faster peak elution needed to be constructed. Probably the biggest challenge for UHPLC was, and still is in many labs, convincing management of the necessity of the required new capital expenses needed to bring this technology into modern laboratories. Fortunately, scheduled depreciation and instrument replacement have caught up to manning most industry laboratories with the capability of executing UHPLC methods.

Prior to the UHPLC "revolution," liquid chromatography had gotten stale. By this, we mean liquid chromatography had become so routine and ubiquitous that research in this area and in many universities has diminished. Methods used in industrial laboratories became sloppy; void volumes were seldom considered, plumbing was only worried about when a method transfer would not work, and injection volumes and detection were solely considered for linearity.

First and foremost, chemists had to decide if their system was capable of running the smaller particle sizes that UHPLC called for. In fact, the integration of this new technology forced chemists to evaluate and decide if their system was capable of running the smaller particle sizes that UHPLC called for. Manufacturers of liquid chromatography instrumentation had to accelerate their design plans to meet the new demands of UHPLC. Column manufacturers had to defend their choice of stationary-phase packings. Superficially porous particles became in vogue again as they were capable of producing "UHPLC like" separations at lower pressures and thus, reducing the need for a rush to new capital expenses in instrumentation. UHPLC led chromatography to become a focus of industrial and academic research again.

1.2 Technology and Chromatography

For many analytical chemists today, chirality was only a concept introduced to us in our sophomore organic chemistry classes.

Now for those of us in pharmaceutical laboratories, chirality is routinely included in determination of purity. The idea of using chromatography to the ability to selectively resolve a compound's capability to rotate a plane of light is, at first, daunting. Yet, after spending some time with the theory and techniques, one comes to realize that chiral separation is just another application of the separation theories we use every day. In other words, it is just an evolving application of the technology and a notion of evolving capabilities is something that chemists must sustain during their career.

It is important to note that instrumental technologies will continue to develop. Thin-layer chromatography (TLC) changed to low-pressure column chromatography, which evolved to high-performance liquid chromatography (HPLC), which has now grown into UHPLC. However, in all these advancements, we still have the same concept of stationary phases, mobile phases, and plate height. Chromatographic theory is not abandoned; it has just progressed with the changing technology. If one overlooks the now outdated columns in stationary-phase packings in use at the time, Snyder and Kirkland's work on liquid chromatographic development is just as relevant today as it was when it was published [1]. In other words, regardless whether if it is TLC, flash chromatography or supercritical fluid chromatography (SFC), the principles of normal-phase chromatography apply.

1.2.1 New Goal of Method Development: Speed

The primary goal of chromatographic method development has always been focused on resolution. One may argue that it is plate height; however, plate theory revolves around achieving the necessary resolution for the intended separation [2]. Sensitivity requirements were added as a focus and industry standards for toxicity of impurities came to light [3, 4].

Today, such method requirements are commonplace, and the requirements for selectivity and sensitivity are expected. Method development is about speed. Every industry is "on the clock" to deliver results. This notion has trickled down to the analytical lab in the urgency to get "results out the door" and "get the product released" for its intended purpose. Thus, the pressure is on to develop methods quickly and robustly for their intended use. Trial-and-error approaches have given way to automated method development schemes that typically yield a fit-for-purpose method in much less time than previously required. However, the fundamental theories behind chromatography have not changed. Instead, the focus on speed in method development has come from the application of automated technologies and approaches. Method development will continue to evolve in this regard, and more suitable technology developed to accompany that evolution.

1.3 The Chromatography Market

Chromatographic instrumentation and consumables is a huge market. Separations technologies play a vital role in the chemical, pharmaceutical, life science, food and beverage, and environmental industries. The global chromatography instruments market is projected to be valued at USD 7.86 Billion in 2017 and is expected to grow at a compound annual growth rate (CAGR) of 6.9% to reach to USD 10.99 billion by 2022 [5]. Market growth is fostered by ever increasing "food safety concerns, increasing use of chromatography tests in the drug approval process, and growing popularity of hyphenated chromatography techniques" [5]. Liquid chromatography sales is expected the reach USD 4.13 billion by 2021 [6]. This forecast is followed by a gas chromatography market prediction of USD 4.13 billion by 2021 [7] and a global flash chromatography market to reach US$ 294.1 million in 2022 [8].

1.4 Chromatographic Method Development

The goal of this text was not to reintroduce theory. We leave theory to instrumental textbooks. This book focuses on how methods for each of the major separation techniques in use in industry are currently being developed. Each chapter addresses how a segment of chromatographic science, currently in routine application in chemical, pharmaceutical, and environmental industrial laboratories is being applied.

1.4.1 Software for Chromatographic Method Development

Modeling technologies are prevalent throughout the industry today and the analytical lab is not an exception. Chromatography theories along with design of experiments (DOE) approaches are now available to automate, evaluate, and implement guided method development approaches. Use of these software programs has led to significant efficiency gains in reducing the time required for developing chromatographic methods. Aoi Ariyasu and Eiki Maeda of Takeda discuss the software approaches available to industrial laboratories for chromatographic method development.

1.4.2 Flash Chromatography Method Development

In modern synthesis laboratories, flash chromatography is a valuable tool reaction monitoring and characterization. Flash chromatography has all but eliminated thin-layer chromatography in this regard and is one of the core techniques used by chemists to isolate synthesized or extracted compounds. Robert Bickler of Biotage presents how flash chromatographic methods are developed in this regard.

1.4.3 Orthogonality in Column Chromatography

Whenever a chromatographic method is being used for separations, the same general questions always arise: (1) "How do I know all my components are resolved?" and (2) "how do I know all my components have eluted?" The traditional approach for answering these questions is to use orthogonal separation methods. While mass spectrometric detection has often been used to address the first question, it cannot address the second. Thus, the use of a differing separation theory/technology is applied to track sample components. In addition, for reversed-phase chromatography the notion of "orthogonality" has evolved into the application of different stationary phases expected to yield significant changes in separation. Jeff and Winton Caldwell of Princeton Laboratories and Gregory Webster of AbbVie discuss how the notion of orthogonality is being used in today's chromatographic method development.

1.4.4 Unified Approach to Reversed-Phase and Normal-Phase Chromatography

As the most commonly used separation technique, liquid chromatography is the furthest along in the theory of method development and using technology to reduce development time. For most laboratories today, reversed-phase chromatographic method development is governed by an automated approach. Normal-phase chromatography is typically performed by SFC. When components of a sample are too polar for efficient reversed-phase resolution, hydrophilic interaction liquid chromatography (HILIC) is often applied. Gregory Webster of AbbVie discusses the automated approach for method development using each of these techniques.

1.4.5 Polysaccharide-Derived Chiral Stationary Phases for the Separation of Enantiomers

Chiral chromatography has evolved rapidly from its introduction as a routine separation technique in the late 1980s. When it was found that differing enantiomers of a compound often had significantly different biological activities, the pharmaceutical industry felt it was best to address safety concerns by implementing this technique as part of a compounds quality control. By far, the most applied stationary phases for chiral separations have been on polysaccharide platforms. Tong Zhang and Pilar Franco of Chiral Technologies discuss stationary-phase technology for chiral separations.

1.4.6 Chiral Chromatography: Method Development

As with routine reversed-phase chromatography, chiral chromatography began with a trial-and-error approach based on the structure of the chiral stationary phase. Method development for chiral separations is based on first finding a stationary phase that exhibits chiral recognition to the sample(s) of interest. After pinpointing an appropriate stationary phase, further method development proceeds in a typical reversed-phase or normal-phase path depending on which phase is being used. This was often a time-consuming approach. Industrial laboratories have found that screening approaches often lead to successful column hits. Laila Kott of Xenon and Gregory Webster of AbbVie discuss current automated approaches to chiral method development.

1.4.7 Aqueous Normal-Phase Chromatography Using Type-C Silica Columns

Type-C silica stationary phases appear to be similar to HILIC stationary phases and the two techniques are often confused. Aqueous normal-phase executed on Type-C materials has a silicon hydride (Si-C) support. HILIC is executed on Type-A or Type-B material which has a silica (Si-O) support. Joseph Pesek and Maria Matyska of San Jose State University distinguish the mechanism of separation for these two technologies and discuss how method development proceeds with aqueous normal-phase chromatography.

1.4.8 Ion Chromatography Columns

While less commonly found in industrial laboratories than gas or liquid chromatography, ion chromatography provides the analytical chemists with a unique tool for detecting ions. Unique stationary phases and capabilities for these columns are less understood by the everyday chemist. Chris Pohl of Thermo Scientific provides the background on the stationary phases used in ion chromatography.

1.4.9 Ion Chromatography: Method Development

This chapter provides the process for developing an ion chromatographic method. Method parameters for ion chromatography are a little different than found in other liquid chromatographic techniques and include selection of eluent systems, detectors, sample preparation, gradient separations, and choice of column chemistries. Laila Kott of Xenon discusses a process for optimizing all the unique ion chromatography parameters.

1.4.10 Fundamentals, Properties, and Applications of Stationary Phases of Gas Chromatography Method Development

Historically, gas chromatography is often the first instrumental separation technique most chemists are trained on. Stationary phases for gas chromatography have evolved from wide bore packed columns made of glass and/or steel to the higher efficiency capillary columns that are mostly used today. Gas chromatography is still the primary way to separate and detect volatile components of a sample. Residual solvent testing in the pharmaceutical industry still primarily relies on this technique as well. Omprakash Nacham and Jared Anderson of Iowa State University provide an updated review of the stationary phases in use today for gas chromatography and some new ionic liquid phases.

1.4.11 Gas Chromatography Method Development

Separation in gas chromatography is driven by volatility differences in analytes as well as, like liquid chromatography, the interactions of the analytes with the stationary phase. Despite the separation occurring in the gas phase, gas chromatography can be used with a variety of analytes, be that solid, liquid, or gas at room temperature. In the pharmaceutical industry, GC is often used for residual solvent analysis and the detection and quantification of genotoxic impurities. Tien Ho and Cody Vinci of AbbVie discuss the various detectors and parameters that can be optimized to make GC a useful tool.

1.4.12 Method Transfer from HPLC to UHPLC

With the inception of UHPLC, many laboratories were tasked with transferring methods from traditional HPLC to UHPLC capable systems. As the UHPLC systems require different conditions, software approaches were developed for this transfer based on traditional column chromatography theory. Susanne Fabel of Thermo Fisher Scientific reviews the software algorithms applied for identifying transferring HPLC method conditions to UHPLC.

1.4.13 Chromatographic Method Validation

For most industries, simply developing a separations method is not good enough. Once developed, the method must show that it can be executed in a repeatable and robust manner in the original and other laboratories. Industry guidelines are commonly applied to validate that the separations method is routinely capable of meeting its intended purpose. Michael Oberlander and his team from Abbvie discuss the industry-accepted guidelines to validate suitability of the developed chromatographic method.

1.4.14 Method Development Strategies for Ion Exchange Chromatography Using pH Gradient Separation for mAbs, Antibody Drug Conjugates (ADCs), and Other Complex Bio-Pharmaceuticals

Factors that affect ion exchange chromatography for both monoclonal antibodies (mAbs) and antibody–drug conjugates (ADCs) are salt concentration and gradient, pH gradients, the pI of the protein, temperature, ion concentration, columns and choice of separation mode. Norman L. Fischer, Scott Allen, and Meg Ruesch from Pfizer show the importance of these parameters, how they are optimized, and how they can affect ion exchange chromatography of both mAbs and ADCs.

1.4.15 Size Exclusion Chromatography Method Development for Therapeutic Proteins

Therapeutic proteins are not necessarily single components. They can exist as aggregates and monomers. While size exclusion chromatography (SEC) was initially developed to characterize polymers, it has found major application in the biopharmaceutical industry. SEC is now commonly used in research and quality control to demonstrate the proteins in the intended product have maintained their same character as the ones used in clinical trial studies. Alexandre Goyon, Jean-Luc Veuthey, Davy Guillarme, and Szabolcs Fekete of the University of Geneva provide the platform commonly used in industry to develop size exclusion methods for proteins.

1.4.16 Ionic Liquids and Counter-Current Chromatography: Is This the Future of Process Purifications?

Current chromatography (CCC)/extraction (CCE) can be considered to be an instrumentally created stationary liquid phase, rather than a stationary phase brought about by chemical bonding of a liquid phase to a solid support. By using high percentages of ionic liquid in biphasic eluents in CCC/CCE (termed ionic liquid-liquid chromatography or ILLC), several dimensions of choice of mobile and stationary phases have been added to the technology. The applications of molecular and ionic solvent systems in CCC/CCE/ILLC include chiral and achiral organics, transition metals, precious metals, actinides, lanthanides, bio-chemicals and natural products, pharmaceuticals, and radionucleotides. In CCC/CCE, compounds can also be partitioned between two mobile phases (for example, in dual flow operation modes), and used to purify one or more target compounds.

Method development in laboratory preparative and process liquid-liquid, counter-current chromatography (CCC)/extraction (CCE) follows the same criteria as for various other solid-liquid forms of chromatography, such as solid-phase extraction (SPE), flash chromatography (FC), and high-performance liquid chromatography (HPLC). These techniques partition target compounds between a stationary phase (SP) and a mobile phase (MP), or two mobile phases in dual flow. SPE/FC/HPLC separations could, in the reverse-phase mode, be considered a form of liquid-liquid partitioning, since many bonded phases can be considered to behave like liquid phases, which are chemically bonded to a solid stationary phase. In normal-phase SPE/FC/HPLC separations, liquid boundary layers also have a major effect on the separations that can be achieved. Leslie Brown and Martyn J. Earle of AECS-QuikPrep present the current technology for CCC/CCE using ionic liquids.

Disclosure

Laila Kott is an employee of Xenon. Gregory Webster is an employee of AbbVie. The study contains no proprietary Xenon or AbbVie data. Xenon and AbbVie jointly participated in the writing, review, approval, and financial support of this publication.

References

1. L. R. Snyder, J. J. Kirkland, J. L. Glajch, *Practical HPLC Method Development,* 2nd ed., Wiley-Interscience: Hoboken, NJ. 1997.
2. G. K. Webster, A. R. Diaz, D. S. Seibert, B. S. Weekley, J. D. Jackson, Plate number requirements for establishing method suitability, *J. Chromatogr. Sci.*, 2005, 43, 67–72.
3. "Impurities in New Drug Products," Q3B(R2), International Conference on Harmonization of Technical Requirements for Registration of Pharmaceuticals for Human Use (ICH), 2 June 2006.
4. "Assessment and Control of DNA Reactive (Mutagenic) Impurities in Pharmaceuticals to Limit Potential Carcinogenic Risk," M7(R1), International Conference on Harmonization of Technical Requirements for Registration of Pharmaceuticals for Human Use (ICH), 31 March 2017.
5. "Chromatography Instruments Market," July 2017, *marketsetandmarkets.com*. Report Code AST 3789.
6. "Global Flash Chromatography Market," August 2017, 2007 Stratview Research, Telibandha, Raipur, 492001 India. Report Code SRAM158.
7. "Gas Chromatography (GC) Market Analysis," August 2017, Grand View Research, San Francisco, CA 94105. Report ID 1-68038-970-8.
8. "High-performance Liquid Chromatography Market: Global Forecast until 2021," February 2017, ReportLinker, Lyon, 69002 France. ID 4753174.

Chapter 2

Software for Chromatographic Methods Development

Aoi Ariyasu and Eiki Maeda

Pharmaceutical Sciences, Takeda Pharmaceutical Company Limited, 17-85, Jusohonmachi 2-chome, Yodogawa-ku, Osaka 532-8686, Japan

aoi.ariyasu@takeda.com, eiki.maeda@takeda.com

2.1 Introduction

Analytical testing and its control strategy play critical roles during the entire life cycle of the drug product in the pharmaceutical industry. Chromatography, especially high performance liquid chromatography (HPLC), ultra-high performance liquid chromatography (UHPLC), and gas chromatography (GC), represent the key techniques employed throughout many aspects of pharmaceutical analyses, including the determination of assays, dissolution, and impurity profiles, process control testing, and drug metabolism studies [1–3]. In general, the use of the trial-and-error approach has been common practice in developing analytical methods for many years. However, this method usually requires large numbers of

Chromatographic Method Development
Edited by Gregory K. Webster and Laila Kott

ISBN 978-981-4800-53-2 (Hardcover), 978-0-429-20172-1 (eBook)
www.jennystanford.com

experiments, and the final outcome is often not optimal, resulting in the requirement for further development in later development stages of the drug product. Thus, a more systematic approach to analytical method development would enable identification of the optimal method parameters, and contribute to faster and efficient method development [4].

In this context, the application of quality by design (QbD) principles, which plan or design quality into an analytical method, can avoid such time-consuming processes and can enable the implementation of a systematic approach to method development. Beyond its traditional use in the pharmaceutical manufacturing process, the concept of QbD has been extended to analytical method development and its control strategy, and is known as analytical QbD (AQbD). In many cases, design of experiments (DOE) with DOE modeling software and/or chromatographic simulation software is implemented and utilized from the aspect of QbD and/or AQbD to achieve optimal method performance [5, 6]. A key benefit of implementing the QbD approach is the development of an understanding of multidimensional combinations, the interaction of input variables, and their impact on quality. In the case of AQbD for chromatographic method development and its control strategy, all analytical method parameters (i.e., input variables) that have a strong influence on chromatographic retention and selectivity (i.e., the quality/performance of the method) should be studied in combination, thus defining perfectly known multidimensional correlations. In addition, an improved understanding of the correlation between analytical method parameters and method performance can aid in the recognition of optimal space of the method parameters, i.e., Design Space. The Design Space, as defined in the ICH Q8(R2) [7], is the "multidimensional combination and interaction of input variables that have been demonstrated to provide assurance of quality." Determination of the optimal space of the method parameters is therefore considered a key factor in quality control (QC), method transfer, and future method changes.

Furthermore, the software, chromatographic simulation software, and DOE modeling software are important for determining the correlation between the analytical method parameters and the method quality, and also in the identification of the optimal space of the parameters by chromatogram simulation, experimental design based on DOE, multivariate analysis calculation, and other functions. Thus, the software enables the smooth and efficient application of the AQbD approach in analytical method development.

In this chapter, we summarize the application of general commercially available software for HPLC/UHPLC or GC, i.e., chromatographic simulation software (e.g., DryLab®, Molnár Institute, Berlin, Germany [8, 9], Fusion QbD® software, S-Matrix Corporation, Eureka, California [10] ACD software, ACD/Labs, Toronto, Canada [11], ChromSword®, ChromSword, Riga, Latvia [12]) and DOE modeling software [13], which has the potential for application in analytical method development. We expect that this information will assist analysts who wish to employ AQbD, and that it will aid in the effective development of optimized and/or robust analytical methods.

2.2 General Features of Software

2.2.1 Typical Procedure for the Systematic Development of Analytical Methods and the Effect of the Software Employed

Software that is considered potentially useful in the development of HPLC/UHPLC or GC analytical methods can be characterized into two main types, namely chromatographic simulation software (hereinafter referred to as chromatographic software) dedicated to the modeling and simulation of retention times of target peaks, and DOE modeling software (hereinafter referred to as DOE software) for the modeling of multidimensional correlations between analytical method parameters (input variables) and method performance (output). Generally, chromatographic software can simulate chromatograms through fewer experiments when compared to DOE software, as its function is specialized for the development of liquid chromatography (LC) or GC methods. However, DOE software can be applied not only for the optimization of various other analytical techniques, but also for demonstrating the multidimensional correlation between sample preparation procedures, equipment models and operators, and outputs.

Thus, the typical strategy for analytical method development is similar to that shown in Fig. 2.1. First, selection of input variables to meet the desired goal of the analytical method takes place, followed by experimental design and execution, response modeling, and further optimization of the analytical method. Determination of the goal is the key step during such method development, as the desired analytical method performance must be defined upon commencing method development. Goal examples include, the complete separation of all peaks, a short analytical time, and analytical robustness. Deep-dive steps employed during chromatographic method development are listed in Table 2.1. Steps 1–3 are always conducted by analysts, based on their knowledge, previous experience, and the results obtained during primary method development. Subsequently, steps 4–7 can be carried out using certain chromatographic software in combination with online LC measurements, while step 5 often requires manual data input into the chromatographic software (i.e., offline measurements). Although DOE software can be used in steps 4, 6, and 7, step 5 must always be conducted by an analyst.

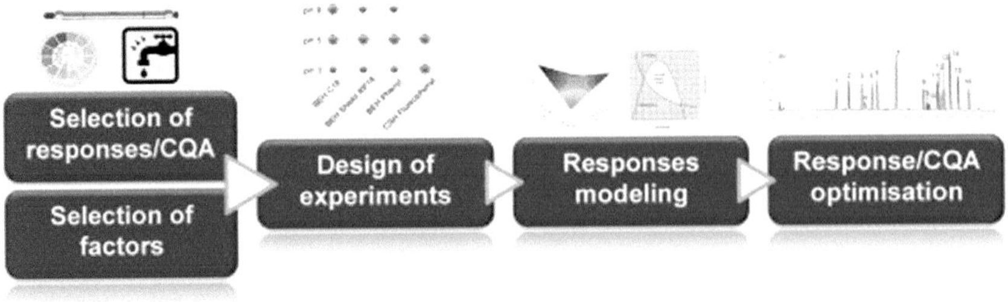

Figure 2.1 Typical steps of the chromatographic method development procedure employed in the pharmaceutical industry. Reprinted from [14], with permission from Elsevier (color altered).

Over the entire pharmaceutical development stages, analytical methods are normally developed and optimized through repetition of steps 2–7. In addition, a phased approach is often adopted for systematic method development, and involves initiating the investigation with a wide range of method parameters, and later narrowing down the investigated range to finally yield the optimal space or point in a phased manner, as shown in Fig. 2.2.

Table 2.1 Steps of chromatographic method development using software

Step	Practical procedure	Process
Step 1	Determination of the desired method performance that is affected by variations in method parameters	Goal determination
Step 2	Determination of method parameters whose variations are evaluated	Experimental design
Step 3	Determination of the investigated range of those parameters	
Step 4	Generation of experimental design to examine the impacts of parameter variation	
Step 5	Execution of experiments based on the determined experimental design, and collection of the method performance results for each experiment	Execution of experiments
Step 6	Model development by correlation analysis between the investigated method parameters and the obtained method performance	Analysis and method optimization
Step 7	Determination of the optimal space or points of method parameters that result in acceptable method performance	

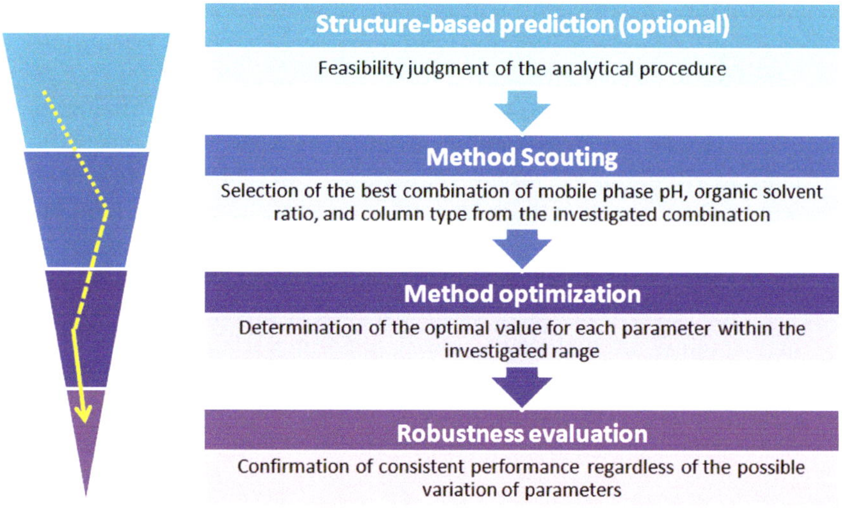

Figure 2.2 Phased approach for analytical method development.

2.2.2 General Software Features and Applicability to the Various Stages of Analytical Method Development

Analytical method development generally consists of two main stages, namely method scouting and method optimization. In the method scouting stage, the column/eluent combination considered preferable for the analytical method is selected through rough screening. This combination does not have to be the best strategy but acts as a guide in the later optimization stages. During the initial method scouting stage, software capable of roughly estimating the impact of method parameters on method performance or simulating analyte behavior is sufficient. Therefore, chromatographic software that can run the experiments automatically under the predesigned conditions can be employed. In contrast, in the method optimization stage, software that assists the precise evaluation of the influence of the method parameters on the target method performance, while aiding in efficient optimization through reducing the number of experiments based on DOE is preferable. In this stage, critical method parameters and their combinations are optimized by experimental design strategies using chromatographic or DOE software, where the chromatographic software addresses method development in a variety of manners using similar resolution maps to that shown in Fig. 2.3 to clearly identify the optimal conditions [15]. In this context, the simultaneous optimization of many factors and seamless variation in the critical resolution are of particular importance.

In addition to these two stages, structure-based prediction is often applied prior to method scouting in early stage analytical method development, while robustness evaluation is conducted following method optimization in late stage analytical method development. During all stages, steps 2–7 are repeated as necessary.

The different functions of each type of software that are useful in analytical method development are summarized in Table 2.2. For example, chromatographic software can be employed in structure-based retention prediction and chromatogram simulation, in addition to method scouting, optimization, and robustness evaluations. Alternatively, DOE software can be used in method scouting, optimization, and robustness evaluations based on offline analysis, which contrasts with chromatographic software that can be used for these purposes based on online analysis. Further details of these functions can be found in later sections and on each vendor website.

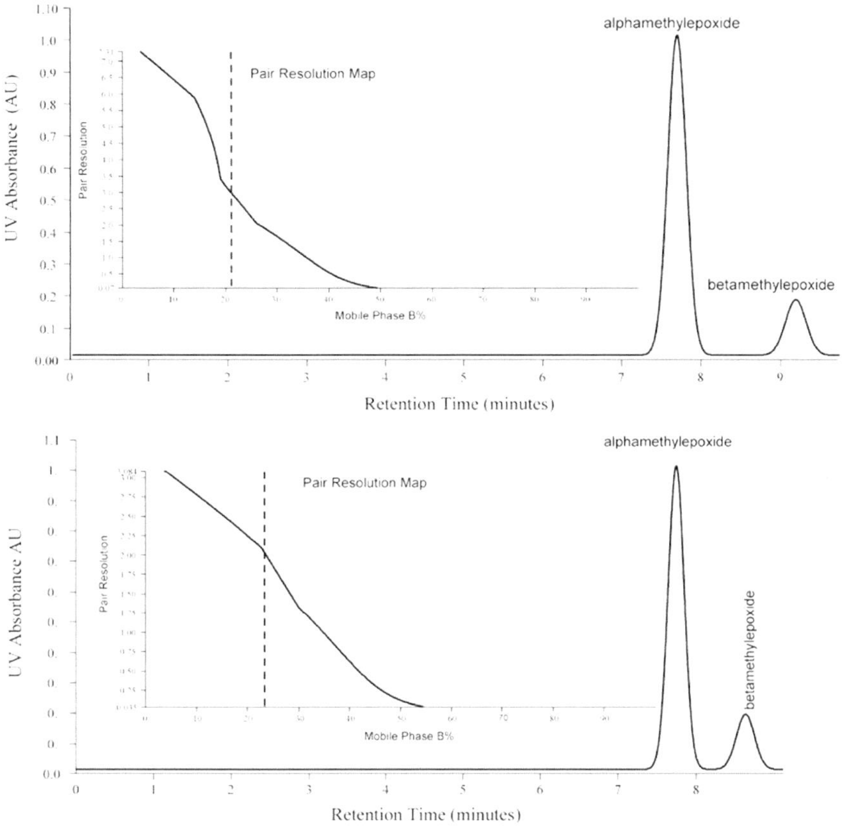

Figure 2.3 Simulated resolution maps and chromatograms from ChromSword® Auto. In the resolution map, the dotted line represents the cursor, which can be moved horizontally across the resolution map to obtain the predicted resolution (y-axis) at the corresponding mobile phase concentration (% B, x-axis). The chromatograms represent simulated separation. Reprinted from [15], with permission from Elsevier.

Although a number of different chromatographic software packages are available (e.g., DryLab®, Fusion QbD® software, ACD software, and ChromSword®), the basic concept is common for all products. However, each software program has a different modeling algorithm and exhibits characteristic functions as developed by each individual company. Users should therefore select the software that exhibits the desired functions for their requirements. In addition, many other examples of DOE software exist, and analysts can adopt their preferred software for their desired analytical method development. For example, Unicorn™, Design Expert®, MODDE, MATLAB, and JMP® are common examples of DOE software. In general, each software is capable of planning experiments and carrying out multivariate analysis to determine the correlation between many input variables and a single output. Indeed, the experimental planning function based on DOE has been introduced in both chromatographic and DOE software, while the modeling and simulation techniques differ between various chromatographic and DOE software.

Table 2.2 General functions of the software employed for chromatographic method development

			(DOE-related)		
Function	**Structure-based retention prediction**	**Chromatogram simulation**	**Method scouting/ screening**	**Method optimization**	**Robustness evaluation**
Chromato-graphic simulation software	Offline	Online /Offline	Online	Online /Offline	Online
DOE modeling software	(n/a)	(n/a)	Offline	Offline	Offline

Online: Separation is conducted automatically for predesigned plans (combination of parameters), thus no data input prior to simulation/calculation is required.

Offline: Separation is conducted independently of software, thus data input for simulation/calculation is required.

2.3 Application of Software to Analytical Method Development

In the following sections, the various features of the software employed for analytical method development (see Table 2.2) are introduced, and examples of their applications are provided.

2.3.1 Structure-Based Retention Prediction

To speed up analytical method development, knowledge and information should be preserved and re-used. Thus, when developing an analytical method from scratch, analysts must consider the implementation of a structure-searchable database that integrates existing screening results, method information, and vendor applications. Such databases have been developed and improved for both reverse-phase LC (RPLC) [16] and ion chromatography (IC) [17]. In the case of RPLC methods, enormous amounts of accumulated data and experience are available, since LC methods have been key analytical methods in the pharmaceutical industry for many years. By searching structure similarity, one can quickly identify a good starting point (such as column material, pH, and mobile phase) for target analytes prior to analytical method development. In addition, the physicochemical properties of a compound (e.g., pKa, log P, and log D values) serve as a rule of thumb to aid in selecting appropriate chromatographic techniques and method conditions, such as the stationary phase, buffer pH, and mobile phase additives for targeted and focused screening.

For example, Wang et al. adopted a structure-searchable database tool, a physicochemical prediction tool, and an LC simulator to create a holistic workflow for efficient HPLC method development and lifecycle management for analytical method development [18]. In particular, they established web-accessible databases that are searchable by structure using the commercially available ACD/Web Librarian software (ACD/Labs, Toronto, Canada). The database includes achiral and chiral separation methods obtained from application notes provided by column manufacturers, literature resources, and analytical methods generated within the company. Such analytical methods for a structurally similar compound can provide a promising starting point for method development when an analytical method for the exact compound is not available. In addition, an LC Simulator (ACD/Labs, Toronto, Canada) can provide calculated pKa values based on the target structure, and as such, can predict optimal separation conditions and retention times using experimental data and physicochemical property predictions.

ChromSword® is a further example of a software package that can provide analysts with first guess analytical methods predicted from structural formulae and column properties. ChromSword® incorporates a database of commercially available columns with retention and selectivity characteristics, which have been experimentally calibrated in acetonitrile-water and methanol-water mobile phases. As such, analysts can determine alternative columns for their target compounds using the ChromSword® column database. ACD/Column Selector (ACD/Labs, Toronto, Canada) and ColumnMatch® (Molnár-Institute, Berlin, Germany) [19] have similar functions. Following determination of the tentative stationary and mobile phases using database predictions, a few trial runs are performed to experimentally verify the in silico selectivity predictions, and chromatographic software is employed to simulate chromatographic separation and aid in the rapid development of faster, greener, and more robust methods.

2.3.2 Chromatogram Simulation

Chromatographic software dedicated to LC can be employed for the simulation of chromatograms. Software tools allow the scientist to model chromatographic separations based upon retention data from a limited number of experiments (two or more runs). In addition, preferred separation conditions can be predicted and virtual chromatograms for user-selected conditions can be simulated using such modeling software. This approach avoids labor-intensive trial-and-error experiments, and potentially results in significant improvements in the efficiency of analytical method development and in the quality of the final analytical method. Furthermore, to gain a better understanding of the method, detailed investigations can be carried out using chromatographic software, which is capable of simulating optimal conditions through a combination of DOE and chromatogram simulations. The depth of the evaluation depends on the goal or intended purpose of the analytical method and the period provided for the method development. Such method optimization using chromatographic software is explained in greater detail in Section 2.3.4.

The simulation feature of chromatographic software such as ChromSword®, DryLab®, ACD/LC and GC Simulators, and Osiris are based on the linear solvent strength (LSS) theory [20, 21]. More specifically, ChromSword® analyzes the LC retention data, builds retention models, and simulates and predicts the separation. It allows for the simultaneous optimization of two variables, such as buffer pH and organic modifier concentration, or temperature and organic modifier concentration. Zheng et al. reported its capability in developing the ion-pair chromatographic method for both the active pharmaceutical ingredients (API) and all known related compounds in final pharmaceutical tablets [22]. Simultaneous optimization of buffer pH and organic modifier (acetonitrile) concentration was performed using ChromSword®. Based on the obtained six chromatograms of the six isocratic runs for the aqueous mobile phases at three different pH values and at two different acetonitrile concentrations, a pair resolution map was constructed with respect to changes in pH and organic modifier concentration, and the simulation was performed based on the map. Baseline separation for the API and all key related compounds was achieved and the experimental results matched well with the predicted results.

DryLab® has a similar function to ChromSword®. Kochling et al. generated the resolution map shown in Fig. 2.4 by only four experimental runs while varying both temperature and gradient time [23]. At any point within this map, the simulated chromatogram can be generated using this software. In this case, they selected the final analytical method parameters based on this resolution map, general knowledge, and past experiences. In addition, Monks et al. generated 3D resolution models for a constant sample using three different columns, with the aim of developing a robust method with known tolerances (see Fig. 2.5). For this purpose, four critical parameters were examined, including gradient time, temperature, pH of eluent A, and stationary phase [4]. Following construction of the resolution cubes and experimental confirmation of the model precision, working points were selected according to the triple criteria of the critical resolution, robustness, and run time. As shown in Fig. 2.6, good correlations between the predicted and experimental results were observed.

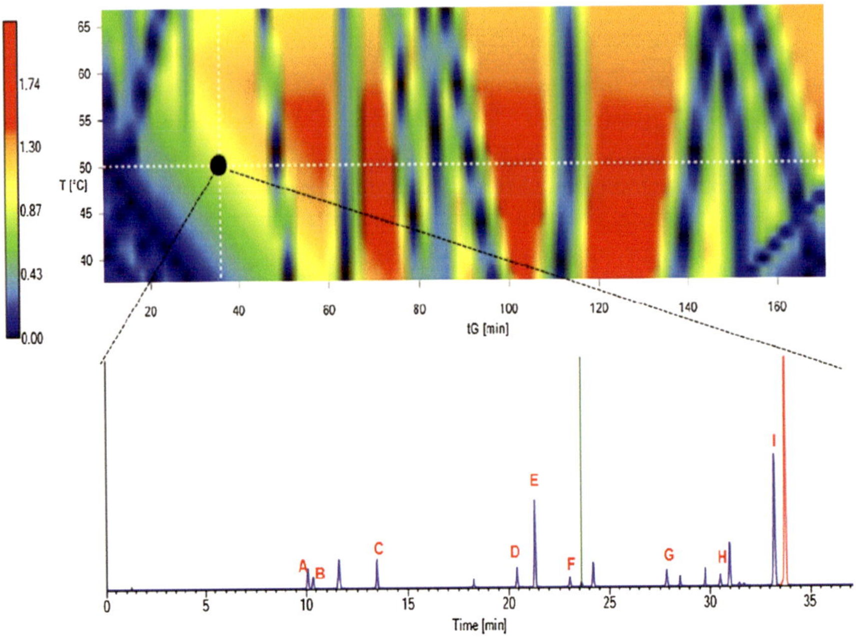

Figure 2.4 A heat map generated using DryLab® software which shows the relationship between time and resolution (top). A simulated chromatogram at the spot marked by the black dot in the upper image (bottom). Reprinted from [23], with permission from Elsevier.

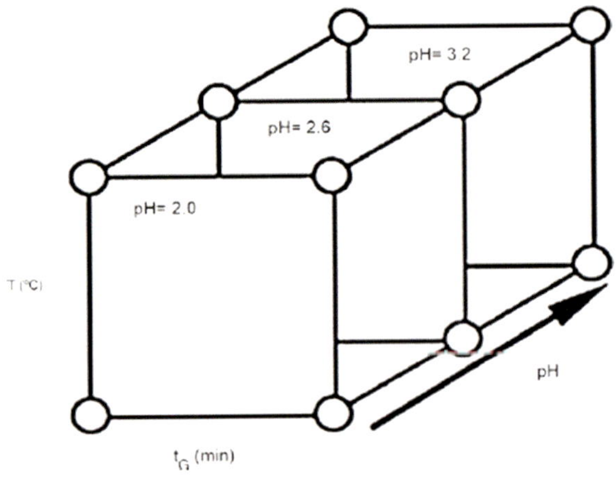

Figure 2.5 Experimental design for a 3D HPLC method optimiation. Reprinted from [4], with permission from Elsevier.

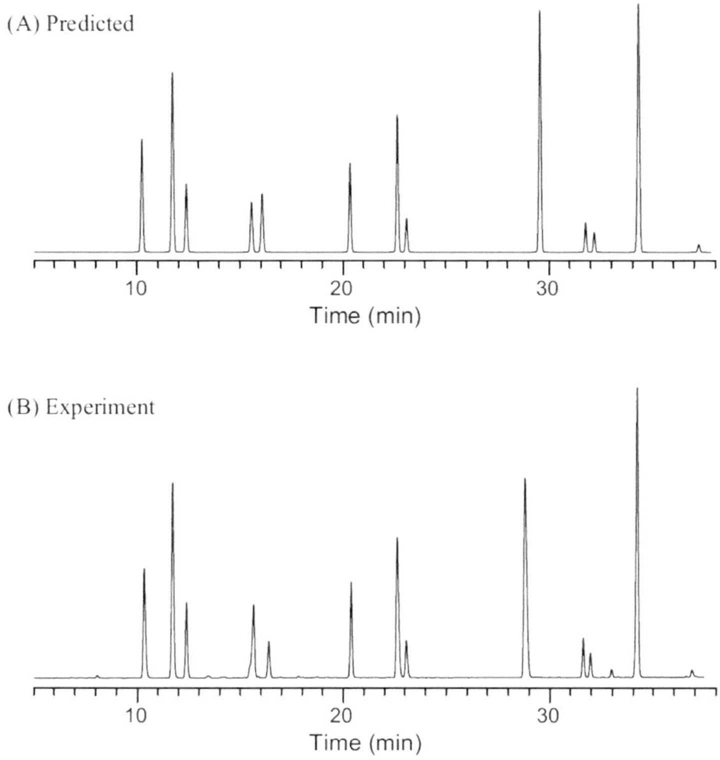

Figure 2.6 Predicted (A) and experimental (B) chromatograms. Reprinted from [4], with permission from Elsevier.

Alternative LC method development software packages (e.g., ACD/LC and GC Simulators, and Osiris) have similar functions, and can be employed to enhance the efficiency of LC method development in an analogous way [11, 24, 25]. However, simulation using chromatographic software has some limitations, such as its inapplicability to methods employing hydrophilic interaction chromatography (HILIC) columns. As a number of software packages apply the linear solvent strength theory or linear free energy relationships, they are unable to simulate chromatograms analyzed using an HILIC column, as in such cases, retention is based on the mixed mode mechanism, i.e., reverse phase and normal phase. However, Tyteca et al. attempted to develop the model to predict the retention behavior in HILIC mode and reported its applicability [26].

2.3.3 Method Scouting/Screening

The first step of analytical method development, i.e., determination of the critical parameters, is carried out either online (auto) [12, 15, 27] or offline (manual) [28] to identify the factors and interactions that have significant effects on the responses among a large set of factors. This step is used to select critical chromatographic parameters that affect retention and selectivity. When scouting is performed online, the chromatographic software or its combination with LC measurements is employed, which allows control of the separation equipment (e.g., HPLC/UHPLC) up to peak integration/identification and following simulation and/or calculation without any manual data input by analysts. In contrast, the offline method involves the use of software alone to carry out simulations and/or calculations, i.e., analysts must input the chromatographic parameters or results by hand. In both cases, the assistance of computer simulation software allows the interrelationships between critical parameters and their impact on the analytical method performance to be studied without the requirement for extensive laboratory experiments. As such, preliminary optimized conditions are obtained through a combination of analytical method parameters.

2.3.3.1 Online method scouting

ChromSwordAuto® (ChromSword, Riga, Latvia) is software used in LC separation methods, and carries out the defined tasks line by line, including automatically switching appropriate columns, solvents, and buffers for scouting purposes. Hewitt et al. investigated the utility of fully automated HPLC method development by applying a variety of compounds, and also examined the utility of "method scouting" using two different approaches [12]. In their first approach (Approach One, Fig. 2.7), the starting conditions were defined by analysts based on previous knowledge and experience, and included selection of a column, buffer, and organic solvent for the target sample (i.e., start without method scouting). For their second approach (Approach Two, Fig. 2.7), 42 combinations of columns, aqueous buffers, and organic modifiers were screened automatically by ChromSword®, the starting condition was selected subjectively by analysts, and the selected condition was applied within approximately 24 h without any

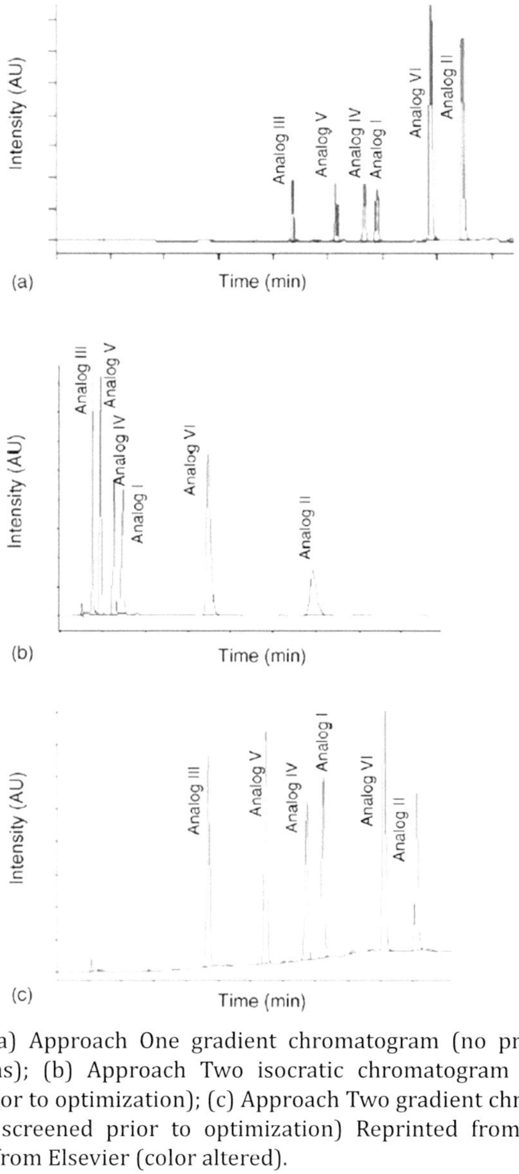

Figure 2.7 (a) Approach One gradient chromatogram (no pre-screening of conditions); (b) Approach Two isocratic chromatogram (conditions screened prior to optimization); (c) Approach Two gradient chromatogram (conditions screened prior to optimization) Reprinted from [12], with permission from Elsevier (color altered).

user assistance (i.e., start with method scouting). Subsequently, automated method development using ChromSwordAuto® full optimization was performed for both series. The first approach identified a gradient method that sufficiently separated the analogs with adequate resolution (Fig. 2.7a) but in the absence of an appropriate isocratic method. In the second approach, the software was successful in identifying both an isocratic and a gradient method of separation for all analogs (Fig. 2.7b,c). The data suggest that the software provides a consistently faster and more efficient method development option, while screening various starting conditions (method scouting) prior to the implementation of software optimization proved valuable in obtaining a suitable method.

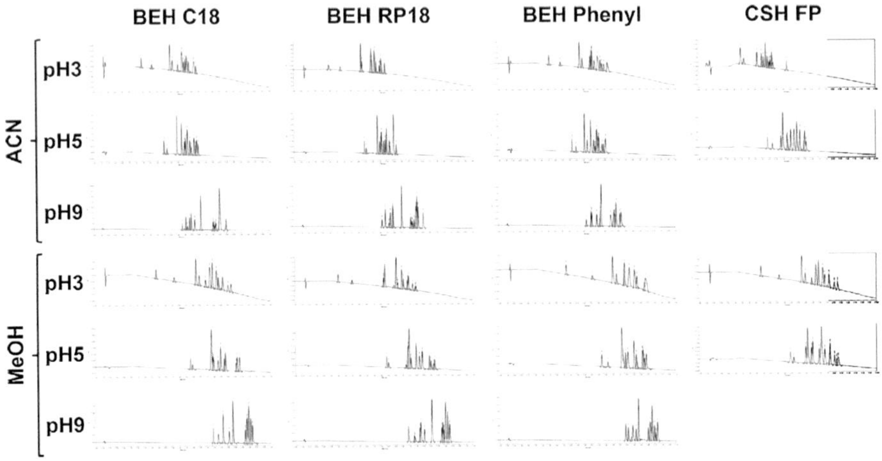

Figure 2.8 Chromatograms recorded during the screening design using Fusion AE™ for four columns (i.e., C18, C18 shield, fluorophenyl, and phenyl), three mobile phase pH values (i.e., 3, 5, and 9) and two organic modifiers (i.e., acetonitrile ACN, and methanol MeOH). Reprinted from [14], with permission from Elsevier (color altered).

The Fusion® software package also incorporates the method scouting function. More specifically, Fusion AE™ software (S-Matrix Corporation, Eureka, California), which is based on DOE and multiple linear regression (MLR), is able to automatically build programs in Empower and screen multiple columns and mobile phases using templates. Indeed, DOE and MLR had already been proven to be good alternatives to LSS for chromatographic method development [29, 30]. Thus, Debrus et al. applied Fusion AE™ software to obtain 22 chromatograms in the separation of 16 compounds for their selection step to evaluate the stationary phases of four columns, three mobile phase pH values, and two organic modifiers [14]. All runs were performed using a generic gradient from 5% to 95% organic modifier over 4 min, and complete screening was performed overnight. The most promising combination was obtained (see Fig. 2.8) where 15 peaks were automatically integrated. Under this condition, the elution window was narrow with a good distribution of compound, giving the opportunity to easily improve the separation during the next optimization step.

2.3.3.2 Offline method scouting

Boussès et al. reported the utilization of Nemrod-W (LPRAI, Marseille, France) and JMP 11 (S.A.S. Institute Inc., Cary, NC, USA) for offline method screening in UHPLC method development to determine a single API [28]. The target analytical method performance, e.g., method selectivity, impurity efficiencies, and environmentally friendly properties, was initially determined, after which, the analytical method parameters whose variability had an impact on method performance and their investigated range were chosen. In this case, a Plackett–Burman design composed of only eight runs for four selected method parameters was chosen, and the experiments were executed. The considered responses were retention time, number of theoretical plates of impurities, and total volume of mobile phase. Coefficient significance was evaluated using a *t*-test comparing the coefficient estimate to its standard deviation. Graphical analysis of the effects (see Fig. 2.9) indicated that flow rate and column temperature had a significant negative effect on retention time, whereas pH had a significant positive effect on these responses. In the context of impurity efficiencies, pH was the only significant factor that imparted a positive effect. Finally, as anticipated, solvent consumption was significantly affected by the flow rate (positive effect) and gradient slope (negative effect). Hence, all four factors had an effect on at least one response of interest. These four critical parameters were therefore further studied by response surface methodology (RSM) in subsequent method optimization studies.

A further example of the application of DOE software for method scouting using Plackett–Burman DOE is introduced below. Wang et al. applied the DOE for analytical method development for the solid phase extraction, high performance liquid chromatography, and ultraviolet/evaporative light scattering detection (SPE-HPLC-UV/ELSD) technique [31]. The combination of sample preparation and chromatographic separation was then examined, and the effects of 11 parameters on the resolutions and signal-to-noise ratios of the chromatographic peaks were investigated by 12 experimental runs. The resulting multiple regression and Pareto ranking analysis (Fig. 2.10) indicated that the six factors (sorbent mass, sample volume, flow rate, column temperature, evaporator temperature, and gas flow rate) were statistically significant (p < 0.05) in this method. Subsequently, the four parameters among the six factors were further applied to DOE using the Box–Behnken design for method optimization.

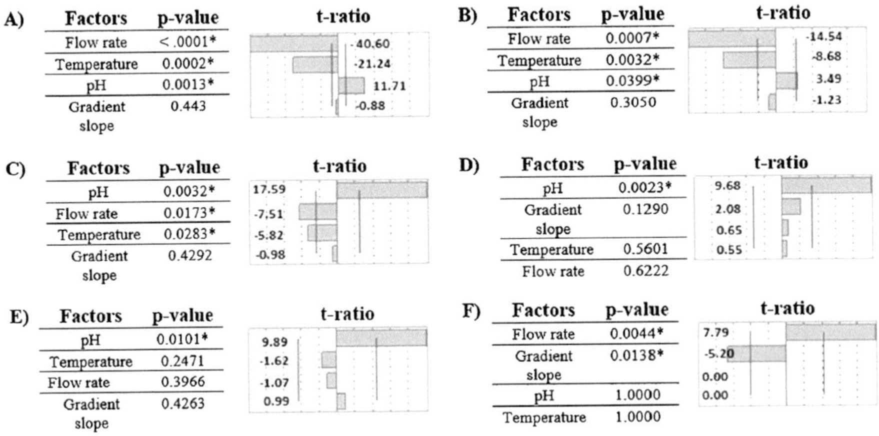

Figure 2.9 Effects of the critical method parameters on the selected method responses for experiments (a)–(f). Representation of the effects (sorted by order of importance) of the 4 factors examined (i.e., flow rate, temperature, pH, and gradient slope) and p-values ($Pr(>t)$) of their significance tests for the responses. Reprinted from [28], with permission from Elsevier (color altered).

2.3.4 Method Optimization

Method optimization is normally conducted following the identification of critical analytical method parameters by method scouting/screening as illustrated previously in Fig. 2.2. During this process, the method parameters are optimized to maximize the analytical method performance and to develop fit-for-purpose and robust methods. In the method optimization stage, DOE is often adopted to investigate the relationship between analytical method parameters (input variables) and method performance (output variables) in a systematic and efficient way. In addition, chromatographic software utilizes the chromatogram simulation. Although both chromatographic software and DOE software can be applied for method optimization, the discussion is ongoing regarding which calculates the optimized analytical method parameters more accurately. A few reports [23] state that chromatographic software, such as DryLab®, is significantly more effective for method optimization than full or partial factorial DOEs; however, further discussion is required based on accumulated case studies.

Upon utilization of the software for method optimization, analysts must consider the investigated range of each analytical method parameter and the run time of each separation to achieve reasonable analysis times, together with an acceptable chromatographic performance. Analysts must therefore set the method development target based on preferred retention times, peak resolution, and an understanding of the physicochemical properties of the analytes and the main degradants of the analytes. As a result of method optimization, it is often difficult to obtain a complete resolution for all peaks. In such cases, the resolution between critical peaks, e.g., the main degradants and substances which increase over time, takes precedence of that of minor peaks such as small peaks and substances which was observed in the early development stage but removed in the later development stage. Thus, even when software is utilized for method development, the planning and checks carried out by analysts prior to and following method development are both necessary and essential.

In this context, both types of method optimization, i.e., using chromatographic and DOE software, are introduced in the following subsections, and a number of applications are exemplified.

Run	Factors[a]											Experimental responses			
	F_1 (mg)	F_2 (mL)	F_3 (mL)	F_4 (mL/min)	F_5 (%)	F_6 (mL/min)	F_7 (°C)	F_8 (°C)	F_9 (°C)	F_{10} (L/min)	F_{11}	d_{R1}	d_{R2}	S_2	S_3
1	60	10	1	1.0	0.1	0.40	30	60.0	50.0	2.00	2	0.81	1.00	18.0	10.5
2	10	5	2	0.2	0.1	0.40	30	110.0	80.0	2.00	2	0.99	1.00	12.2	8.2
3	60	10	1	0.2	0	0.40	30	110.0	80.0	1.00	4	0.11	0.83	56.1	29.1
4	10	10	1	1.0	0.1	0.20	40	110.0	80.0	1.00	2	0.23	1.00	43.7	20.8
5	60	5	1	0.2	0.1	0.20	40	110.0	50.0	2.00	4	0.00	1.00	24.3	13.5
6	60	5	2	1.0	0.1	0.20	30	60.0	80.0	1.00	4	0.80	0.73	29.8	11.1
7	10	5	1	1.0	0	0.40	40	60.0	80.0	2.00	4	0.96	0.29	15.6	9.0
8	60	5	2	1.0	0	0.40	40	110.0	50.0	1.00	2	0.00	1.00	42.0	22.8
9	10	10	2	0.2	0.1	0.40	40	60.0	50.0	1.00	4	0.36	0.00	41.1	14.7
10	10	5	1	0.2	0	0.20	30	60.0	50.0	1.00	2	0.96	0.66	34.4	19.3
11	60	10	2	0.2	0	0.20	40	60.0	80.0	2.00	2	0.49	1.00	60.1	37.9
12	10	10	2	1.0	0	0.20	30	110.0	50.0	2.00	4	0.57	0.53	49.0	28.1

[a] F_1, sorbent mass F_2, sample volume; F_3, elution volume; F_4, elution speed; F_5, formic acid in mobile phase; F_6, flow rate of mobile phase; F_7, column temperature; F_8, evaporator temperature; F_9, nebulizer temperature; F_{10}, gas flow rate; F_{11}, PMT gain

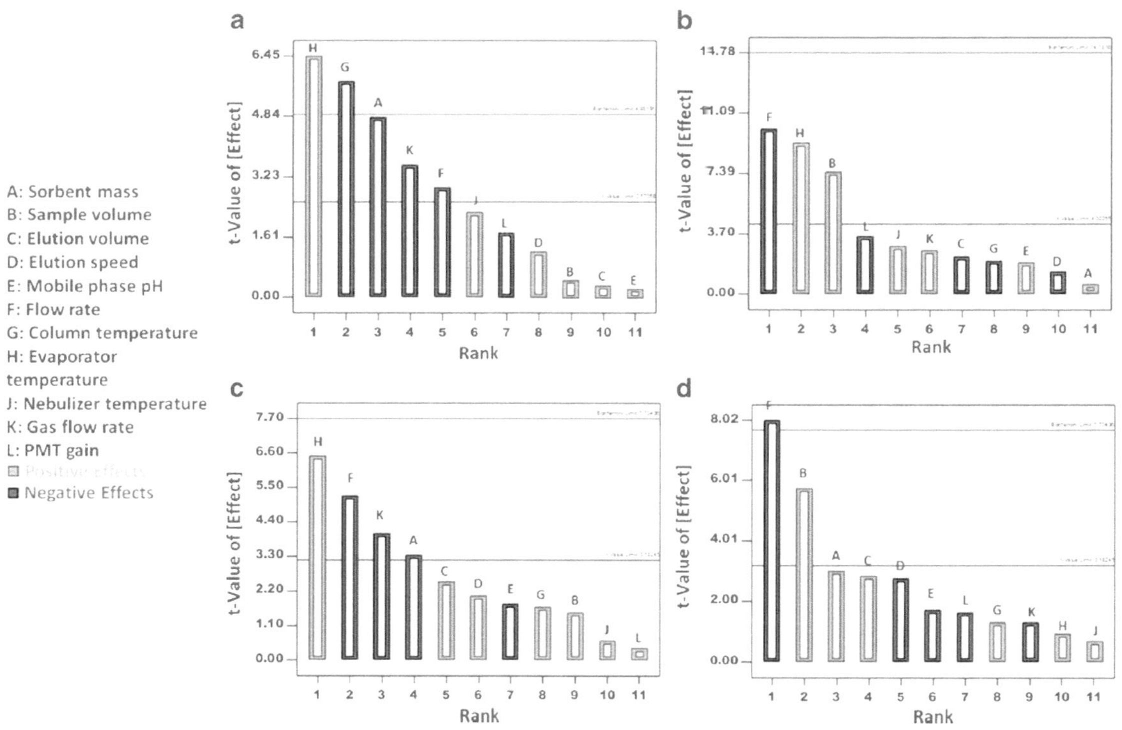

Figure 2.10 Experimental design using the Plackett–Burman design and the resulting Pareto charts of factorial analysis showing the effects of various factors on the modeled responses. Reprinted from [31], with permission from Springer (color altered).

2.3.4.1 Method optimization using chromatographic software

A number of chromatographic software packages adopt the combined DOE and chromatogram simulation function, where various kinds of method optimization exist using the chromatographic software, through on/offline optimization combined with on/offline measurements. Although few chromatographic software packages contain the online method optimization function, ChromSword® is the one that does have this capability. Following determination of the investigated method parameters and their subsequent input into the software by analysts, the software constructs the experimental design and automatically runs LC measurements (online measurements). Based on the obtained chromatograms, the software can generate the critical resolution map and propose the optimal method parameters (offline optimization), or it can search the optimal parameters by the repetition of experiments in a stepwise fashion (online optimization). Such online optimization is a trial-and-error approach carried out with the aid of results from offline optimization. Example results are shown in Fig. 2.7.

In contrast, DryLab® is a software package that combines offline measurements with offline optimization. DOE studies are conducted offline by analysts using full factorial design for the predetermined range, and the obtained results (e.g., retention times and peak areas) are input into the software, leading to the simulation of chromatograms and the generation of a resolution map for the points within the investigated range. Figure 2.11 shows the 3D resolution map representing the simultaneous influence of three parameters on the selectivity and critical resolution of the chromatogram. The minimum resolution value (i.e., the value of critical resolution) at each point was calculated by DryLab® and plotted. In this study, Molnár et al. observed more than one point located in the middle of each robust sphere (red region) representing regions where the critical resolution was >1.5, indicating that all peaks were well separated [32]. These points were therefore taken as the potential sets of the optimal analytical method parameters.

In addition, both software packages incorporate the gradient optimization function. This is especially important, as LC methods for obtaining impurity profiles often require the application of a gradient program to separate the numerous analyte peaks. For example, based on the obtained results of two linear gradients (i.e., increasing organic solvent concentration in the mobile phase at varying times), the optimized gradient program can be calculated by the software. The gradient program presented as the dotted line in each panel of Fig. 2.12 indicates the optimized gradient program calculated by ChromSword® in this manner.

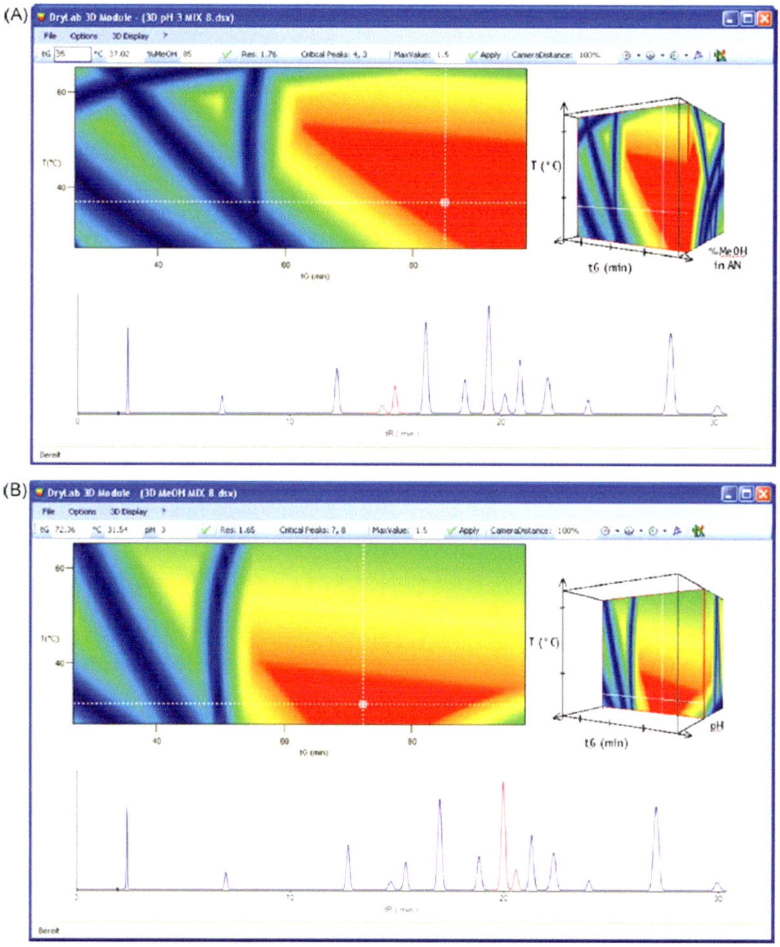

Figure 2.11 3D resolution spaces generated by DryLab®. (A) Gradient time–temperature–ternary eluent composition and (B) gradient time–temperature–pH models. Reprinted from [32], with permission from Elsevier.

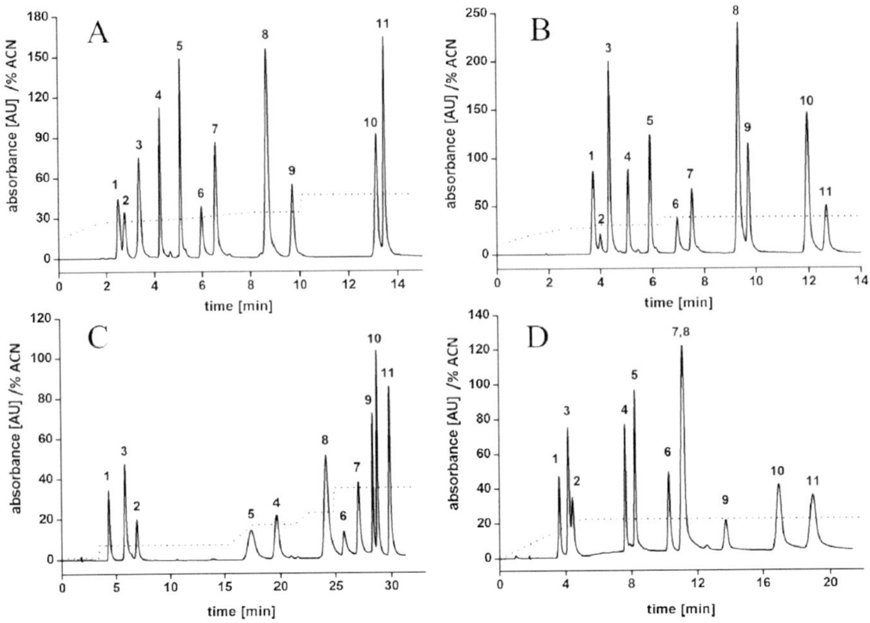

Figure 2.12 Examples of the gradient optimization by ChromSword®, indicating the optimized gradient separations of flavonoids using different columns. Gradient profiles were provided by ChromSword® and are indicated by dotted lines: (A) Phenyl, (B) Phenoxy, (C) Phenyl-amide, (D) Phenyl-amine. Reprinted from [33], with permission from Elsevier.

2.3.4.2 Method optimization using DOE software

When general DOE software is applied to analytical method development, analysts must select a suitable DOE type depending on the stage of analytical method development. Common types of DOE are summarized in Table 2.3 and illustrated in Fig. 2.13. Box–Behnken and central composite designs are two of the most widely used designs for method optimization, and enable the generation of reliable contour plots or surface response plots as outputs using fewer experiments than full factorial design. While the contour plot presents the variation in analytical method performance (output variables) with the contour line as a function of two analytical method parameters (input variables), the surface response plot is a 3D plot showing the relationship between the method performance index and three method parameters.

Table 2.3 Commonly used types of DOE and their outcomes

Types of DOE	Experimental design	General application	Response surface plot	Factor interaction investigation
(1) Full factorial	N^k design	Robustness confirmation	Possible (from >2 level design)	Possible
(2) Central composite	2^k full factorial design + center point + star points (including edge points)	Robustness confirmation, method optimization	Possible	Partly possible
(3) Box–Behnken	Three-level incomplete factorial design (not including the edge points)	Method optimization	Possible	Partly possible
(4) Plackett–Burman	Economical design	Method screening	Not possible	Not possible

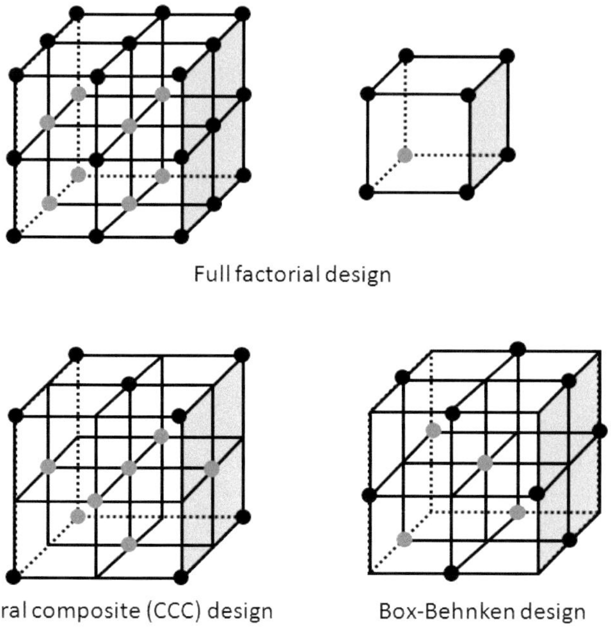

Figure 2.13 Schematic representations of the experimental design images.

Furthermore, the Plackett–Burman design is also commonly employed for method scouting/screening and for primary method development, as this type of DOE can identify and rank the critical analytical method parameters that have a large impact on the target method performance. This again requires fewer experiments than Box–Behnken or central composite designs (see Section 2.3.3.2, Offline method scouting). The selected critical method parameters based on the results of DOE using Plackett–Burman design can then be investigated further in subsequent method optimization studies.

Figure 2.14 shows one example of the contour plots obtained as a result of DOE studies for method optimization.

DOE studies for method optimization. Boussès et al. investigated the resolution between various sets of peaks, the efficiency of critical peaks, the mobile phase volume, and the volume of ethanol as the target analytical method performance affected by the four critical analytical method parameters identified in the previous method screening stage [28]. The relationship between these method parameters and method performances was examined using central composite design DOE with a mobile phase pH range of 4.0–5.0, a gradient slope of 2.0–4.0%/min, a column temperature of 30–50°C, and a flow rate of 0.2–0.4 mL/min. The multivariate correlation analysis of the obtained method performance results with method parameters using the selected software produced the contour plots shown in Fig. 2.14. The white regions in the plots indicate that the results of all target method performances met an acceptable value. Furthermore, the method operable design regions were identified as a combination of parameters with ranges where any combination of parameters could assure an acceptable method performance. The regions are also known as the optimal space of method parameters.

The above example was obtained following the application of DOE for an LC separation method. As previously mentioned, DOE can also be applied for other chromatographic separation methods, such as SFC methods. In this context, Landagaray et al. developed the central composite design with three factors (i.e., flow rate, pressure outlet, and percentage of ethanol) to optimize the performance of the SFC method [34].

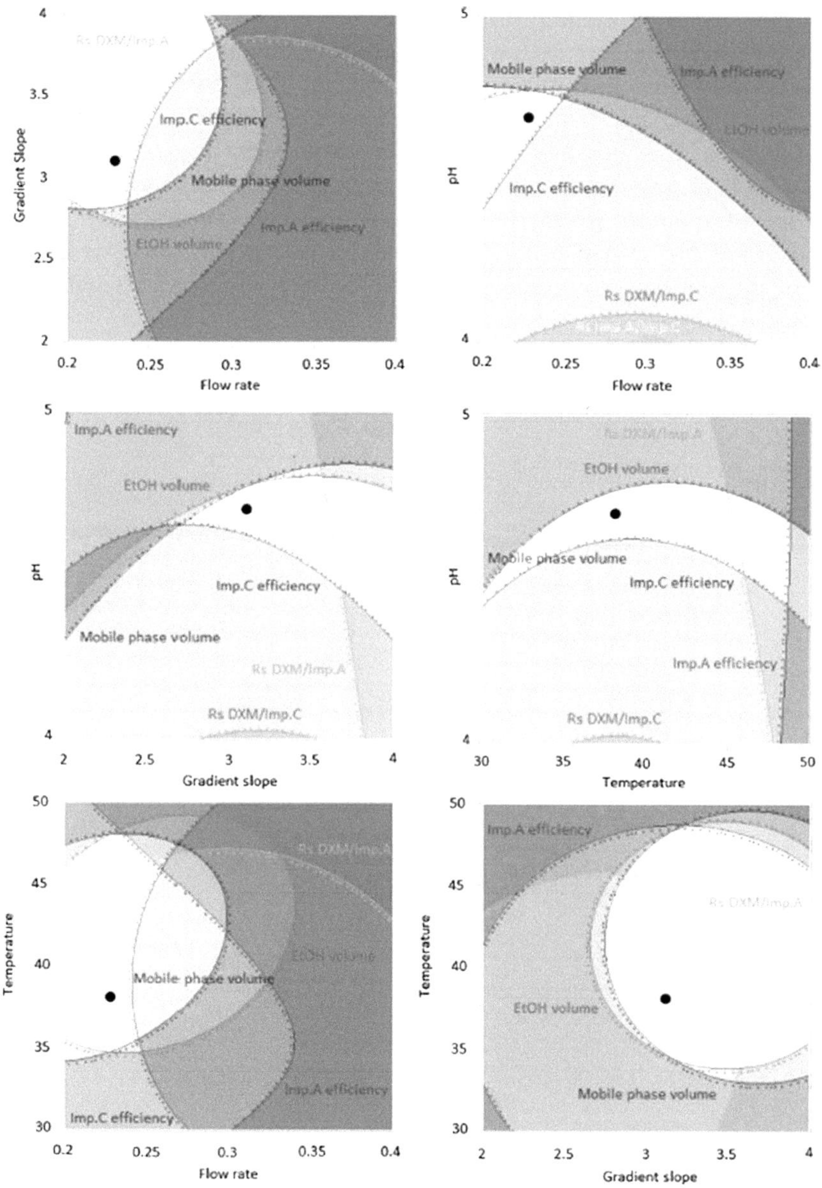

Figure 2.14 Contour plots of the seven method responses of interest as a function of the 4 studied method parameters. The white zone corresponds to the optimal space of method parameters where all responses fulfilled the requirements, with (•) representing the optimum conditions. Reprinted from [28], with permission from Elsevier (color altered).

The results obtained through this experimental design were input into the software, and the surface response plot outlining the relationship between the three method parameters and the method performance was calculated as shown in Fig. 2.15. Based on these results, analysts can either select the appropriate analytical method parameters or allow DOE software to select a set of method parameters to maximize the output analytical method performance. In this research, the optimization module of the software was utilized and the optimal method parameters were selected.

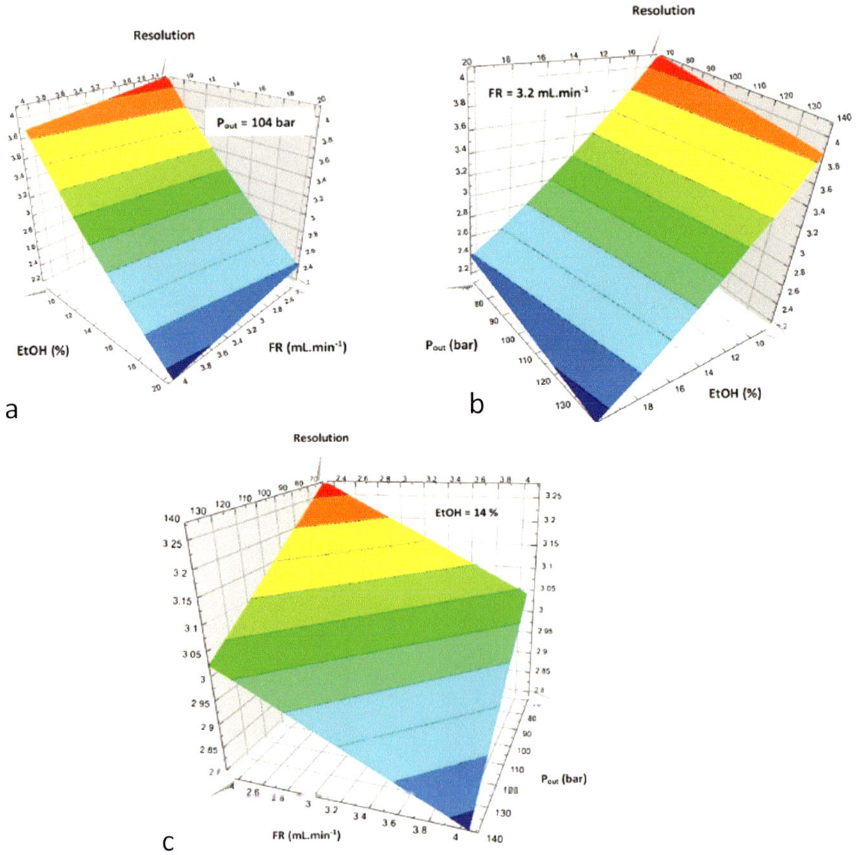

Figure 2.15 Surface response modeling of the resolution for the SFC method as a function of (a) the flow rate, (b) the percentage of ethanol in the mobile phase, and (c) the outlet pressure. Reprinted from [34], with permission from Elsevier.

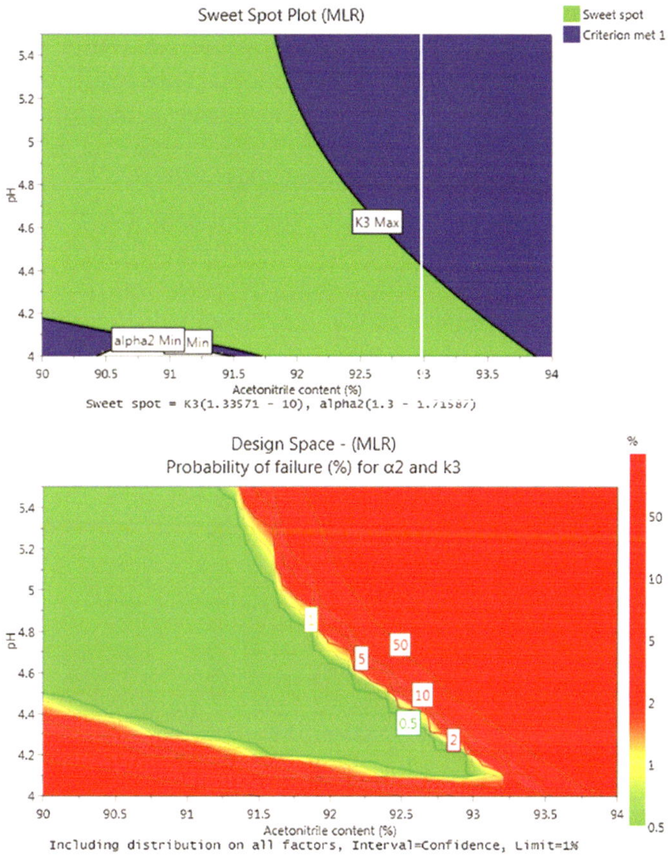

Figure 2.16 Example 1 of the optimal space of method parameters considering the probability concept. (Left) Sweet Spot plot obtained by plotting acetonitrile content vs. aqueous phase pH as defined by the requirements. Regions where only one criterion is met are colored in blue, while regions where both criteria are met are colored in green. (Right) Design space defined for the acetonitrile content and pH of the aqueous phase in consideration of the probability. Reprinted from [35], with permission from Elsevier.

Recently, the probability concept has also been considered when determining the optimal space of analytical method parameters, with one example being depicted in Fig. 2.16. The left-hand panel shows the Sweet Spot plot, which was obtained by multivariate analysis based on the obtained results, followed by the construction of a prediction model by the DOE software.

However, this model does not take into consideration the variation in precision of the investigated analytical method parameters and model errors. The border lines were therefore reconsidered to assure a method performance with a probability of 99%, and the resulting optimal/design space is presented as the green region of the right-hand panel.

Hubert et al. also adopted the probability concept for determination of the optimal space using Monte-Carlo simulations [36]. The determined optimal space is presented as the area within the blue rectangle in Fig. 2.17C, with the working point indicated by the red circle. Based on this result, the robustness was evaluated for the conditions within the optimal space in terms of the preset target method performance.

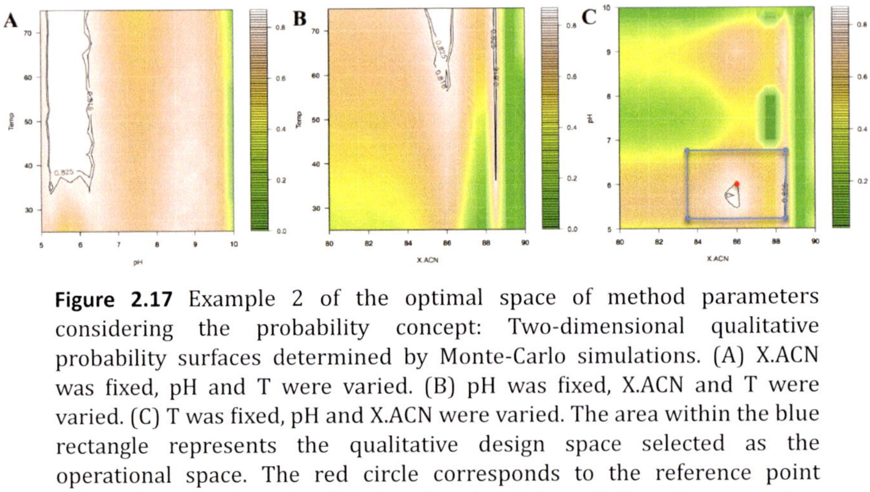

Figure 2.17 Example 2 of the optimal space of method parameters considering the probability concept: Two-dimensional qualitative probability surfaces determined by Monte-Carlo simulations. (A) X.ACN was fixed, pH and T were varied. (B) pH was fixed, X.ACN and T were varied. (C) T was fixed, pH and X.ACN were varied. The area within the blue rectangle represents the qualitative design space selected as the operational space. The red circle corresponds to the reference point selected for the formal validation. Reprinted from [36], with permission from Elsevier.

2.3.5 Robustness Evaluation

The final stage of analytical method development involves evaluation of the robustness of the optimized method. The robustness of an analytical method is particularly important in the case of methods employed over a number of years, such as release testing methods for commercial products. As stated in ICH Q2(R1) [37], robustness is a measure of reliability of an analysis with respect to deliberate variations in method parameters and should be evaluated during method development phase. Indeed, robustness should be evaluated and confirmed at least in the final stage of the method development phase. The conventional approach for robustness evaluation is the one parameter approach, i.e., the effect of changing a single analytical method parameter on method performance is evaluated while the other parameters remain unchanged. However, it is not possible to examine the interaction effects of various factors using this approach. As such, the application of multivariate analysis has recently become more common, as its procedure for robustness evaluation is similar to DOE studies for method optimization, with the only differences being the investigated range of method parameters and DOE type.

A number of methods are available to carry out robustness evaluations. For example, full factorial DOE is a basic technique used to confirm that all edge points of an optimal space meet the acceptance criteria. In the case of an analytical method developed and optimized based on DOE studies that was consequently well understood, and where the identification of the optimal space of method parameters within the contour plot or surface response plot had been carried out, the application of the partial factorial DOE or Plackett–Burman design may be acceptable for the subsequent robustness evaluation studies.

Furthermore, the functions of chromatographic software packages have recently been expanded, including the inclusion of automatic robustness evaluation. For example, some chromatographic software packages can conduct DOE modeling and analysis combined with the simultaneous operation of chromatographic apparatus, i.e., steps 4–6 in Table 2.1, can be performed automatically. For example, ChromSword® software coupled with Agilent LC apparatus has the capability to carry out such an evaluation. Following determination of the parameters and ranges to be investigated for robustness evaluation, the various values are input into the software, and the following procedures are conducted automatically. After designing all experiments and carrying out the corresponding LC measurements for all planned experiments, the obtained results of the robustness evaluation are plotted as a contour plot. Furthermore, robustness confirmation can also be obtained based on simulated chromatograms, with one example including the Robustness Module of DryLab®. Where a DOE study has previously been conducted to investigate the impact of the target analytical method parameters on method performance over a wide investigation range, the robustness of the method performance within the space of a narrow range of method parameters can be evaluated without additional experiments. Finally, the chromatograms are simulated for all points of the full factorial design experiments, and the robustness is evaluated based on the simulated data.

2.4 Advantages and Disadvantages of Using Available Method Development Software Programs

As discussed above, the implementation of chromatographic software and/or DOE software into analytical method development is extremely beneficial. As software capabilities advance, such application will be enhanced in future decades. However, at present, the pros and cons of the software packages and their implementation in analytical method development can be summarized as follows:

Pros

- ✓ Possibility of significantly reducing experimental/analytical time, resources, and number of experiments required
- ✓ Systematic approach enabling the identification and statistical analysis of the optimal point or space of the analytical method parameters
- ✓ Feasibility of chromatogram simulation without carrying out any physical experiments (only for chromatographic software)
- ✓ Easy to understand through a visualized outcome
- ✓ Easier transfer for knowledge and technical skills via better understandings of analytical methods

Cons

- × Possibility of inconsistency between the predicted (simulated) and obtained chromatogram (Prediction accuracy)

- Necessity of training to use the software
- Inability to characterize the effect of injection solvent mismatch and injection solvent volume overload [38].

Among the various points outlined above, a complementary relationship can be established between the number of necessary experiments and the prediction accuracy, with an increase in the number of experiments increasing the prediction accuracy, and vice versa. Therefore, each function of the software is employed at a different development stage depending on the stage in question. This relationship is summarized in Table 2.4. For example, in the early development stage, a number of parameters and multidimensional factors must be widely addressed. However, the prediction accuracy is not critical at this stage, as no or few experimental results are required. In contrast, in the late development stage, only critical parameters over a narrower range should be considered, and the prediction should be accurate and based on a large amount of previous and/or ongoing development outcomes.

Table 2.4 Comparison of software features for chromatographic method development

Development stage	Early stage ⟷ Late stage				
Type of function	Structure-based retention prediction	Chromato-gram simulation	Method scouting/ screening	Method optimization	Robustness evaluation
Parameters/ interactions to be investigated	Various ⟷ Focused				
Prediction accuracy	Not critical ⟷ Critical				
Experiments	No or a few experiments ⟷ Many experiments				

2.5 Future Perspective

As described in this chapter, the implementation of software into the analytical method development process provides a number of advantages in terms of saving time and resources, and reducing the number of experiments required. Over time, the functions and features of software packages have been developed and gradually improved according to scientific advances and recent trends toward AQbD approaches. As the use of software has become more common, a number of future perspectives for the application of such software in chromatographic method development have been outlined below.

In the middle of the analytical method life cycle, the parameters of the developed method are often changed to improve separation, reduce run times, achieve the desired robustness and accuracy at later developmental stages, and due to the unavoidable changes in column or critical reagents. Indeed, both major and minor changes are often made even in post-approval stages. For such cases, the AQbD approach established for chromatographic method development by the appropriate software is the key for such changes; an example would be changes in the analytical method parameters within the developed optimal space, i.e., working point shifts within the established ranges. If the method is developed without the AQbD approach, even such a minor change requires additional experiments and method validation. In contrast, when the method is developed using the AQbD approach assisted by software allowing the determination of an optimal space, further studies are generally not required following such a change.

A further example is the unavoidable change in column vendor. As the column vendor is not a continuous parameter such as pH or ion strength, establishing an optimal space that incorporates multiple column vendors may be difficult without the aid of software. However, if the analytical method for the original column is developed based on the AQbD approach with the support of appropriate software, and that for the new column is developed in a similar way, the two analytical methods can be easily compared, and any differences between them identified. Such a comparison can be performed either before or after adopting the method. Figure 2.18 shows an example of the latter case, where the 2D planes of the 3D resolution cubes are presented with a color gradient from blue to red, which corresponds to low- to high-resolution values of two peaks for each of the three columns. The similarities of the optimal space between the first (original) and second columns can be confirmed through a similar color distribution pattern, while the differences between the third column and the other two columns can be identified through different patterns. This may suggest the possibility of easily shifting the optimal space for the first column to that of the second column, and the difficulty of shifting to the third column. Thus, such an AQbD approach should ideally be adopted at the beginning of method development, as it can permit simple and efficient changes in methods in the later stages of development.

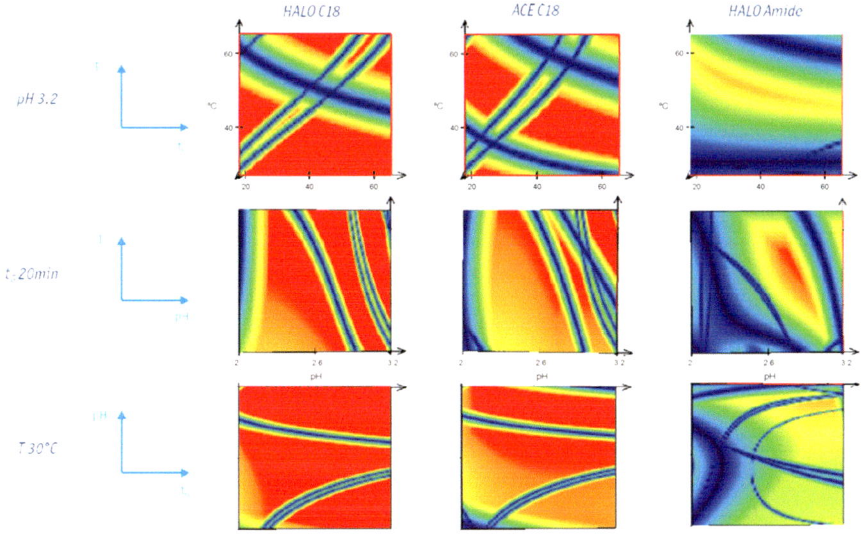

Figure 2.18 Comparison of three different 2-D planes of the 3-D resolution spaces generated using DryLab® for each column. Reprinted from [4], with permission from Elsevier.

Thus, in summary, in recent years, the chromatogram simulation capability of chromatographic software has been advanced, and the utilization of software for analytical method development has become more widespread. In addition, there have been increasing reports of the QbD approach being adopted for analytical method development (AQbD). This AQbD approach is combined with analytical method development using software or through DOE studies. It may therefore be possible that the simulated or calculated results generated by the software can be utilized for robustness evaluations or other areas of commercial authorization application documents for analytical methods, thus avoiding the necessity for obtaining actual experimental data through a more traditional development strategy. It is therefore expected that the utilization of chromatographic or DOE software may be accepted by the regulatory authorities in the future through the increased implementation of the QbD approach in the pharmaceutical industry, i.e., representing a science- and risk-based approach to pharmaceutical analytical method development.

References

1. Lunn, G. (2005). *HPLC Methods for Recently Approved Pharmaceuticals.* (Wiley, USA).
2. Snyder, L. R., Kirkland, J. J., and Glajch, J. L. (1997). *Practical HPLC Method Development.* (Wiley, USA).
3. Swadesh, J. K. (1996). *HPLC: Practical and Industrial Applications.* (CRC Press, USA).
4. Monks, K. E., Rieger, H.-J., and Molnár, I. (2011). Expanding the term "Design Space" in high performance liquid chromatography (I). *Journal of Pharmaceutical and Biomedical Analysis*, **56**, pp. 874–879.
5. Rozet, E., Lebrun, P., Debrus, B., Boulanger, B., and Hubert, P. (2013). Design spaces for analytical methods. *Trends in Analytical Chemistry*, **42**, pp. 157–167.
6. Snyder, L. R., Dolan, J. W., and Quarry, M. A. (1987). High-performance liquid chromatographic method-development using computer simulation. *Trends in Analytical Chemistry*, **6**, pp. 106–111.
7. ICH Q8(R2). Pharmaceutical Development.
8. Molnar, I. (2002). Computerized design of separation strategies by reversed-phase liquid chromatography: Development of DryLab software. *Journal of Chromatography A*, **965**, pp. 175–194.

9. Snyder, L. R., Dolan, J. W., and Lommen, D. C. (1989). Drylab® computer simulation for high-performance liquid chromatographic method development: I. Isocratic elution. *Journal of Chromatography A*, **485**, pp. 65–89.

10. Karmarkar, S., Yang, X., Garber, R., Szajkovics, A., and Koberda, M. (2014). Quality by design (QbD) based development and validation of an HPLC method for amiodarone hydrochloride and its impurities in the drug substance. *Journal of Pharmaceutical and Biomedical Analysis*, **100**, pp. 167–174.

11. Meng, M., Rohde, L., Čápka, V., Carter, S. J., and Bennett, P. K. (2010). Fast chiral chromatographic method development and validation for the quantitation of eszopiclone in human plasma using LC/MS/MS. *Journal of Pharmaceutical and Biomedical Analysis*, **53**, pp. 973–982.

12. Hewitt, E. F., Lukulay, P., and Galushko, S. (2006). Implementation of a rapid and automated high performance liquid chromatography method development strategy for pharmaceutical drug candidates. *Journal of Chromatography A*, **1107**, pp. 79–87.

13. Maeda, E., Urakami, K., Shimura, K., Kinoshita, M., and Kakehi, K. (2010). Charge heterogeneity of a therapeutic monoclonal antibody conjugated with a cytotoxic antitumor antibiotic, calicheamicin. *Journal of Chromatography A*, **1217**, pp. 7164–7171.

14. Debrus, B., Guillarme, D., and Rudaz, S. (2013). Improved quality-by-design compliant methodology for method development in reversed-phase liquid chromatography. *Journal of Pharmaceutical and Biomedical Analysis*, **84**, pp. 215–223.

15. Xiao, K. P., Xiong, Y., Liu, F. Z., and Rustum, A. M. (2007). Efficient method development strategy for challenging separation of pharmaceutical molecules using advanced chromatographic technologies. *Journal of Chromatography A*, **1163**, pp. 145–156.

16. Talebi, M., Schuster, G., Shellie, R. A., Szucs, R., and Haddad, P. R. (2015). Performance comparison of partial least squares-related variable selection methods for quantitative structure retention relationships modelling of retention times in reversed-phase liquid chromatography. *Journal of Chromatography A*, **1424**, pp. 69–76.

17. Park, S. H., Shellie, R. A., Dicinoski, G. W., Schuster, G., Talebi, M., Haddad, P. R., Szucs, R., Dolan, J. W., and Pohl, C. A. (2016). Enhanced methodology for porting ion chromatography retention data. *Journal of Chromatography A*, **1436**, pp. 59–63.

18. Wang, L., Zheng, J., Gong, X., Hartman, R., and Antonucci, V. (2015). Efficient HPLC method development using structure-based

database search, physico-chemical prediction and chromatographic simulation. *Journal of Pharmaceutical and Biomedical Analysis*, **104**, pp. 49–54.

19. Monks, K., Molnár, I., Rieger, H. J., Bogáti, B., and Szabó, E. (2012). Quality by Design: Multidimensional exploration of the design space in high performance liquid chromatography method development for better robustness before validation. *Journal of Chromatography A*, **1232**, pp. 218–230.
20. Horváth, C., Melander, W., and Molnár, I. (1976). Solvophobic interactions in liquid chromatography with nonpolar stationary phases. *Journal of Chromatography A*, **125**, pp. 129–156.
21. Nikitas, P., and Pappa-Louisi, A. (2009). Retention models for isocratic and gradient elution in reversed-phase liquid chromatography. *Journal of Chromatography A*, **1216**, pp. 1737–1755.
22. Zheng, J., and Rustum, A. M. (2010). Rapid separation of desloratadine and related compounds in solid pharmaceutical formulation using gradient ion-pair chromatography. *Journal of Pharmaceutical and Biomedical Analysis*, **51**, pp. 146–152.
23. Kochling, J., Wu, W., Hua, Y., Guan, Q., and Castaneda-Merced, J. (2016). A platform analytical quality by design (AQbD) approach for multiple UHPLC-UV and UHPLC-MS methods development for protein analysis. *Journal of Pharmaceutical and Biomedical Analysis*, **125**, pp. 130–139.
24. Euerby, M. R., Hulse, J., Petersson, P., Vazhentsev, A., and Kassam, K. (2015). Retention modelling in hydrophilic interaction chromatography. *Analytical and Bioanalytical Chemistry*, **407**, pp. 9135–9152.
25. Goga-Remont, S., Heinisch, S., and Rocca, J. L. (2000). Use of optimization software to determine rugged analysis conditions in high-performance liquid chromatography. *Journal of Chromatography. A*, **868**, pp. 13–29.
26. Tyteca, E., Guillarme, D., and Desmet, G. (2014). Use of individual retention modeling for gradient optimization in hydrophilic interaction chromatography: Separation of nucleobases and nucleosides. *Journal of Chromatography A*, **1368**, pp. 125–131.
27. Alden, P., Yurach, D., and Warren, P. (2010). An automated workflow for accelerated development of a robust LC method within a QbD framework. *The Column*, **6**, pp. 15–19.
28. Boussès, C., Ferey, L., Vedrines, E., and Gaudin, K. (2015). Using an innovative combination of quality-by-design and green analytical chemistry approaches for the development of a stability indicating

UHPLC method in pharmaceutical products. *Journal of Pharmaceutical and Biomedical Analysis*, **115**, pp. 114–122.

29. Debrus, B., Lebrun, P., Ceccato, A., Caliaro, G., Rozet, E., Nistor, I., Oprean, R., Rupérez, F. J., Barbas, C., Boulanger, B., and Hubert, Ph. (2011). Application of new methodologies based on design of experiments, independent component analysis and design space for robust optimization in liquid chromatography. *Analytica Chimica Acta*, **691**, pp. 33–42.
30. Debrus, B., Lebrun, P., Kindenge, J. M., Lecomte, F., Ceccato, A., Caliaro, G., Mavar Tayey Mbay, J., Boulanger, B., Marini, R. D., Rozet, E., and Hubert, Ph. (2011). Innovative high-performance liquid chromatography method development for the screening of 19 antimalarial drugs based on a generic approach, using design of experiments, independent component analysis and design space. *Journal of Chromatography A*, **1218**, pp. 5205–5215.
31. Wang, L., and Qu, H. (2016). Development and optimization of SPE-HPLC-UV/ELSD for simultaneous determination of nine bioactive components in Shenqi Fuzheng injection based on quality by design principles. *Analytical and Bioanalytical Chemistry*, **408**, pp. 2133–2145.
32. Molnár, I., Rieger, H. J., and Monks, K. E. (2010). Aspects of the "Design Space" in high pressure liquid chromatography method development. *Journal of Chromatography A*, **1217**, pp. 3193–3200.
33. Janas, P., Bocian, S., Jandera, P., Kowalkowski, T., and Buszewski, B. (2016). Separation of flavonoids on different phenyl-bonded stationary phases-the influence of polar groups in stationary phase structure. *Journal of Chromatography A*, **1429**, pp. 198–206.
34. Landagaray, E., Vaccher, C., Yous, S., and Lipka, E. (2016). Design of experiments for enantiomeric separation in supercritical fluid chromatography. *Journal of Pharmaceutical and Biomedical Analysis*, **120**, pp. 297–305.
35. Terzić, J., Popović, I., Stajić, A., Tumpa, A., and Jančić-Stojanovic, B. (2016). Application of analytical quality by design concept for bilastine and its degradation impurities determination by hydrophilic interaction liquid chromatographic method. *Journal of Pharmaceutical and Biomedical Analysis*, **125**, pp. 385–393.
36. Hubert, C., Houari, S., Rozet, E., Lebrun, P., and Hubert, Ph. (2015). Towards a full integration of optimization and validation phases: An analytical-quality-by-design approach. *Journal of Chromatography A*, **1395**, pp. 88–98.

37. ICH Q2(R1). Validation of Analytical Procedures: Text and Methodology.
38. Jeong, L. N., Sajulga, R., Forte, S. G., Stoll, D. R., and Rutan, S. C. (2016). Simulation of elution profiles in liquid chromatography-I: Gradient elution conditions, and with mismatched injection and mobile phase solvents. *Journal of Chromatography A*, **1457**, pp. 41–49.

Chapter 3

Flash Chromatography Method Development

J. Robert Bickler

Biotage LLC,
10430 Harris Oaks Blvd., Suite C, Charlotte, North Carolina 28269, USA

bob.bickler@biotage.com

3.1 Introduction

Flash chromatography (a.k.a. flash purification) is a rapid separation and purification technique (hence the term flash) used by organic, medicinal, and natural product chemists to isolate one or more compounds in crude chemical mixtures. The term Flash Chromatography was first coined by W. Clark Still et al. in his seminal paper "Rapid Chromatographic Technique for Preparative Separations with Moderate Resolution," published in 1978 [1].

Flash chromatography requires only a few items—a solid, inert sorbent (e.g., silica), a sturdy, inert column (glass, polypropylene),

Chromatographic Method Development
Edited by Gregory K. Webster and Laila Kott

ISBN 978-981-4800-53-2 (Hardcover), 978-0-429-20172-1 (eBook)
www.jennystanford.com

solvents, and a pump or vacuum source. Today, though, most chemists use automated flash systems with built-in pumps, gradient making capability, in-line detectors, and fraction collectors, all controlled by relatively simple-to-use software.

3.2 Goal Setting

As with any chromatographic technique, the results obtained are dependent on several factors so in order to insure success goals should be set. Chemists use flash chromatography for one reason—to purify and isolate the compound synthesized.

Medicinal and organic chemists utilize flash chromatography to purify synthesized intermediates while natural product chemists use flash chromatography to separate extracts into as many individual compounds as possible. In both cases, having set goals and expectations will help in determining the best route or routes to satisfactorily complete the task.

Typical goals include the following:

- Purity level
- Yield amount
- Throughput (mg/h or g/h)
- Scalability (from bench to pilot to production)

Note that even though a detector is frequently employed, quantitation is not performed in this technique and typical goals for analytical techniques such as precision and sensitivity are not given any priority.

3.3 Starting Out

Flash chromatography itself is not an overly difficult technique; however, it is the sample complexity that challenges most chemists. Having the right tools and performing some preliminary tests using thin-layer chromatography (TLC) with a variety of different solvents will supply the information required for an efficient flash purification. Most chromatography techniques rely on a series of scouting runs to determine optimal conditions for a separation. Because this application is to purify a newly synthesized compound only one time, TLC results are used as a substitute for scouting runs.

3.3.1 Thin-Layer Chromatography

Thin-layer chromatography can be (and should be) a chemist's most important tool for purifying crude mixtures. With TLC, chemists are able to scout various solvent mixtures that help optimize both the separation and load amount.

TLC data is directly transferrable to flash chromatography. A compound's retardation factor (Rf) on TLC is inversely proportional to the number of column volumes (CV) required for its elution from a flash column using the identical solvent blend (Table 3.1):

$$CV = 1/Rf \tag{3.1}$$

Table 3.1 The relationship between TLC Rf values and the number of required column volumes for elution from a flash column

Rf	CV
0.1	10
0.2	5
0.3	3.33
0.4	2.5
0.5	2
0.6	1.67
0.7	1.43
0.8	1.25
0.9	1.11
1	1

In Fig. 3.1, we see how that relationship looks in terms of both a TLC plate and a flash column.

For those familiar with other chromatography techniques such as HPLC or GC, think of CV as k′. However, the separation performance with TLC and flash is based on the difference between any two adjacent compounds' CV. This value is known as ΔCV:

$$\Delta CV = CV_2 - CV_1 \tag{3.2}$$

A large ΔCV indicates a good separation and a high loading capacity for any given cartridge or column.

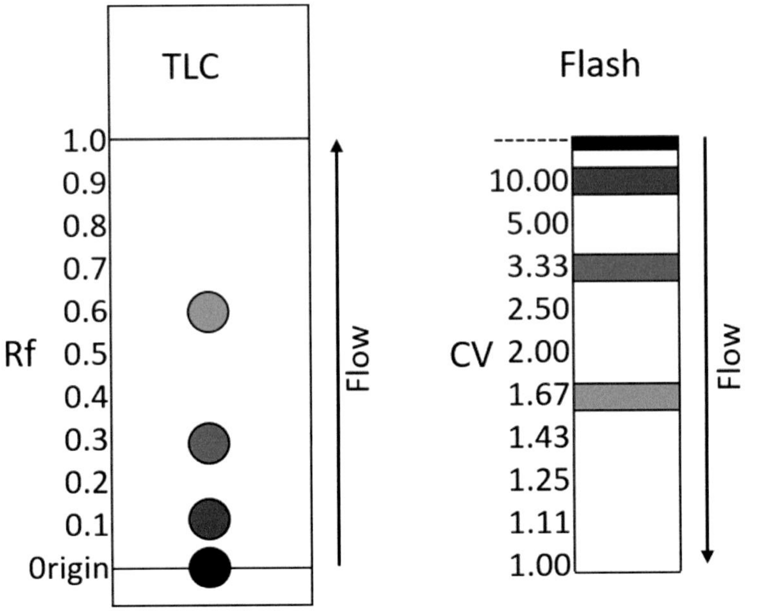

Figure 3.1 The relationship between Rf from a TLC plate and the number of CV required to elute the compound is illustrated here. The relationship is reciprocal allowing the chemist to determine the required solvent volume to elute each compound in the mixture being separated.

3.3.2 Using TLC to Scout Suitable Elution Solvents

When first developing a flash purification method an investigation into suitable solvents is recommended. Solvents are classified by their chemistry into selectivity classes and relative elution strengths versus silica and/or alumina. These are separate variables which need exploration separately, starting with selectivity.

Table 3.2 provides a list of commonly used solvents and their selectivity class. The key to using this data is to assist with solvent evaluation. Create solvent blends from different selectivity groups to evaluate their impact on the separation. As an example, choose hexane and ethyl acetate as well as a mix of hexane and ether or hexane and DCM. More likely than not, a different separation will be seen using TLC for each solvent blend.

Table 3.2 Common flash solvents and their selectivity group [1]

Solvent	Selectivity group
Ether	I
MeOH	II
EtOH	II
PrOH	II
THF	III
DCM	V
Acetone	VI a
EtOAc	VI a
MeCN	VI b
Toluene	VII
Chloroform	VIII
Hexane	—
Heptane	—
Isooctane	—

Once you have found a solvent or solvent blend that best separates your target compound from the other impurities and by-products it is time to modify the blend's strength. Strength adjustment ensures your product elutes with maximum resolution from the impurities in minimal time using minimal solvent.

Table 3.3 provides a listing of common flash solvents and their strength relative to silica [2].

Elution strength is adjusted by manipulating the solvent blend ratio so your target compound has a TLC Rf of between 0.10 and 0.4 (10 CV to 2.5 CV). This is true for both isocratic and gradient elution. Usually it will not matter if the other compounds in the crude sample are not separated from each other as long as your targeted compound is separated from everything else.

To decrease Rf values, add a weaker solvent such as hexane as it—and other saturated hydrocarbon solvents—has essentially no elution strength. Once you have attained a suitable separation from TLC, you can directly transfer the conditions to your flash system.

Table 3.3 Common flash solvents and their strength relative to silica

Solvent	Strength
MeOH	0.70
EtOH	0.68
PrOH	0.60
THF	0.53
MeCN	0.51
Acetone	0.50
EtOAc	0.43
Ether	0.40
DCM	0.32
Chloroform	0.26
Toluene	0.22
Hexane	—
Heptane	—
Isooctane	—

3.4 Determining Cartridge Load Capacity

Before you can purify your crude material, you need to determine the amount of media (silica, C18, etc.) needed in order to fully separate your targeted compound from the others in the mix. To do this, you can use your TLC's Rf data for your target compound and its two nearest neighbors—the one directly above it and the one directly below it.

Convert each compound's Rf value to a CV value using Eq. 3.1 and then calculate the ΔCV for each pair using Eq. 3.2. Use the lower ΔCV value for determining load capacity from Table 3.4.

For example, let us say you have TLC data showing three spots with Rf values of 0.6, 0.3, and 0.1. Using Eq. 3.1, we can convert each Rf into its CV, determine the ΔCV for each adjacent pair.

$$CV_1 = 1/0.6 = 1.67$$

$$CV_2 = 1/0.3 = 3.33$$

$$CV_3 = 1/0.1 = 10$$

$$CV_2 – CV_1 = 1.66$$

$$CV_3 – CV_2 = 6.67$$

Based on this data, the lower ΔCV is 1.66 and the suggested load amount for any cartridge size is found in the column headed by <2 in Table 3.4. For instance, if the sample size is 0.1 g, the appropriate cartridge size is 10 g.

Table 3.4 Flash cartridge load estimate (in grams) based on TLC DCV data

	ΔCV		
Cartridge size (g)	**<2**	**2–3.9**	**>4**
5	<0.05	0.05–0.25	0.25–0.5
10	<0.1	0.1–0.5	0.5–1.0
25	<0.25	0.25–0.75	0.75–2.5
50	<0.5	0.5–2.5	2.5–5.0
100	<1.0	1.0–5.0	5.0–10
250	<2.5	2.0–7.5	7.5–25
500	<5	5–25	25–50
750	<7.5	6.0–22.5	22.5–75
1000	<10	10–50	50–100
1500	<15	12–45	45–150

Frequently chemists will take a shortcut and simply apply the rule-of-thumb that for every 1 g of crude material to be separated, 100 g of silica is needed; this is known as the 1% rule.

While this approach can work, it ignores the fact that the ΔCV's usefulness is in determining the optimal amount of silica for a particular separation.

3.5 Determining Elution Style

There are three basic elution styles used in flash chromatography:

- Isocratic
- Linear gradient
- Step gradient

Each of these typically provides a suitable separation, but gradients elute the compounds in more highly concentrated bands and often with better resolution between them.

3.5.1 Isocratic Elution

Isocratic elution just means that the solvent ratio remains constant throughout the purification process (Fig. 3.2).

Your TLC provides you with the solvent ratio to use and the Rf data. Converting the Rf values to CV and then ΔCV provides separation quality information you can use to determine cartridge size and loading. Although isocratic elution can be quick and simple, it usually results in lower loading capacity and less than ideal separations.

3.5.2 Linear Gradient Elution

With linear gradient elution, the mobile phase ratio changes over time from a weak solvent blend to a stronger solvent blend. The theory is that poorly retained compounds will retain and separate better starting the elution with a weak solvent blend. The better-retained compounds will then separate and elute by increasing the solvent strength at a constant rate over a period of time (Fig. 3.3).

A major benefit to linear gradient elution is its ability to reduce the bandspreading seen with isocratic elution methods. This concentration effect helps with both compound detection and recovery. Load capacity is typically improved as well. Most chemists today use linear gradients for flash purification.

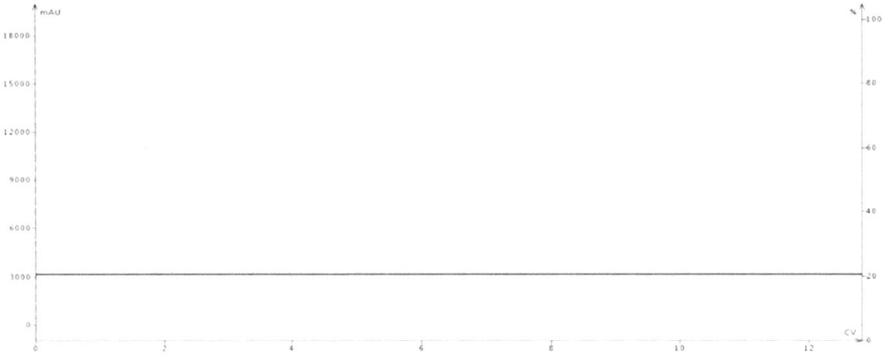

Figure 3.2 An example of isocratic flash chromatography. The solvent ratio remains constant throughout the chromatography run and the compounds elute at volumes consistent with the predicted values from TLC.

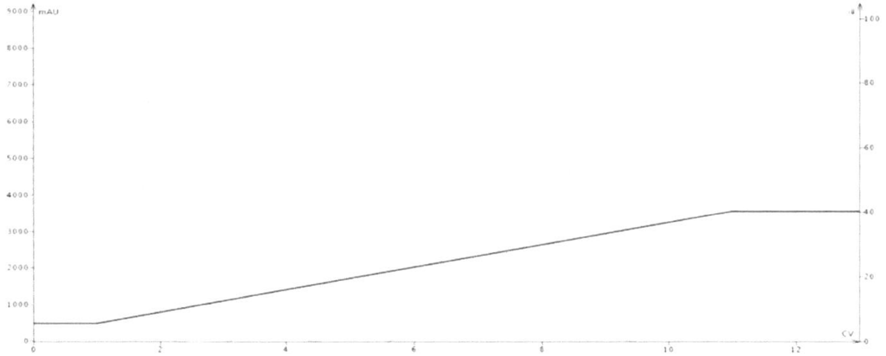

Figure 3.3 An example of linear gradient elution. The ratio of strong solvent to weak solvent changes during the separation enabling poorly retained compounds to separate better and well retained compounds to elute in higher concentration.

The keys to a good linear gradient are the starting and ending solvent ratios and the gradient slope. Many chemists use a gradient that runs from 0% strong solvent to 100% strong solvent over a specific time period or volume. While this approach can work, it is not based on TLC data and often results in poor separations using either too much or too little solvent.

A better approach is to base your linear gradient on your TLC data using the algorithm below.

¼ TLC%strong solvent for 1 CV
¼ TLC%strong solvent to 2× the TLC% strong solvent in 10 CV
2× the TLC%strong solvent for 2 CV

As an example, the linear gradient used in Fig. 3.2 was created based on TLC using 20% ethyl acetate in hexanes. Following the algorithm above, the gradient started with 5% ethyl acetate for 1 CV. This was followed by a linear gradient from 5% to 40% ethyl acetate over 10 CV. The gradient ended with 40% ethyl acetate being held for 2 CV. Gradients based on TLC data using the algorithm above will elute all compounds with Rf values from 0.1 to 1.

3.5.2.1 Linear gradients from 0% to 100%

It is common practice, but not supported by this author, among many chemists to use linear gradients that start at 0% strong solvent and run to 100% strong solvent over a span of 10–40 column volumes. This approach skips any efforts to optimize the separation with the assumption that by running over the entire range of solvent blends, the compound of interest will come out at some unpredictable point along the chromatogram.

The lack of predictability, waste of solvent, and necessity of guessing the amount of silica that should be used for separation make this shortcut approach quite undesirable.

3.5.3 Step Gradient Elution

The most powerful elution method is the step gradient. It can also be the most difficult to create and optimize. Step gradients are more powerful because each compound in the sample mix can have its separation parameters optimized. This has been found to be true in many cases, and Fig. 3.4 is a nice example using the same five-component mix used in Figs. 3.2 and 3.3.

By using an optimized step gradient, the elution of each compound is determined by step-wise solvent ratio changes. Some commercially available automated flash chromatography systems have capability to create an optimized step gradient from a pair of TLC plate data. In Fig. 3.4, Rf data from 20% and 30% ethyl acetate in hexane TLC runs were used.

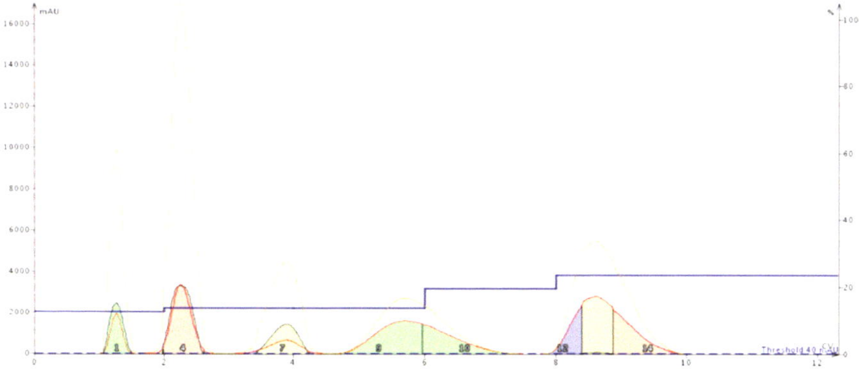

Figure 3.4 An example of step gradient elution. The ratio of strong solvent to weak solvent changes step-wise during the separation. If properly created, the amount of separation between each compound can be optimized leading to higher fraction purity and increased loading.

3.6 Choosing between Normal- and Reversed-Phase

Polar and ionizable compounds typically are not amenable to normal-phase chromatographic methods using organic solvents and polar sorbents such as silica. For these molecules, reversed-phase chromatography is employed. The difference between normal- and reversed-phase is that with reversed-phase the solvents are polar and the sorbent is lipophilic, just the opposite of normal phase.

In reversed-phase, compounds are retained and separated primarily based on hydrophobicity. Reversed-phase method development is also different. While TLC is an excellent method development tool in normal phase, the results from reversed-phase TLC plates do not reliably predict an optimal flash chromatography method. This is due to stationary phase wettability issues and mass-transfer kinetics which differ between reversed-phase TLC plate development and reversed-phase column chromatography.

Instead, methods and gradient shapes are developed by reversed-phase HPLC using columns packed with the same sorbent used in the flash cartridge or on a small flash cartridge since they are re-usable.

A good practice with reversed-phase method development is to start with a solubility test. If the crude material is water soluble, start your gradient in 10% organic solvent (typically methanol or acetonitrile) and run a linear gradient to 100% organic in 10 column volumes (CV). If the compounds elute too fast, repeat the test but set the final organic concentration for 50%. This process can be continued until an acceptable separation is achieved.

If the crude sample is most soluble in organic solvent (methanol or acetonitrile), start your method development at 50% organic and run a 10 CV linear gradient to 100% organic. Make changes to either the starting solvent blend% or the final blend% to obtain the desired separation.

Once a suitable gradient is determined, a loading study can be conducted to determine the method's limitations. This is best accomplished by increasing the amount of sample loaded on the flash cartridge. Maximizing sample concentration in the most polar solvent available will help maximize load. Dissolution solvents such as DMSO, DMF, and water perform best as they are quite polar.

3.7 Conclusion

Flash chromatography is a useful technique for the separation of a broad variety of compounds. It is one of the workhorse techniques used by chemists who need to isolate compounds that are recently synthesized, or extracted from a wide variety of matrices.

References

1. Biotage Flash Consumables brochure (2015) p. 27.
2. http://www.sanderkok.com/techniques/hplc/eluotropic.html.

Chapter 4

Orthogonality in Chromatographic Methods

Zachary S. Breitbach and Gregory K. Webster

AbbVie Inc., Research and Development, 1 N. Waukegan Rd., North Chicago, Illinois 60064, USA

zachary.breitbach@abbvie.com

4.1 Introduction

Throughout chromatographic method development, challenges often arise when peaks elute with the void volume, co-elute with each other, or do not elute at all. In these cases, it is beneficial to evaluate and/or develop orthogonal separation methods. Chromatographic orthogonality simply refers to a separation technique, mode, or phase that offers unique selectivity. Early on, this was most frequently seen as using alternate methods with dissimilar mechanisms of separation to evaluate the selectivity profile of the original method. This was often performed by using a normal-phase method to evaluate a reversed-phase profile but can also be done using TLC, GC, or capillary electrophoresis techniques. Coupling mass spectrometric detection provides a

Chromatographic Method Development
Edited by Gregory K. Webster and Laila Kott

ISBN 978-981-4800-53-2 (Hardcover), 978-0-429-20172-1 (eBook)
www.jennystanford.com

great deal of selectivity information for the original method, but is not considered an "orthogonal" evaluation.

4.2 Traditional View of Orthogonal LC Methods

When developing new LC methods, it is important to consider what mode of chromatography is best employed. Table 4.1 lists several unique modes of LC. Each column of the table (i.e., normal-phase, reversed-phase, and others) can be considered orthogonal to each other, while each mode within each column of the table can also be considered orthogonal. For example, normal-phase separations are always considered orthogonal to reversed-phase separations. Yet, within the normal-phase mode, there is the opportunity of unique selectivity when testing HILIC vs. SFC for example. Further, within a given mode, there are selections of columns which are intentionally developed to be orthogonal to each other. As discussed later in the chapter, different approaches of qualifying or quantifying orthogonality have been developed. While no model is perfect and cannot overcome the need for column screening, the systems described herein give the practitioner direction in choosing columns or groups of columns to screen.

Table 4.1 Example of orthogonal modes of LC

Normal-phase	Reversed-phase	Others
Traditional normal-phase	Traditional reversed-phase	Ion-exchange
Supercritical fluid	Mixed mode	Size exclusion
HILIC	Ion-pairing	Hydrophobic interaction
Polar organic		

4.3 Column Orthogonality in Reversed-Phase Chromatography

4.3.1 Introduction

For the last two decades, researchers have developed alternative approaches to method development. The focus changed from applying different chromatographic methods to evaluate impurity methods. Instead, the primary method developed was challenged by columns that offer differing reversed-phase performance. The second "orthogonal" method is used to identify overlapping peaks or missing peaks found in the initial method. Figure 4.1 illustrates the method development scheme proposed from Pellet and co-workers [1].

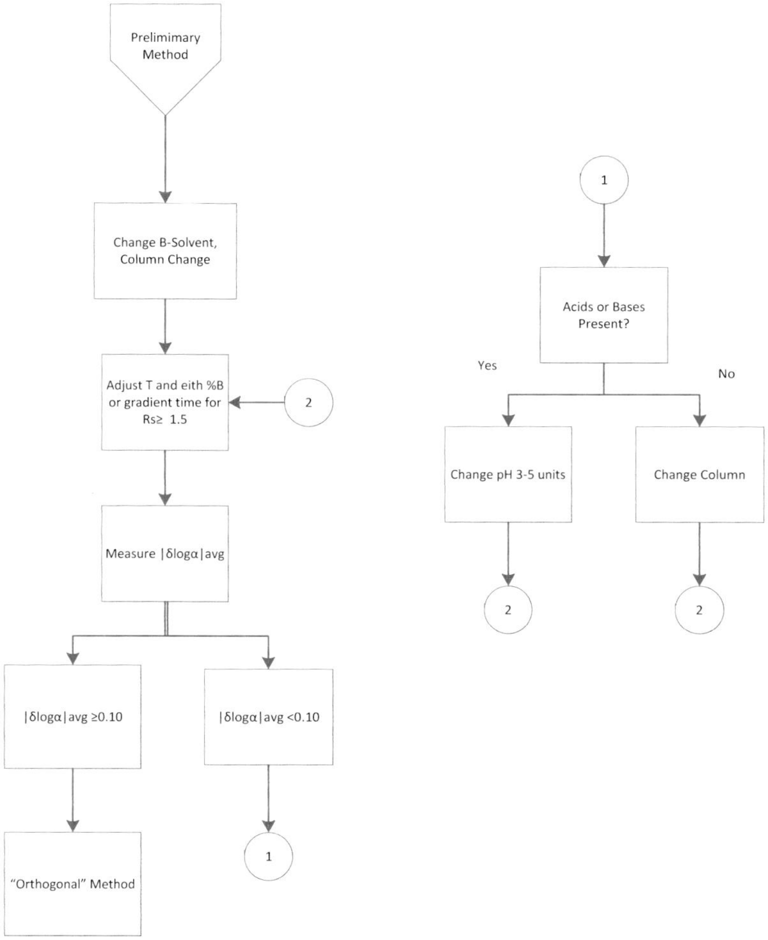

Figure 4.1 Orthogonal column scheme for methods development.

This scheme requires that the change in column selectivity is large enough to resolve previously overlapping peaks. This requires a change in the absolute value of log α ($|\delta \log\alpha|$) of $\geq$ 0.05 units. Larger values of $|\delta \log\alpha|$avg increase the likelihood of resolving overlapped peaks.

4.3.2 Hydrophobic Subtraction Theory

Several groups worked to develop a system of comparing reversed-phase columns that ultimately came to be called the "Hydrophobic Subtraction Theory" (HST) [1–7]. Other systems, such as linear solvation energy relationships (LSER), have been developed but, the HST system is currently most widely used. The HST system for the determination of the separation factor is based on Eq. 4.1:

$$\log\alpha = \eta' H - \sigma' S^* + \beta' A + \alpha' \mathrm{B} + \kappa' C, \tag{4.1}$$

where the solute properties and the column properties are as follows:

Solute Properties

η' = solute hydrophobicity
σ' = solute bulkiness
β' = solute hydrogen-bond basicity
α' = solute hydrogen-bond acidity
κ' = charge on the solute molecule

Column Properties

H = hydrophobicity
S^* = steric interaction
A = column hydrogen-bond acidity
B = column hydrogen-bond basicity
C = cation-exchange activity

The HST system applies five column-selectivity parameters to correlate selectivity of commercial reversed-phase columns. The parameters used are based on (a) steric interaction, (b) column hydrogen-bond acidity, (c) column hydrogen-bond basicity, (d) water sorption and (e) cation-exchange activity. Hydrophobicity, H, is a measure of the hydrophobic interaction between the analyte and the stationary phase, which is the traditional model of the reversed-phase separation. Steric interaction, S^*, is defined by Snyder et al. [7] as a measure of the resistance to insertion of bulky solute molecules into the stationary phase. This parameter is similar to shape selectivity. Hydrogen-bond acidity, A, is characteristic of the non-ionized silanols present in the packing while hydrogen-bond basicity, B, comes from water sorbed in the alkyl silica of Type B stationary phases. Finally, column cation-exchange activity, C, comes from ionized silanols. This last parameter can vary with mobile phase pH.

For an ideal C18 column, H would be considered equal to 1.00, while S^*, A, B, and C would be equal to 0.00. Solute parameters for Eq. 4.1 have been derived from various authors and manufacturers for many commercially available reversed-phase columns and characteristic analytes of varying properties.

4.3.2.1 HST for column comparison

The wide availability of the column data has enabled an efficient way to compare reversed-phase columns without making an injection. The authors of the HST system further developed a simple equation to compare columns called the "Column Selectivity Function" (Eq. 4.2):

$$F_s = [(H_2 - H_1)^2 + (S_2^* - S_1^*)^2 + (A_2 - A_1)^2 + (B_2 - B_1)^2 + (C_2 - C_1)^2]^{1/2} \quad (4.2)$$

Scientists do not have to derive these parameters for a majority of the commercially available reversed-phase columns on the market today. Databases are found in USP and many manufacturer websites.

4.3.2.2 HST values for commercial reversed-phase columns

It is not unusual for laboratories to look for equivalent columns to act as a backup for their methods. Sometimes, the original method column is no longer in production and must be replaced. Using a current website for HST comparisons [8], it is relatively simple to find comparable columns. It is generally accepted that comparable columns can be found when F_S for two columns is less than or equal to three.

For example, three commercial core-shell columns have column properties seen in Table 4.2.

Table 4.2 Core-shell column HST column properties at pH = 7.0

Column	Phenomenex Kinetex C18	Supelco Ascentis Express C8	Macherey Nagel Nucleoshell RP 18
H	0.96	0.91	1.11
*S**	0	0.0100	0.360
A	−0.130	−0.110	0.0770
B	−0.0100	0	−0.0400
C	0.120	0.170	0.266

Thus, when looking for an equivalent column to your method using a Phenomenex Kinetex C18 at pH = 7.0, the Supelco Ascentis Express C8 is likely to yield an equivalent method profile. A Macherey Nagel Nucleoshell RP 18 likely will not. Of course, once picked, this must be experimentally confirmed and validated (Table 4.3).

Table 4.3 Core-shell column HST column selectivity function

Column	Column selectivity function (F_S)
Phenomenex Kinetex C18, 100 A	0.00
Supelco Ascentis Express C8	1.95
Macherey Nagel Nucleoshell RP 18	10.94

4.3.2.3 HST for orthogonal reversed-phase separations

For impurity methods development and validation, the HST system can be used to find orthogonal sets of columns using large differences in the column selectivity function (F_S). Some researchers consider this to be an $F_S > 10$. Using the same site as in the earlier section, this time the goal is to find columns that have large differences in the selective function. In Table 4.4, an alternative column for a separation developed with a YMC Pack Pro column is presented along with another commercial phase that has significantly different column properties (Column X). From this data, F_S is determined to be 59.93. As illustrated in Fig. 4.2, the said orthogonal column resolves a previously unresolved peak seen between peak “e” and “f” on Column X. In this example, the HST system yields a true orthogonal separation.

Table 4.4 Core-shell column HST column properties at pH = 7.0

Column	YMC Pack Pro C18 RS	Column X
H	1.11	0.73
S^*	0.0500	−0.190
A	−0.0600	−0.340
B	−0.0500	0.020
C	−0.220	0.64

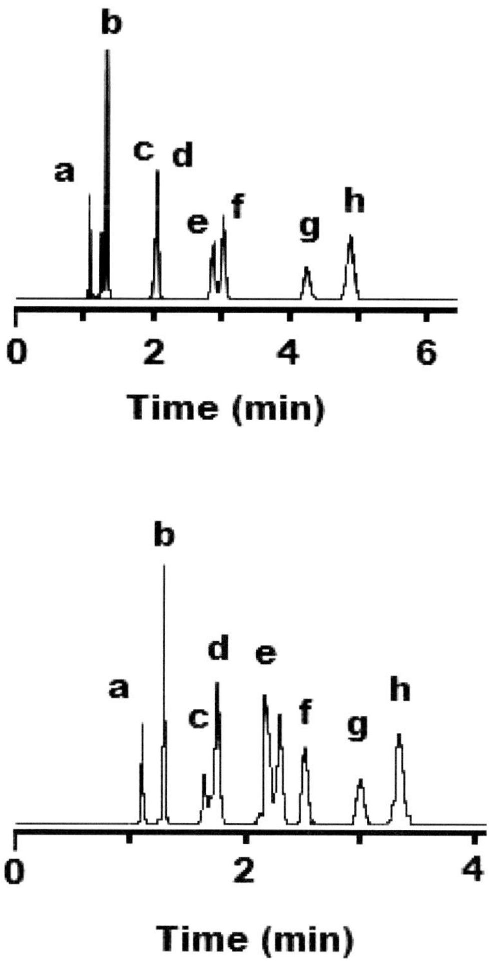

Figure 4.2 Orthogonal column profiles.

The challenge of using large differences in column selectivity function (F_S) is that column cation-exchange activity, C, can have a disproportionate effect. This means ionized analytes in the sample can have a significant effect on F_S. Thus, columns that appear to have high orthogonality potential may not see a change in neutral compounds. For separation of neutral compounds, it is suggested to drop the cation-exchange term, C, from the F_S equation.

4.4 Orthogonality in Traditional Normal-Phase LC

The use of traditional normal-phase for analytical methods has found diminished utility in recent years, in part because of the versatility of reversed-phase HPLC (*vide supra*), but also because supercritical fluid chromatography (SFC) has developed into a greener, safer, and cheaper direct replacement. As such, there have been nearly no advancements in normal-phase achiral columns in recent years. Nonetheless, there are a handful of traditional normal-phase columns (e.g., bare silica, diol, amino, and cyano) that are available to offer orthogonality during method development.

The most common means to quantify orthogonality amongst traditional normal-phase columns has been through the use of LSERs. Equation 4.3 describes a typical LSER. Solute parameters (known values for a set of carefully chosen and well defined probe analytes) are given in capital letters. Once retention factors for the set of probe analytes are determined, multiple linear regression analysis can be employed to solve for the stationary phase descriptors which can be used to compare properties of stationary phases to one and other.

$$\text{Log } k' = c + eE + sS + aA + bB + vV, \tag{4.3}$$

where the solute properties and the column properties are as follows:

Solute Properties

E = excess molar refraction

S = dipolarity/polarizability
A = solute hydrogen-bond acidity
B = solute hydrogen-bond basicity
V = molecular volume

Column Properties

e = interaction with n- and π-electrons
s = dipolar interactions
a = hydrogen-bond basicity
b = hydrogen-bond acidity
v = dispersive and cavity effects

One attempt to study silica using LSERs concluded that the LSER methods typically used to classify differing column interactions are not appropriate in predicting solute retention [9]. One reason for this is the difficulty in defining the bulk stationary phase in a bare silica column [9]. Other reports proved successful in evaluating bonded normal-phase columns using the same LSER approach [10] but did report negative values for dispersive and cavity formation effects. Table 4.5 lists LSER coefficients which have been determined for a series of traditional normal-phase columns. This information can be useful in determining the controlling factors that lead to differences between columns, as well as, give an idea of which columns of similar phase type are most unique. As seen in Table 4.5, interactions through dipolarity and polarizability seem to be fairly consistent among the tested columns (note, e values are not reported in the table as they were noted to be insignificant on each test stationary phase). Interactions through hydrogen bonding give the greatest differences in the tested columns. For example, this could be particularly useful in determining which cyanopropyl stationary phases are most unique. In this case the Spherisorb CN appeared to be orthogonal to the other cyanopropyl columns. Of course it is important to note that mobile phase components can greatly alter interaction types of normal-phase columns and the apparent differences in selectivity from these evaluations may not always remain.

Table 4.5 LSER coefficients for bonded normal-phase columns (values determined in hexane containing 1% (v/v) methanol)

Stationary phase	Ligand	Particle Size (μm)	Surface area (m^2/g)	*c*	*s*	*a*	*b*	*v*
Hypersil silica	Bare Silica	5	170	−1.21	1.06	2.23	1.56	−0.83
Hypersil APS NH_2	Aminopropyl	5	170	−1.14	0.94	2.94	1.20	−0.72
Lichrospher Diol	Dihydroxypropyl	5	250	0.94	1.07	2.37	1.47	−0.85
Lichrospher CN	Cyanopropyl	5	350	−0.89	0.99	1.94	1.04	−0.64
Ultrasphere CN	Cyanopropyl	5	200	−1.10	1.03	1.97	1.08	−0.60
Hypersil CN	Cyanopropyl	5	170	−1.16	0.95	1.86	1.15	−0.61
Spherisorb CN	Cyanopropyl	5	220	−0.99	1.00	2.13	1.69	−0.84

Source: Adapted from reference [10].

4.5 Orthogonality in Supercritical Fluid Chromatography

As noted earlier, SFC is an excellent replacement for normal-phase chromatography and has seen continued growth in utility in recent years as commercial instrumentation has become quite robust and reliable. With the growing popularity of SFC, the number of task-specific SFC stationary phases has increased. Typically SFC columns are polar phases (e.g., bare silica, 2-ethylpyridine, picolylamine) but some orthogonal separations have been observed on C18 stationary phases in SFC.

As such, LSERs have been applied to dozens of stationary phases in SFC to determine and quantify orthogonality [11–15]. The resulting system constants were used to plot different phases on a five-sided diagram (see Chapter 5). An improved LSER was developed (see Eq. 4.4) to also include electrostatic interactions [16]. With the added anionic and cationic (d^- and d^+) stationary phase descriptors, over 30 stationary phases have been evaluated and their resulting polarity has been displayed on a seven-sided diagram as shown in Fig. 4.3.

Figure 4.3 Spider diagram based on LSER models calculated with seven descriptors and Eq. 4.4, with the retention data measured for the 85 neutral species and 24 ionizable species, for 31 columns. Chromatographic conditions: CO_2–MeOH 90:10 (v/v), 25°C, 150 bar, 1 or 3 mL/min depending on column dimensions. Reproduced with permission from [16].

$$\text{Log } k' = c + eE + sS + aA + bB + vV + d^-D^- + d^+D^+, \tag{4.4}$$

where the solute properties and the column properties are as follows:

Solute Properties

E = excess molar refraction
S = dipolarity/polarizability
A = solute hydrogen-bond acidity
B = solute hydrogen-bond basicity
V = molecular volume
D^- = negative charge
D^+ = positive charge

Column Properties

e = interaction with n- and π-electrons

s = dipolar interactions
a = hydrogen-bond basicity
b = hydrogen-bond acidity
v = dispersive and cavity effects
d^- = electrostatic interaction with anions
d^+ = electrostatic interaction with cations

Further, using over 100 probe analytes, a hierarchical cluster map of 31 stationary phases was produced using normalized retention data as shown in Fig. 4.4. This figure is a more practical representation of column uniqueness and is useful in understanding the orthogonality of such columns. In Fig. 4.4, columns which are further away from each other should result in the largest difference in selectivity.

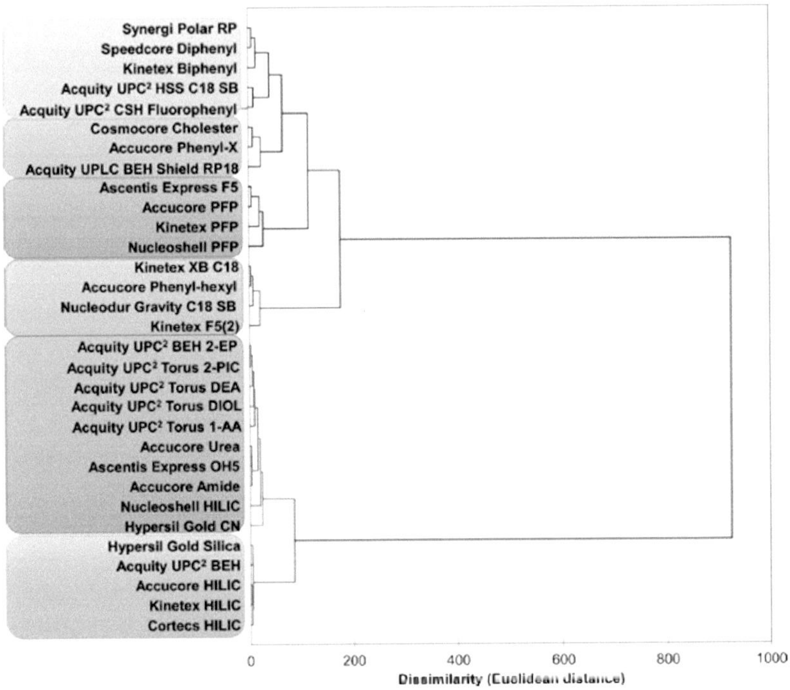

Figure 4.4 Hierarchical cluster analysis (HCA) on the normalized retention data (centered and reduced log k values) measured for 109 analytes. Retention dissimilarity in the abscissa reflects Euclidean distance between the stationary phases. Chromatographic conditions: CO_2–MeOH 90:10 (v/v), 25°C, 150 bar, 1 or 3 mL/min depending on column dimensions. Reproduced with permission from [16].

4.6 Orthogonality in HILIC

HILIC has been used for the separation of polar analytes for decades. Traditionally, bare silica was the main stationary phase used for HILIC; however, in the past 10–15 years, a variety of new HILIC columns have been introduced and have shown orthogonal utility when compared with silica and other traditional polar phases (Table 4.6).

Irgum and co-workers used a series of carefully chosen probe analytes to evaluate the orthogonality of 22 HILIC stationary phases by characterizing the phases' ability to interact with analytes via hydrophilic, hydrophobic, electrostatic, H-bonding, dipole-dipole, and π-interactions as well as their interplay with the eluent [17]. Principle component analysis was applied to the column/analyte library and four main functional groups were determined for HILIC column selectivity.

Table 4.6 Groups of HILIC stationary phases as determined in ref. [17].

Group	Stationary phase
Cation-exchange	Silica
Anion-exchange	Amino
Dipole-dipole and multi-point H-bonding	Polymeric sulfobetaine and poly(2-sulfoethyl aspartamide)
Low specific interaction	Hydroxyl, diol, amide, monomeric zwitterionic

A similar but simplified version of evaluating HILIC stationary phase orthogonality was developed by Lucy and co-workers using just five specifically chosen analytes [18]. The ratio of benzyltetramethylammonium chloride's (BTMA) retention factor to cytosine's retention factor yields a measure of a HILIC columns ion-exchange character. Similarly, the retention factor ratios of cytosine/uracil and adenosine/adenine represent the hydrophilicity and H-bonding abilities, respectively.

The following selectivity maps were assembled and are excellent guides in understanding orthogonality between different HILIC phases as well as orthogonality within a given group of HILIC phases (Fig. 4.5).

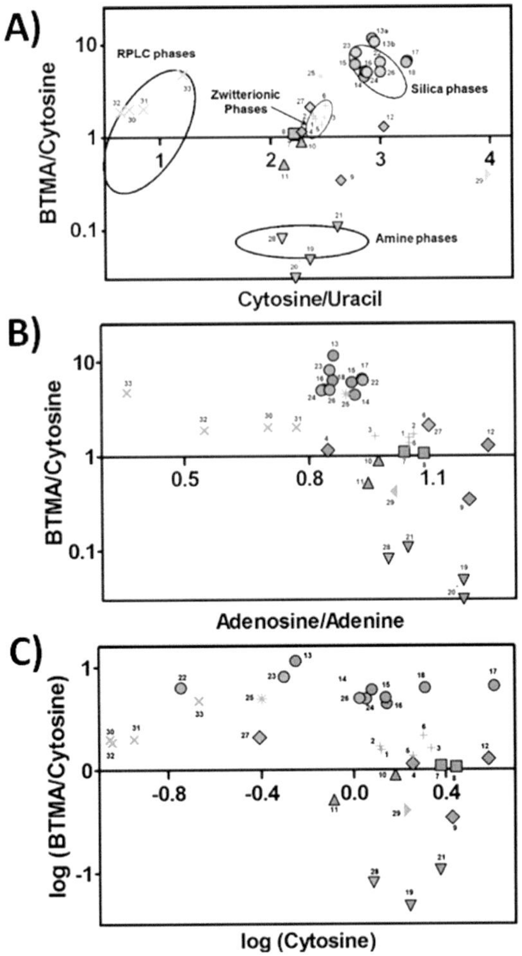

Figure 4.5 (A) Hydrophilicity vs. ion-exchange selectivity plot of HILIC phases, (B) Ion-exchange vs. H-bond formation capability selectivity plot of HILIC phases, and (C) The HILIC-Phase Selectivity Chart. Columns: 1 ZIC-HILIC, 2 ZIC-HILIC, 3 ZIC-HILIC, 4 ZIC-pHILIC, 5 Nucleodur HILIC, 6 PC HILIC, 7 TSKgel Amide 80, 8 TSKgel Amide 80, 9 PolyHYDROXYETHYL A, 10 LiChrospher 100 Diol, 11 Luna HILIC, 12 PolySULFOETHYL A, 13 Chromolith Si, 14 Atlantis HILIC Si, 15 Purospher STAR Si, 16 LiChrospher Si, 17 LiChrospher Si 60, 18 Cogent Type C Silica, 19 LiChrospher 100 NH2, 20 Purospher STAR NH2, 21 TSKgel NH2-100, 22 Atlantis HILIC, 23 Onyx silica monolith, 24 Zorbax HILIC plus, 25 Silica monolith coated with AS9-SC, 26 Zorbax RRHD HILIC plus, 27 Acclaim Trinity P1, 28 Cosmosil HILIC, 29 Acclaim HILIC-10, 30 Zorbax Eclipse XDB-C18, 31 XBridge C18, 32 YMC Pro C18, and 33 Zorbax SB-aq. Adapted from [18].

4.7 Orthogonality in Chiral Chromatography

Chiral stationary phases (CSPs) for HPLC/SFC generally perform best in either normal-phase/SFC mode or in reversed-phase and polar organic mode. Polymeric and Pirkle-type columns tend to be screened in normal-phase and SFC, whereas macrocyclic glycopeptides, cyclofructans, and cyclodextrins are often best screened in reversed-phase and polar organic modes. These groups of columns are somewhat orthogonal in terms of the mechanism that drives the enantiomeric recognition capabilities (e.g., π–π interactions on Pirkle-type columns vs. H-bonding and inclusion complexation on cyclodextrin based CSPs).

Because of the unique, multi-point, interaction mechanism needed for chiral recognition, quantifying orthogonality of CSPs is difficult and yet to be accomplished. Utilizing large libraries of chiral compounds and columns tested under many conditions has been the best way to qualitatively discover orthogonal CSPs [19–20]. Ideally, one would have a system for reversed-phase and polar organic mode screening and another for SFC screening, each system equipped with unique CSPs.

4.8 Orthogonality in Multi-Dimensional Chromatography

The heart of 2D-LC is having two orthogonal phases in the primary and secondary dimensions. In this way, one can maximize the 2D separation space and take full advantage of the multiplicative gains in peak capacity 2D separations can offer. See Fig. 4.6, where panel A shows no orthogonality. Figure 4.6A describes the relationship between the data (i.e., retention of analytes) in the *x*- and *y*-axis (i.e., first and second dimension column) as a straight line; when this is the case, the two columns are orthogonal to each other. When two columns behave the same or similarly, this line has a correlation coefficient close to unity. The further the correlation coefficient deviates from one, the more orthogonal the dimensions are. This is shown in panel B, which illustrates theoretically perfect orthogonality, and panel C shows a realistic representation of an ideal system.

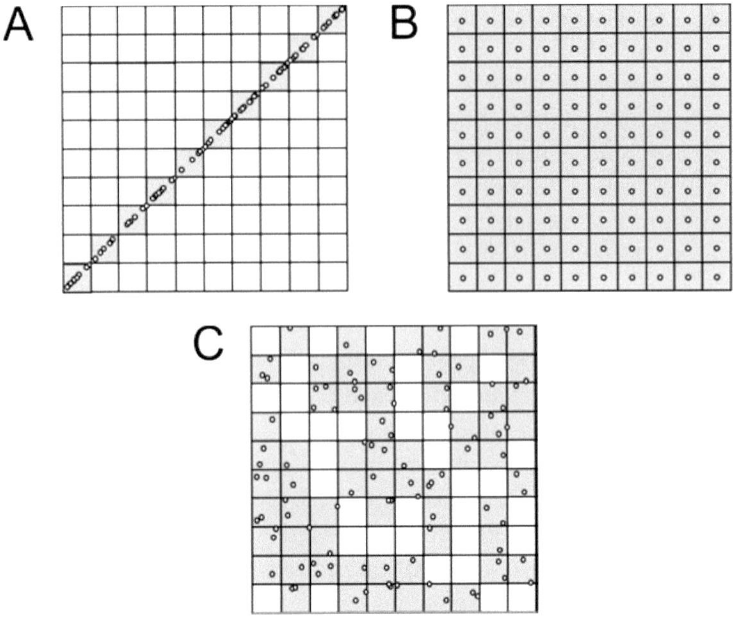

Figure 4.6 Geometric orthogonality concept. Hypothetical separation of 100 analytes In 10 × 10 normalized separation space. (A) Nonorthogonal system, 10% area coverage represents 0% orthogonality. (B) Hypothetical ordered system, full area coverage. (C) Random, ideally orthogonal, system, area coverage is 63% representing the 100% orthogonality. Reproduced with permission from [21].

A full mathematical understanding of orthogonal 2D-LC phases is underdeveloped. However, coupling different columns from the unique modes described throughout this chapter is the best chance for success. That being said, using a similar phase but different mobile phase conditions can also give rise to the desired orthogonality. Giliar et al. used a geometric approach (as described in Fig. 4.6) to examine orthogonal 2D configurations for peptide analysis. In Fig. 4.7, SFC-RP, HILIC-RP, and SCX-RP all yielded excellent utilization of the 2D separation space, as did a C18-C18 configuration when operated at high pH in one dimension at low pH in the other. Needless to say, finding optimal 2D configurations remains a matter of trial and error.

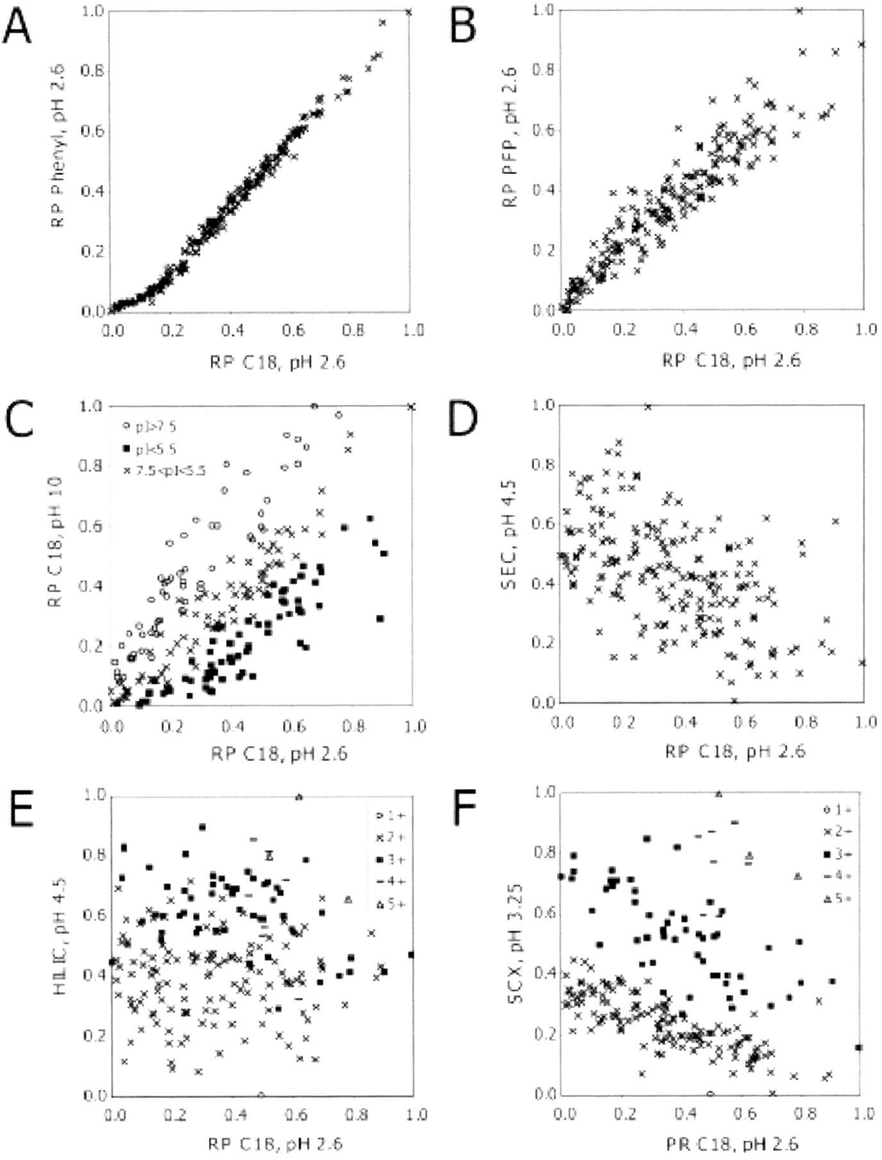

Figure 4.7 Normalized retention time plots for selected 2D-LC systems. Reproduced with permission from [21].

4.9 Conclusion

Separation techniques, modes of chromatography, and stationary phase material all can be orthogonal. Many models for characterizing orthogonality can be studied and used to make decisions on screening column sets for method development (as discussed in Chapter 5). It should be noted that column vendors specifically choose to commercialize sets of columns that are orthogonal for their internal test analytes libraries. Thus, if a column company has two HILIC columns available, they believe them to be orthogonal. This, however, may not prove true in all cases when applied to real world sample or using unique mobile phases. At the end of the day, using the tools and maps described herein can only guide educated guesses on column orthogonality; it is still up to the practitioner to find the right column for the job as is outlined in the proceeding chapter.

Disclosure

Zachary S. Breitbach and Gregory K. Webster are employees of AbbVie. AbbVie provided no additional financial support for this work. AbbVie participated in the writing, reviewing, and approving the publication. The presentation contains no proprietary AbbVie data.

References

1. Pellett, J., Lukulay, P., Mao, Y., Bowen, W., Reed, R., Mac, M., Mungerc, R. C., Dolan, J. W., Wrisley, L., Medwide, K., Toltl, N. P., Chan, C., Skibic, M., Biswas, K., Wells, K. A., and Snyder, L. R., 'Orthogonal' separations for reversed-phase liquid chromatography, *J. Chromatogr. A*, **2006**, 1101, 122–135.
2. Wilson, N. S., Nelson, M. D., Dolan, J. W., Snyder, L. R., Wolcott, R. G., and Carr, P. W., Column selectivity in reversed-phase liquid chromatography I. A general quantitative relationship, *J. Chromatogr. A*, **2002**, 961, 171–193.
3. Wilson, N. S., Nelson, M. D., Dolan, J. W., Snyder, L. R., Wolcott, R. G., and Carr, P. W., Column selectivity in reversed-phase liquid chromatography II. Effect of a change in conditions, *J. Chromatogr. A*, **2002**, 961, 195–215.

4. Wilson, N. S., Nelson, M. D., Dolan, J. W., Snyder, L. R., Wolcott, R. G., and Carr, P. W., Column selectivity in reversed-phase liquid chromatography III. The physico-chemical basis of selectivity, *J. Chromatogr. A*, **2002**, 961, 217–236.
5. Dolan, J. W., Maule, A., Bingley, D., Wrisley, L., Chand, C. C., Angodd, M., Luntee, C., Kriskoe, R., Winstone, J. M., Homeiere, B. A., McCalley, D. V., and Snyder, L. R., Choosing an equivalent replacement column for a reversed-phase liquid chromatographic assay procedure, *J. Chromatogr. A*, **2004**, 1057, 59–74.
6. Marchand, D. H., Snyder, L. R., and Dolan, J. W., Characterization and applications of reversed-phase column selectivity based on the hydrophobic-subtraction model, *J. Chromatogr. A*, **2008**, 1191, 2–20.
7. Snyder, L. R., Dolan, J. W., and Carr, P. W., A new look at the selectivity of RPC columns, *Anal. Chem.*, **2007**, 79, 3254–3262.
8. http://www.hplccolumns.org/database/compare.php.
9. Cheong, W. J., and Choi, J. D., Linear solvation energy relationships in normal-phase liquid chromatography based on retention data on silica in 2-propanol/kexane eluents, *Anal. Chim. Acta,* **1997**, 342, 51–57.
10. Park, J. H., Yoon, M. H., Ryu, Y. K., Kim, B. E., Ryu, J. W., and Jang, M. D., Characterization of some normal-phase liquid chromatographic stationary phases based on linear solvation energy relationships, *J. Chromatogr. A*, **1998**, 796, 249–258.
11. West, C., and Lesellier, E., Characterization of stationary phases in subcritical fluid chromatography by the solvation parameter model I. Alkylsiloxane-bonded stationary phases, *J. Chromatogr. A*, **2006**, 1110, 181–190.
12. West, C., and Lesellier, E., Characterisation of stationary phases in subcritical fluid chromatography by the solvation parameter model II. Comparison tools, *J. Chromatogr. A*, **2006**, 1110, 191–199.
13. West, C., and Lesellier, E., Characterisation of stationary phases in subcritical fluid chromatography with the solvation parameter model III. Polar stationary phases, *J. Chromatogr. A*, **2006**, 1110, 200–213.
14. West, C., and Lesellier, E., Characterisation of stationary phases in subcritical fluid chromatography with the solvation parameter model IV. Aromatic stationary phases, *J. Chromatogr. A*, **2006**, 1115, 233–245.
15. West, C., and Lesellier, E., Characterisation of stationary phases in subcritical fluid chromatography with the solvation parameter

model V. Elaboration of a reduced set of test solutes for rapid evaluation, *J. Chromatogr. A*, **2007**, 1169, 205–219.

16. West, C., Lemasson, E., Bertin, S., Hennig, P., and Lesellier, E., An improved classification of stationary phases for ultra-high-performance supercritical fluid chromatography, *J. Chromatogr. A*, **2016**, 1440, 212–228.
17. Dinh, N. P., Jonsson, T., and Irgum, K., Probing the interaction mode in hydrophilic interaction chromatography, *J. Chromatogr. A*, **2011**, 1218, 5880–5891.
18. Ibrahim, M. E. A., Liu, Y., and Lucy, C. A., A simple graphical representation of selectivity in hydrophilic interaction liquid chromatography, *J. Chromatogr. A*, **2012**, 1260, 126–131.
19. Akin, A., Antosz, F. J., Ausec, J. L., Greve, K. F., Johnson, R. L., Magnusson, L. E., Ramstad, T., Secreast, S. L., Seibert, D. S., and Webster, G. K., An orthogonal approach to chiral method development screening, *Curr. Pharm. Anal.*, **2007**, 3, 53–70.
20. Barhate, C. L., Joyce, L. A., Makarov, A. A., Zawatzky, K., Bernardoni, F., Schafer, W. A., Armstrong, D. W., Welch, C. J., and Regalado, E. L., Ultrafast chiral separations for high throughput enantiopurity analysis, *Chem. Commun.*, **2017**, 53, 509–512.
21. Gilar, M., Olivova, P., Daly, A. E., and Gebler, J. C., Orthogonality of separation in two-dimensional liquid chromatography, *Anal. Chem.*, **2005**, 77, 6426–6434.

Chapter 5

Unified Approach to Reversed-Phase and Normal-Phase Chromatographic Method Development

Gregory K. Webster

AbbVie Inc., Research and Development, 1 N. Waukegan Rd., North Chicago, Illinois 60064, USA

gregory.webster@abbvie.com

5.1 Introduction

Method development for liquid chromatography has not changed substantially from the groundbreaking work of Snyder and Kirkland [1] and updated texts [2, 3]. We still need selectivity and sensitivity in validatable and transferable methods. The requirements from the USP [4] and ICH [5] have not changed. What has changed in this new millennium is the need for efficiency (speed) over "academic" methods. The pharmaceutical industry develops thousands of methods a year to support rapid drug development. Methods need to meet their intended goal in their role of moving the drug development pipeline forward. The

Chromatographic Method Development
Edited by Gregory K. Webster and Laila Kott

ISBN 978-981-4800-53-2 (Hardcover), 978-0-429-20172-1 (eBook)
www.jennystanford.com

theory of chromatographic method development has not been radically transformed; however, the complexity of the applications, number of samples running at any given moment, and instrument technology certainly have. Textbook analytical methods have given way to fast turnaround.

5.2 Unified Approach to Method Development

There is no set standard approach to developing methods in liquid chromatography. Older chemists tend to rely on experience as a way to muscle through the process. However, with today's complex regulations regarding detection and selectivity of impurities, many labs have adopted a more automated approach. The principles set out by Lloyd Snyder and Joseph Kirkland [1] are still currently relevant in method development. While column and system technology has greatly advanced over the years, the fundamentals have not changed and still hold true.

5.2.1 "Like Dissolves Like"

Fundamental to liquid chromatographic method development is the understanding of how the process works. If one recalls the differences between normal and reversed-phase chromatography and the concept of "like dissolves like" taught in general chemistry, you are 70% there. We have to recall that normal-phase chromatography uses a polar stationary phase and a non-polar mobile phase, and in this case the least polar compounds elute first. As we increase the polarity of the mobile phase using alcohol or another additive, there comes a point when the polar compounds have equal to greater affinity for the mobile phase. At this point, they traverse down the column. It is this changing affinity from stationary to mobile phase that results in selectivity. Nonpolar compounds are less strongly held by the polar stationary phase and elute off the column first. Elution proceeds as the compounds increase in polarity.

Conversely, reversed-phase chromatography uses a non-polar stationary phase with a polar mobile phase. Non-polar compounds tend to retain ("dissolve") on the nonpolar stationary phase. As we decrease the polarity of the mobile phase using acetonitrile or another additive, there again comes a point with non-polar compounds having equal or greater affinity for the mobile phase, and again the resulting traversal down the LC column results in selectivity based on hydrophobicity. Polar compounds are less strongly held by the non-polar stationary phase and elute off the column first. Elution proceeds as the compounds decrease in polarity (Table 5.1).

Table 5.1 LC elution

LC type	Stationary phase	Mobile phase
Normal phase	Polar	Non-polar → More polar
Reversed-phase	Non-polar	Polar → Less polar

The advantage of looking at method development in this regard allows chromatographers to standardize a method development approach regardless of developing a normal or reversed-phase method. Method development simplifies down to choosing the right stationary phase that will initially retain the sample analytes and then manipulating the mobile phase so that the analytes will traverse the column in a controlled and reproducible manner.

The scenario in Table 5.1 is true when the analytes of interest are not charged. To address this, chromatographers can manipulate the mobile phases in the hopes of eliminating charge. This is done by operating below the pKa, above the pKb, or through the use of ion paring reagents. This is also why even when analytes are suspected to be non-polar, a pH adjustment to mobile phase is routinely applied in reversed-phase chromatography.

5.2.2 Information and Samples Required

Ideally, chromatographic method development begins with a targeted KPSS, or key predictive sample set. Chromatographers need to know what they intend on separating. Not long ago, most chromatographers could look at the molecule and know the conditions that would likely get them 80% done with the separation. Today, however, regulations often make this more complex. The chromatographer has to know if the goal of the separation is only to simply look at the profile of the compound, to determine the potency, detect degradation products ("stability indicating") or to profile purity while being selective to synthetic impurities, degradants, and/or toxic or medically significant impurities. Besides the actual sample, the method development chemist would be helped by knowing the number of compounds likely present (Table 5.2). Additionally with each possible component in the mixture, key information, such as molecular weight, structure, pKa, solubility and concentration range, helps establish whether a normal or reversed-phased method is appropriate and what detection scheme is needed.

The choice of detection scheme defines what mobile phases and additives can be used. Examples are given in Table 5.3. The traditional phosphate buffers and perchloric acid ion pairing reagents are used much less often in modern laboratories simply because of their lack of compatibility with mass spectrometric detection (MS). In regulated laboratories, a mass spectrometric compatible LC method is desired even for UV quality control methods. This is because, should an investigation be necessary, no additional validation would be required for this method on the MS detector. Knowing what detectors will be needed ultimately limits what mobile phases can be used.

Table 5.2 What information do I ask for?

Structural information	Structure/functional groups pKa Hydrophobicity Chirality
Synthesis Information	Synthesis route(s) Chirality
Samples	Drug Precursors/Intermediates Reaction-by products Stressed samples
Initial methods	Sample preparation information LC conditions Detectors
Goal for method	Potency/purity Stability indicating

Table 5.3 Detection schemes

Detector technology	Mobile phase restriction(s)
Absorption (UV-Vis, fluorescence)	Minimize background chromophore
Refractive index (RI)	No gradient
Evaporative light scattering (ELSD) Mass spectrometry Charged aerosol detection (CAD)	Volatile mobile phases and additives
Electrochemical	Charge carrying ions present
Chemical luminescence nitrogen detection (CLND)	No ACN or other nitrogen containing additives

5.3 Reversed-Phase Chromatography

5.3.1 Control of Analyte Retention

For reversed-phase liquid chromatography method development, it has been shown that changes in organic modifier content yield the most significant changes in separation [1–3, 13]. Choice and concentration of the organic modifier is of primary importance. Separations are affected by other conditions as well, but these usually yield only small changes in resolution. Knowing this, systematic method development can be used to minimize the time it takes to develop methods suitable for their intended purpose. When starting a method development project, many of the common considerations are listed in Table 5.4.

Table 5.4 RPLC method considerations

Stationary phase	Start with C18, unless early methods indicate otherwise
Organic modifier	Use ACN, MeOH and 1:1 ACN/MeOH; can look at THF/EtOH/IPA as a second tier
Buffer	pH control needed for ionizable compounds
Avoid using TFA/TEA in early screens	Keep separation mechanism consistent. TFA/TEA add ion pairing capabilities that may or may not be needed

(*Continued*)

Table 5.4 (*Continued*)

Temperature	Start with 30°C to maintain column above ambient conditions
Gradient steepness	Robustness of method
Delay volume on system	Needs to be known for transfer

5.3.2 Hydrophobic Subtraction Theory/Columns

The combined use of solvent- and column-type selectivity may be useful for the separation of extremely difficult samples. Historically, solvent-type selectivity is first investigated for a C8 or C18 column. If the resulting profile is suitable, no additional development is needed. If further selectivity is needed, a phenyl or cyano (CN) columns can be applied. However, with a change of column a new optimum in mobile phase may be needed. RPLC retention for non-polar, non-ionic compounds generally follows:

$$Rt_{C18} \approx Rt_{C8} > Rt_{Phenyl} > Rt_{C4} > Rt_{C2} > Rt_{CN} >> Rt_{silica}$$

A chromatographer told me once when changing a column for method development that "it is just dumb to pick up a random C18 column and expect it to perform differently." Choices of screening columns should maximize their differences as discussed in Chapter 4. Start with a column you have history with and then select stationary phases known to behave differently for the chemistry of the intended sample set.

5.3.3 Solvent Strength

Reversed-phase chromatography uses a non-polar stationary phase with a polar mobile phase. Traditionally, this meant a C18 stationary phase bonded to a silica support. Today, this definition has expanded to C1–C30, fluorinated, phenyl and porous carbon stationary phases with varying endcapping and polar embedding. Yet, these stationary phases remain relatively non-polar. Thus, for reversed-phase chromatography, strong solvents are organics such as acetonitrile (ACN), methanol (MeOH), or tetrahydrofuran (THF) to "dissolve" the retained analyte from the stationary phase into the mobile phase. Weak solvents would be solvents which do not induce the analyte into the mobile phase, such as water (Table 5.5).

Table 5.5 Strong and weak solvents in RPLC

Reversed-phase strong solvents	Reversed-phase weak solvents
Primary:	Water
Acetonitrile (ACN)	Aqueous buffers and ion pairing solutions
Methanol (MeOH)	
Tetrahydrofuran (THF)	
Secondary:	
Ethanol (EtOH)	
Isopropanol (IPA)	

It should be noted that stationary phases that enhance selectivity by π–π bonding interactions, such as phenyl phases, need the presence of solvents such as MeOH to promote these interactions.

5.3.4 Screening

If early methods are suitable to meet the separation goal, use them. Otherwise, we need to do some injections! Since method development is about critical pair resolution [6], not plate number [7], we need to have the critical pair(s) in mind going into development. All too often, advocates of computer-assisted method development are too busy encouraging the use of their software and they forget to promote that all development starts with a few injections. We need initial information about the elution characteristics of the sample and the critical pairs. Depending on the chromatographer's experience, there are several starting points for this. However, they all must include looking at extremes in pH and mobile phase. Most compounds tend to have a region of pH where they have optimal resolution.

5.3.4.1 System considerations: UHPLC vs. Traditional

With chromatographic system technology changing at a fast pace, industry has a mix of system capabilities at play. Older systems still operate with a maximum pressure of ~400 bar while modern ultrahigh performance liquid chromatography (UHPLC) systems operate well over 1000 bar. While the maximum pressure of a system delegates the size of columns, stationary phase particles and system settings for the method conditions, the theory of method development is independent of columns size, or system pressures, with the notable exception that smaller stationary phase packings operating at higher pressures generally exhibit higher plate numbers and better resolution. Scaling methods to older platforms generally work well (Chapter 13). The reverse, however, generally does not [8, 9].

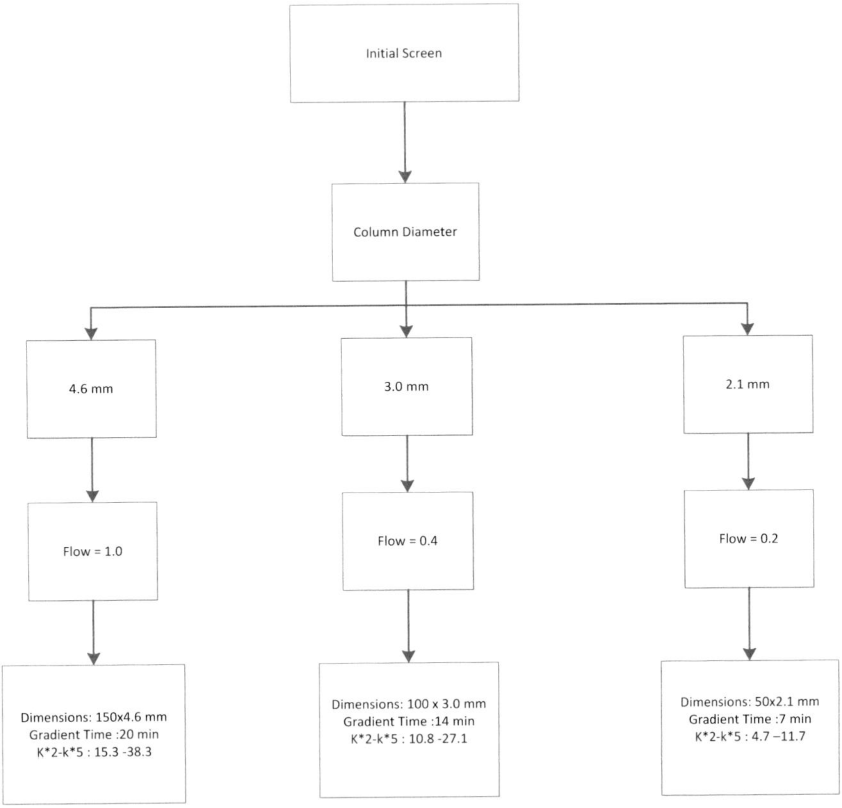

Scheme 5.1 System requirements.

No matter how much one believes in their method development software, the process demands actual injections on real liquid chromatographs! The conditions in Table 5.6 are generally where most method developers start.

Table 5.6 Representative starting conditions

Column – C18	C18
Detection[1]	UV Detection at 201 nm, 254 nm
Organic Component	ACN MeOH 1:1 ACN/MeOH or THF
Buffer	pH = 3, pH = 6.8, pH = 9
Gradient[2]	Initial: 95% weak solvent/5% strong solvent Final: 5% weak solvent/95% strong solvent

[1]Assuming existence of a chromophore, otherwise a CAD or ELSD detector with volatile mobile phase adjustments can be used.
[2]Based on Scheme 5.1.

As a general rule led by the pharmaceutical industry, reversed-phase liquid chromatography method development is based on the notion of column orthogonality (Chapter 4). Pellet et al. [10] came up with a process based on Scheme 5.2. In this scenario, the developer optimized the preliminary method through a series of column and solvent changes to optimize resolution. The process is governed by the $|\delta \log \alpha|_{avg}$ term discussed in Chapter 4. When this term reaches a value of 0.10 or greater, an orthogonal method has been reached.

The Pellet et al. process can be tedious when done manually. For this reason, the traditional trial and error approach has been replaced through automation to achieve shorter development times. Seen in many labs today is a (a) traditional approach, or (b) varying automated approaches to reversed-phase method development.

5.3.4.2 Traditional approach

The traditional approach to LC method development is to systematically vary one factor, such as mobile phase composition or selection of a column, with all other parameters being held constant. This "trial and error" approach (seen in Figure 4.1 in the last chapter) is rather slow, depending on the experience of the method developer and knowledge of the KPSS. This approach is repeated until an adequately performing instrument method is obtained.

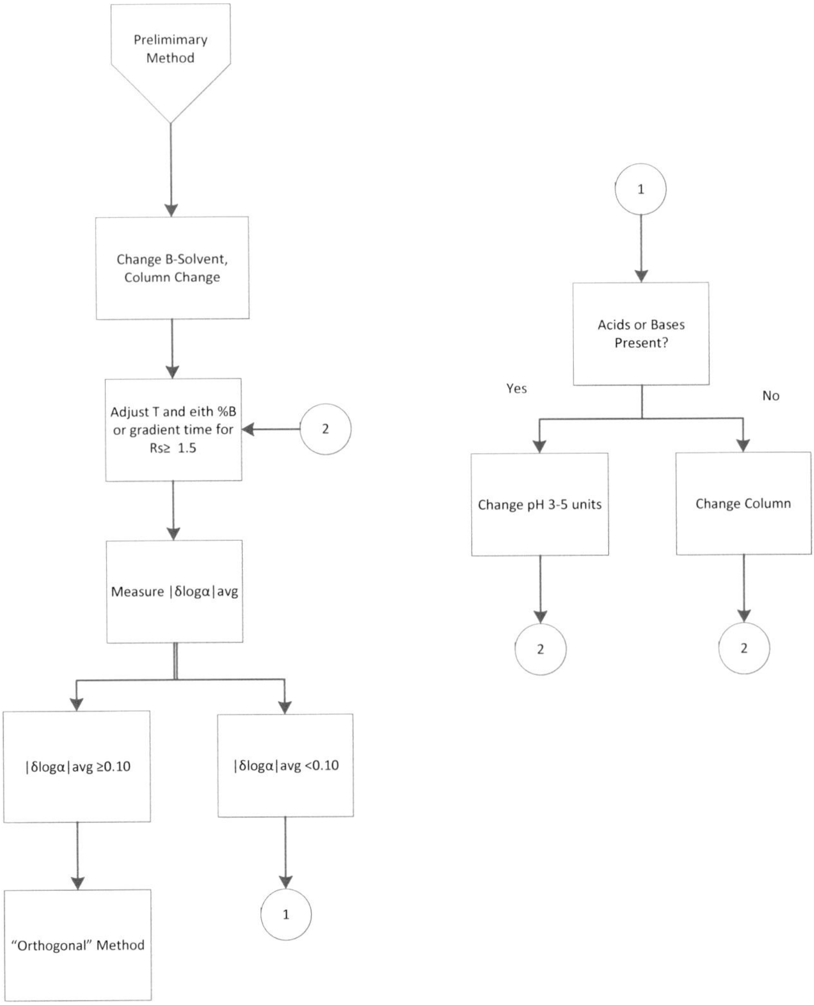

Scheme 5.2 Preliminary orthogonal method development.

This approach generally starts with high levels of acetonitrile in the mobile phase to ensure elution of all the analytes in the KPSS as well as a low wavelength in tandem with MS to monitor for impurities. The level of the organic modifiers (or gradient slope) is then backed off to look for improved critical pair resolution. This would be repeated for other organic modifiers, pH and column stationary phase. Temperature would be the last challenge to the system, as it has been shown to have the least dramatic effect on resolution for routine separations.

Oftentimes, ion pair reagents (TFA, TEA) are introduced too early in the method development process. This will be discussed in a later section of this chapter.

The application of this traditional approach has worked historically. The issue is time. Modern laboratories want robust methods with adequate selectivity for the intended purpose, but the development of these methods is not commonly given time on the product's critical path to market. This leads not only to potential delays but, often also comes with poor development documentation and process understanding.

5.3.4.3 Automated approach

With the advent of computer controlled instrumentation, the traditional approach is being replaced with programmed method development approaches to rapidly develop and optimize separation parameters. Use of these automated approaches can reduce development time from weeks to days with full development report documentation.

5.3.4.3.1 Model 1: Resolution optimization

Resolution optimization models use linear and nonlinear regression to predict an analyte's retention time (t_r) as a function of multiple method and instrument parameters. An example of this approach is using Drylab®, or ChromSword® (see Chapter 2). The software continuously solves the equations for the critical pairs to predict their resolution [2]:

$$\ln k = aV^{2/3} + b(\Delta G) + c, \tag{5.1}$$

where k is the retention factor of a solute, V the molecular volume of a solute, ΔG the energy of interaction of the solute with water, and a,b,c the characteristic parameters of the reversed-phase column in the eluent being used.

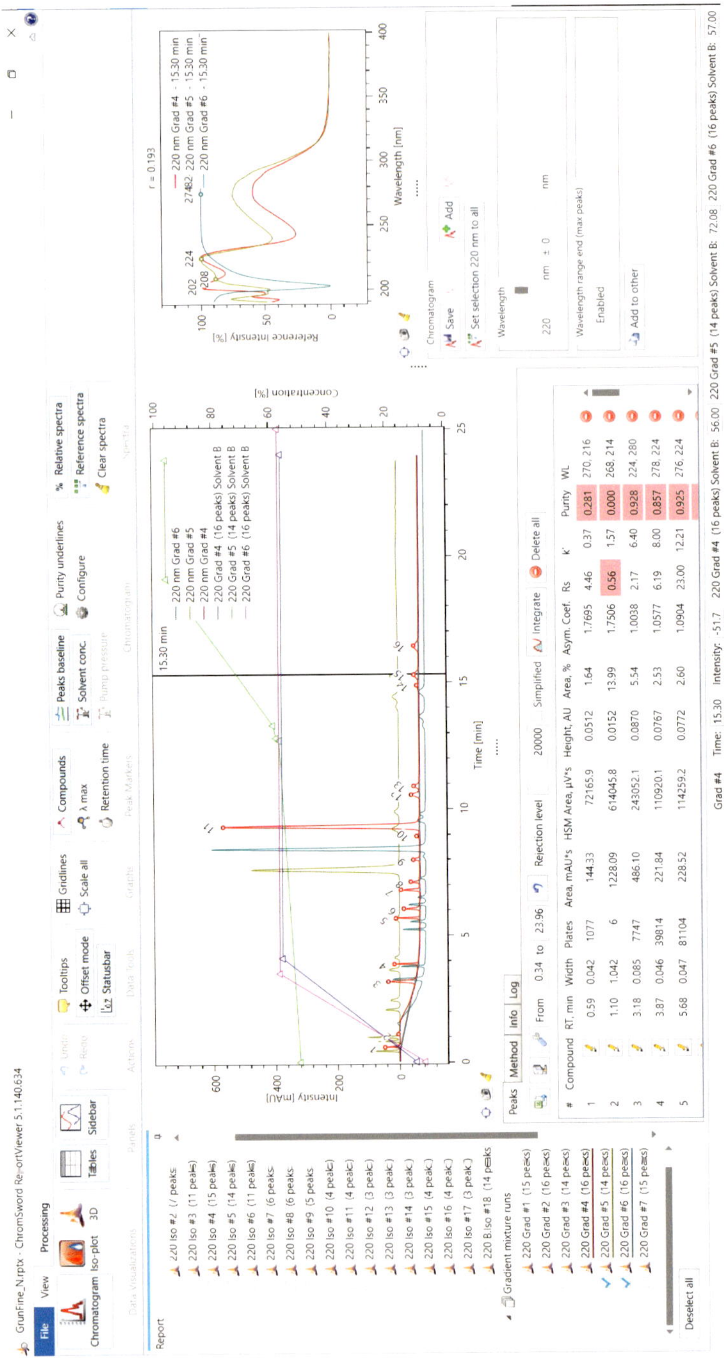

Figure 5.1 Model 1 approach with Chromsword®. Figure used with permission.

The process has to start somewhere, so initial runs to adapt the theoretical coefficients for prediction of proceeding experimental conditions need to be run. The major limitation to this approach is that the equation cannot be modified. It must be independently applied to solve for other variables. Often it is simply assumed that variables such as pH and gradient slope do not interact. This application is detailed in Chapter 2.

5.3.4.3.2 Model 2: Design of experiments (DOE) optimization

The DOE approach screens columns based on the optimization of factors such as pH, gradient slope, and organic solvent type (Table 5.7). This practice has been well accepted in the pharmaceutical industry [11, 12]. An example of this would be using Fusion QBD®. Here a statistical experiment design is used to systematically vary the factors together with experimental runs to explore a multi-factor design space. The interaction effects of these parameters are optimized with their linear and curvilinear effects. Proprietary DOE designs allow a method development approach to simultaneously look at multiple columns, pH, organic mobile phase type and composition as well as temperature. A second DOE tier can be added to investigate other parameters to further optimize method performance.

Table 5.7 DOE optimizations

Tier 1 DOE	Tier 2 DOE
Colum screening	Temperature
pH	Run time
Organic modifier	Flow rate
Gradient	Use of ion pair reagents

5.3.4.3.3 pH control

Automated systems for method development minimize the number of mobile phases needed by taking advantage of the Henderson-Hasselbalch equation to control buffer pH. Table 5.8 illustrates a six-level mass spectrometry detection-friendly pH screen from pH 2.75 to 6.25 using formic acid and ammonium formate. For sample sets requiring a low UV detection wavelength, a similar series is easily generated with phosphate buffers. Other pH series can be found in the literature and product websites.

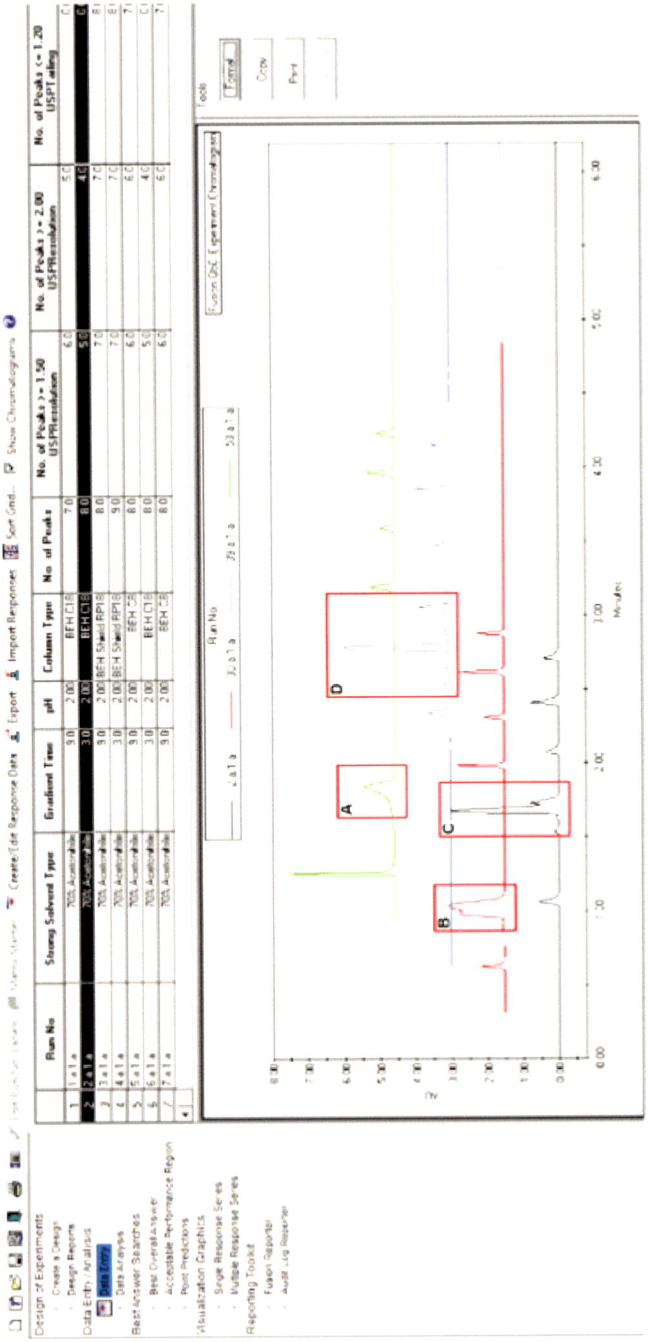
A
B
C
D

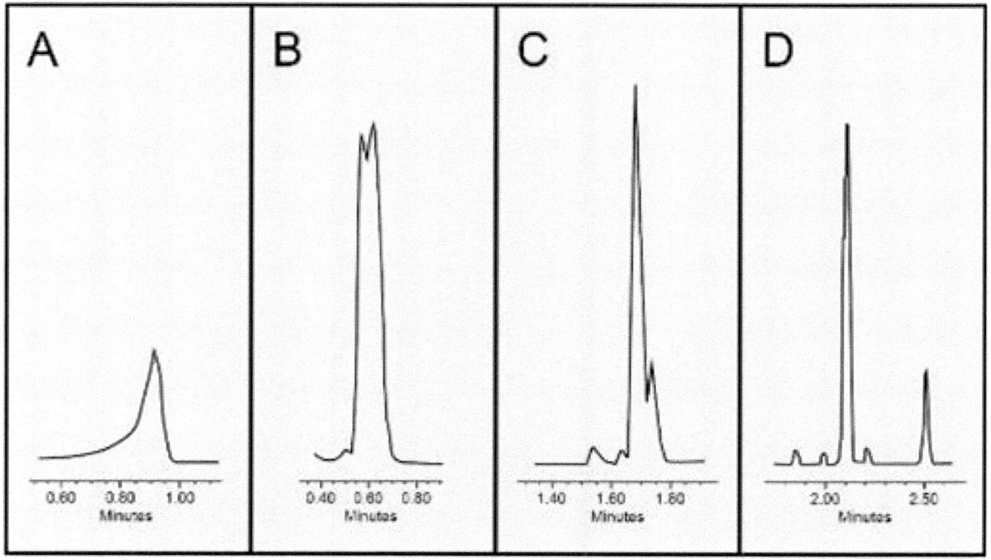

Figure 5.2 Model 2 approach with Fusion®. Fusion QbD-LC Method Development White Paper, V3.2, S-Matrix Corporation. Used with permission.

Table 5.8 Automated pH control

pH	20.0 mM formic acid (%)	20.0 mM ammonium formate (%)
2.75	100	0
3.16	80	20
3.70	50	50
4.34	20	80
5.42	5	95
6.24	0	100

5.3.5 Ion Pair Reagents

A decade ago, if method development chemists saw aromatic amine functionality in their KPSS, they would often simply default to the use of an ion pairing reagent, usually perchloric acid or trifluoroacetic acid, in the aqueous portion of their mobile phase. Yet, many times simply buffering the mobile phase pH to a specific range was all that was needed, as evidenced by the future use of automated method development systems. Buffering mobile phases are commonly easier to transfer, more robust and present less fouling of mobile phase lines and columns than the use of ion pairing reagents memory effects.

So, when should ion pairing reagents be introduced? A traditional view of the mechanism for using ion paring reagents was to cover the bonded phase surface to minimize the ion exchange retention by absorbed analyte and the uncovered stationary phase. If the reagent concentration of the ion pairing reagent is sufficient to provide coverage of the bonded phase surface, column selectivity becomes relatively unaffected and will become similar for different columns [13]. However, the use of polar embedded agents and endcapping in current stationary phases have greatly reduced the need for this application. In fact, much of today's reversed-phase methods no longer require ion paring reagents. The application can be minimized to only applications to control acids or bases on a case-by-case basis.

5.3.6 Reversed-Phase Screen Conditions

Table 5.9 is an example of a preliminary set of conditions for an orthogonal reversed-phase LC screen.

Table 5.9 Example preliminary reversed-phase screen conditions

Column phases (Chapter 4)	1. Symmetry C18 2. Xterra C8 RP 3. Thermo CN 4. Discovery HS F5 5. ZirChrom-EZ 6. Bondclone C18	
Organic phases	1. ACN 2. MeOH 3. ACN/MeOH	
Buffer	Formic acid/ammonium formate pH screen	
Instrument settings	Traditional 150 × 4.6 mm Columns (3 μm fully porous particles)	5 to 95% organic 1 mL/min, 30°C Gradient Time: 20 min K*2-k*5 : 15.3–38.3.
	Traditional 100 × 3.0 mm Columns (2.6–3 μm particles)	5 to 95% organic 0.4 mL/min, 30°C Gradient time: 14 min K*2–k*5 : 10.8–27.1
	UHPLC 50 × 2.1 mm Columns (≤2.7 μm particles):	5 to 95% organic 0.2 mL/min, 30°C Gradient time: 7 min K*2–k*5 : 4.7–11.7.
Detection	UV @ 254 nm, MS	

5.4 Normal-Phase Chromatography

The traditional mode of normal-phase chromatography using hexane, heptane, isooctane etc. is no longer a major player in the world of chromatographic separations. It has been replaced by modern supercritical fluid chromatography (SFC) [18].

Supercritical carbon dioxide behaves very similar to hexane in a normal-phase mode. However, its use has many advantages such as operation as a "green" technique, easier recovery of analytes in preparative applications and greater chromatographic efficiency in analytical applications. Because of this, our discussion of normal-phase method development will be limited to achiral SFC in this chapter and chiral applications in Chapter 7.

5.4.1 Control of Analyte Retention

For normal-phase liquid chromatography method development, it has again been shown that changes in organic modifier content yield the most significant changes in separation [1–3]. That is, choice and concentration of the organic modifier remains of primary importance. When starting a method development project, many of the common considerations for normal-phase separations are listed in Table 5.10.

Table 5.10 Achiral SFC method considerations

Stationary phase	Start with silica unless early methods indicate otherwise
Organic modifier	Use MeOH, EtOH or IPA and 1:1 MeOH/EtOH; can look at $MeCl_2$ as a second tier
Buffer	Buffers not used. Acetic acid, ammonium acetate, TFA and alkyl amines for ionizable compounds
Avoid using TFA/TEA in early screens	Keep separation mechanism consistent. TFA/TEA add ion pairing capabilities that may or may not be needed
Temperature	Need to maintain column above supercritical or evaluate sub-critical fluid conditions.
Gradient steepness	Robustness of method
Delay volume on system	Needs to be known for transfer

5.4.2 Hydrophobic Subtraction Theory

Because normal-phase columns are less in demand, formal orthogonal column guides are generally not available online or on vendor websites. Start with a column you have history with and then select stationary phases known to behave differently for the chemistry of the intended sample set.

5.4.3 Solvent Strength

Normal-phase chromatography uses a polar stationary phase with a non-polar mobile phase. Commonly used strong and weak solvents for normal phase chromatography are listed in Table 5.11 [14].

Table 5.11 Strong and weak solvents in NPLC

Normal-phase strong solvents	Normal-phase weak solvents
Primary: Isopropanol (IPA) Methanol (MeOH) Ethanol (EtOH) *Secondary*: Acetonitrile (ACN) Methylene chloride (CH_3Cl_2)	Hexane/heptane Supercritical fluid carbon dioxide (CO_2) Supercritical fluid nitrous oxide (N_2O)

Traditionally, this meant a silica stationary phase. Today, this definition has expanded to cyano, diols and most importantly, alky pyridines [15]. For normal-phase chromatography, strong solvents are alcohols such as methanol (MeOH), ethanol (EtOH) or isopropanol (IPA) to "dissolve" the retained analyte from the stationary phase into the mobile phase. Weak solvents would be solvents which do not induce the analyte into the mobile phase, such as supercritical fluid carbon dioxide or hexane.

5.4.4 Ionic Samples

Since buffers are not effective in normal phases, the control of ionization is typically controlled using acid and basic modifiers. Acetic and trifluoroacetic acid are popular acidic modifiers with ammonium acetate and alkyl amines for higher pHs. In addition, low levels of water can be used as an acidic modifier in SFC. Unlike with reversed-phase chromatography, trifluoroacetic acid and alky amines are used in combinations for some normal-phase applications.

5.4.5 Screening

5.4.5.1 Mobile phases

Elise et al. recently studied different mobile phase additives for the achiral SFC separation of 160 proprietary drugs [16]. Using methanol without additives as a reference, they studied the basic, acidic and salt additives to methanol. The basic additives studied were ammonium hydroxide, diethylamine, diethanolamine and iso-propylamine. The acidic additives selected were trifluoroacetic acid and water. The salt additives for this study were ammonium acetate and ammonium formate. Each additive was evaluated for their effect upon elution capability, peak shapes, UV baseline drift, and UV and mass responses (signal-to-noise ratios) for the sample set.

The study used Derringer desirability functions to rank the performance of the selected additives. The authors concluded that ammonium acetate at 20 mM was the most effective additive in terms of chromatographic efficiency and detection. Diethylamine, trifluoroacetic acid and water yielded surprisingly poor results. The study also found that ammonium hydroxide provided useful MS responses and low baseline drift with UV detection.

In a different screening study, Hicks et al. proposed that diethylamine or formic acid were effective additives for their sample sets [17].

5.4.5.2 Columns

In an adjacent study, Elise et al. also studied column orthogonality with SFC for the achiral SFC separation of 160 proprietary drugs [18]. They evaluated 1.7–2.5 μm fully porous or 2.6–2.7 μm superficially porous particles, with a variety of stationary phase chemistries. Because of the varying column dimensions used, the linear speed and gradient conditions were adjusted to be nearly equivalent for each of the columns.

In this study, a Derringer desirability functions was used to rank the selected columns with regard to retention, retention window and peak shapes. The results enabled the column set to be segregated in four groupings: (i) 1-aminoanthracene and 2-ethylpyridine phases, (ii) diol-type phases and the sole sulfobetaine phase, (iii) diethylamine and 2-picolylamine phases and (iv) the three bare silica phases and C18-bonded phase. For screening, these groupings may be considered orthogonal to each other.

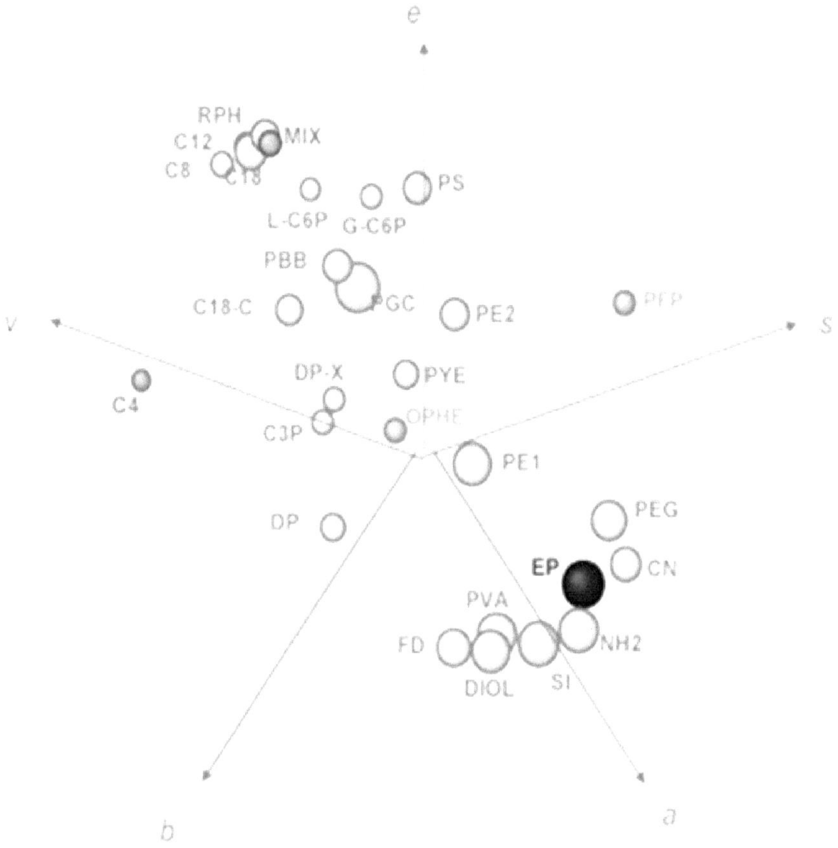

Figure 5.3 LSER classification of SFC columns. Figure used with permission from ref. [15].

West and Lesellier [19–24] built a database for commercially available SFC columns using a QSRR (quantitative structure-retention relationship) approach. They used a five-dimensional classification diagram (Fig. 5.3) that was based on a linear solvation energy relationships (LSER) to represent column selectivity similarity and differences. This work suggested that three major groups of stationary phases can be defined: nonpolar, moderately polar and very polar [19]. Alkyl-bonded stationary phases with no polar functionalities represent the nonpolar group. SFC behaves somewhat like the reversed-phase retention mode in this grouping. The moderately polar group consists of stationary phases with intermediate polarity, such as ODS phases containing a polar-embedded group or a hydrophilic end-capping group, as well as aromatic stationary phases. SFC behaves like a mixed mode that is in between reversed-phase retention and normal-phase retention. Finally, the polar group consists of more traditional normal phases such as bare silica gel (Si), 3-aminopropyl bonded silica (NH2), 3-cyanopropyl-bonded silica (CN) and propanediol-bonded silica (DIOL) in addition to fluoroalkyl phase (FD), polymer-coated phases (PEG and PVA) and alkylpyridines. SFC behaves in a more traditional normal-phase retention mode with the polar phase. It is recommended that the initial column screening is started with picking one or two columns from each group. These stationary phase combinations would provide a high probability to find appropriate starting conditions for method optimization.

5.4.5.3 Automated Normal-Phase Screening Systems

Due to a smaller market demand, there are no commercially available systems for automated normal-phase screening. However, the data could be entered into Model 1 or Model 2 software systems and evaluated from there.

5.4.5.3.1 Achiral SFC screen conditions

Table 5.12 is an example of a preliminary set of conditions for an orthogonal achiral SFC screen.

Table 5.12 Example preliminary achiral SFC screening conditions

Column phases	1. Ethylpyridine 2. Diol 3. Sulfobetaine 4. 2-picolylamine 5. Silica 6. C18

Additives	1. Acetic Acid 2. Ammonium acetate 3. neat MeOH	
Instrument settings	Columns with 100 × 3.0 mm dimensions (1.7–2.5 μm fully porous particles)	CO_2 with 5 to 50% MeOH in 10 min, flow rate at 1 mL/min, temperature at 25°C and outlet pressure at 150 bar
	Columns with 150 × 4.6 mm dimensions (2.6 μm superficially porous particles)	CO_2 with 5 to 50% MeOH in 15 min, flow rate at 2.35 mL/min, temperature at 25°C and outlet pressure at 150 bar
	Columns with 150 × 3.0 mm dimensions (2.7 μm fully porous particles):	CO_2 with 5 to 50% MeOH in 15 min, flow rate at 1 mL/min, temperature at 25°C and outlet pressure at 150 bar
Detection	UV @ 254 nm, MS	

5.5 Hydrophilic Interaction Liquid Chromatography

Hydrophilic interaction chromatography (HILIC) may be viewed as a variation of normal-phase chromatography. Because the mechanism for HILIC is not well understood, the technique is being cataloged in its own section for this chapter. The technique has become popular in its effectiveness to retain polar analytes that often elute in the solvent front of reverse phase methods. Oftentimes, polar impurities can be retained while the active drug elutes early.

Many believe that the primary means of separation is through the partitioning of analytes into a water layer held on the column surface. This has been demonstrated by experimental and simulation investigations [25]. In addition to simple partitioning, other influences such as multipoint hydrogen bonding and electrostatic interactions with bonded ionic groups have been proposed.

5.5.1 Control of Analyte Retention

In normal-phase chromatography, the mobile phase is 100% organic. In HILIC, the initial mobile phase is highly organic and increasing levels of water are used to elute analyte from water/polar packing surface. Water in the mobile phase attracts the polar analytes bound to the stationary phase to be eluted. Analytes are eluted in order of increasing hydrophilicity. Considerations for starting a HILIC method development are listed in Table 5.13.

Table 5.13 HILIC method considerations

Stationary phase	Start with silica unless early methods indicate otherwise
Organic modifier	Use ACN/H_2O
Buffer	Buffers used to control ionization
Avoid using TFA/TEA in early screens	Keep separation mechanism consistent. TFA/TEA add ion pairing capabilities that may or may not be needed
Temperature	Typical held above ambient conditions.
Gradient steepness	Robustness of method
Delay volume on system	Needs to be known for transfer

5.5.2 Hydrophobic Subtraction Theory

As with SFC, formal orthogonal column guides are generally not available online or on vendor websites. Start with a column you have history with and then select stationary phases known to behave differently for the chemistry of the intended sample set.

5.5.3 Solvent Strength

Hydrophilic interaction chromatography uses a polar stationary phase with weak polar organic non-polar mobile phase.

Traditionally, this meant a silica stationary phase. Today, this definition has expanded to bonded chemistries where the column remains quite polar. Most HILIC methods have been developed using acetonitrile and will not work if significant levels of alcohols such as methanol (MeOH) or isopropanol (IPA) are used. The strong solvent to "dissolve" the retained analyte from the stationary phase into the mobile phase is water. Weak solvents would be solvents which do not induce the analyte into the mobile phase, such as acetonitrile or acetone (Table 5.14).

Table 5.14 Strong and weak solvents in HILIC

HILIC strong solvents	HILIC weak solvents
Water	*Primary*:
Aqueous buffers and ion pairing solutions	Acetonitrile (ACN)
Methanol (MeOH)	*Secondary*:
	Acetone
	Dioxane

5.5.4 Ionic Samples

Buffers or ion pairing additives are used to control ionization in HILIC. Formic and trifluoroacetic acid are popular modifiers as they maintain mass spectrometry compatibility of the mobile phase. At pH 6, ionic effects from the stationary phase silanols begin to influence selectivity [25].

5.5.5 Screening

5.5.1.1 Mobile phases

Traditionally, acetonitrile is used as the organic phase in hydrophilic interaction chromatography. Use of small levels of methanol and isopropanol has illustrated mixed results [25, 26]. Thus, the primary alternative to acetonitrile is the use of acetone.

Periat et al. found that for the 82 drugs they studied using HILIC, pH 3 and pH 6 buffers at 10 and 50 mM were effective [26]. McCalley found additional selectivity with using trifluoroacetic or heptafluorobutyric acid additives [27]. However, as these additives can produce memory effects in columns and system tubing, they may be best suited for a second tier screen.

5.5.1.2 Columns

While literature method development studies [25–28] have looked to optimize columns, Periat et al. went a step further looking for column orthogonality for HILIC separations [26]. Their approach proposed four columns with sub-2 μm particles which seems like an excellent starting point for early HILIC screening: Agilent Zorbax RRHD (bare silica), Acquity BEH (hybrid silica), Thermo Syncronis HILIC (zwitterionic phase) and Acquity BEH amide (hybrid silica bonded with amide group).

5.5.1.3 Automated HILIC phase screening systems

Again, due to a smaller market demand, there are no commercially available systems for automated HILIC screening. However, Tumpa et al. [28] demonstrated a design of experiments (DOE) approach using Mode1 (Umetrics, Umea, Sweden).

5.5.1.4 HILIC screen conditions

Table 5.15 is an example of a preliminary set of conditions for an HILIC screen [26, 27].

Table 5.15 Example preliminary HILIC screen conditions

Column phases	1. Agilent Zorbax RRHD (bare silica) 2. Acquity BEH (hybrid silica) 3. Thermo syncronis HILIC (zwitterionic phase) 4. Acquity BEH amide (hybrid silica bonded with amide group)	
Mobile phases	1. ACN/ammonium formate, (pH 3.0, 10 mM) 2. ACN/ammonium formate, (pH 6.0, 10 mM) 3. ACN/trifluoroacetic Acid, (0.1%)	
Instrument settings	Columns with 50 × 2.1 mm dimensions (1.7 μm fully porous particles):	Initial isocratic hold at 5% ACN for 1 min, then to 65% ACN in 3 min and then 1 min at 65% ACN and re-equilibration during 3 min, flow rate at 0.5 mL/min, temperature at 30°C
Detection	UV @ 254 nm, MS	

5.6 Conclusion

Reversed-phase chromatography is once again rapidly evolving with the advent of small particle columns with high pressure systems optimized for minimal void volumes and fast detection. Yet, the basics of method development for this mode is still very much in accordance with the standards set by Snyder and Kirkland years ago [1]. The principles set forth by this classic work have simply been updated to reflect technology capabilities. In addition, automated system opportunities have enabled statistical and DOE algorithms to facilitate finding optimal separations conditions.

Supercritical fluid chromatography has, to date, been very successful in normal-phase applications, particularly in pharmaceutical analysis. SFC provides several advantages over traditional normal-phase liquid chromatography in terms of resolution, throughput, and consumption of hazardous/expensive solvents. However, these advantages have not been fully utilized in achiral analysis. There is little doubt SFC will become more and more popular in achiral analysis in the near future.

HILIC has found a very successful niche in chromatographing polar impurities that are difficult to resolve by reversed-phase separations. In particular, HILIC conditions are especially amenable to mass spectrometric detection for low level sensitivity and high selectivity.

Disclosure

AbbVie provided no financial support outside of Dr. Webster being an employee of AbbVie. AbbVie participated in writing, reviewing, and approving the publication. The chapter contains no proprietary AbbVie data.

References

1. L. R. Snyder, J. J. Kirkland, J. L. Glajch: *Practical HPLC Method Development,* 2nd ed., Wiley-Interscience: Hoboken, NJ. 1997.
2. *HPLC Made to Measure: A Practical Handbook for Optimization*, S. Kromidas, ed., Wiley-VCH, Weinheim, 2006.

3. *HPLC for Pharmaceutical Scientists*, Y. Kazakevich and R. Lobrutto, ed., John Wiley & Sons, Hoboken, NJ, 2007.
4. <621> Chromatography, United States Pharmacopeia and National Formulary (USP 40-NF 35). Rockville, MD: United States Pharmacopeia Convention, 2017.
5. The International Council for Harmonization of Technical Requirements for Pharmaceuticals for Human Use (ICH), 1996, Validation of Analytical Procedures: Text and Methodology.
6. G. K. Webster and C. L. Basel, Critical Pairs in Column Chromatography: A Primer for Pharmaceutical Method Validations, *LCGC*, 2003, 21(3), 286–294.
7. G. K. Webster, A. R. Diaz, D. S. Seibert, B. S. Weekley, J. D. Jackson, Plate Number Requirements for Establishing Method Suitability, *J. Chromatogr. Sci.*, 2005, 43, 67–72.
8. G. K. Webster and A. Elliott, Direct Method Scaling from UHPLC to HPLC: Is This Feasible for Pharmaceutical Methods? *Am. Pharm. Rev.*, 2011, 14, 32–40.
9. G. K. Webster and M. A. Gragg, Scaling LC Methods Using Superficially Porous Particle Stationary Phases, *LCGC North Am.*, 2018, 36 (3), 184–193.
10. J. Pellett, P. Lukulay, Y. Mao, W. Bowen, R. Reed, M. Mac, R. C. Mungerc, J. W. Dolan, L. Wrisley, K. Medwide, N. P. Toltl, C. Chan, M. Skibic, K. Biswas, K. A. Wells and L. R. Snyder, 'Orthogonal' Separations for Reversed-Phase Liquid Chromatography, *J. Chromatogr. A*, 2006, 1101, 122–135.
11. S. Karmarkar, R. Garber, Y. Genchanok, S. George, X. Yang, and R. Hammond, Quality by Design (QbD) Based Development of a Stability Indicating HPLC Method for Drug and Impurities, *J. Chromatogr. Sci.*, 2011, 49, 439–446.
12. B. Debrus, D. Guillarme, S. Rudaz, Improved Quality-by-Design Compliant Methodology for Method Development in Reversed-Phase Liquid Chromatography, *J. Pharm. Biomed. Anal.*, 2013, 84, 215– 223.
13. N. S. Wilson, M. D. Nelson, J. W. Dolan, L. R. Snyder and P. W. Carr, Column Selectivity in Reversed-Phase Liquid Chromatography II: Effect of a Change in Conditions, *J. Chromatogr. A*, 2002, 961, 195–215.
14. Webster (ed.), *Supercritical Fluid Chromatography: Advances and Applications in Pharmaceutical Analysis*, G. K. Pan Stanford: Singapore, 2014.

15. J. W. Caldwell, W. B. Caldwell, G. K. Webster and Z. Wang Method Development for Achiral SFC, Chapter 4, in *Supercritical Fluid Chromatography—Advances and Applications in Pharmaceutical Analysis*, Pan Stanford Publishing, 2014.
16. E. Lemassona, S. Bertin, P. Hennigb, H. Boiteuxc, E. Leselliera, C. West, Development of an Achiral Supercritical Fluid Chromatography Method with Ultraviolet Absorbance and Mass Spectrometric Detection for Impurity Profiling of Drug Candidates. Part II. Selection of an Orthogonal Set of Stationary Phases, *J. Chromatogr. A*, 2015, 1408, 217–226.
17. M. B. Hicks, Y. Zhang, D. Moore, A. Apedo, Advancing SFC Methods Development with a Multi-Column Supercritical Fluid Chromatography with Gradient Screening, *Am. Pharm. Rev.,* 2011 14(6), 52, 54, 56, 58–60. http://www.americanpharmaceuticalreview.com/Featured-Articles/37308-Advancing-SFC-Method-Development-with-a-Multi-Column-Supercritical-Fluid-Chromatography-with-Gradient-Screening/.
18. E. Lemassona, S. Bertin, P. Hennigb, H. Boiteuxc, E. Leselliera, C. West, Development of An Achiral Supercritical Fluid Chromatography Method with Ultraviolet Absorbance and Mass Spectrometric Detection for Impurity Profiling of Drug Candidates. Part I: Optimization of Mobile Phase Composition, *J. Chromatogr. A*, 2015, 1408, 227–235.
19. West, C., Lesellier, E. A Unified Classification of Stationary Phases for Packed Column Supercritical Fluid Chromatography, *J. Chrom. A*, 2008, 1191, 21–39.
20. West, C., Lesellier, E. Characterization of Stationary Phases in Supercritical Fluid Chromatography with the Solvation Parameter Model, *Adv. Chromatogr.*, 2010, 48, 195–253.
21. West, C., Lesellier, E. Characterization of Stationary Phases in Subcritical Fluid Chromatography with the Solvation Parameter Model IV. *J. Chrom. A*, 2006, 1115, 233–245.
22. West, C., Lesellier, E. Characterization of Stationary Phases in Subcritical Fluid Chromatography by the Solvation Parameter Model, *J. Chrom. A*, 2006, 1110, 181–190.
23. West, C.; Lesellier, E. Characterisation of Stationary Phases in Subcritical Fluid Chromatography with the Solvation Parameter Model. III. Polar Stationary Phases, *J. Chrom. A*, 2006, 1110, 200–213.
24. West, C., Lesellier, E. Orthogonal Screening System of Columns for Supercritical Fluid Chromatography, *J. Chrom. A*, 2008, 1203, 105–113.

25. A. Kumar, J. C. Heaton, D. V. McCalley, Practical Investigation of the Factors that Affect the Selectivity in Hydrophilic Interaction Chromatography, *J. Chromatogr. A*, 2013, 1276, 33–46.
26. A. Periat, B. Debrus, S. Rudaz, D. Guillarme, Screening of the Most Relevant Parameters for Method Development in Ultra-High Performance Hydrophilic Interaction Chromatography, *J. Chromatogr. A*, 2013, 1282, 72–83.
27. D. V. McCalley, Study of Retention and Peak Shape in Hydrophilic Interaction Chromatography Over a Wide pH Range, *J. Chromatogr. A*, 2015, 1411, 41–49.
28. A. Tumpa, A. Stajić, B. Jančić-Stojanović, M. Medenicaba, Quality by Design in the Development of Hydrophilic Interaction Liquid Chromatography Method with Gradient Elution for the Analysis of Olanzapine, *J. Pharm. Biomed. Anal.*, 2017, 134, 18–26.

Chapter 6

Polysaccharide-Derived Chiral Stationary Phases for the Separation of Enantiomers

Tong Zhang and Pilar Franco

Chiral Technologies Europe, Parc Innovation,
160 Bd. Gonthier d'Andernach, F-67400 Illkirch, France

tzhang@chiral.fr

6.1 Introduction

Based on the research work from Prof. Okamoto's group [1–3], polysaccharide-derived and silica-supported chiral stationary phases were introduced in the 1980s. They are considered as the first choice of chiral stationary phases (CSPs) for enantiomer separations by liquid chromatography (LC) and supercritical fluid chromatography (SFC). A considerable number of publications over three decades have reported and reviewed the applications of these CSPs by scientists working in the field of enantioseparations [4–24].

Chromatographic Method Development
Edited by Gregory K. Webster and Laila Kott

ISBN 978-981-4800-53-2 (Hardcover), 978-0-429-20172-1 (eBook)
www.jennystanford.com

Currently many polysaccharide-derived CSPs are commercially available, either as coated or immobilized phases, with particle sizes ranging from 20 μm for preparative separations over 5 μm for conventional method development to 3 μm and sub-2 μm for ultrafast analysis of enantiomers [13, 25, 26]. The most advanced series includes the CSPs by the optimized immobilization technologies: CHIRALPAK® IA, CHIRALPAK® IB, CHIRALPAK® IC, CHIRALPAK® ID, CHIRALPAK® IE, CHIRALPAK® IF, CHIRALPAK® IG and CHIRALPAK IH (Fig. 6.1). These immobilized CSPs combine the benefits of polysaccharide-based chiral selectors—that is their broad application scope and their preparative potential—with the advantages of the immobilization process, which leads to strong CSP robustness and practically unrestricted solvent compatibility.

The second series contains the classical coated CSPs, bearing many different substitution patterns: CHIRALPAK® AD, CHIRALPAK® AS, CHIRALPAK® AY, CHIRALPAK® AZ, CHIRALCEL® OD, CHIRALCEL® OJ, CHIRALCEL® OZ, and CHIRALCEL® OX among others (Fig. 6.2).

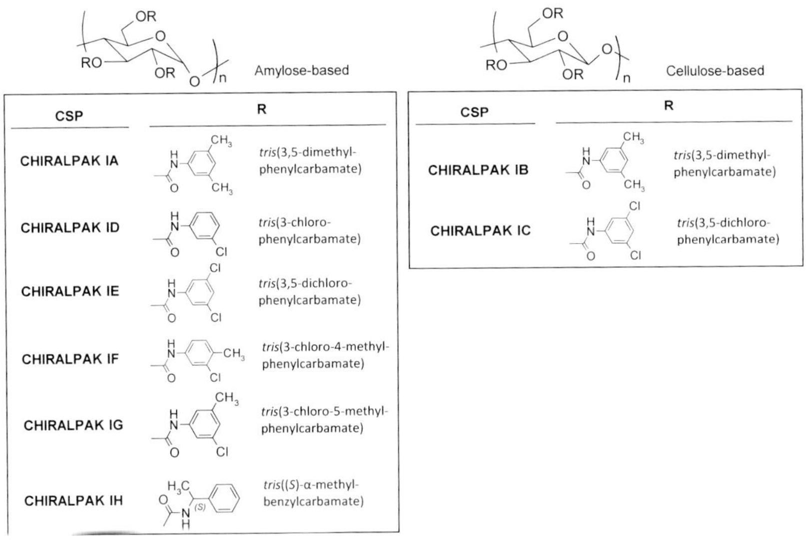

Figure 6.1 Structures of the immobilized polysaccharide-derived chiral selectors.

One essential feature of these polysaccharide derivatives is their broad application domain. Accordingly, a judicious selection of CSPs allows the design of effective screening kits with a limited number of columns and a high success rate. For many years this has been the rational choice of four coated CSPs as primary screening tools in the research laboratories and in pharmaceutical industry. Concerning the immobilized CSP selection, primary screening kit will contain the selectors with the broadest application domains. The remaining selectors would be included in a secondary screening.

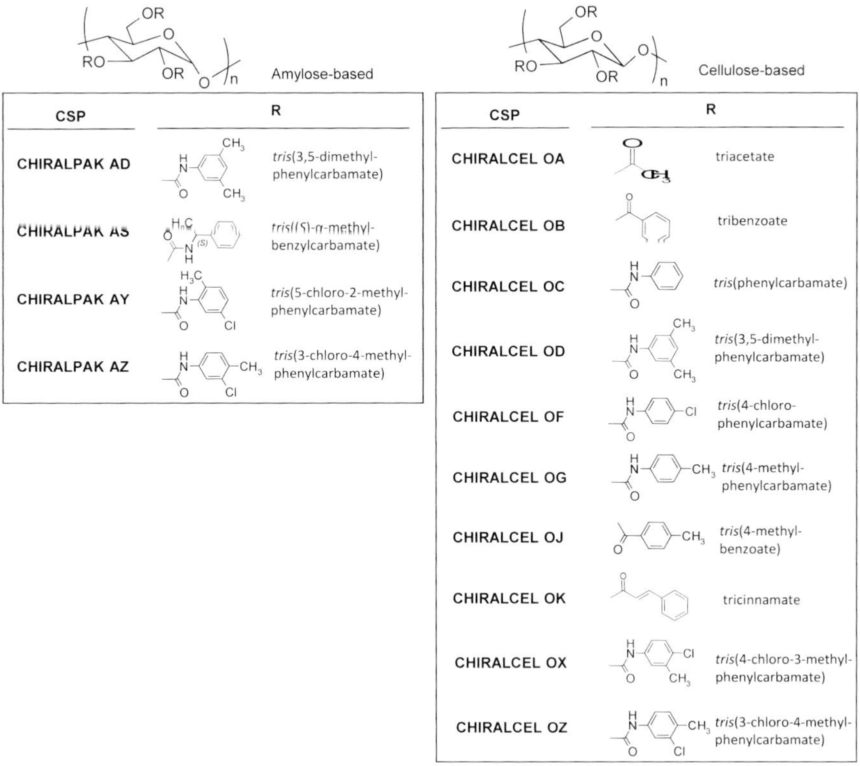

Figure 6.2 Structures of the coated polysaccharide–derived chiral selectors.

In this chapter, we overview and compile the screening strategies or approaches for the separations of enantiomers on these types of chiral supports. The main issues to be addressed include the configuration of the column set for sample screening, the separation technique (HPLC or SFC) and the choice of mobile phase in different separation modes, i.e., normal phase (NP) mode, polar organic (PO) mode and reversed phase (RP) mode to achieve efficient, fast, and appropriate method which is matched to the requirements of the project and the molecule.

6.2 Column and Mobile Phase Selection

The standard screening protocols adopted by laboratories or research groups worldwide usually use three or four analytical columns in the primary screening. Two main options for setting up the primary screening, with either the four traditional coated columns (CHIRALPAK AD and AS, CHIRALCEL OD and OJ) or alternatively the immobilized phases for enhanced applicability and robustness [27–34], including CHIRALPAK IA, IB, IC, and IG. Such configurations with polysaccharide-derived columns, tested on a randomly composed sample pool, and with four solvent mixtures, can afford complete enantiomer resolution for over 90% of the racemates examined and enantiorecognition above 95%.

Although most separations can be achieved with a reduced number of columns and mobile phases, there remains a need for unique selectivities that allow the resolution of certain challenging isomers or their impurities. This is where the "secondary" column set can step in. It should be noted that continuous research effort over the last decades has been devoted to the development new and powerful selectors showing enantiomeric recognition ability complementary to the primary column sets used for screening. A common and typical characteristic of these new CSPs are cellulose and amylose derivatives bearing methyl- and chloro-substituted phenylcarbamate groups, instead of the dimethyl- or dichloro- present in the standard systems either in coated or immobilized fashion (Figs. 6.1 and 6.2 for structural features). Some of these materials were initially developed in 20 μm particle size and involved in our CSP library application solely for preparative purposes but following their high success rates for nearly a decade, they have been introduced in the product line in 20, 5, 3, and now sub-2 μm formats for LC and SFC separations.

Method screening with polysaccharide-derived CSPs should consider several practical aspects and has to be designed based on the needs of each particular application. Polysaccharide-derived

CSPs exhibit chiral recognition abilities in organic and aqueous mobile phases in LC, as well as under supercritical fluid (SFC) conditions. Their enantioselectivity, retentivity and resolution degree are differently modulated in those modes, as solvent eluting strength, solvation effect, supra-molecular structure of the polymer and interaction mechanism (such as hydrogen bonding, dipole–dipole interaction, p-p stacking, steric hindrance) will not be equivalent in all of them. Therefore, the choice of the chromatographic mode will be an essential element in the experimental design. Moreover, when setting up a screening strategy, it is important to consider the separation scale (analytical or semi-preparative or preparative) that has to be achieved, the time constraints, the compatibility of mobile phase with the CSP and compound as well as environmental restrictions.

On this basis, the present section will aim to give a general overview of the main principles to be applied, although it is recommended to adapt these recommendations to the individual needs. Three main areas will be covered:

1. Use of organic mobile phases in HPLC
2. Use of water compatible mobile phases in HPLC
3. SFC conditions

6.2.1 Method Screening with Organic Mobile Phases in HPLC

Given the increasing number of chiral compounds in drug development and other domains, efficient and automated HPLC screening using columns packed with various CSPs has been a challenging subject and discussed by several authors [27–33, 35]. The strategies to achieve enantiomer separations are quite variable from one laboratory to another depending on the type of substances in screening, the equipment available and the R&D background. However, the motivation for the development of efficient screening approaches remains common. We all search for the easiest, the fastest, the most successful and reliable way to analyze a series of molecules with diverse structures and chemical properties in a reasonable time frame. This is the standpoint for chiral compound screening.

6.2.1.1 Immobilized CSPs under organic mobile phase conditions

Since their introduction in the market in 2004, a certain number of research teams have published and described the use of immobilized polysaccharide CSPs in HPLC for new and diverse applications involving enantioseparations [16–18, 25, 26, 32–34, 36–81]. As already mentioned, immobilized polysaccharide-derived CSPs have several benefits that should be considered in the screening process, but some guidelines may be needed to direct the choice of solvents from their very broad range to compose the mobile phase [16, 17, 32–38].

Based on exhaustive investigations on immobilized columns, it was concluded that, when used as the mobile phase or mobile phase components, a group of solvents and their mixtures can lead to better selectivity values for most of the compounds tested. These organic conditions could be considered as a primary set of mobile phases for screening with these CSPs. Primary solvent configurations for the analytical screening system are shown in Table 6.1. In practice, these four mobile phase systems can be set up using a quaternary HPLC screening system.

It is well known that the mobile phase may have major effects on the enantiomeric separations. The chromatograms in Fig. 6.3a demonstrate how significantly the enantiomeric separation can be dependent on the mobile phase composition. On a CHIRALPAK IA-U column (1.7 μm, 100 × 3.0 mm i.d.), the resolution of N-CBZ-DL-Ala enantiomers is partial using the mobile phase of hexane/EtOH/FA 80/20/0.1 (FA: formic acid). The simple replacement of EtOH with isopropyl alcohol (2-PrOH) at the same volume percentage leads to a much larger separation of the same enantiomers. Figure 6.3b shows a case of reversal of elution order of enantiomers on the same CHIRALPAK IA-U column, induced by the change from 2-PrOH to EtOH as the polar mobile phase component.

Chlorinated solvents are usually submitted to a more rigorous control due to environmental restrictions. Nevertheless, their use at analytical level may be allowed in most laboratories, provided that their waste residues are treated appropriately. If the use of dichloromethane (DCM) is indeed constrained, tetrahydrofuran (THF)-based mixtures could be implemented instead.

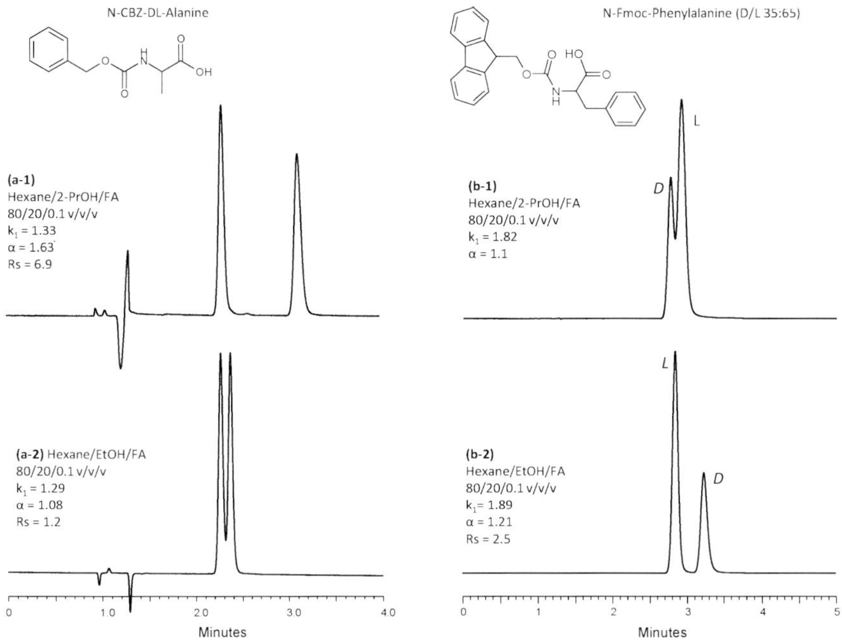

Figure 6.3 Effect of mobile phase on chiral separations. Column: CHIRALPAK IA-U (1.7 μm, 100 × 3.0 mm i.d.); temperature: 25°C; flow rate: 0.425 mL/min.

If additional screening of mobile phases was required, the use of the following mixtures as a secondary set is recommended: heptane or hexane/ethyl acetate EtOAc, heptane or hexane/THF, acetonitrile (ACN) and methanol (MeOH) (Table 6.1).

Different alcohols, with shorter or longer aliphatic chains, e.g., EtOH, 1-propanol (1-PrOH) and 2-PrOH, can play a "dual" role to mix either with non-polar alkanes or with polar solvents (ACN, MeOH), though alkane is immiscible to ACN and MeOH. However, the solvent nature varies little; hence, the choice stays quite limited if the viscosity factor is considered. In contrast, solvents such as THF, EtOAc, MtBE, DCM, etc., belong to different solvent families with distinct properties. Unlike heptane, methanol, and acetonitrile, those solvents are all of mid-polarity and miscible with both non-polar and polar organic solvents. This feature is beneficial for fine-tuning composition of mobile phase, as it allows easy adjustment in retention, selectivity and resolution.

Table 6.1 Mobile phases for sample screening with immobilized CSPs

	Solvent mixture	Starting condition (%)	Typical optimization range (%)
Primary mobile phases	Alkane/EtOH	80/20	99/1 to 50/50
	Alkane/2-PrOH	80/20	99/1 to 50/50
	Alkane/DCM/EtOH	50/50/2	85/15/0 to 0/80/20
	Alkane/MtBE/EtOH	0/98/2	80/20/0 to 0/40/60
Secondary mobile phases	Alkane/EtOAc	50/50	80/20
	Alkane/THF	70/30	95/5 to 0/100
	MeOH	100	
	ACN	100	

The proposed mobile phase systems were selected to maximize complementary behavior among the CSPs of the set and also between solvents on the same column. It is important to note that elution order of enantiomers may change in certain cases due to the mobile phase (Fig. 6.3b) and due to the different selectors. This phenomenon can also be observed for separation of certain target sample and its impurities. Such systematic screening can be a useful tool to choose the most suitable CSP-mobile phase combination for the specific separation aims.

The first objective is to set the right conditions to start the screening, leading to appropriate retention times and successful chiral separations. As an example, and based on an important number of experimental data, the proportions indicated in Table 6.1 can be taken as starting point and possibly adjusted according to the first results and the needs for optimization. The retention factor of the compound will steer the adjustment of the mobile phase for eluting strength. With these first data, one can decide the best solvent(s) or solvent ratios to achieve the desired separation.

EtOAc, MtBE, THF, DCM, and chloroform ($CHCl_3$) can be used as mobile phases in their undiluted form or even combined with alcohols. However, neat THF appears to be the mobile phase with the highest eluting strength among them. Its use in its pure form often leads to short or no retention of solutes, although we have already seen certain exceptional resolutions in THF/alcohol mixtures.

If an exhaustive investigation is to be carried out with a given compound, or the results from the screening with the first set of mobile phases are not satisfactory, it may be worth considering expanding the solvent range. For example, acetone, 1,4-dioxane, toluene, methylal (dimethoxymethane) or ethylal (diethoxymethane) have proved to be useful as mobile phase components for successful chiral separations in certain cases.

It is important to point out that the conditions described in Table 6.1 are directly applicable for neutral (bearing no charges) compounds. If acidic or basic racemates have to be separated into enantiomers, addition of an acidic or basic additive to these mobile phases may be necessary, although the same principles as described for the bulk solvent composition will remain valid. The role and particular aspects of the additives will be discussed later in this chapter.

No particular advice needs to be given for the transition between different organic mobile phases, as long as miscibility issues are taken into account. However, when switching from very different mobile phase conditions, equilibration times have to be adapted (e.g., at least 30 min at 1 mL/min for analytical columns sized 150 × 4.6 mm i.d.) in order to ensure method reproducibility as a result of complete equilibration of the column.

Taking into account the different nature and proportion of the solvents described earlier, the use of a UV diode array detector may not be highly recommended. Certain mobile phases containing toluene, acetone and high proportions of DCM or EtOAc may cause UV detection problems due to their high UV cut-off. In this case, hyphenation of other type of detectors may be necessary, such as CAD (charged aerosol detector), ELSD (Evaporative light scattering detector), polarimeter, CD (circular dichroism), RI (refractive index) or MS (mass spectrometry) detector.

The use of gradients could also be envisaged to select appropriate solvent composition in terms of retention and selectivity [27]. It would then be important to make sure that the HPLC system in use is able to guarantee proper and accurate metering and mixing of different mobile phase components, as this is not always the case with DCM, EtOAc or THF mixtures.

Sample solubility in the mobile phase, enantioselectivity and loading capacity are key factors in chiral preparative chromatography to render a preparative separation practically feasible and/or economically attractive. However, sample solubility and stability in the injection solvent are also important at analytical level. Such a limitation may be circumvented by using immobilized CSPs. In contrast to the coated CSPs, the immobilized ones allow method development directly guided by sample solubility with the possibility to improve the enantioselectivity and separation. From the method development viewpoint, a compromise of sample solubility, enantioselectivity, resolution and retention time is often needed.

6.2.1.2 Coated CSPs under organic mobile phase conditions

The approach of method development and optimization on the coated CSPs can be based on a similar strategy to the one used for the immobilized-type phases. However, the choice of solvents must be restricted to those being compatible with these supports. This means that all solvents or solvent combinations that could potentially dissolve, even partially, the polysaccharide derivative must be avoided in the mobile phase. The appropriate solvent set for screening will be then composed of alkane/EtOH, alkane/2-PrOH, alcohol-based and ACN-based mixtures (Table 6.2). In principle, these mobile phase systems can be set up using the same HPLC screening system. However, it is recommended that ACN or its mixtures with alcohols will be carefully managed, preferably separately, in order to avoid any potential accidental wrong mixing between alkane and ACN. If such an event happens, the coated columns could be irreversibly damaged. Obviously, a good column flushing with a transition solvent having a broad miscibility (e.g., EtOH [82]) must be inserted between the two mobile phases which are not miscible with each other.

Table 6.2 Mobile phases for sample screening with coated CSPs

Solvent mixture	Starting condition (%)	Typical optimization range (%)
Alkane/EtOH	80/20	99/1 to 50/50
Alkane/2-PrOH	80/20	99/1 to 50/50
MeOH	100	
ACN	100	

6.2.2 Method Screening with Aqueous or Water Compatible Mobile Phases in HPLC

The polysaccharide type CSPs seem to be most frequently used with organic eluents (alkane-alcohol mixtures or the polar solvents discussed above). However, resolution of enantiomers on these columns by aqueous eluents has a history almost as long as their applications with organic mobile phases [11]. Method development in RP-mode on polysaccharide-derived CSPs has been discussed in several articles [11, 83–85] and some new applications have been recently developed [48, 54, 65].

The choice of RP-mode is often dictated by the need for direct injection of the samples of biological extracts. It is also an alternative to certain compounds with low retention, or limited solubility or lack of enantioselectivity with organic mobile phases. An additional and increasing interest in RP methods is their undeniable suitability for LC-MS applications.

The chiral recognition mechanism of polysaccharides in the presence of water differs from the one observed in organic mobile phase. The solvation effect on the stationary phase induced by water may be strong and will modify the retention and enantiorecognition mechanisms. Therefore, reverse phase chromatography is often complementary to normal phase separation mode and offers the possibility of enhancing the resolution success.

6.2.2.1 Immobilized CSPs in aqueous or water compatible solvent mixtures

As a logical sequence of the recent investigations with organic mobile phases in HPLC and SFC, an intensive study on the chromatographic behavior of the immobilized CSPs towards various mobile phase systems was undertaken in our laboratories. This has allowed the development of simple and straightforward strategies for efficient method screening [18, 32, 33].

The nature of the molecules to be analyzed should be considered for mobile phase selection. Under this primary guideline, a "decision tree" can be defined to start the method development (Fig. 6.4). The selected mobile phases do not really differ from the proposals for achiral reverse phase screening.

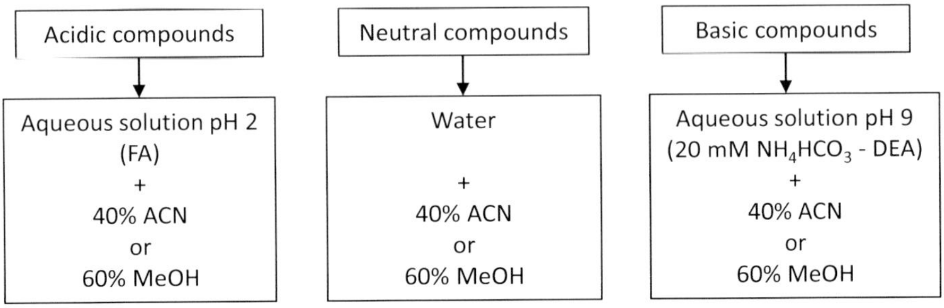

Figure 6.4 HPLC screening strategy on polysaccharide-derived columns using aqueous mobile phase.

The selection of the mobile phase for neutral compounds is straightforward: plain water is suitable to be used as the aqueous component. The most common and effective pH value of the aqueous solution for the resolution of acidic molecules is in the range 2.0–2.5 in which the ionization of most organic acids will be suppressed. FA is a good choice where LC-MS compatibility is required while other acids (trifluoroacetic acid (TFA), acetic acid, H_3PO_4) may be valuable for other applications [18, 33]. Based on the same reasoning, basic aqueous solutions should normally be in use to suppress the ionization of basic solutes. For this purpose, certain buffers or aqueous media (phosphates, borates, ammonium acetate or KPF_6 aqueous solutions) at basic pH may also be useful for enantiorecognition [18]. Where RP methods suitable for MS detection for basic compounds are required, ammonium bicarbonate solution adjusted to pH 9 is a good option.

The organic modifier plays an important role in regulating retention and modulating enantioselectivity. The first choice of modifier to be combined with the aqueous solution is ACN and MeOH. In this context, the RP-mode could be sometimes considered as an extension of the PO-mode when retention is not sufficient in the absence of water.

MeOH can be a useful modifier in RP-mode but with weaker eluting strength and leading to much more viscous aqueous mobile phases. ACN has the advantages of low viscosity, low UV cut-off, adapted eluting strength on the polysaccharide-type CSPs and the outstanding ability to induce good enantioselectivity. The use of other organic modifiers, such as EtOH, 1-PrOH and 2-PrOH can also be considered. In our practice, however, these modifiers would be tried only if no successful hit is observed with ACN and MeOH in the mobile phase.

Additionally, and specifically to the immobilized supports, THF can also be employed to compose the RP mobile phase. It may sometimes be a good alternative to ACN and MeOH, but in a general way it seems to be a less versatile organic modifier. The relative eluting strength of these three organic modifiers can be ranked as follows: THF > ACN > MeOH. Moreover, in case of need, sample can be dissolved in DMSO, DMF or acetone for injection on the columns packed with the immobilized CSPs.

As in NP-mode, the key point for RP method development would be setting the right conditions to start the sample screening. Based on a significant number of experimental data, the mobile phase compositions indicated in Fig. 6.4 can be recommended as the starting point. The typical starting conditions represent actually the mobile phases of upper-middle eluting strength (containing enough organic modifier and enabling elution of most of the compounds in a reasonable analysis time frame). In many cases, successful separations can be directly obtained issuing the screening procedure. However, certain compounds or families of compounds may be either extremely retained or too fast eluted in these conditions. In this case, an optimization step should be considered for the individual samples, by playing the nature and the percentage of the organic modifiers. Gradient elution can certainly be used for sample and mobile phase screening [86], but isocratic elution mode seems to be more convenient as for method reproducibility or robustness and for straightforward system-to-system or lab-to-lab method transfer.

Another interesting feature of immobilized polysaccharide-derived supports is the possibility to use the "strong" solvents (such as THF and DMF) for column flushing/cleaning and for efficient transition from RP- to NP-mode and vice versa. The flushing protocol is described in detail on the instruction manual of the immobilized columns (discussed in Section 6.7 in this chapter) but will not be compatible with the previously existing coated CSPs.

It is worth mentioning the recent reports about chiral separations carried out on the immobilized CSPs by incorporating a low percentage of water (<40% in volume) in ACN or alcohol (MeOH or EtOH) as mobile phase [87–90]. In these studies, HILIC (hydrophilic interaction liquid chromatography) or HILIC-like mode is suggested for effective retention and enantiomer separation of certain polar and hydrophilic compounds.

6.2.2.2 Coated CSPs in aqueous or water compatible solvent mixtures

Method development in RP-mode in the case of the coated supports can follow exactly the same strategy as described in Fig. 6.4. However, apart from alcohols and acetonitrile, no other alternative organic solvents can be used, as they may irreversibly damage the columns.

6.2.3 Mobile Phases in SFC

SFC applications in the separation of enantiomers are a growing field [91–95] and many applications using polysaccharide-based columns have been described [17, 33, 34, 40, 96–113].

6.2.3.1 Immobilized CSPs in SFC

Experiments to date with polysaccharide-derived CSPs show that success rates can be achieved by using alcohol co-solvents [17, 34, 96–104]. In particular, MeOH, EtOH, and 2-PrOH are very good initial co-solvent choices (Fig. 6.5). ACN seems to be less successful, but it may be an option for a secondary screening. If the elution of the analytes is not fast enough when ACN is used, an alcohol can be added to the CO_2/ACN mixture. The screening can be performed in isocratic and gradient mode [33, 101, 103, 104].

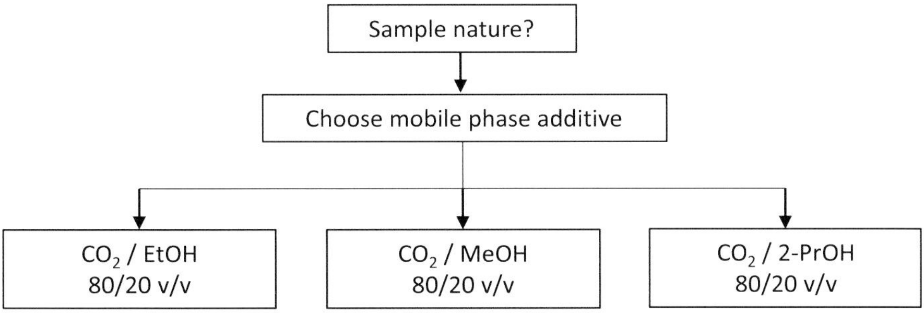

Figure 6.5 Primary screening strategy on polysaccharide-derived columns in SFC mode.

To resolve basic compounds, add a small amount (normally 0.5–1% in the co-solvent) of a basic additive (typically isopropylamine (iPAm), diethylamine (DEA) or trimethylamine (TEA)). For acidic compounds, an acidic additive (i.e., TFA, FA or acetic acid) may be helpful for better resolution, although the acidity of CO_2 could in certain cases be sufficient for good enantiomer resolution acidic compounds [104].

Moreover, immobilized polysaccharide-derived CSPs are compatible with a much wider range of solvents than are the coated phases [114]. The use of such "extended range" solvents was investigated for SFC screening in a similar manner as for LC separations. Initially these solvents were used to enhance sample solubility, to solve difficult separations or to process compounds being unstable in alcohols. Subsequently we have found that these solvents can also offer exceptional enantioselectivity profiles. THF, MtBE, DCM, or EtOAc, for example, can be used as the CO_2 modifier. In some cases, pure DCM, MtBE and EtOAc may not be strong enough to elute certain compounds. In these cases, the addition of small percentages of an alcohol mixed with the co-solvent is advised (Fig. 6.6) [17, 33, 103].

As already mentioned, success rate is already very high with the alcohol modifiers. Therefore, it has become established practice to use either the immobilized columns or alternatively the coated ones—in the column set for screening with the alcohol modifiers as first strategy. Then the solvent range can be broadened on the immobilized columns if no satisfactory separation is achieved or sample solubility/stability becomes an issue. SFC and LC are complementary techniques and comparable success rates are achieved with the same set of columns. Nevertheless, recognition patterns are sometimes different and leading to different retentive and selectivity profiles. These two factors plus the likely sample solubility issues will be determinant in the choice of one technique or the other for separation of a given compound.

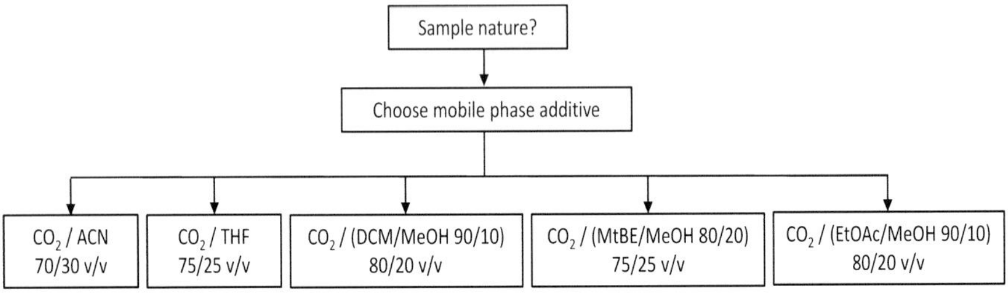

Figure 6.6 Secondary screening strategy on polysaccharide-derived columns in SFC mode (only compatible with immobilized CSPs, except ACN).

Analytes are preferably dissolved in the solvent mixture that is used as the co-solvent in SFC. If this is not possible, choose a better sample diluent (e.g., ethanol) and monitor any potential perturbance of the baseline due to the injection solvent. For sample insoluble in alcohols or ACN, it is possible to dissolve these in DCM or DCM mixtures for injection when using immobilized phases. In such cases, caution should be taken to avoid on-line precipitation of the compounds, but this should not be done with coated phases without compromising life-time of the column.

6.2.3.2 Coated CSPs in SFC

Coated CSPs are only compatible with alcohols and acetonitrile mixtures in SFC. As discussed in the previous section, normally the screening starts with alcohol screening and acetonitrile would be only screened as a secondary option, either in pure form or mixed with alcohols to enhance elution.

6.3 The Role of Additives

For basic and acidic samples, it may be necessary to incorporate an additive in the mobile phase in order to get reasonable elution of the compounds with good peak shape and optimum resolution degree. Thus, basic samples may require a basic additive (DEA, butylamine (BA), 2-aminoethanol (AE), ethylenediamine (EDA)) and acidic compounds the addition of an acid (TFA, FA, or acetic acid). The percentage needed is typically 0.1% (in volume) and should normally not exceed 0.5% (see examples described in the following sections) [16, 17, 33]. In SFC this amount is also respected, although as added in the co-solvent it may be 0.5–1% to keep the overall percentage <0.5% in the final mobile phase. In RP-conditions, buffers will be used as discussed in the corresponding section [18].

It has been found that certain amines, such as EDA and AE induce much better behavior for certain compounds than the more commonly used DEA. Resolution degree and peak symmetry can be dramatically improved with this type of additives, as exemplified in Fig. 6.7. For practical purposes, as EDA and AE may have limited miscibility in certain organic solvents in the absence of alcohol, it is better running screenings with DEA, but it will be important in the optimization step to consider this fact.

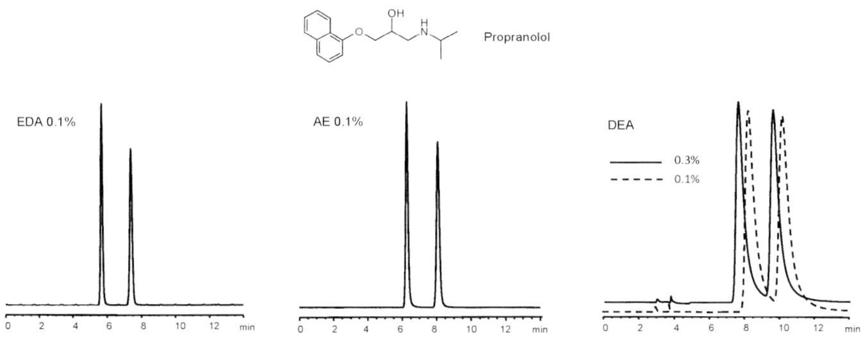

Figure 6.7 Additive effect on enantiomer resolution. Column: CHIRALPAK IB (5 μm, 250 × 4.6 mm i.d.); mobile phase: hexane/2-PrOH/additive, 90/10/0.1 (or 0.3%) v/v/v; flow rate: 1.0 mL/min; temperature: 25°C.

6.4 Column Temperature

Column temperature is a critical factor impacting the thermodynamic process and determining the kinetic aspects of separations by chromatography. This is also true for enantiomer separation using CSPs. The column temperature can induce multiple effects: it can affect retention, enantioselectivity, column efficiency, resolution degree, column backpressure, analysis speed, sample, and column stabilities and even elution order of the enantiomers [115–124]. For CSPs based on polysaccharide derivatives, temperature could also alter the conformation of the polymeric chiral selectors and trigger different interaction mechanisms between the enantiomers and the chiral selector.

The effect of column temperature on chromatographic retention and selectivity can be expressed by van't Hoff equations:

$$\ln k = -\Delta H°/RT + \Delta S°/R + \ln Ø \quad (6.1)$$

$$\ln \alpha = -\Delta(\Delta H°)/RT + \Delta(\Delta S°)/R, \quad (6.2)$$

where $\Delta H°$ and $\Delta S°$ are, respectively, the standard enthalpy and entropy changes for the analyte molecule transfer from mobile phase to stationary phase; Ø is the phase ratio defined as the ratio of the volume occupied by the stationary phase to the volume of the mobile phase contained in a column; $\Delta(\Delta H°)$ and $\Delta(\Delta S°)$ are, respectively the differences in the changes in standard enthalpy and entropy of the enantiomers in separation; R is the gas constant; T is the absolute temperature expressed in K (Kelvin). The plots of ln k and ln α versus $1/T$ will reveal the mechanistic information of the retention and enantiomer separation. The case study below demonstrates the effects of temperature and the approaches for method optimization.

Racemic Praziquantel was well resolved into enantiomers on CHIRALPAK IA (5 μm, 250 × 4.6 mm i.d.) using a mobile phase composed of n-heptane/EtOH 45/55 v/v at 1.0 mL/min (Fig. 6.8). The temperature was varied in the range of 25–50°C in 5°C increments. As shown in Fig. 6.8c–d, van't Hoff plots of the capacity factors and selectivity were linear in the temperature range of 50°C down to 30°C ($1000/T$ = 3.09 to 3.30 K^{-1}) but started to bend at around $1000/T$ = 3.3 (30°C). As defined by van't Hoff equations, enantioselectivity is the balanced parameter between differences in enantiomeric binding enthalpy and disruptive entropic effects. A transition point breaking the linear relationship between ln k or ln α and I/T in van't Hoff plot indicates an evolution of the retention and/or selective mechanisms governing the separation. Very probably, a change in the conformation of the chiral selector occurred under the given mobile phase condition at approximately 30°C (or 3.30 1/°K), inducing the mechanistic modification in the case of Praziquantel. Typically, the conformational change of the chiral selector is thermodynamically reversible if T is kept in the reasonable range, ensuring the good reproducibility of the separation under given chromatographic conditions. As an overall trend, the enantioselectivity is slightly increasing with the decrease in temperature. This is actually the scenario observed in enantiomer resolution of Praziquantel on CHIRALPAK IA.

As far as the kinetic aspect is concerned, the increase in temperature normally accelerates the mass transfer process, leading to enhanced column efficiency (Fig. 6.8e) and sharp shortening of the analysis time (Fig. 6.8f). As a balance of the temperature effect on column efficiency, on enantioselectivity and retention factor, the resolution degree of Praziquantel enantiomers followed a bell-shaped curve against temperature and reached the highest Rs value at 40°C (Fig. 6.8f).

The experimental evidence suggests that adjustment and control of the temperature may be a very effective tool for optimizing chiral separations, although the achievement of the maximized resolution degree for a given pair of enantiomers would need optimization efforts through a systematic study on temperature effect. When combined with other parameters such as the reduced particle size of the CSP, reduced column size and high flow rate, the increase in temperature can accelerate the analysis to sub-minute (fast or ultrafast analysis). This is the case of enantiomer resolution of Indapamide on a short CHIRALCEL OD-3 column (50 × 4.6 mm i.d.) packed with 3 μm CSP particles. As shown in Fig. 6.9a, the enantiomers were well resolved within 30 s of time at 25°C with MeOH at 3 mL/min. By increasing the temperature to 40°C, the analysis time could be shortened even further to about 20 s. The chromatograms in Fig. 6.9b illustrate the temperature effect on peak shape improvement. Raising the temperature from 25 to 40°C led to drastic peak narrowing and sharpening of Mephobarbital on CHIRALPAK AD-3, most remarkably for the second eluting enantiomer. In this case, the temperature increase brought forth multiple benefits: ultra-fast analysis, enhanced resolution degree (Rs from 4.7 up to 7.7) without mentioning the potential gain in detection sensitivity (or signal-to-noise ratio) of the compound.

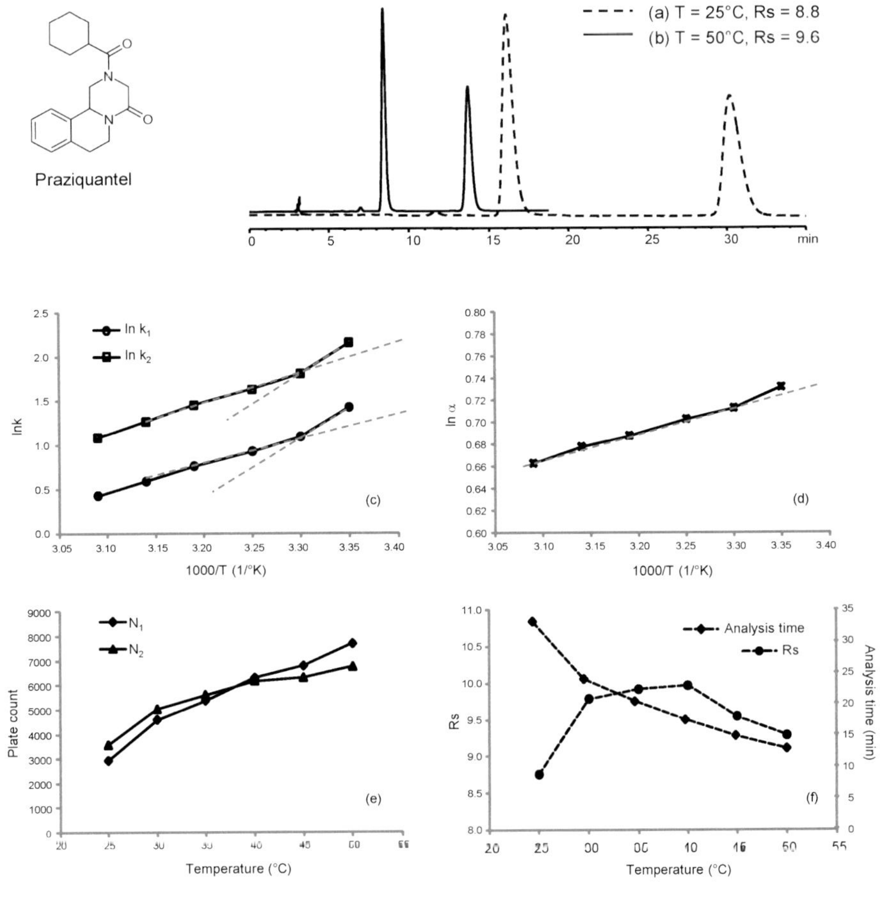

Figure 6.8 Effect of temperature on enantiomer separation of Praziquantel. Column: CHIRALPAK IA (5 μm, 250 × 4.6 mm i.d.); mobile phase: n-heptane/EtOH 45/55 v/v; flow rate: 1.0 mL/min.

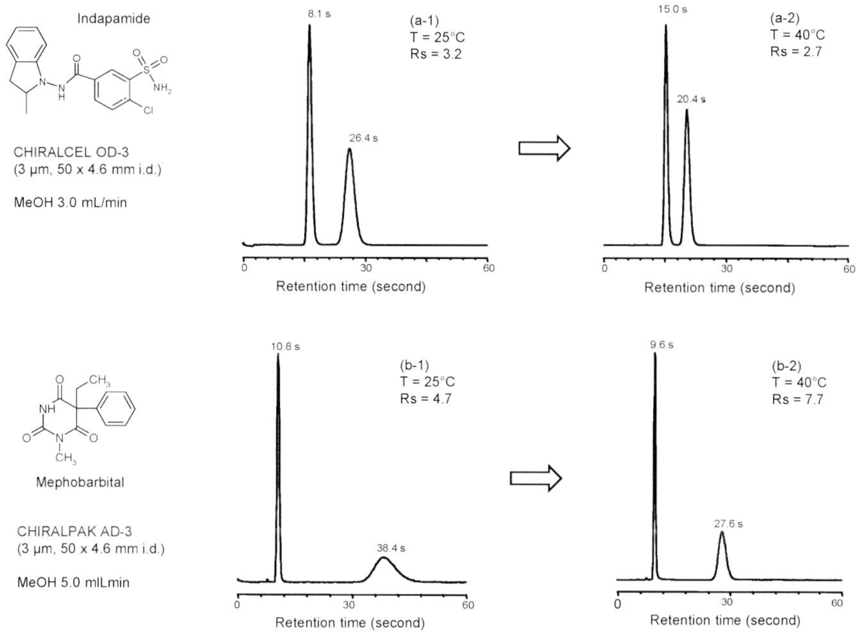

Figure 6.9 Method optimization by temperature.

In practice, the use of temperature higher than 50°C is not advised in a general way with the polysaccharide-based columns for chiral separations. From one side, raising the temperature up to 40°C can usually fulfill the purpose of method optimization without causing on-line sample degradation. On the other side, the frequent use of some very volatile solvents (e.g., hexane) or solvents of low boiling points (e.g., DCM and MtBE) in the mobile phase will create inconvenient chromatographic conditions and working environment if $T > 40$°C.

It should be noted that temperatures below ambient, typically 10–15°C, may also be envisaged in an attempt to reach good enough separation of some difficult compounds by possibly increase the enantioselectivity. This may be also a good option when the thermal stability issue is raised for the sample to be analyzed. As consequences of lowing temperature, one should expect higher viscosity of the mobile phase, higher column back pressure, longer retention times and reduced column efficiency which may counter balance the gain in selectivity for the separation.

6.5 Effects of Flow Rate and Particle Size

In rate theory of chromatography, flow rate is a variable affecting the longitudinal diffusion and mass transfer process of the analyte molecules [125], therefore the column efficiency (represented by the plate count *N*) for a given separation on a column of given dimension. Its role in column efficiency is well defined by Van Deemter equation:

$$\text{HETP} = \text{A} + \text{B}/\mu + \text{C}\mu \tag{6.3}$$

HETP refers to the height equivalent to the theoretical plate of the column. The smaller the HETP, the higher the column efficiency, μ is the linear velocity of the mobile phase which can be determined by the division of flow rate over the column cross-sectional area. Coefficient A is the multi-path term for packed columns in which the analyte molecules move through different paths. The contribution of this term to chromatographic band spreading can be reduced if the column packing is improved and/or the column is packed with smaller particles. B is the longitudinal diffusion term linked to the analyte diffusion from the center to the edges of the chromatographic band. Increasing the linear velocity will normally attenuate the longitudinal diffusion of the analyte molecule as a consequence of its shorter stay in the column. The C-term in Van Deemter equation is related to the mass transfer kinetics, more precisely to the speed of adsorption-desorption process of the analyte towards the stationary phase. Too high linear velocity will enforce the analyte to elute too fast from the column without reaching or approaching the mass transfer equilibrium. Thus, the higher the velocity, the severer the band broadening becomes.

Figure 6.10 depicts the Van Deemter plots of trans-Stilbene oxide on the pair of columns CHIRALPAK AD-H (particle size: 5 μm) and CHIRALPAK AD-3 (particle size: 3 μm) in NP-mode, as well as those of 5,5-Diphenyl-4-methyl-2-oxazolidinone on CHIRALPAK IB (particle size: 5 μm) and CHIRALPAK IB-3 (particle size: 3 μm) in RP-conditions. They were traced with the data of the first eluting peak, but similar scenarios were obtained for the second eluting peak. These plots confirm the clear trend that columns packed with smaller particles afford higher column efficiency (smaller A-term, represented by the lower position of the curves) and have more favorable mass transfer kinetics (the reduced C-term, represented by the flatter slope of the curves in the extended linear area).

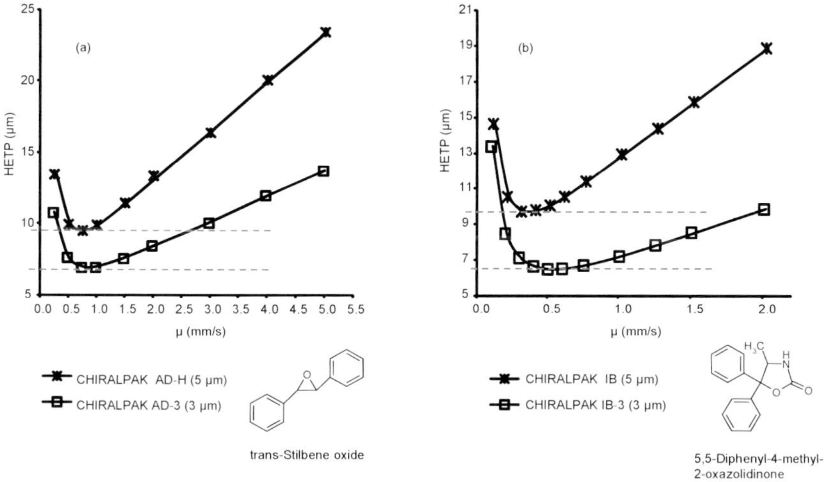

Figure 6.10 Van Deemter plots: Column conditions: T = 25°C; column size: 150 × 4.6 mm i.d; mobile phase: (a) hexane/2-PrOH 90/10 v/v, (b) ACN/H_2O 40/60 v/v.

These are characteristics of columns packed with particles of small size and offer significant benefits of running the analysis at high velocity without compromising significantly the column efficiency [25]. Figure 6.11 shows the enantiomer separation of trans-Stilbene oxide on 50 × 3 mm i.d. columns packed, respectively with 5, 3, and 1.6 μm CSPs containing the same chiral selector: immobilized amylose tris(3,5-dimethylphenylcarbamate). It can be observed that, in comparison with the 5 μm CHIRALPAK IA column, (ultra) faster analysis could be obtained over the 3 μm (CHIRALPAK IA-3) to sub-2 μm (CHIRALPAK IA-U) by increasing the flow rate (or μ) without compromising the resolution degree of the enantiomers. Additional examples of sub-1 min separations

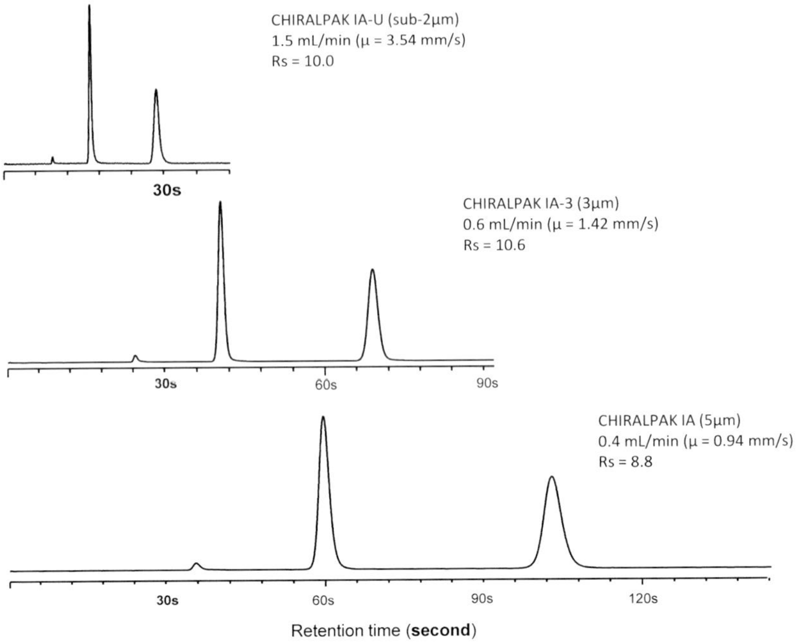

Figure 6.11 Effect of particle size: Solute: trans-Stilbene oxide; column size: 50 × 3 mm i.d; mobile phase: hexane/2-PrOH 90/10 v/v; temperature: 25°C; instrument: UHPLC (X-LC, Jasco).

are shown in Fig. 6.12. It has to be noted that the instrument void should be optimized when working with columns packed with small CSP particles in order to maximize the gain in resolution afforded by the high column efficiency. For this purpose, ultra-high-performance of liquid chromatography (UHPLC) system is strongly recommended to run separations with sub-2 μm packed columns.

For optimization of analytic method, the adjustment of linear velocity (or the flow rate on a column with a given internal diameter) may be helpful in search of the optimal column efficiency, therefore the improved resolution degree. This is particularly critical in the attempt to transfer a partial resolution to a full one. Such an approach is normally more effective with CSPs of larger particles than smaller particles due to the steeper slope of the linear region in Van Deemter plot for the larger particle size. As demonstrated in Fig. 6.13 for enantiomer resolution of

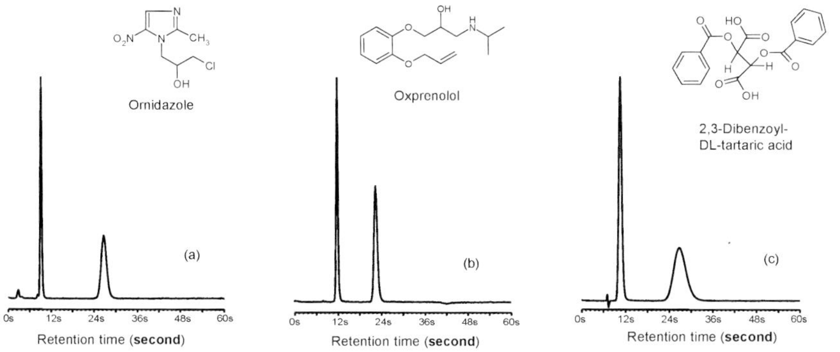

Figure 6.12 Fast analysis at high flow rate on 3 μm-packed columns. Conditions: Flow rate = 5.0 mL/min; T = 25°C; column size: 4.6 × 50 mm; particle size: 3 μm; instrument: UPLC (I-class, Waters); column and mobile phase: (a) CHIRALPAK IA-3, DCM/MeOH/2-AE 98/2/0.1 v/v/v; (b) CHIRALPAK IB-3, hexane/THF/2-AE 70/30/0.1 v/v/v; (c) CHIRALPAK IC-3, hexane/2-PrOH/TFA 80/20/0.1 v/v/v.

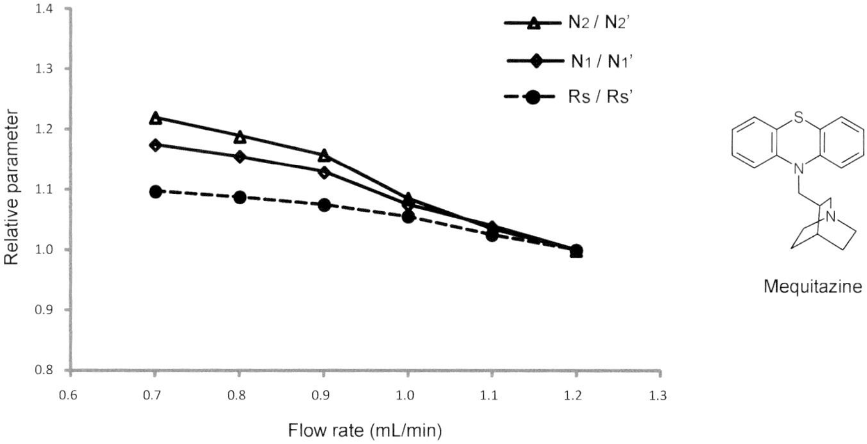

Figure 6.13 Effect of flow rate on relative column efficiency and resolution degree of Mequitazine enantiomers, CHIRALPAK ID (5 μm, 250 × 4.6 mm i.d.). Mobile phase: hexane/2-PrOH/DEA 80/20.0.1 v/v/v; T = 25°C; detection UV 254 nm; N1, N2, and Rs: the values obtained at 0.7–1.1 mL/min; N1′, N2′, and Rs′: the values obtained at 1.2 mL/min.

Mequitazine on CHIRALPAK ID (5 μm), the reduction of the flow rate from 1.2 to 0.7 mL/min allowed increasing the plate counts at the level of 20%, resulting in enhancement of the resolution degree at the level of 10%. It should be mentioned that the gain of 10% in resolution is at the expense of much longer (about 80%) analysis time. In some cases, playing the mobile phase composition to optimize the selectivity seems to be a more attractive alternative for method optimization as the lack of column efficiency can often be largely compensated by a high selectivity [126].

6.6 Influence of Analyte Structure and the Substituent Effect

It has been observed in experimental work that a given single column can be peculiarly versatile for chiral separation of whole series of structure-related compounds [51, 127–131]. Still, some efforts would often be needed for optimizing the methods for the given compound by varying the chromatographic conditions (e.g., temperature, mobile phase composition) and even by changing the chromatographic mode. This is undoubtedly related to the complexity of the chiral recognition mechanism operating on the polysaccharide-derived CSPs.

The interactions between analyte molecules and a chiral stationary phase are multiple and of diverse natures, including mainly hydrogen bonding, π–π associations, dipole–dipole stacking, hydrophobic interactions and the steric fit of the asymmetric portions of the analyte structure to the chiral environment, which is defined by the molecular and supramolecular structure of the polymeric chiral selector [132]. Depending on the analyte molecular structure, the mobile phase in use and the solvation state of the polymer, each of the interactions mentioned above may play a role of higher or lower importance in retention and in recognition of the enantiomers. Such a complexity makes the prediction of chiral separation difficult and the trial-and-error approach has been constantly adopted for method development on polysaccharide-based CSPs. Even in the case where the column and the chromatographic conditions are kept the same, the slightest difference in the analyte molecular structure can induce drastic change in chromatographic results. This is the case, for instance, of the compound pair of Verapamil and Methoxyverapamil.

Structurally, Verapamil and Methoxyverapamil differ from each other merely from the additional methoxy group on the benzeneacetonitril moiety in Methoxyverapamil. However, such a small structural variation brought about quite different chromatographic results on CHIRALPAK IE with Hexane/2-PrOH/DEA 80/20/0.1 v/v/v as the mobile phase. As shown in Fig. 6.14a, the first eluting peak of both compounds had very similar retention time (ca. 19 min). In contrast, the second peak of Methoxy-verapamil was eluted much later than that of Verapamil, leading to significant enhancement in enantioselectivity and in resolution degree of the Methoxyverapamil enantiomers. It seems that the presence of the third bulky methoxy group at 5-position of the phenyl group would intensify the steric selective effect but not disturb interactions of all the other types between the analytes and the CSPs. N-CBZ-Valine and N-CBZ-Norvaline are isobaric amino acid derivatives. Their structural difference is in the aliphatic side chain: isopropyl in the former; n-propyl in the later. It seems that the isopropyl group has more favorable steric effect than n-propyl group on IA-U column, resulting to much larger enantioselectivity for N-CBZ-DL-Valine than for N-CBZ-DL-Norvaline enantiomers (Fig. 6.14b).

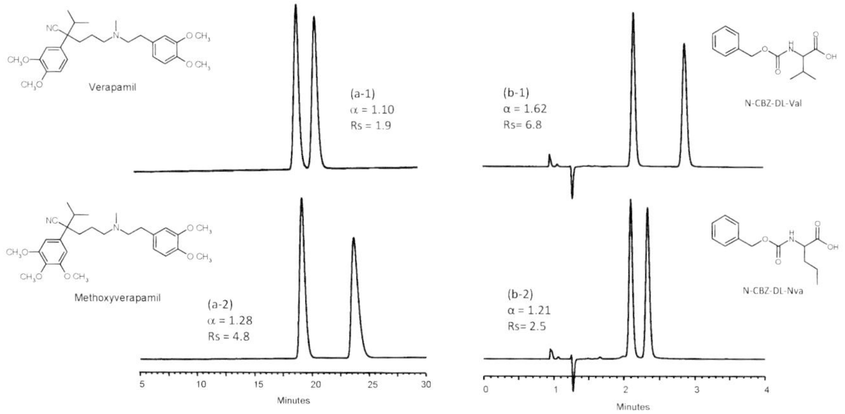

Figure 6.14 Substituent effect of the analytes on enantiomer separations. (a) CHIRALPAK IE (5 μm, 250 × 4.6 mm i.d.); mobile phase: hexane/2-PrOH/DEA 80/20.0.1 v/v/v; flow rate: 1.0 mL/min; *T* = 25°C; (b) CHIRALPAK IA-U (1.7 μm, 100 x 3.0 mm i.d.); mobile phase: hexane/EtOH/FA 80/20.0.1 v/v/v; flow rate: 1.425 mL/min; *T* = 25°C.

If the substituent effects in the above examples could be quite easily assessed in the enantiomer recognition, the consequence of the structural change at the substituent level is not always singular. In most cases, the substituent effect is more complex and can hardly be predictable. In a systematic study on the substituent effect, a series of ß-blockers was injected on a CHIRALPAK ID column using a single mobile phase (Hexane/2-PrOH/DEA 80/20/0.1 v/v/v). As summarized in Table 6.3, the chromatographic data scattered very much for these compounds.

Table 6.3 Enantiomer separations of β-blockers on CHIRALPAK ID

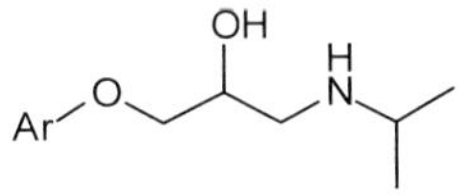

	Ar	t_1	t_2	k_1	α	Rs
Bunolol	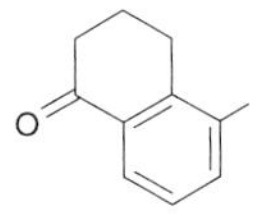	9.13	9.78	1.97	1.11	1.44
Pindolol	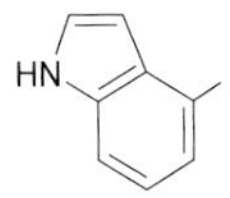	6.56	6.94	1.13	1.11	1.26
Propranolol	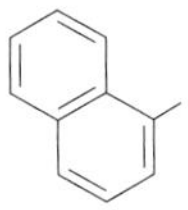	5.09	5.75	0.66	1.32	3.50
Carazolol	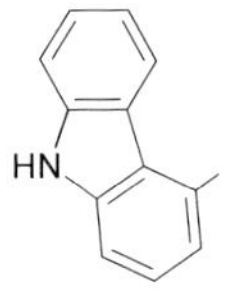	7.51	7.93	1.44	1.09	1.12
Atenolol	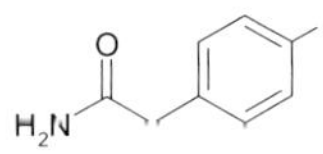	19.19	20.16	5.24	1.06	0.86
Alprenolol	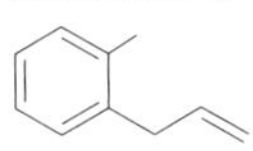	4.82	4.82	0.57	1.00	0.00

	Ar	t_1	t_2	k_1	α	**Rs**
Oxprenolol	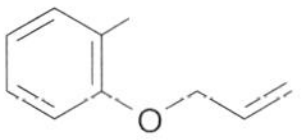	5.87	13.34	0.91	3.67	13.59
Tiprenolol	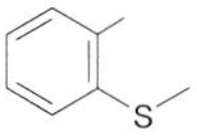	6.51	7.88	1.12	1.40	4.91
Acebutolol	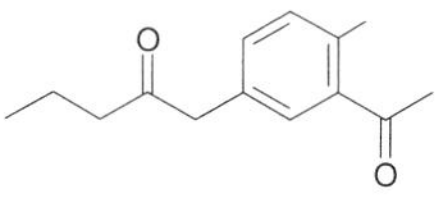	15.70	17.14	4.11	1.11	1.44
Metoprolol	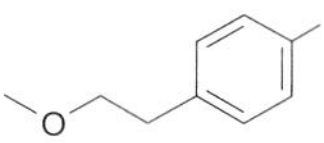	7.14	11.87	1.32	2.16	11.57
Bisoprolol	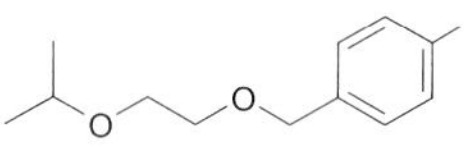	6.98	9.22	1.27	1.57	6.20

Note: Column size: 250 × 4.6 mm i.d., Particle size: 5 μm. Mobile phase: Hexane/2-PrOH/DEA 80/20/0.1 v/v/v. Flow rate: 1.0 mL/min, *T*: 25°C.

Although bearing the same general or basic structure, these ß-blockers had very large variation in retention factor (k_1 ranging from 0.57 to 5.24) and in enantioselectivity (α spreading from 1.00 to 3.67) in function of the aromatic scaffold. The most striking difference in enantiomer separation was observed between Alprenolol and Oxprenolol. Both molecules have a single substituent at position-2 of the phenoxy group. The substituent is an allyl group in Alprenolol but an allyloxy group in Oxprenolol. The structural difference is very subtle but the chromatographic results were extremely distinctive. Chromatographically, racemic Alprenolol was eluted very fast in a single peak (Fig. 6.15a), while racemic Oxprenolol was resolved into enantiomers to a very large degree with slightly longer retention time for the first peak (Fig. 6.15b). There is no doubt that the allyloxy group in Oxprenolol played an essential role in the enantioselective mechanism under the given chromatographic conditions. This phenomenon with the same compounds may not be as significant as described here if the chromatographic conditions are not the same (e.g., changes in mobile phase and/or in column).

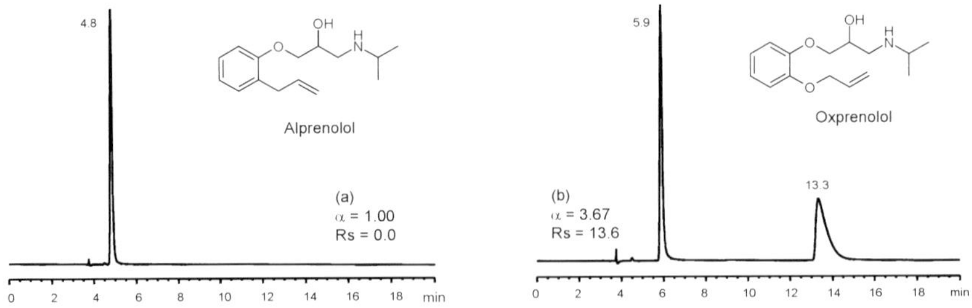

Figure 6.15 Substituent effect of the ß-blockers on enantiomer separations. Column: CHIRALPAK ID (5 μm, 250 × 4.6 mm i.d.); mobile phase: hexane/IPA/DEA 80/20/0.1 v/v/v; flow rate: 1.0 mL/min; *T*: 25°C.

6.7 Column Maintenance, Cleaning, and Regeneration

The column lifetime depends on daily care and the operator's awareness of the cautions to be taken for the columns in use. The factors affecting the column performance in the time course are multiple. Among them, solvent compatibility would be the most important, especially when it comes to the coated-type columns based on polysaccharide derivatives.

As previously described, in the coated CSPs the polysaccharide derivative is not chemically bonded but physically deposited by special technology onto the silica matrix. Therefore, all the mid-polar solvents, such as THF, DCM, chloroform, MtBE, acetone, and EtOAc cannot be used as mobile phase or mobile phase components as they are able to dissolve off the polymers and damage irreversibly the columns. The solvents compatible to the coated polysaccharide-CSPs are typically the mixtures of non-polar solvents (e.g., hexane or heptane) with polar solvents (e.g., EtOH and 2-PrOH) for NP-mode. For PO-mode, alcohols (e.g., MeOH and EtOH), ACN and their mixtures can be efficiently used. When running the columns under RP conditions, all the above-mentioned polar solvents can be employed as the organic modifier for chiral separations. Any immiscible mixtures, for instance hexane and ACN, must be avoided because they won't allow getting a homogeneous eluent medium and could also be destructive to the coated-type columns.

In contrast, immobilized CSPs have no constrains in the choice of the solvents, if all the solvents in use are miscible [16, 34–38]. Although the application of the non-miscible solvents does not damage the immobilized columns, the solvent miscibility should always be considered for the sake of reliable and reproducible chromatographic conditions.

The columns can become "dirty" or the column inlet frit is "clogged" if the non-cleaned sample solutions are injected, or some samples are too strongly retained on the column, or on-column precipitations of the sample or the salt additives occurred. "Dirty" column will lead to the baseline instability, need prolonged column equilibration time and compromise the method performance. "Clogged" column will induce higher and higher backpressure and alter the column efficiency. In these cases, column cleaning may be needed.

Ideally, the solvent affording the best solubility to the "dirty" samples and the impurities should be the best for column cleaning. However, the natures of the cumulated dirtying species are not always clear and the free choice of the cleaning solvent is not feasible with the columns of coated-type. According to our experience, the most efficient cleaning method for columns of coated-type is to flush (or carefully back-flush) the column with pure EtOH at reduced flow rate for several hours (≥3 h). When used in RP-mode, the column can be cleaned with salt-free and water-enriched mobile phase, followed by ACN flushing at reduced flow rate.

For columns of the immobilized-type, the column cleaning may be more efficient by choosing the right solvent without worrying about the CSP stability. A general protocol is proved to be effective for column cleaning packed with the immobilized CSPs. It consists of three steps: (1) EtOH for 30 min; (2) THF for 2 h; (3) EtOH for 30 min. If the column cleaning is not successful enough with this protocol, the column flushing can be repeated but using DMF for 3 h instead of THF for 2 h. Again, the column flushing is recommended to be carried out at reduced flow rate. For columns of 4.6 mm internal diameter, for instance, the flow rate can be set at 0.5 mL/min for EtOH and THF, at 0.3 mL/min for DMF.

Another potential issue occasionally happening could be a change in the polymer conformation in the column. When dealing with immobilized-type phases, although the polysaccharide derivatives cannot be stripped off by solvents, especially the solvents of mid-polarity in use for the mobile phase and sample solvent, the inter- and intra-molecule interactions inside the bulk structure of the polymer may be weakened or modified, the polymer then swells to adopt a specific conformation by following the lowest energy rule in the given solvent medium. As a result, the new polymer conformation created may have an influence in certain sensitive applications, by changing selectivity, resolution and/or efficiency.

Figure 6.16 shows an example of conformation change on the separation of a pair of enantiomers on CHIRALPAK IA (5 μm, 250 × 4.6 mm i.d.). In comparison with the separation initially obtained (chromatogram (a)) using hexane/EtOH 90/10 v/v as mobile phase, the resolution degree of racemic (4-fluorophenyl)-γ-butyrolactone was very much compromised after the application of DCM-containing mobile phases (chromatogram (b)) and EtOAc-containing mobile phases (chromatogram (c)). In both cases, the separation deterioration was commonly characterized by a significant reduction of enantioselectivity (the second peak eluted earlier). In the case of (c), the reduction of selectivity was accompanied by the broadened peak shape. As a part of the typical characteristics of the immobilized CSPs, the column performance deviation can be remedied by a specific regeneration method. Once regenerated, the column can afford perfectly reproducible separations regardless of its mobile phase background (chromatogram (d)).

CSP regeneration could be particularly valuable while switching from mobile phases containing the mid-polar solvents to the ones composed of alkane-alcohol mixtures or of polar solvents. Having the ability to dissolve the free polysaccharide derivatives, all the solvents of mid polarity can condition the CSP by themselves, erase the memory imprinted by other solvents and allow the polymer to develop a specific solvated state or chain conformation which, in its turn, determines the specific enantioselective properties of the CSP.

The protocols to regenerate immobilized columns are quite like the general column cleaning method described above. Thus, for easy protocol handling, the column regeneration and column cleaning can in principle be combined for these columns. If we consider in detail the whole series, some differences can be observed between cellulose and amylose derivatives. The amylose derivatives (CHIRALPAK IA, CHIRALPAK ID, CHIRALPAK IE, CHIRALPAK IF, and CHIRALPAK IH) have rather loose and flexible structures due to the α-1,4 linkage between the glucose units. It is easier to get the amylose-based CSPs changed but also faster to get the column performance regenerated. On the other side, the columns based on cellulose derivatives (CHIRALPAK IB and CHIRALPAK IC) usually undergo much less pronounced performance change than the amylose-based ones, even with prolonged exposure to the non-standard mobile phases. This may be attributed to the intrinsic properties of the cellulose materials. In cellulose, the glucose units are bonded together by β-1,4 linkage, which keeps the polymer chains from coiling. The straight and relatively rigid chain conformation in cellulose or cellulose derivatives favors, both in intensity and in strength, the inter- and intra-chain interactions by multiple attracting forces. It is this capacity for self-assembling that firmly holds the polymer chains together against the exertions for modification of the supra-molecular structure. For this reason, the regeneration of CHIRALPAK IB and CHIRALPAK IC is rather rarely needed.

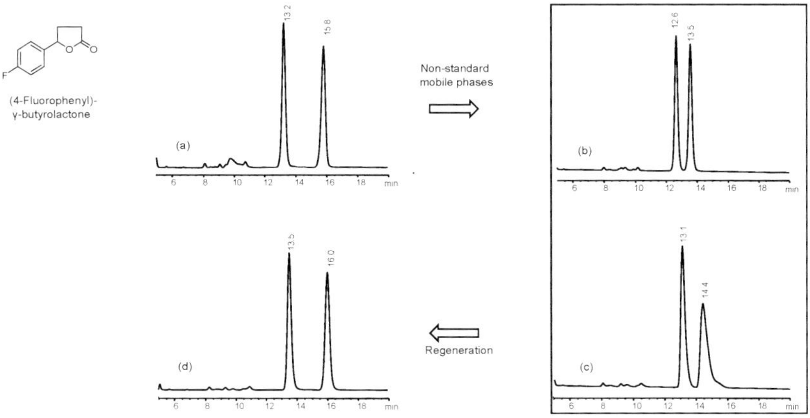

Figure 6.16 Effect of the non-standard mobile phases and the regeneration: CHIRALPAK IA (5 μm, 250 × 4.6 mm i.d.); flow rate: 1.0 mL/min; mobile phase: hexane/EtOH 90/10 v/v. *T*: 25°C. (a) Initial separation. (b) Separation obtained after application of DCM-containing mobile phase. (c) Separation obtained after application of EtOAc-containing mobile phase. (d) Separation obtained after column regeneration.

6.8 Conclusions

The technology advancement in preparing polysaccharide-based chiral stationary phases has provided an impetus to the field of chiral separation by chromatography. It gives rise to CSPs with higher performance, broader application scope, higher versatility and enhanced robustness. It also makes the approaches of method development evolve to reach higher success rate in analytical and preparative separations of chiral compounds.

The establishment of the screening protocols by choosing the appropriate chiral columns and the suitable mobile phase is essential for efficient method development. The option to use 3 to 4 immobilized polysaccharide-derived CSPs in combination with the 4 most versatile mobile phase systems fits well the current challenge of fast screening for an increased number of compounds in a limited time frame. The choice in chromatographic mode (SFC or HPLC, NP or RP, etc.) can be considered in connection with the instrument availability, the application purpose and specific aspects of the sample source and sample properties.

Based on the screening results, the optimization to obtain the best suited method can be accomplished by fine-tuning the mobile phase composition, by using the most appropriate additives in amount and in nature, by rational use of sample solvent, by adjustment in temperature and in flow rate, by changing the column dimension, by considering the particle size in the column and, of course, without forgetting to try the alternative columns which may exhibit complementary properties for chiral separation. It is important to note that many methods issuing the first sample screening can directly meet the separation objective with no need of further optimization efforts. Fast or ultra-fast chiral analysis can be achieved by running short columns packed with 3 μm or sub-2 μm particles on optimized chromatographic systems such as UHPLC or ultra-high-performance SFC.

References

1. Okamoto, Y., Kawashima, M., Yamamoto, K., and Hatada, K. (1984) *Chem. Lett.*, **13(5)** p. 739.

2. Okamoto, Y., Kawashima, M., and Hatada, K. (1984) *J. Am. Chem. Soc.* **106** p. 5357.
3. Okamoto, Y., Aburatani, R., and Hatada, K. (1987) *J. Chromatogr. A* **389** p. 95.
4. Dingenen, J. (1994) Polysaccharide phases in enantioseparations in *A Practical Approach to Chiral Separations by Liquid Chromatography*, ed., Subramanian, G., Chapter 6 (VCH, Weinheim).
5. Francotte, E. (1997) Chromatography as a separation tool for the preparative resolution of racemic compounds in *Chiral Separations, Applications and Technology*, ed., Ahuja, S., Chapter 10 (American Chemical Society, Washington).
6. Okamoto, Y., and Kaida, Y (1994) *J. Chromatogr. A* **666** p. 403.
7. Francotte, E. (1994) *J. Chromatogr. A* **666** p. 565.
8. Okamoto, Y., and Yashima, E. (1998) *Angew. Chem. Intl. Ed. Engl.* **37** p. 1020.
9. Maier, N. M., Franco, P., and Lindner, W. (2001) *J. Chromatogr. A* **906** p. 3.
10. Yashima, E. (2001) *J. Chromatogr. A* **906** p. 105.
11. Tachibana, K., and Ohnishi, A. (2001) *J. Chromatogr. A* **906** p. 127.
12. Franco, P., Senso, A., Oliveros, L., and Minguillón, C. (2001) *J. Chromatogr. A* **906** p. 155.
13. Francotte, E. (2001) *J. Chromatogr. A* **906** p. 379.
14. Aboul-Enein, H. Y. (2001) *J. Chromatogr. A* **906** p. 185.
15. Cox, G. (ed.) (2005) *Preparative Enantioselective Chromatography* (Blackwell Publishing, Oxford).
16. Zhang, T., and Franco, P. (2007), *Chiral Separation Techniques: A Practical Approach* (3rd revised and updated Edition) ed., Subramanian, G., Chapter 3 (VCH, Weinheim).
17. Franco, P., and Zhang, T. (2008) *J. Chromatogr. B* **875** p. 48.
18. Zhang, T., Nguyen, D., and Franco, P. (2010) *J. Chromatogr. A* **1217** p. 1048.
19. Lee, J., Lee, J. T., Watts, W. L., Barendt, J., Yan, T. Q., Huang, Y., Riley, F., Hardink, M., Bradow, J., and Franco, P. (2014) *J. Chromatogr. A* **1374** p. 238.
20. Khater, S., Lozac'h, M.-A., Adam, I., Francotte, E., and West, C. (2016) *J. Chromatogr. A* **1467** p. 463.
21. Francotte, E. (2016) *LCGC Eur.* **29** p. 194.

22. Baudelet, D., Schifano-Faux, N., Ghinet, A., Dezitter, X., Barbotin, F., Gautret, P., Rigo, B., Chavatte, P., Millet, R., Furman, C., Vaccher, C., and Lipka, E. (2014) *J. Chromatogr. A* **1363** p. 257.
23. Rizzo, S., Menta, S., Faggi, C., Pierini, M., and Cirilli, R. (2014) *J. Chromatogr. A* **1363** p. 128.
24. Lämmerhofer, M. (2010) *J. Chromatogr. A* **1217** p. 814.
25. Zhang, T., and Franco, P. (2008) *LC-GC Europe* **21**(9) p. 430.
26. Cirilli, R., Ferreti, R., Gallinilla, B., La Torre, F., Mai, A., and Rotili, D. (2006) *J. Sep. Sci.* **29** p. 1399.
27. De la Puente, M. L., White, C. T., Rivera-Sagredo, A., Reilly, J., Burton, K., and Harvey, G. (2003) *J. Chromatogr. A* **983** p. 101.
28. De la Puente, M. L. (2004) *J. Chromatogr. A* **1055** p. 55.
29. Andersson, M. E., Aslan, D., Clark, A., Roeraade, J., and Hagman, G. (2003) *J. Chromatogr. A* **1005** p. 83.
30. Wetli, H. A., and Francotte, E. (2007) *J. Sep. Sci.* **30** p. 1255.
31. Matthijs, N., Maftouh, M., and Heyden, Y. V. (2006) *J. Chromatogr. A* **1111** p. 48.
32. Zhang, T., and Franco, P. (2010) *LC-GC Europe* **23**(6) p. 302.
33. Franco, P., and Zhang, T. (2013) *Chiral Separations, Methods and Protocols* (Second edition), ed., Scriba, G., Chapter 6 (Humana Press), pp. 113–126.
34. Zhang, T., Nguyen, D., and Franco, P. (2008) *J. Chromatogr. A* **1191** p. 214.
35. Younes, A. A., Ates, H., Mangelings, D., and Vander Heyden, Y. (2013) *J. Pharm. Biomed. Anal.* **75** p. 74.
36. Zhang, T., Kientzy, C., Franco, P., Ohnishi, A., Kagamihara, Y., and Kurosawa, H. (2005) *J. Chromatogr. A* **1075** p. 65.
37. Zhang, T., Nguyen, D., Franco, P., Murakami, T., Ohnishi, A., and Kurosawa, H. (2006) *Anal. Chim. Acta* **557** p. 221.
38. Zhang, T., Nguyen, D., Franco, P., Isobe, Y., Michishita, T., Murakami, T. (2008) *J. Pharm. Biomed. Anal.* **46** p. 882.
39. Zhang, T., and Franco, P. (2004) *Int. Lab.* **34-5** p. 24.
40. Cox, G. B. (2008). A screening process for development of enantioselective supercritical fluid chromatography separation methods, *LC.GC Application notes.*
41. Amoss, C., Cox, G. B., Franco, P., and Zhang, T. (2008) CHIRALPAK® IC™—An Immobilized Polysaccharide Chiral Stationary Phase with a Unique Chiral Selector, *LC.GC Application Notes.*

42. Zhang, T., Schaeffer, M., and Franco, P. (2005) *J. Chromatogr. A* **1083** p. 96.

43. Cox, G. B., Maier, N. M., Zhang, T., and Franco, P. (2009) *LC GC Application Notes*.

44. Zhang, T., Nguyen, D., and Franco, P. (2006) *J. Sep. Sci.* **29** p. 1517.

45. Ghanem, A., and Aboul Enein, H. Y. (2005) *Anal. Chim. Acta* **548** p. 26.

46. Ghanem, A., and Naim, L. (2006) *J. Chromatogr. A* **1101** p. 171.

47. Ali, I., Naim, L., Ghanem, A., and Aboul Enein, H. Y. (2006) *Talanta* **69** p. 1013.

48. Cirilli, R., Simonelli, A., Ferretti, R., Bolasco, A., and Climenti, P. (2006) *J. Chromatogr. A* **1101** p. 198.

49. Cirilli, R., Orlando, V., Ferretti, R., Luciana Turchetto, L., Silvestri, R., De Martino, G., and La Torre, F. (2006) *Chirality* **18** p. 621.

50. Andersson S. (2007) *Chiral Separation Techniques: A Practical Approach* (3rd revised and updated Edition), ed., Subramanian, G., Chapter 17 (VCH, Weinheim).

51. Cirilli, R., Ferretti, R., Gallinella, B., De Santis, E., Zanitti, L., and La Torre, F. (2008) *J. Chromatogr. A* **1177** p. 105.

52. Cirilli, R., Ferretti, R., Gallinella, B., Billa, A. R., Vincieri, F. F., and La Torre, F. (2008) *J. Sep. Sci.* **31** p. 2206.

53. Cirilli, R., Ferretti, R., De Santis, E., Gallinella, B., Zanitti, L., and La Torre, F. (2008) *J. Chromatogr. A* **1190** p. 95.

54. Ferretti, R., Gallinella, B., La Torre, F., Zanitti, L., Turchetto, L., Mosca, A., and Cirilli, R. (2009) *J. Chromatogr. A* **1216** p. 5385.

55. Pawar, R. S., Grundel, E., Mazzola, E., White, K. D., Krynitsky, A. J., and J. I. Rader, J. I. (2009) *J. Sep. Sci.* **33** p. 200.

56. Wang, C., Armstrong, D. W., and Chang, C. D. (2008) *J. Chromatogr. A* **1194** p. 172.

57. García, P., Franco, P., Alvarez, R., and De Lera., U. R. (2011) *J. Sep. Sci.* **34** p. 999.

58. Zhang, T., Franco, P., Nguyen, D., Hamasaki, R., Miyamoto, S., Ohnishi, A., and Murakami, T. (2012) *J. Chromatogr. A* **1269** p. 478.

59. Ashok, S., Varma, M. S., and Swaminathan, S. (2012) *J. Chromatogr. Sci.* **50** p. 799.

60. Zhou, J., Du, Q. Z., Zhao, S. Z., Sun, F., Li, X. Y., and Zhang, Z. Z. (2014) *J. Chromatogr. Sci.* **53** p. 959.

61. Geryka, R., Kalíkováa, K., Vozkaa, J., Plecitáa, D., Schmidb, M. G., and Tesărová, E. (2014) *J. Chromatogr. A* **1363** p. 155.

62. Peluso, P., Mamane, V., Aubert, E., and Cossu, S. (2014) *J. Sep. Sci.* **37** p. 2481.
63. Harrison, A., Melchionna, M., Franco, P., and Hlavac, J. (2014) *J. Chem.* **38** p. 5491.
64. Rizzo, S., Menta, S., Benincori, T., Ferretti, R., Pierini, M., Cirilli, R., and Sannicolo, F. (2015) *Chirality* **27** p. 888.
65. Geryk, R., Kalíková, K., Vozka, J., and Tesarŏvá, E. (2015) *Chromatographia* **78** p. 909.
66. Liu, N., Dong, F., Xu, J., Liu, X. G., Chen, Z. L., Tao, Y., Pan, X. L., Chen, X. X., and Zheng, Y. Q. (2015) *J. Agric. Food Chem.* **63** p. 6297.
67. Islam, M. F., and Lee, W. (2015) *J. Chosun Natural Sci.* **8** p. 111.
68. Rao, R. N., and Kumar, K. N. (2015) *J. Chromatogr. Sci.* **53** p. 295.
69. Menta, S., Carradori, S., Secci, D., Faggi, C., Mannina, L., Cirilli, R., and Pierini, M. (2015) *J. Org. Chem.* **80** p. 11932.
70. López-Ram-de-Víu, P., Gálvez, I., and Díaz-de-Villegas, M. D. (2015) *J. Chromatogr. A* **1390** p. 78.
71. Qian, Y., Zhao, X. Q., Zhao, L., Cui, L. L., Liu, L., Jiang, X. L., Liu, Y. J., Gao, L. P., and Xia, T. (2015) *J. Chromatogr. B* **1006** p. 1.
72. Cirilli, R., Ferretti, R., Gallinella, B., Turchetto, L., Zanitti, L., and La Torre, F. (2015) *J. Pharm. Biomed. Anal.* **50** p. 9.
73. Zanitti, L., Ferretti, R., Gallinella, B., La Torre, F., Sanna, M. L., Mosca, A., and Cirilli, R. (2015) *J. Pharm. Biomed. Anal.* **52** p. 665.
74. Sola, I., Viayna, E., Gómez, T., Galdeano, C., Cassina, M., Camps, P., Romeo, M., Diomede, L., Salmona, M., Franco, P., Schaeffer, M., Colantuono, D., Robin, D., Brunner, D., Taub, N., Hutter-Paier, B., and Muñoz-Torrero, D. (2015) *Molecules* **20** p. 4492.
75. Moldovan, R. C., Dascal, G. S., Mirel, V., Bodoki, E., and R. Oprean, R. (2015) *Pharmacia* **63** p. 909.
76. Adhikari, S., Kang, J. S., and Lee, W. (2016) *Chirality* **28** p. 789.
77. Srinivasu, G., Nagesh Kumar, K., Thirupathi, C., Lakshmi Narayana, C., and Parameswara Murthy, C. (2016) *Chromatographia* **79** p. 1457.
78. Saduttoa, D., Ferretti, R., Zanitti, L., Casulli, A., and Cirilli, R. (2016) *J. Chromatogr. A* **1445** p. 166.
79. Gallinella, B., Ferretti, R., Zanitti, L., Sestili, I., Mosca, A., and Cirilli, R. (2016) *J. Pharm. Anal.* **6** p. 132.
80. Kim, S. J., Nam, K. W., Park, B., Md. F. Islam, Md. F., and Lee, W. (2016) *Korean Soc. Biotechnol. Bioeng. J.* **31** p. 186.

81. Lotter, C., Poehler, E., Heiland, J., Mauritz, L., and Belder, D. (2016) *Lab Chip.* **16** p. 4648.

82. Török, G., Goetelen, L., Luyckx, R., and Van Broeck, P. (2005) *J. Pharm. Biomed. Anal.* **39** p. 425.

83. Huybrechts, T., Török, G., Vennekens, T., Sneyers, R., and Vrielynck, S. (2007) *LCGC Europe* **20** p. 319.

84. Ishikawa, A., and Shibata, T. (1993) *J. Liq. Chromatogr.* **16** p. 859.

85. Zhou, L., Welch, C., Lee, C., Gong, X., Antonucci, V., and Ge, Z. (2009) *J. Pharm. Biomed. Anal.* **49** p. 964.

86. Amoss, C., Coryell, B. R., Cox, G. B., and Maier, N. M. (2010) Poster at the HPLC2010, Boston, USA.

87. Cirilli, R., Ferretti, R., Gallinella, B., and Zanitti, L. (2013) *J. Chromatogr. A* **1304** p. 147.

88. Materazzo, S., Carradori, S., Ferretti, R., Gallinella, B., Secci, D., and Cirilli, R. (2014) *J. Chromatogr. A* **1327** p. 73.

89. Gallinella, B., Bucciarelli, L., Zanitti, L., Ferretti, R., and Cirilli, R. (2014) *J. Chromatogr. A* **1339** p. 210.

90. Ferretti, R., Zanitti, L., Casulli, A., and Cirilli, R. (2016) *J. Sep. Sci.* **39** p. 1418.

91. Petersson, P., Lundell, N., and Markldes, K. E. (1992) *J. Chromatogr.* **623** p. 129.

92. Wang, R. Q., Ong, T. T., Tang, W. H., and Ng, S. C. (1012) *Trends Anal. Chem.* **37** p. 83.

93. Kalíková, K., Slechtová, T., Vozka, J., and Tesărová, E. (2014) *Anal. Chim. Acta* **821** p. 1.

94. Regalado, E. L., and Welch, C. J. (2015) *J. Sep. Sci.* **38** p. 2826.

95. Lemasson, E., Bertin, S., and West, C. (2016) *J. Sep. Sci.* **39** p. 212.

96. Cox, G., Stringham, R., and Matabe A. (2002) *LC GC North America, The Application Notebook*, **August,** p. 24.

97. Maftouh, M., Granier-Loyaux, C., Chavana, E., Marini, J., Pradines, A., Vander Heyden, Y., and Picard, C. (2005) *J. Chromatogr. A* **1088** p. 67.

98. Diehl, G., Meishammer, A., Huynh, D., and Francotte, E. (2006) Presentation at SPICA 2006, Innsbruck, Austria.

99. Zhang, T., Nguyen, D., and Franco, P. (2007) Presentation at the HPLC2007, Ghent, Belgium.

100. Cox, G. B. (2007) *Presentation at the 1st International Conference on SFC*, Pittsburgh, Pennsylvania, USA (September 2007).

101. Miller, L., and Potter, M. (2008) *J. Chromatogr. B* **875** p. 230.

102. Zhang, T., Nguyen, N., Franco, P., and Vollmer, M. (2011) Posters at Chirality ISCD 23, July 2011 (Liverpool, UK) and at SFC 2011, July 2011 (New York, USA).

103. Zhang, T., Nguyen, N., Franco, P., and Vollmer, M. (2011) Agilent application note (2011): http://www.chem.agilent.com/Library/applications/5990-9315EN.pdf.

104. Zhang, T., Nguyen, N., Franco, P., and Vollmer, M. (2011) Agilent application note (2011): http://www.chem.agilent.com/Library/applications/5990-9459EN.pdf.

105. Ye, Y. K., Lynam, K. G., and Stringham, R. W. (2004) *J. Chromatogr. A* **1041** p. 211.

106. Hamman, C., Wong, M., Hayes, M., and Gibbons, P. (2011) *J. Chromatogr. A* **1218** p. 3539.

107. Bhavyasri, K., Rambabu, D., Prasad, P. S. S., and Murali Balaram, V. (2013) *Am. J. Anal. Chem.* **4** p. 51.

108. Kasten, S. A., Zulli, S., Jones, J. L., Dephillipo, T., and Cerasoli, D. M. (2014) *Chirality* **26** p. 817.

109. Khater, S., Zhang, Y., and West, C. (2014) *J. Chromatogr. A* **1363** p. 278.

110. DaSilva, J. O., Coes, B., Frey, L., Mergelsberg, I., McClain, R., Nogle, L., and Welch, C. J. (2014) *J. Chromatogr. A* **1328** p. 98.

111. Khater, S., and West, C. (2014) *J. Chromatogr. A* **1373** p. 197.

112. Khater, S., and West, C. (2015) *J. Pharm. Biomed. Anal.* **102** p. 321.

113. Jin, L. X., Gao, W. L., Yang, H. Y., Lin, C. M., and Liu, W. L. (2011) *J. Chromatogr. Sci.* **49** p. 739.

114. Miller, L. (2012) *J. Chromatogr. A* **1256** pp. 261–266.

115. O'Brien, T., Crocker, L., Thompson, R., Toma, P. H., Conlon, D. A. Feibush, B., Moeder, C., Blacker, G., and Grinberg, N. (1997) *Anal. Chem.* **69** p. 1999.

116. Wang, F., O'Brien, T., Dowing, T., Bicker, G., and Wyvratt, J. (2002) *J. Chromatogr. A* **958** p. 69.

117. Sanaie, N., and Haynes, C. A. (2006) *J. Chromatogr. A* **1104** p. 164.

118. Lao, W., and Gan, J. (2006) *J. Chromatogr. A* **1131** p. 74.

119. Yao, B., Zhan, F., Yu, G., Chen, Z., Fan, W., Zeng, X., and Weng, W. (2009) *J. Chromatogr. A* **1216** p. 5429.

120. Schlauch, M., and Frahm, A. W. (2001) *Anal. Chem.* **73** p. 262.

121. Gasparrini, F., Marini, F. F., Pierini, M., and Villani, C. (1999) *Enantiomer* **4** p. 325.

122. Fulde, K., and Frahm, A. W. (1999) *J. Chromatogr. A* **858** p. 33.

123. Persson, B. A., and S., Andersson, S. (2001) J. *Chromatogr. A* **906** p. 195.

124. Balmer, K., Lagerström, P. O., and Persson, B. A. (1992) *J. Chromatogr.* **592** p. 331.

125. Majors, R. E. (2006) *LC-GC Europe* **19** p. 352.

126. Butchart, K., Foster, G., and Temesi, D. (2007) *Chromatogr. Today* **1** p. 3.

127. Aboul-Enein, H. Y., and Serignese, V. (1994) *Chirality* **6** p. 378.

128. Aboul-Enein, H. Y., and Bakr, S. A. (1997) *Chirality* **9** p. 10.

129. Toribio, L., del Nozal, M. J., Bernal, J. L., Jimenez, J. J., and Alonso, C. (2004) *J. Chromatogr. A* **1046** p. 249.

130. Toribio, L., del Nozal, M. J., Bernal, J. L., Alonso, C., and Jimenez, J. J. (2005) *J. Chromatogr. A* **1091** p. 118.

131. Myrdal, P. B., Angersbach, B. S., Karlage, K., and Kuehl, P. J. (2006) *J. Chromatogr. A* **1132** p. 315.

132. Gaffney, M. H., Stiffin, R. M., and Wainer, I. W. (1989) *Chromatographia* **27** p. 15.

Chapter 7

Chiral Chromatography: Method Development

Laila Kott[a] and Gregory K. Webster[b]

[a]*Xenon, 200-3650 Gilmore Way, Burnaby, British Columbia V5G 4W8, Canada*
[b]*Abbvie, 1 N. Waukegan Rd, North Chicago, Illinois 60064, USA*

lkott@lu.lv, gregory.webster@abbvie.com

7.1 Introduction

In achiral chromatography, there are many understood interactions and equilibria that take place between the analytes, the solid phase of the column and the liquid (mobile) phase. These interactions can be modeled (see Chapter 2) and often fall within well-established theories and equations of chromatographic separations. The same is not true for chiral separations. Chiral separations depend initially on how well (points of contact) the chiral molecules interact with the chiral selector on the solid phase of the column. The packing of the column, the number of chiral selectors and

Chromatographic Method Development
Edited by Gregory K. Webster and Laila Kott

ISBN 978-981-4800-53-2 (Hardcover), 978-0-429-20172-1 (eBook)
www.jennystanford.com

how they are attached (i.e., coated vs. immobilized) have effects on the chiral separations and dictate how much method development space there is within given chromatographic parameters. Many chiral methods have very small design spaces as the chiral selectors work under quite specific conditions. For example, in an achiral assay, if one had two closely eluting peaks, one could adjust the column (longer, smaller particle size, smaller internal diameter) to increase the resolution. These parameters will all work for chiral method development as well. In reverse-phase chromatography, one can almost always slow down elution by slowing the change from polar to non-polar liquids. This theoretically can be done ad infinitium, as the peaks will continue to spread but nonetheless, what happens to the peaks is highly predictable. Conversely, in chiral chromatography one may only have a window of 5% to 10% organic where the interactions with the chiral phase favor chiral separation. Going below or above that range may cause the two enantiomers to, once again, merge into one peak. Overall, the goal of chiral chromatography is to achieve a resolution (R_S) of at least 1.5 between enantiomer peaks. In some cases when the 1.5 resolution is not achievable, the method can still be used, if the small enantiomer peak elutes prior to the parent peak. This order of elution is a second (and sometimes achievable) goal when developing a chiral method. Once an enantioselective stationary phase has been found, then much of the chiral method development follows the same processes as developing an achiral method.

These goals lead to the first and most useful part of chiral chromatographic method development—column screening. This chapter presents a roadmap for the development of a chiral purity method.

7.2 Start at the End, with the Detectors

One tool available to aid in chiral analyses is detectors that differentiate between enantiomeric pairs and provide unique, distinguishable signals for each enantiomeric compound. In the absence of a detector that can differentiate between enantiomers, the analyst is left with only a diode array detector or a mass detector to identify chiral pairs. The danger in relying solely on the

UV spectral data is that this type of data may not be selective to one compound, whereas there is more confidence in mass spectral data being unique to a compound, but not all chiral separation buffers are mass spectrometry "friendly." Many compounds can give a similar UV spectrum, therefore trying to discern enantiomeric pairs can be challenging in the presence of other impurities of similar structure.

There are several detectors that can differentiate between enantiomers and this is highlighted in the chromatographic traces by the presence of positive and negative peaks. Currently there are two types of detectors on the market that can be incorporated into chromatographic systems: polarimeters and circular dichroism (CD) detectors. Both detector types use Drude's equation, which describes the variation of optical rotation as a function of wavelength.

7.2.1 Polarimeters

Normal light waves vibrate in many planes; however, plane polarized light is generated when normal light is passed through an optical polarizing filter. This effect results in a light beam emerging that vibrates in a single plane (linearly polarized). Linearly polarized light is rotated when passing through a compound which is optically active. The degree of rotation depends on both the concentration of a chiral compound and its molecular structure. Every optically active substance has its own specific rotation (degree of rotation in polarized light). The specific rotation of the molecule, not the absorption characteristics, is what determines the signal strength using the polarimeter.

Drude's equation highlights the effect of wavelength on the specific rotation $[\alpha]_\lambda$, such that

$$[\alpha]_\lambda = \frac{\sum A_i}{(\lambda_2 - \lambda_i)}, \tag{7.1}$$

where A_i is a molecular constant and λ_i is a constant wavelength [1]. The equation also points out that the degree of rotation, as a function of wavelength, is greatest at shorter wavelengths. Therefore, to optimize the chiral response in a molecule, lower source wavelengths should yield stronger responses.

Interestingly, at lower wavelengths right- and left-handed circularly polarized light[1] propagate at different velocities and are absorbed by molecules to a different extent. (This phenomenon is known as circular dichroism; see below.) When this happens, it causes a deviation from Drude's equation, known as the "Cotton effect."

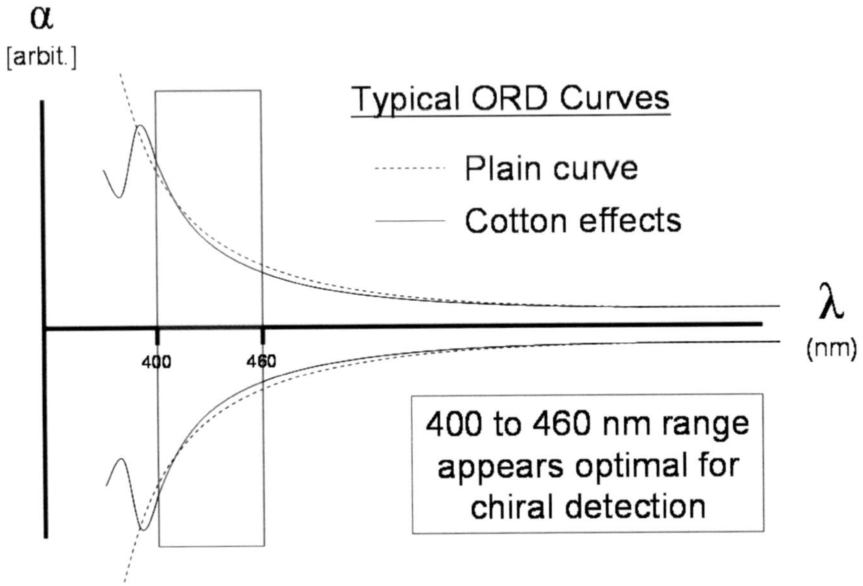

Figure 7.1 Graphical representation of Drude's equation showing the Plain curve, the regions where there is the possibility of the Cotton effect, and the theoretically optimal wavelength range. Reprinted from [2], with permission from Elsevier.

The plain optical rotation dispersion (ORD) curve (or Plain curve) shows the relationship between the magnitude of the optical rotation versus wavelength (i.e., a representation of Drude's equation). The plain curve is affected by the Cotton effect and Fig. 7.1 shows the optimum regions for analysis as well as the flat region (a section of curve where there is little change in $[\alpha]$) but is a region with few optical interferences.

[1]Circularly polarized light: The two beams of linearly polarized light that are of equal amplitude and are a quarter wave out of phase.

7.2.2 Circular Dichroism Detector

When an optically active compound preferentially absorbs right or left circularly polarized light, the difference between the right and left absorbances, which is often a very small value, is denoted as the circular dichroism (CD) signal. CD wavelengths are measured in the UV wavelength range. As with UV absorbance, the CD signal is wavelength dependent. From the above discussion of the Cotton effect, one could assume that a molecule should have a chromophore with absorption in the range of 200 to 420 nm to have strong CD signal. This broad generalization was somewhat disproven by Kott et al. [2], but considering that CD is an absorbance based technique, chromophores will only add to the signal. There are molecules that have CD spectra outside of this wavelength range (i.e., porphyrins [3]), but most small active pharmaceutical ingredients (APIs) fall within this wavelength range.

Molecules have a distinctive CD spectrum, similar to having unique UV spectra and for the best results, the CD detector should be set to the enantiomeric pair's maximum/minimum wavelength, as one would choose a high absorbing wavelength for UV detection in liquid chromatography.

If both the UV and CD data are collected, the anisotropy or *g*-factor can be used in achiral chromatography to determine the enantiomeric purity of a compound. Enantiomers do not have different responses in the UV, therefore a 0.1 mg/mL sample of an R-enantiomer, 0.1 mg/mL of an S-enantiomer of the same compound and a 0.1 mg/mL solution of a racemic mix, will yield the same UV signal intensity at a given wavelength. The CD signal, however, for the R-enantiomeric solution will give the maximum positive signal at the optimum CD wavelength, the S-enantiomeric solution will give the maximum negative signal and racemic mix will give no signal at all. The difference in these CD signals versus the UV signal can also allow for the simultaneous UV detection and the determination of the ratio of enantiomers.

Beer's law states that the absorbance of a compound is dependent on its molar absorptivity (ε), solution concentration (c) and path-length of the cell (l):

$$A = \varepsilon \times c \times l \tag{7.2}$$

and it has been discussed that the circular dichroism (CD) signal is the difference between the absorbed right ($A_{(r)}$) and left ($A_{(l)}$) circularly polarized light:

$$CD = A_r - A_1 \tag{7.3}$$

Therefore, the anisotropy or g-factor is the ratio of the two signals:

$$g\text{-factor} = CD/A, \tag{7.4}$$

which can be reduced to a ratio of the difference in CD and absorbance extinction coefficients:

$$\Rightarrow \frac{(\varepsilon_r - \varepsilon_l) \times c \times l}{\varepsilon \times c \times l} = \frac{\Delta\varepsilon}{\varepsilon} \tag{7.5}$$

which results in an expression independent of concentration and path-length, provided they are constant for both measurements. A g-factor of zero would then be indicative of a racemic mix.

7.2.3 When Chromatography Does Not Work: An Alternative Technique

Vibrational circular dichroism (VCD) is similar to CD, but is a spectroscopic technique which detects differences in absorption of left and right circularly polarized infrared (IR) and near-IR light, where vibrational transitions in molecules are observed. VCD is an absorption measurement and its intensity is linearly proportional to concentration and path-length in accordance to Beer's law. But unlike its parent IR, the VCD intensity is also linearly proportional to enantiomeric purity of the sample, where the signals for enantiomers are equal but opposite in sign. This feature of VCD allows direct measurement of percent desired enantiomer without a need for chromatographic separation [4].

The mid-IR region from 2000 to 800 cm^{-1} (5000 to 12500 nm) is typically were the small VCD signal is seen. Therefore, VCD is fundamentally an IR measurement, which has an abundance of spectroscopic bands, and works well with chemometric analyses.

For this type of analysis, the VCD spectra of mixtures of known enantiomeric ratios are collected and analyzed. Chemometrics allows for the use of many more data points than a standard linear curve and, as such, allows one to use a defined wavelength range instead of just a single wavelength to develop models. With the extra data incorporated into the analysis, the error, also known as the variability, is also calculated and included. This error allows for a better understanding of the fit of the model and how predictive it can be in the future, which is an indication of how accurate and reliable these measurements are. Using chemometrics for these percent enantiomer models leads to a robust and sensitive approach to analysis of complex data sets and is an alternative method of determining chiral purity in the absence of chromatographic separation [4].

7.3 Next Step—the Type of Chromatography

7.3.1 Choice of Phase

A decision on the type of chromatography used is made based on the solubility and stability of your molecule. The main choices are the traditional normal phase, supercritical fluid chromatography (SFC) or reverse phase. Polar organic chromatography has also been used with varying degrees of success, as such it is not highlighted in this chapter [5–7]. Other options include chiral GC for small volatile compounds with no chromophore and chiral CE for small charged molecules like peptides.

Choice of chromatographic separation often rests on the solubility of the compound. More and more reverse-phase columns are being developed (see Chapter 6), such that it is becoming the first choice for chiral separations, as often the same or similar preparation can be used for both the achiral and chiral assays. If your compound is soluble in water, reverse phase is a good place to start. If your compound is soluble in methanol, supercritical fluid chromatography (SFC) may be useful staring point, due to its quick throughput. If you are looking at a chiral synthetic intermediate, which may be more soluble in strong organic solvents or your compound is unstable under aqueous conditions, then one tends to favor using normal-phase chromatography. Normal-phase columns traditionally take a long time to equilibrate, although gradients are also possible.

Recently there have been some major improvements in analytical SFC. SFC, using mainly compressed liquid carbon dioxide as one of the mobile phases, is a faster way to screen for chiral separations over normal phase, providing one has the equipment. Normal-phase chromatography can be run on almost any LC system, whereas SFC requires a dedicated system. However, if available, using SFC is much easier as the columns equilibrate much faster and due to the low viscosity of liquid CO_2 the flow rates are much faster (typically between 1 and 2.5 mL/min for 4.6 mm id columns). SFC has different selectivity, is a greener technology, is more compatible with mass spectrometry and scales easily to preparation scale.

Bottom line, the first step in developing a chiral method is to screen different columns and modes of chromatography to find the best enantiomeric separation(s) [8–10].

7.3.2 Normal-Phase Screening

Normal phase has traditionally been more selective for isomers and has been the first choice for many as a chiral screen. Typically, initial screens use hexane(s) or heptane as the weak eluting phase, coupled with alcohols such as ethanol or isopropyl alcohol (IPA). When using a combination of normal phase and SFC or converting between the two, IPA is the preferred alcohol to use as elution with IPA in normal phase correlates better to what is seen under similar conditions in SFC than ethanol does. Unlike the reverse-phase systems, gradients tend to be shallower or not run at all, as the columns take a longer time to re-equilibrate under normal-phase conditions. Steep gradients can be run as one needs to elute the compounds, but care must be taken to ensure that the systems properly re-equilibrate, for if the system has not fully equilibrated, the chromatography achieved may not be reproducible. Many normal-phase methods often are screened using a gradient and when the optimum separation conditions are found, they are converted to isocratic.

If both an analytical and preparatory chromato-graphic methods are required, developing a normal-phase method yields fractions with easily removable solvents for easier compound recovery. For a non-exhaustive list of columns currently used in the pharmaceutical industry, see the Introduction section of Chapter 6.

7.3.3 Supercritical Fluid Chromatography (SFC) Screening

SFC is a normal-phase technique where supercritical fluid CO_2 replaces the traditional alkane leading to much more efficient chromatography [11]. Supercritical CO_2 has a higher diffusivity and a lower viscosity than traditional liquid chromatography mobile phases [11], which allows for high flow rates and rapid re-equilibration. These two factors lend well to SFC column screening. Like normal-phase chromatography, SFC requires screening several columns, although the mobile phase conditions are usually restricted to the use of carbon dioxide and methanol. Ethanol, isopropanol, and acetonitrile can also be used [12]. Table 7.1 compares example screening protocols based on the column dimensions available. Note that other method parameters (temp, back pressure regulator (BPR) settings, etc.) for method screening should be based on standard settings for instrument used.

Under SFC conditions, polysaccharide stationary phases have the potential to provide separations for most compounds representative of molecules in development by pharmaceutical laboratories. Table 7.2 is a non-exhaustive list of currently used chiral SFC columns used by the pharmaceutical industry.

Note that many of these may also be used in normal and reverse-phase chromatography; however, it is strongly recommended that once a column is used under a particular chromatographic condition (reverse phase, normal phase, SFC) that column is only to be used for that type of chromatography. Again, as with the additives, the columns can become less reproducible when forced to switch back and forth between conditions and different solvents.

Table 7.1 Comparison of SFC screening methods based on column dimensions

	Column dimensions		
Conditions	**4.6 × 250 mm**	**4.6 × 250 mm**	**4.6 ×150 mm**
Particle size (μm)	10	5	3
Flow rate (mL/min)	1.5	2	3
Starting % MeOH	4	4	1
Hold (min)	4	4	0
Ramp time to 40% (min)	18	9	5
Hold (min)	3	2	1
Post equilibration time (min)	5	1	0
Run cycle time (min)	30	16	6

Source: Reproduced with permission from [12].

Table 7.2 Chiral SFC columns used by the pharmaceutical industry

Manufacturer	Type of phase	Column name	Particle sizes (μm)	Typical dimensions (mm)
Waters (Trefoil series)	polysaccharide	Cell-1 Cell-2 Amy-1	2.5	3.0 × 100 (150)
Phemomenex (Lux series)	polysaccharide	Amylose-1, -2 Cellulose-1, -2, -3, -4 iCellulose-5 iAmylose-1	3, 5	3.0 × 100 (150)
Chiralcel	polysaccharide	See Figures 6.1 and 6.2 in Chapter 6	Sub 2, 3, 5	2.1 (3.0) × 150
Regis	Pirkle type	Whelko (S,S) and (R,R) Ulmo (S,S) and (R,R)	1.8 (Whelko 01), 3.5, 5, 10	1.6 (3.0) × 250 (150)

7.3.4 Reverse-Phase Screening

To make the most out of reverse-phase chiral screening it is recommended that the screen spans the pH range and employs multiple organic solvents for elution. A typical screen would include a low pH (~2), a neutral pH (6.8 to 7.4) and a high pH (~9). For this type of screen to be successful, verify that the columns used can operate within this pH range. If not, then adjust the pH range of columns.

As a first pass, it is recommended that methanol and acetonitrile be the organic eluents. These solvents can be used with either coated or immobilized chiral phases. If neither of these phases proves to work, other organic solvents that are miscible with water can be used (i.e., isopropyl alcohol, THF); however, it should be verified whether such solvents are compatible with the chosen chiral phases. Immobilized phases have a larger solvent range than coated phases and certain solvents can strip the (very expensive) chiral phase off coated columns.

For setting up a screen, it is best to use a system with a column-switching valve (the more, the better; two will take a long time to screen many columns) and, if possible, pumps with more than the standard four lines. Ideally one would want a system that can handle at least the three different pH aqueous phases and two organic phases. Having lines to flush the columns between pH changes helps prevent salting out and other damage that sudden pH changes can cause. To expedite the screen process, it is recommended to use a gradient, as one can essentially perform several "isocratic" screens in one gradient run. Keep in mind that when methanol and water mix, the mobile phase heats up, but when acetonitrile and water mix the endothermic enthalpy of mixing causes the mobile phase temperature to decrease. If using high salt buffers, the aqueous solubility of the buffer decreases with mobile phase temperature; care must be taken when using acetonitrile to keep its maximum gradient concentration low enough to not precipitate the salt in buffers. Table 7.3 highlights typical gradient screening programs that can be adjusted as needed.

Table 7.3 Typical reverse-phase gradient starting conditions for reverse-phase chiral screening

Parameter	Settings		
pH's screened	1 to 2	Neutral (6.8)	9.0
Organic phase, first pass	Methanol	Acetonitrile	
Gradient	90 (aqueous):10 (methanol) to 20:80	90 (aqueous):10 (ACN) to 30:70	
Column	4.6 × 100 or 150 mm [3 μm particle size]	3.0 × 100 or 150 mm [3 μm particle size]	2.1 × 100 or 150 mm [3 μm particle size]
Flow rate (ml/min)	1.0	0.75	0.5
Gradient time	20 to 40 min		
Detection	Collect all wavelengths, if possible, otherwise 220, 254, 280 nm		
Detection (optional)	Circular dichroism at estimated appropriate wavelengths		
Detection (optional)	Mass spectrometry if aqueous phases used are volatile		

If sufficient column chemistries have been screened, often there are 'hits' which show some degree of separation, which can then be taken to a standard system and optimized. Unfortunately there are times when no resolution is achieved. In this case, as sometimes happens with strongly acidic or basic compounds, it may be beneficial to use an additive in the mobile phase. In general, additives are not the first choice to use on a screening system as many of the additives used in chiral screen have memory effects. These compounds may react/functionalize/or in some way alter the stationary phase with their use and as such the stationary phase does not return to its original condition. This then throws into question all subsequent screens on a system with that column. A separation may work on the screening system but not offline with a fresh column not functionalized in the same way, the column on the screen system is. Figure 7.2 shows a chart for choosing an additive for the chiral mobile phases, if needed.

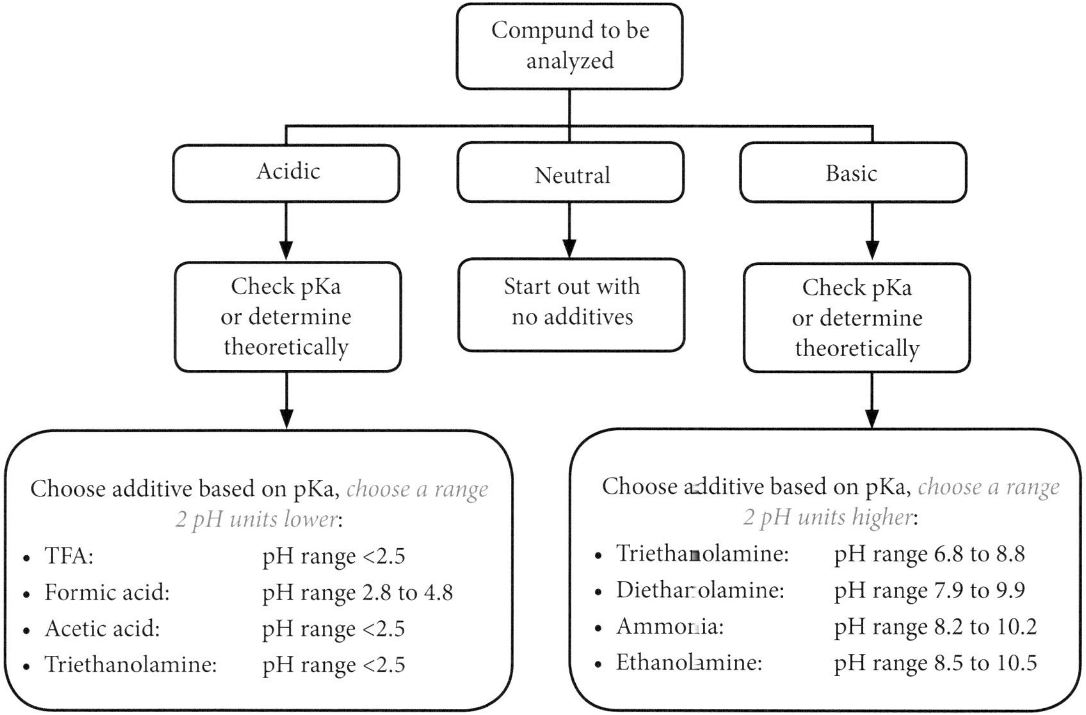

Figure 7.2 Flow chart to aid in choosing an additive for chiral screening.

7.3.5 Screening Success

Screening is used extensively in the pharmaceutical industry, especially once a standard in-house screen has been established. Such screens are used as they can rapidly find enantiomeric separations for 60% to 80% of the compounds in a time frame much faster than trial and error alone. Unlike for the case of achiral method development where some knowledge of the structure can lead to a good suggestion for a chromatographic starting point based on previous knowledge and experience with different stationary phases, the separation of chiral molecules is too complex to intuitively see the structure and choose the best chiral phase. Polysaccharide columns have historically had the best success rates and are the most prevalent columns in most in-house screens. Following are summaries of some of the different column types available.

7.4 The Columns

As discussed above and in Chapter 6, the chiral column is key to the separation of the enantiomers. There are several types of columns to consider; however, the polysaccharide columns (cellulose and amylose) are often the first type of column tried as their technology is continuously improving; these types of columns have become more rugged and have been adapted for use with UHPLC, for instance, the sub 2 µm from Chiralcel (see Chapter 6). Other types of columns available include the Pirkle (or brush type) and antibiotic-based columns.

7.4.1 Polysaccharide-Type Columns

Several manufacturers have polysaccharide-type columns. Currently Chiral Technologies has the greatest selection of phases (see Chapter 6), but other companies such as Phenomenex and Waters are also developing polysaccharide chiral columns. These columns are available as both coated, where the chiral selector is coated on to the silica phase, and as an immobilized phase, where the chiral selector is physically bonded to silica. There are advantages and disadvantages to both types of column. Since the silica is only coated with the chiral phase, choice of chromatographic solvent is key as solvents other than alcohols and acetonitrile can wash the chiral phase off of the silica, destroying the expensive chiral column. Alternatively, immobilized phases are more rugged and can be used with a wider variety of organic solvents but tend to have a lower density of the chiral selector. Since polysaccharide columns are the most prevalent chiral columns currently used by the pharmaceutical industry, they are available in a large range of particle sizes ranging from sub 2 μm for analytical use up to 20 μm for preparative separations, and the corresponding column dimensions [13, 14].

There are traditionally two types of polysaccharide columns: one based on amylose and one based on cellulose. It is of note that it is these sugars that are the true chiral selector and that the different column selectivity is based on the different derivatives of these polysaccharides. For example, *tris* (3-chloro-4-methyl-phenylcarbamate) and *tris* (3,5-dimethyl-phenylcarbamate) are widely used derivatives of amylose and cellulose for chiral separations, but these functional groups themselves are not chiral, the selectivity comes from the combination of the added derivative and the polysaccharide. For a list of currently used derivatives of amylose and cellulose, see Chapter 6.

7.4.2 Pirkle-Type Columns

These phases were developed as π-electron acceptors, π -electron donors and chiral selectors with both, π-donor and π-acceptor attributes. Several studies have been conducted with individual compound and column investigations using Pirkle-type stationary phases [15–23]. Normally, the same mobile phase approaches that work with polysaccharide stationary phases also work with Pirkle-type chiral stationary phases. Because of the commercial success of the polysaccharide columns, many pharmaceutical laboratories simply choose not to screen Pirkle-type columns when looking for enantioseparations. This is unfortunate as unique separation opportunities may be missed. Past performance of stationary phases may not be indicative of future enantioresolutions with new pharmaceutical entities. Thus, maintaining screening capabilities with alternative stationary phase choices is prudent.

Because Pirkle-type chiral phases are covalently bound to the chromatographic support, they exhibit excellent stability. Many polysaccharide phases in use today are non-covalent coatings that can be washed off the support material if the chromatographic conditions are not carefully chosen. Pirkle-type chiral stationary phases are stable from pH 2.5 to 7.5 and can be used in both the normal-phase mode and the reverse-phase modes. An important characteristic of the Pirkle-type chiral stationary phases is their availability in both enantiomeric and/or opposite diasteriomeric forms. This choice of enantiomeric forms allows the chromatographer to manipulate the separation for optimization, as often, for ease of quantitation, the goal is to have the enanatiomer elute in front of the parent peak. For example, in case of an S,S-configured chiral stationary phase, the elution order of separated enantiomers can be reversed simply by switching to the R,R-configured chiral selector.

7.4.3 Antibiotic Type Columns

These phases are based on the chiral selectivity of several macrocyclic antibiotic molecules. The most common antibiotics used as chiral selectors are as follows [24, 25]:

- Vancomycin and its analogs
- Teicoplanin and its analogs, including Teicoplanin aglycone (TAG)
- Ristocetin and its analogs

All these phases can be used in reverse-phase, normal-phase, and polar organic modes. Table 7.4 highlights the selectivities of these three types of phases.

7.4.4 Columns for Ultrafast Chiral Separations

Now that ultra-high-pressure liquid chromatography (UHPLC) is found in most laboratories, chiral column manufacturers have been bringing more UHPLC-type chiral columns on the market, allowing for faster chiral separations [15]. Sub-minute separations are possible using fully porous particles (FPP); however, the bonding of chiral phases to superficially porous particles (SPP) has shown to be even more effective for sub-minute and even sub-second enantiomeric separations [26–28].

Table 7.4 Selectivities of several antibiotic based chiral columns

Chiral selector	Modification(s)	Column name	Selective for
Vancomycin	None	Chirobotic V	Neutral molecules, amides, acids, esters and cyclic amines
Teicoplanin	None	Chirobiotic T	Underivatized amino acids, N-derivatized amino acids, carboxy cyclic acids, phenols, small peptides, aromatic molecules, and aliphatic and cyclic amines
Teicoplanin aglycone	Sugars removed	Chirobiotic TAG	Amino acids and other amphoteric molecules, cyclic amino acids, and neutral molecules such as diazopenes, hydantoin and oxazolidinones
Ristocetin A		Chirobiotic R	Anionic analytes (i.e., α-hydroxy acids, substituted aliphatic acids), chiral alcohols, secondary and tertiary amines

Several instrument settings and configurations need to be considered for such quick separations. From a data collection standpoint, the collection rate on the detector(s) should be at 80 Hz, as a minimum. Extra column volumes need to be reduced which can be accomplished by replacing high volume mixtures and using low volume fittings.

Some columns used for ultrafast separations are commercially available, while others require special functionalization of packing materials. Table 7.5 gives an overview of columns used for ultrafast separations.

Table 7.5 Overview of chiral columns used for ultrafast separations

Column	Phase/chiral separator	Commercially available	Reference
LarihcShell-P	Isopropyl bonded cyclofrutan CF6-P	AZYP LLC (Arlington, TX, USA)	28
CDShell-RSP	2-hydroxypropyl-β-cyclodextrin	AZYP LLC (Arlington, TX, USA)	24, 26
TeicoShell	Teicoplanin	AZYP LLC (Arlington, TX, USA)	28
VancoShell	Vancomycin	AZYP LLC (Arlington, TX, USA)	26
NicoShell	Modified macrocyclic glycopeptide	AZYP LLC (Arlington, TX, USA)	
Chiralpak IA-U and IC-U	Polysaccharide	Chiralcel	See Fig. 6.1, Chapter 6
(S,S) and (R,R) Whelk-O1	Pirkle type	Regis	

7.4.5 Other Phases for Chiral Separations

This chapter and Chapter 6 have explored the more traditional and available chiral columns, but this is not an exhaustive list. There are many older chiral selectors that have fallen out of use but, nonetheless, with ever-complicated separations could be worth looking at if all other columns tried do not lead to a satisfactory separation [29]. Additionally, new and interesting chiral phases are emerging based on new and/or modified technologies [30]. This is especially true for ultrafast chiral chromatography, so keeping on top of new chiral column technologies is very useful for chiral method development.

7.5 Conclusions

As discussed in this chapter and in Chapter 6, the key to quick chiral method development is first and foremost to find the conditions and column to achieve the enantiomeric separation. This can be simplified by screening multiple columns under multiple conditions, and even further simplified with the use of chiral detectors. Once conditions that work have been found, then the method development can be taken off-line from the screening system for optimization, which can be similar to achiral optimization at this point. Different approaches and techniques have been discussed, and all work well for certain compounds. It may take several attempts to find the right conditions for separation of enantiomers, but options are many and usually some separation of enantiomers is found.

Acknowledgments

The authors would like to thank Zachary Breitbach for review of the chapter and providing useful insights.

Disclosure

Laila Kott is an employee of Xenon. Gregory Webster is an employee of AbbVie. The study contains no proprietary Xenon or AbbVie data. Xenon and AbbVie jointly participated in writing, reviewing, and approving and providing financial support for this publication.

References

1. Snatzke. G. (1968) Circular dichroism and optical rotatory dispersion—Principles and application to the investigation of the stereochemistry of natural products, *Angewandte. Chemie International Edition in English*. **7**(1), pp. 14–25.
2. Kott, L., Holzheuer, W. B., Wong, M., Webster, G. K. (2007) An evaluation of four commercial HPLC chiral detectors: A comparison of three polarimeters and a circular dichroism detector, *J. Pharmaceutical and Biomedical Analysis.* **43**(1), pp. 57–65.

3. Huang, X., Nakanishi, K., Berova, N. (2000) Porphyrins and metalloporphyrins: versatile circular dichroic reporter groups for structural studies, *Chirality*. **12**, pp. 237–255.
4. Kott, L., Petrovic, J., Phelps, D., Roginski, R., Schubert, J. (2014) Determination of a low-level percent enantiomer of a compound with no ultraviolet chromophore using vibrational circular dichroism (VCD): Enantiomeric purity by VCD of a compound with three chiral centers, *Applied Spectroscopy*. **68(10)**, pp. 1108–1115.
5. Matthijs, N., Maftouh, M., Heyden, Y. V. (2006) Screening approach for chiral separation of pharmaceuticals IV. Polar organic solvent chromatography. *Journal of Chromatography A*, **1111**, pp. 48–61.
6. Matthijs, N., Maftouh, M., Heyden, Y. V. (2006) Chiral separation strategy in polar organic solve chromatography and performance comparison with normal-phase liquid and supercritical-fluid chromatography, *Journal of Separation Science*. **29**, pp. 1353–1362.
7. Morante-Zarceno, S., Sierra, I. (2012) Comparative HPLC methods for β-blockers separation using different types of chiral stationary phases in normal phase and polar organic phase elution modes. Analysis of propranolol enantiomers in natural waters, *Journal of Pharmaceutical and Biomedical Analysis*. **62**, pp. 33–41.
8. Akin, A., Antosz, F. J., Ausec, J. L., Greve, K. F., Johnson, R. L., Magnusson, L. E., Ramstad, T., Secreast, S. L., Seibert, D. S., Webster, G. K. (2007) An orthogonal approach to chiral method development screening, *Current Pharmaceutical Analysis*. **3**(1), pp. 53–70.
9. Wong, M. M., Holzheuer, W. B., Webster, G. K. (2008) A Comparison of HPLC and SFC chiral method development screening approaches for compounds of pharmaceutical interest, *Current Pharmaceutical Analysis*. **4**(2), pp. 101–105.
10. Holzheuer, W. B., Wong. M. M., G. K. Webster, (2009) Reverse phase chiral method development screening for compounds of pharmaceutical interest, *Critical Pharmaceutical Analysis*. **5**, pp. 345–357.
11. Webster, G. K. (2014) The SFC market: 'Yesterday, today, and tomorrow, in *Supercritical Fluid Chromatography: Advances and Applications in Pharmaceutical Analysis*, ed., Webster, G. K., Chapter 1 (Pan Stanford Publishing, USA), pp. 1–14.
12. Schafer, W., Chandrasekaran, T., Prizada, Z., Zhang, C., Gong, X., Biba, M., Regalado, E. L., Welch, C. J. (2013) Improved chiral SFC screening for analytical method development, *Chirality*. **25(11)**, pp. 799–804.

13. Zhang, T., Franco, P. (2010) Finding the best separation for enantiomeric mixtures, *LCGC North America.* **28**(9), pp. 818–822.
14. Woiwode, U., Neubauer, S., Kaupert, M., Lindner, W., Lammerhofer, M. (2017) Trends in enantioselective high performance liquid chromatography, *LCGC Europe.* (Special issue) **30**(6), pp. 34–42.
15. Kennedy, J. H. (1996) Comparison of chiral separations on polysaccharide chiral stationary phases to an improved Pirkle-type phase, *Journal of Chromatography.* **725**, pp. 219–24.
16. Siluveru, M., Stewart, J. T. (1996) Stereoselective determination of R(-)- and S(+)-prilocaine in human serum using a brush-type chiral stationary phase, *Journal of Pharmaceutical and Biomedical Analysis.* **15**, pp. 389–92.
17. Zhang, X., Ouyang, J., Yang, Y. (2001) A simple method for chiral separation of ephedrines using (R)-1-naphthylglycine and 3,5,-dinitrobenzoic acid as stationary phase, *Analytical Letters.* **34**, pp. 1851 1865.
18. Shao, B.-H., Xu, X.-Z., Wu, Q.-Z., Lu, J.-D., Fu, X.-Y. (2005) Comparative enantioseparation of 2-arylpropionic acid esters on cellulose derivative and (S,S)-Whelk-O 1 columns, *Journal of Liquid Chromatography & Related Technologies.* **28**, pp. 63–80.
19. Shao, B.-H., Xu, X.-Z., Wu, Q.-Z., Lu, J.-D., Fu, X.-Y. (2003) Comparison of enantioseparation and chiral recognition mechanism of racemic naproxen esters on (S,S)-Whelk-O 1 and CDMPC chiral columns, *Acta Chimica Sinica.* **61**, pp. 1635–1640.
20. Madhavan, P., Rao, B. M., Pravin, A., Kumar, P. R., Screenivasulu, M., Chandrasekhar, K. B. (2007) A validated chiral HPLC method for the determination of enantiomeric purity of R-β-amino-β-(4-methoxyphenyl) propionic acid, *Chromatographia.* **65**, pp. 81–84.
21. Szczerba, T. (2011) Summarization of screening hits on the Whelk-O® 1, RegisPack and RegisCell chiral stationary phases (CSPs), *LCGC North America.* (Suppl.), **24** (February) p. 24.
22. Holzheuer, W. B., Wong, M. M., Webster, G. K. (2009) Evaluation of Pirkle-type stationary phases in chiral method development screening for compounds of pharmaceutical interest, *Critical Pharmaceutical Analysis.* **5**, pp. 10–20.
23. McDevitt, T. F., Vicente, G., Webster, G. K., Szczerba, T. (2009) Column coupling to solve a challenging API separation using the Whelk-O® 1 chiral stationary phase, *LC/GC Application Notebook.* September **2009** p. 23.

24. Lee, D., Web, M. L. (2009) Chiral analysis of pharmaceuticals, in *Pharmaceutical Analysis*, eds., Lough, W. J., Chapter 3 (Wiley & Sons, Oxford, UK), pp. 88–90.
25. Sigma-Aldrich (Millipore Sigma), https://www.sigmaaldrich.com/analytical-chromatography/hplc/columns/chiral/chirobiotic.html (accessed November 13, 2017).
26. Patel, D. C, Brietbach, Z. S., Wahab, M. F, Barhate, C. L., Armstrong, D. W. (2015) Gone in seconds: Praxis, performance, and peculiarities of ultrafast chiral liquid chromatography with superficially porous particles, *Analytical Chemistry*. **87**, pp. 9137–9148.
27. Wahab, M. F., Wimalasinghe, R. M., Wang, Y., Barhate, C. L., Patel, D. C., Armstrong, D. W. (2016) Salient sub-second separations, *Analytical Chemistry*. **88**, pp. 8821–8826.
28. Barhate, C. L., Joyce, L. A., Makarov, A. A., Zawatsky. K., Bernardoni, F., Schafer, W. A., Armstrong, D. W., Welch, C. J., Regalado, E. L. (2017) Ultrafast chiral separations for high throughput enantiopurity analysis, *Chemical Communications*. **53**, pp. 509–512.
29. Webster, G. K., Kott, L. (2011) Method development for pharmaceutical chiral chromatography, in *Handbook of Modern Pharmaceutical Analysis*, eds., Ahuja, S., Scypinski, S., Chapter 7 (Elsevier, San Diego), pp. 251–282.
30. Brown, L., Webster, G. K., Kott, L., Rao, N. K. R., Luu, T. A., Shah, R., Henriques, L. (2011) Novel coated cellulose carbamate stationary phase to enhance selectivity of compounds of pharmaceutical interest, *Pharmaceutica Analytica*. **2**, pp. 2–7.

Chapter 8

Aqueous Normal-Phase Chromatography Using Type C Silica Columns

Joseph J. Pesek and Maria T. Matyska

Department of Chemistry, San Jose State University, San Jose, California 95192, USA

joseph.pesek@sjsu.edu

This chapter presents the general principles of aqueous normal-phase (ANP) chromatography, a technique that has features of both reversed-phase (RP) and hydrophilic interaction chromatography (HILIC). It is a unique chromatographic format and while some types of stationary phases exhibit a limited ANP capability, it is silica hydride-based materials that are most successful in this mode and have the broadest range of applications. Thus, this review will focus exclusively on the fundamental aspects of ANP, the uniqueness of the silica hydride material and review applications utilizing both the RP and ANP capabilities of these columns.

Chromatographic Method Development
Edited by Gregory K. Webster and Laila Kott

ISBN 978-981-4800-53-2 (Hardcover), 978-0-429-20172-1 (eBook)
www.jennystanford.com

8.1 Introduction

A variety of approaches and stationary phases are available for the separation of mixtures or analyses of specific compounds based on the chemical and structural properties of the analyte. A large fraction of current chromatographic methods utilize a format that revolutionized separation science more than five decades ago; chemically bonded stationary phases that are fabricated on silica as a support material with hydrophobic moieties attached to the surface. Columns made from these materials are referred to as reversed-phase (RP) and the most common bonded group is the straight chain C18 hydrocarbon. Other chain lengths between C4 and C30 have demonstrated the same type of retention based on hydrophobic interactions with the degree of retention based on the number of carbon atoms, the amount of the bonded phase and the properties of the analytes. Other types of bonded moieties with significant hydrophobic character can also provide RP retention and many have been synthesized over the years but only a relatively small number have merited commercial development. While other contributions to retention can occur on these phases, very often adsorption effects from residual silanols on the silica surface, the hydrophobic effects are usually the dominant factor in these types of chromatographic columns.

However, the analysis of polar (hydrophilic) compounds has always presented a serious challenge due to the low retentivity on typical RP stationary phases. In this case, the typical approach for many analyses was to derivatize the analyte to make it less polar so that it could be retained on a typical RP column. Only a few stationary phases were polar enough to provide significant retention for these types of compounds. The most common being ion-exchange materials which could be used for analytes possessing either a permanent positive or negative charge. This could often be achieved by adjusting the pH of the mobile phase or dealing with the salts of the parent compound. The relatively few good options for polar compounds provided opportunities for new approaches and materials to deal with these challenging types of analytes.

In recent years, there have been two types of chromatographic materials developed more specifically designed for the analysis of hydrophilic compounds: mixed-mode and hydrophilic interaction liquid chromatography (HILIC) stationary phases [1, 2]. In the mixed-mode format, groups are bonded to the surface that have some polar components as part of the bonded group as well as a hydrophobic alkyl chain. Thus, polar compound retention is determined by the nature of the functional group on the bonded moiety. In HILIC, the most common phases are silica or silica with a bonded group containing a polar functionality like cyano, amino or diol. In this case, retention is due to a water layer that forms on the surface of the stationary phase that results in a partitioning effect between the adsorb water layer and the organo/aqueous mobile phase. The mixed-mode phases often have limited applicability and retention and the HILIC materials have other issues that will be described in more detail later when specific properties of the silica hydride phases are discussed.

8.2 Silica Hydride

There have been numerous reports in the literature describing a broad array of stationary phases based on ordinary silica [1, 2]. The general approach to attaching an organic group to a silica surface utilizes an organosilanization reaction. The organosilane reagent can be either monofunctional or trifunctional which leads to a single attachment to the silica surface or the formation of a polymer layer on the support material. While other approaches such as monoliths or different materials like zirconia or titania have been developed over the years, they have not significantly replaced ordinary silica as the predominant source for HPLC stationary phases. Although some desirable properties of these alternative materials have been found, the positive features are often negated by other factors that have precluded them from displacing silica except in a limited number of applications.

However, over the last two decades, one material has been documented in the literature to have significantly different properties than ordinary silica but still retains many of its good features such as high surface area, excellent mechanical strength, and the ability to fabricate the material in a broad range of particle sizes. That material is silica hydride [3–7]. The essential difference between ordinary silica and silica hydride is shown in Fig. 8.1. The underlying structure in both materials is a polymeric series of silicon-oxygen linkages. However, the surface of ordinary silica is populated by polar silanols while silica hydride contains virtually only mildly hydrophobic Si-H moieties. This significant difference can be observed in the chromatographic properties of stationary phases fabricated on these two support materials. However, a simple example of the nature of the silica hydride surface is shown in Fig. 8.2. In this case, simple nonpolar molecules are retained on unmodified silica hydride using simple reversed-phase conditions. These results illustrate the mildly hydrophobic nature on the base material used for the fabrication of stationary phases that are further modified.

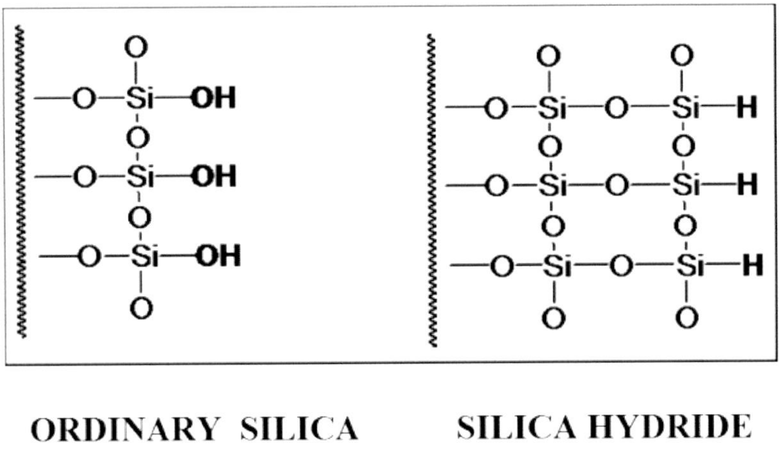

Figure 8.1 Comparison of the surface structures of ordinary silica and silica hydride.

The first important property of silica hydride that parallels a desirable aspect of ordinary silica is its ability to be modified by a simple and well-known chemical reaction. In the case of silica, it is organosilanization. For silica hydride, the comparable reaction is hydrosilation that involves the addition of Si-H to an olefin or alkyne [4]. Thus, bonding to the surface is through a stable silicon-carbon bond that provides durability and long lifetime to these materials. Once the material was fabricated and through many prior investigations, a unique and distinctive property of all silica hydride-based stationary phases is their ability to operate in both the reversed-phase and normal-phase modes [8, 9]. However, in this case normal-phase refers to the retention of polar/charged compounds using a mobile phase containing at least a minimal amount water. Hence, the term aqueous normal phase(ANP) is used to describe this property. This mechanism is in contrast to organic normal phase where the retention of moderately polar compounds occurs using organic solvents in the mobile phase. It should be noted that another interesting feature of silica hydride stationary phases is that they can also function under organic normal-phase conditions [10]. Their slight hydrophobic nature creates a different type of material than the one most often used in this separation mode: ordinary silica.

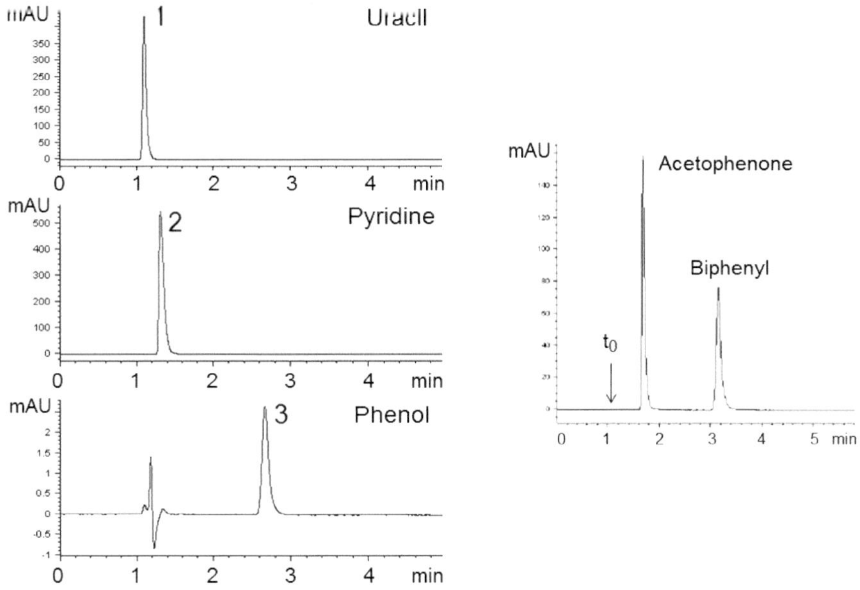

Figure 8.2 Chromatograms showing the retention of small hydrophobic compounds on silica hydride using reversed-phase conditions.

An important clue to the nature of silica hydride has been provided primarily by Jandera and coworkers, who measured the amount of water on the surface of various chromatographic materials [11]. A variety of silica hydride modified stationary phases showed less than one-half of a monolayer of water on the surface. This is contrast to ordinary silica and various modified silicas where there was anywhere from 4 to 10 monolayers of water on the surface. This fundamental difference between ordinary silica and silica hydride materials leads to important variations in chromatographic properties when comparing columns fabricated on the two supports. There are at least two important consequences of the difference in the amount of surface water. In the instances where variations in the water layer occur, such as gradients, this can often be a time-consuming process requiring many column volumes before the stationary phase returns to its original configuration. Since this is not an issue with silica hydride, the time required to re-equilibrate the column is often significantly shorter than with ordinary silica phases. The lack of water on the surface also facilitates operation in the organic normal-phase mode. On silica and other materials used for this chromatographic mode, water can drastically affect retention of analytes. Therefore, mobile phases must often be scrupulously dried in order to achieve reproducible retention due adsorption of water on the surface. Since silica hydride has a very low affinity for water, its presence in the mobile phase has little effect on its retention properties.

The absence of a water layer on silica hydride stationary phases explains many of its unique properties. However, the one feature of these materials that was difficult to explain involved the retention of hydrophilic compounds. Based on a number of different analytical methods, the surface was shown to contain no more than 5% residual silanols [7]. In addition, the chromatographic results described above demonstrate the general hydrophobic nature of the silica hydride surface. The first step to formulating an explanation for this feature of silica hydride materials was to determine the physical state of the surface. This was accomplished by measuring the zeta potential of the stationary phases [12]. The zeta potential determines the net overall charge on the surface and the region within a few nanometers. Surprisingly the measurements showed a very high negative zeta potential, comparable to ordinary silica, for the silica hydride materials. Since the presence of silanols could not explain this property, another theory was sought to account for the substantial negative charge and hence the retention of hydrophilic compounds. Because the overall surface of a silica hydride particle is hydrophobic, a search of the literature determined the best model for this material was an oil droplet suspended in an organo/aqueous liquid environment. A number of investigations have demonstrated that under these conditions the droplet has a strong negative charge on the surface [13–15]. The source of the negative charge was determined to be hydroxide ions from the autodissociation of water [16]. Thus, retention of polar compounds is based on either charge attraction or displacement of hydroxide or other adsorbed mobile phase components like buffers or additives.

The above description of the surface characteristics of silica hydride and in particular the mechanism for the retention of polar compounds, provides a sharp contrast to a currently common method for their analysis, HILIC. In HILIC the compounds are separated due to the water layer on the surface of the stationary phase [17]. These water layers are comparable to those described above for silica-based materials which are the base for many HILIC stationary phases. Even HILIC columns in other formats like polymeric materials still rely on the substantial water layer at the surface. This layer acts as a pseudo stationary phase that is more polar than the bulk mobile phase. Therefore, hydrophilic compounds preferentially partition into this layer and are thus retained on the column. The contrast between an adsorption type mechanism on silica hydride and the partition mechanism in HILIC have some very important consequences. For example, retention generally increases on HILIC phases when the concentration of additives in the mobile phase increases. However, just the opposite effect is noticed for silica hydride, i.e., retention decreases as additive concentration increases. The practical result is particularly relevant when using mass spectrometry for detection. High additive concentrations (some methods require from 20 to 100 mM) in the mobile often lead to deposits forming on the ionization source that must be cleaned more frequently as well as producing inconsistent ionization. Silica hydride almost exclusively operates in the 5 to 15 mM additive concentration range. The other consequence of the difference in mechanisms is the re-equilibration time after gradient runs. Gradient runs are necessary when complex samples with numerous analytes that are frequently encountered in biological, clinical, pharmaceutical, forensic, food and environmental determinations. At the end of the gradient run, it often takes numerous column volumes (sometimes 20 or more) to bring the stationary phase (water layer) back to the original conditions at the start of the gradient. For silica hydride materials, generally only two or three column volumes are necessary to achieve the original conditions at the start of the gradient. Practically, this means that many more analyses can be done over a given time period and considerably less solvents are used in the process.

The features described above de monstrate that silica hydride and silica hydride-based stationary phases are unique separation materials. They are clearly different from the most closely related materials; ordinary silica and silica modified using traditional organosilanization reactions. A number of advantages have already been cited. These will be reiterated to some extent in the examples given in the following sections as well as other features that provide new approaches for the separation and analysis of samples ranging from the determination of a single component to multicomponent mixtures in both simple and complex matrices.

8.3 Applications of Silica Hydride-Based Separation Materials

8.3.1 Reversed-Phase

As shown in the example above, bare silica hydride without any further modification through hydrosilation possesses RP retention properties for typical small hydrophobic molecules. As expected, modification of silica hydride with nonpolar organic moieties further enhances the reversed-phase capabilities of this material. Thus, any silica hydride-based stationary phase can be used for RP applications, but of course, the more hydrophobic attached organic moieties provide the greatest RP retention.

Some typical examples of RP analyses done on a C18 modified silica hydride material are shown in Fig. 8.3. In Fig. 8.3A, two hydrophobic components in a sun screen formulation are analyzed on this C18 column. The structures of these compounds are also shown in the figure. The chromatogram demonstrates the high efficiency and excellent peak shape that can be achieved in the

RP mode with silica hydride-based columns. Another feature of these materials is also highlighted in this figure and that is the high degree of reproducibility from batch to batch in the fabrication of the stationary phase. A second example on this same C18 column is shown in Fig. 8.3B. In this analysis of an acetaminophen formulation, there is thc primary pharmaceutical component and two other minor constituents. Based on the separation shown, both acetaminophen as the primary component as well as caffeine and butalbital could be quantitated as needed to assure the quality of the marketed product. In both examples shown, peak shape is excellent that demonstrates the value of the silica hydride surface being slightly hydrophobic and essentially free of silanols for RP applications.

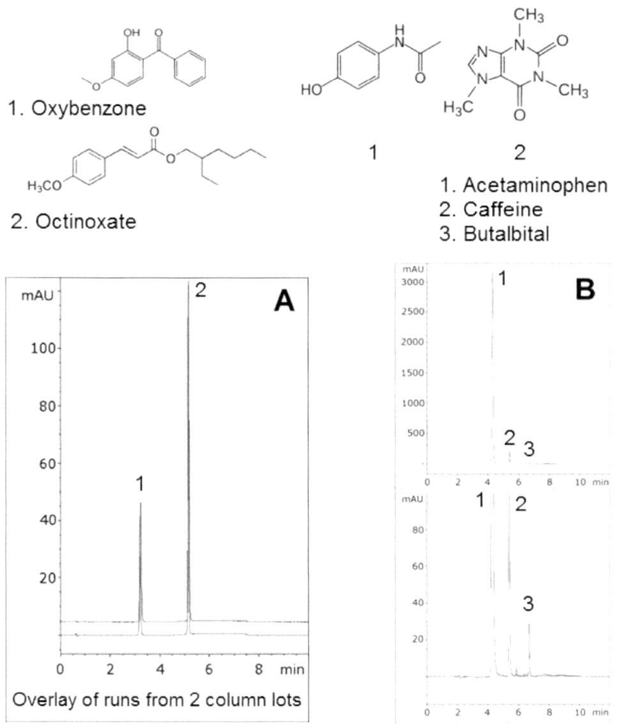

Figure 8.3 Examples of reversed-phase separations on a C18 modified silica hydride stationary phase using an acetonitrile/water mobile phase. (A) Sunscreen components oxybenzone (1) and octinoxate and (B) pharmaceutical formulation containing acetaminophen (1), caffeine (2), and butalbital (3).

A variety of applications have been reported on silica hydride stationary phases that are primarily focused on the analysis of hydrophobic compounds. One example relevant to forensic science is the analysis of drugs of abuse [18]. Two types of silica hydride columns were tested: C18 and phenyl. Both stationary phases were successful in separating 17 common drugs of in less than 9 min. However, selectivity for the two columns was different, so the proper choice for a particular analysis would depend on the target compounds. The same combination of columns was used to analyze the phenolic components in mesquite flour again showing that both were adequate but had different selectivity [19]. The hydride-based phenyl phase has also proved useful in the analysis of mycotoxins [20]. Bean, rice, maize, and wheat samples were stored under conditions conducive to fungal development and the protocol developed could detect a variety of mycotoxins in the matrices. The phenyl column has also been used to analyze for the phenolic composition in pomegranate [21]. Another interesting application is the determination of amoxicillin using a USP method on a silica hydride C18 column [22]. The results showed both improved reproducibility and peak shape using this column format. Food analysis is becoming an increasingly important field for HPLC that can be used to determine quality, nutrients and authentication for certain high cost or rare products. One example is the analysis of capsaicinoids in various hot sauces which are responsible for the spicy flavor in this product as well as in peppers [23]. The determination was made using both 2 μm and 4 μm C18 silica hydride columns. Another example of food analysis using the same type of C18 column was the determination of trans-resveratrol in wine [24]. The silica hydride column performed better than a standard silica C18 column for the separation of the target analyte from matrix components so that good quantitation in real samples was obtained. The silica hydride-based C8 column was used for the analysis of bisphenol A (BPA) in paper products such as carbon paper and store receipts [25]. BPA can be transferred by contact with the surface of these materials and has been shown to be responsible for a number of physiological problems. The protocol reported that included a sample preparation procedure resulted in good retention and excellent reproducibility so that both identification and quantification of BPA can be done utilizing UV detection. An interesting application in the RP mode has been reported using a capillary column format. The analysis of nonsteroidal anti-inflammatory drugs and steroids was reported using 50 and 100 μm columns packed with the C18 silica hydride material [26]. A fast analysis (<3 min) having excellent resolution for multicomponent mixtures was developed for a broad range of these compounds.

One additional aspect of the RP uses for the silica hydride materials is for preparative applications. One such example of a viable format is one that was developed for the preparation of carotenoids [27]. Carotenoids have been used as dietary supplements and have been purported to have significant health benefits. Another potential expansion of silica hydride phases would be for the separation of chiral compounds. A preliminary study utilizing a vancomycin stationary phase proved that enantiomeric separation formats can be developed using silica hydride as the support material [28]. Future investigations could focus on different types of chiral selectors to be bonded to the silica hydride surface.

8.3.2 Use of Aqueous Normal Phase for Polar Compounds

Because there are often limited options for the analysis of polar compounds or those that are available such as HILIC have some substantial drawbacks, ANP utilizing silica hydride-based stationary phases is becoming a more common technique [29]. HILIC, for example, is often limited by reproducibility of retention times on both an intraday and interday basis, column lifetime (robustness of the stationary phase), long equilibration times for gradient methods, and the need for high concentrations of additives to the mobile phase that make MS detection more difficult. ANP on silica hydride materials is virtually free of all of these chromatographic problems.

A dramatic example of the difference between ordinary silica-based materials and silica hydride is shown in Fig. 8.4. The compound analyzed in the chromatogram is methotrexate, an extremely polar drug. The stationary phase used in this analysis is the silica hydride C18. Thus, even though the amount of stationary phase on the surface is similar to those phases produced through organosilanization, it still retains the capability to operate in the normal-phase mode. This is a unique feature of the silica hydride columns. All phases can operate in either the RP or ANP modes, and in this case even if the surface is modified with a very hydrophobic moiety, it still retains normal-phase retention capabilities. The retention for this compound is quite strong since a very steep gradient that ends at 20% water (strong solvent) is need to elute the compound in a reasonable period of time. Under these experimental conditions, peak shape is excellent and quantitative measurements, if needed, could easily be done. A similar result on a stationary phase made on ordinary silica using organosilane chemistry modification would result in little or no retention of this compound. Another highly polar drug, tobramycin, has similar behavior on hydrophobic silica hydride columns, even when using methanol as the weak solvent.

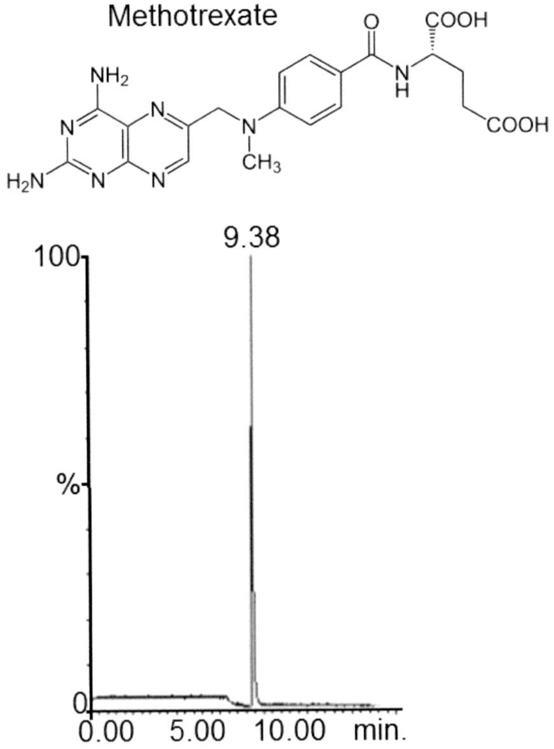

Figure 8.4 Retention of the polar compound methotrexate in the aqueous normal phase mode on a C18 modified silica hydride column.

While these are two interesting examples and illustrate the unique behavior of the silica hydride phases, most analyses of polar compounds utilize other phases with less hydrophobic modifications, or in some cases very little or no modification at all. The charged nature of the silica hydride surface as determined by the zeta potential measurements cited above provide the properties necessary for the retention of hydrophilic compounds. The first significant report on the use of a silica hydride phase in ANP focused on metabolites such as amino acids, organic acids, and carbohydrates [30]. The stationary phase used is a material referred to as the diamond hydride (DH); essentially a silica hydride surface with only about 2% bonded carbon. Mass spectrometry (MS) was used as the means of detection since most clinical and pharmaceutical analyses for these compounds by HPLC are increasingly done by MS. The advantage of MS detection is that only isobaric compounds need to be separated in order to determine the target analytes in the method. The requirements for separation can be further reduced by using accurate mass time-of-flight (TOF) or tandem mass (MS/MS) analyzers. Another study using silica hydride columns reported on the analysis of both hydrophobic and hydrophilic metabolites in serum [31]. An important finding was that ion suppression occurred in a smaller range of k values on the silica hydride columns than for traditional columns whether the analysis was done in a single column or tandem column format. Other metabolite studies have included the monitoring of pathogens in cells [32] as well as the effect of immunosuppressant drugs on cell pathogens [33]. It was also possible to discern difference in metabolites found in the urine and saliva of normal individuals and cancer patients [34]. Metabolic profiling has also been used in the investigation of biomarkers for bladder cancer [35]. Another interesting metabolic study showed that elevated levels of hydrophilic sulfur amino acids could be correlated with an increased risk cardiovascular disease [36]. The ANP with MS/MS detection has excellent reproducibility and quantitation in blood plasma. A sample calibration curve (homocysteine) is shown in Fig. 8.5. An ANP method using the DH column was used to identify aspartate as the source of nitrogen for tuberculosis bacillus [37]. Metabolites associated with malaria have also been studied using an ANP protocol [38]. The DH column was also shown to be useful for global untargeted as well as targeted metabolomic analysis [39]. Microbiota were fingerprinted according to their small molecule constituents using an ANP format. The role of succinate in the metabolism of *Mycobacterium tuberculosis* was determined using an ANP method on the DH column [40]. With this information, a subsequent investigation determined that para-aminosalicylic acid could be used for an effective treatment of tuberculosis [41]. An important class of metabolites that often are difficult to analyze by HILIC and other methods are nucleotides. Both the DH and a weak cation phase, UDA (based on undecanoic acid) have shown to be effective for the retention and separation of these compounds in the ANP mode [42]. A comprehensive comparison of HILIC and ANP for metabolomics was made and the advantages of the silica hydride materials documented [43].

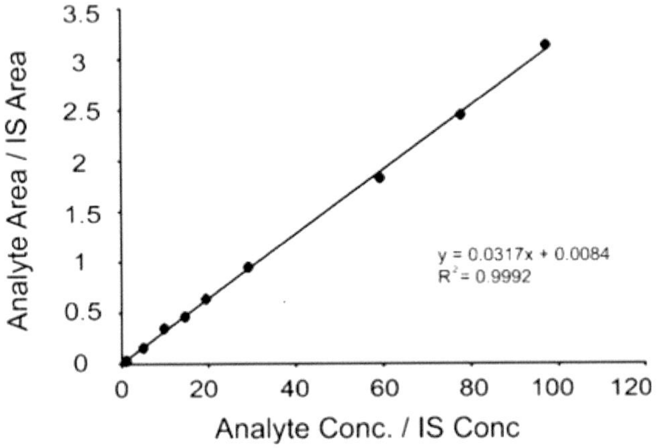

Figure 8.5 Calibration curve for homocysteine in blood plasma using the diamond hydride stationary phase in the aqueous normal phase mode. Reprinted with permission from ref. 36.

While ANP is now an established technology in metabolomics, it has also proven to be effective in a variety of other analyses involving hydrophilic compounds. For example, thiopurines that are not readily analyzed by RP methods have been shown to be potential biomarkers for some physiological conditions. These compounds are easily retained and effective separations can be obtained utilizing ANP formats [44]. Another potential clinical application involves the analysis of the two isobaric compounds symmetric and asymmetric dimethylarginine. These compounds have been associated with a number of renal and cardiac physiological conditions. Using exact mass TOF for detection a separation protocol was developed for their analysis since both compounds typically appear in any physiological sample [45]. Another approach is to use partial separation and a single quad MS with in-source fragmentation to identify and quantitate the two species. The actual separation of the two compounds (shown in Fig. 8.6) done under isocratic conditions has not been reported previously and was accomplished using an experimental silica hydride material with 2-acrylamido-2-methylpropane sulfonic acid (AMPS) bonded to the surface. Ethyl glucuronide (EtG) is used as an indicator of long-term alcohol abuse and clinical testing for it is common for public safety officers, airline pilots, and pregnant women. A simple viable format for EtG was developed using either the DH or UDA phases [46]. Another interesting application is the use of ANP technology for the analysis of various acidic lipids in brain and kidney extracts [47]. Another application in the area of medical analysis is the determination of PEG 400 that is used for intestinal permeability studies [48]. An evaporative light scattering detection protocol was developed due to the high salt content of the perfusate samples.

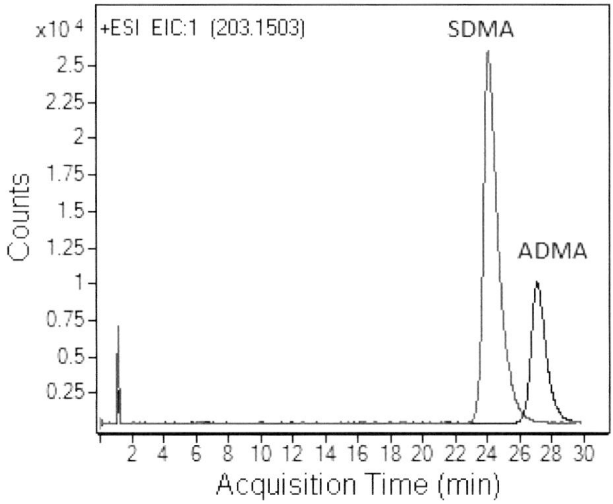

Figure 8.6 Separation of asymmetric and symmetric dimethylarginine on an experimental AMPS column in the aqueous normal phase mode. Reprinted with permission from ref. 45.

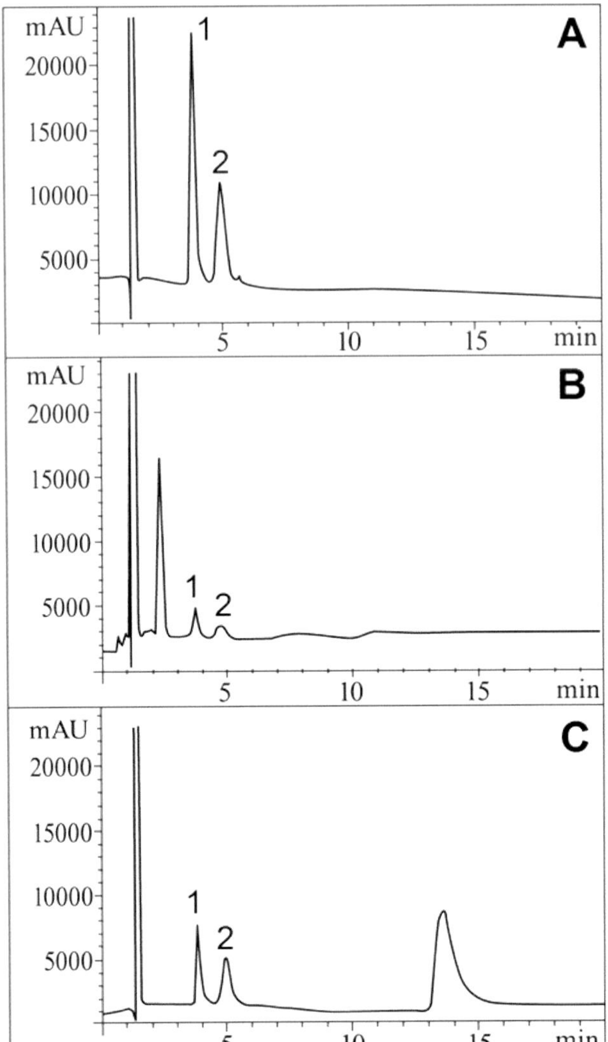

Figure 8.7 Analysis of fructose (1) and glucose (2) in various beverages using the Diamond Hydride column in the aqueous normal phase mode. (A) Cola, (B) red wine, and (C) grape juice. Reprinted with permission from ref. 51.

A growing area that is becoming more reliant on HPLC and LCMS methods is food and beverage analysis. Folic acid is found in many food products, including juices and cereals. A protocol including sample preparation and quantitation for these matrices has been reported using the DH stationary phase [49]. Histamine, a compound that can cause allergic reactions in many people, can be found in wine, tuna, and cheese. A DH format at levels resulting in physiological problems was designed for analyses in each of these matrices [50]. Sugar levels in many food products are an essential measurement for both quality and taste. A simple method for fructose and glucose that is also useful for other sugars using refractive index detection on a silica hydride amide column has been reported [51]. Some beverage analyses are shown in Fig. 8.7. Another type of sample where quality and regulation are increasingly important is dietary supplements. A number of investigations have been reported for these products which are generally available over the counter. The compound 1,3-dimethylamine, a component in dietary supplements, and a number of similar hydrophilic species have been analyzed using the DH column [52]. The substantial number of articles in the food and beverage area utilizing ANP formats were described in a review [53]. A silica hydride material was also used in a sample preparation mode in the analysis of hydrophilic components in dark chocolate [54].

Methods for the analysis of hydrophilic pharmaceutically and related compounds have been reported based on the use of silica hydride phases [55]. The powerful antibiotic, cycloserine, was analyzed on a DH column [56]. The parent compound as well as it degradation product and two impurities were successfully determined with MS detection. The drug zanamivir, used for treating influenza A and B, was analyzed in human serum using several different columns [57]. The DH provided the best retention and results under identical experimental conditions (Fig. 8.8). This investigation also documented the reproducible retention times for the drug on the DH vs. a HILIC column where t_R varied from run to run.

The versatility of the silica hydride phases is illustrated (Fig. 8.9) by another application where even the hydrophobic Zoloft with a log P value of 2.9 is retained under ANP conditions with a gradient going from 95% to 50%, acetonitrile, over 6 min [58]. This same effect is seen in the analysis of peptides on the DH column. A successful separation has been reported for a mixture containing both hydrophobic and hydrophilic peptides [59]. Therefore, it should be possible to analyze an enzymatic digest in a single run rather than the two-dimensional approaches currently in use. A potential clinical analysis of the peptide bradykinin in saliva has been demonstrated [60]. Bradykinin has been shown to reduce blood pressure and a simple analysis in saliva, an easily obtained physiological fluid, would make clinical determinations more expedient. A series of collagen crosslinks having a complex amino acid-like structure were analyzed on the DH column [61] (Fig. 8.10). These collagen crosslinks were analyzed in skin, bone, and tendons. The studies could be a precursor to understanding normal and diseased tissue in a variety of abnormalities including cancer as well as providing biomarkers for clinical analyses.

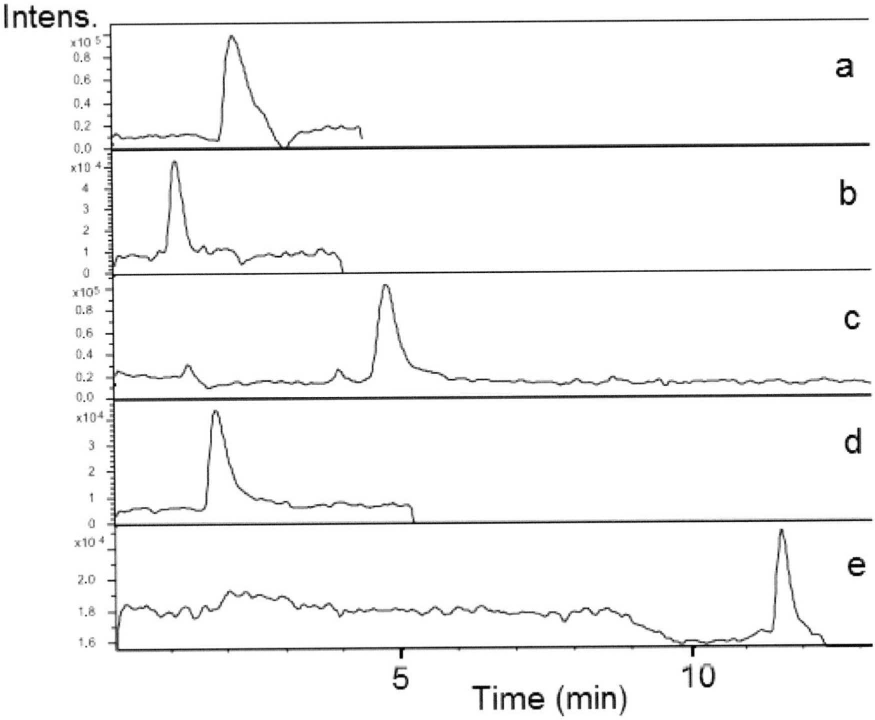

Figure 8.8 Retention of the drug zanamivir under optimal conditions on several commercially available HPLC columns. (a) Scherzo SM C18, (b) Agilent Zorbax SB-Aq, (c) Luna HILIC, (d) Cogent Bidentate C18 and (e) Cogent Diamond Hydride. Reprinted with permission from ref. 57.

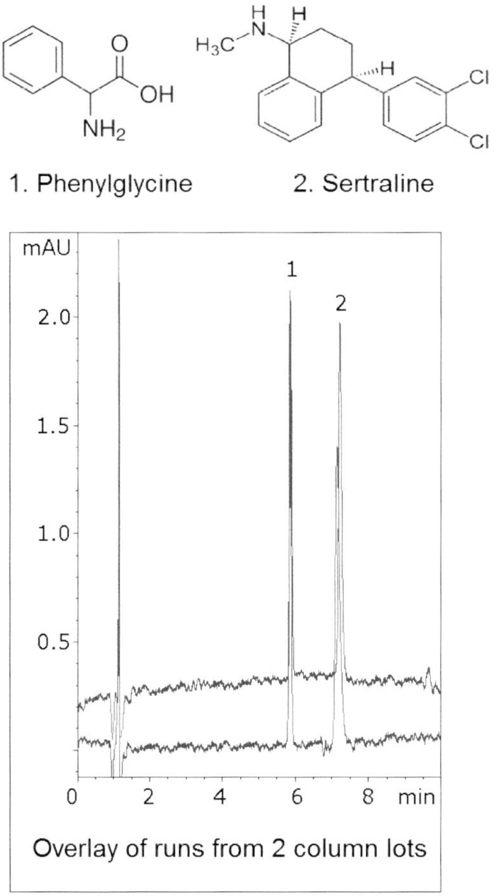

Figure 8.9 Simultaneous retention of a hydrophobic (sertraline) and a hydrophilic (phenylglycine) compound under aqueous normal chromatography conditions (linear gradient from 95% to 50% acetonitrile) on the diamond hydride column.

8.3.3 Organic Normal Phase on Silica Hydride

This aspect of silica hydride technology has not been extensively explored to date but does show promise to alleviate some of the problems associated with conventional organic normal-phase chromatography. The main problem in conventional organic normal-phase chromatography, which typically uses ordinary silica as the stationary phase, is associated with water. Because silica

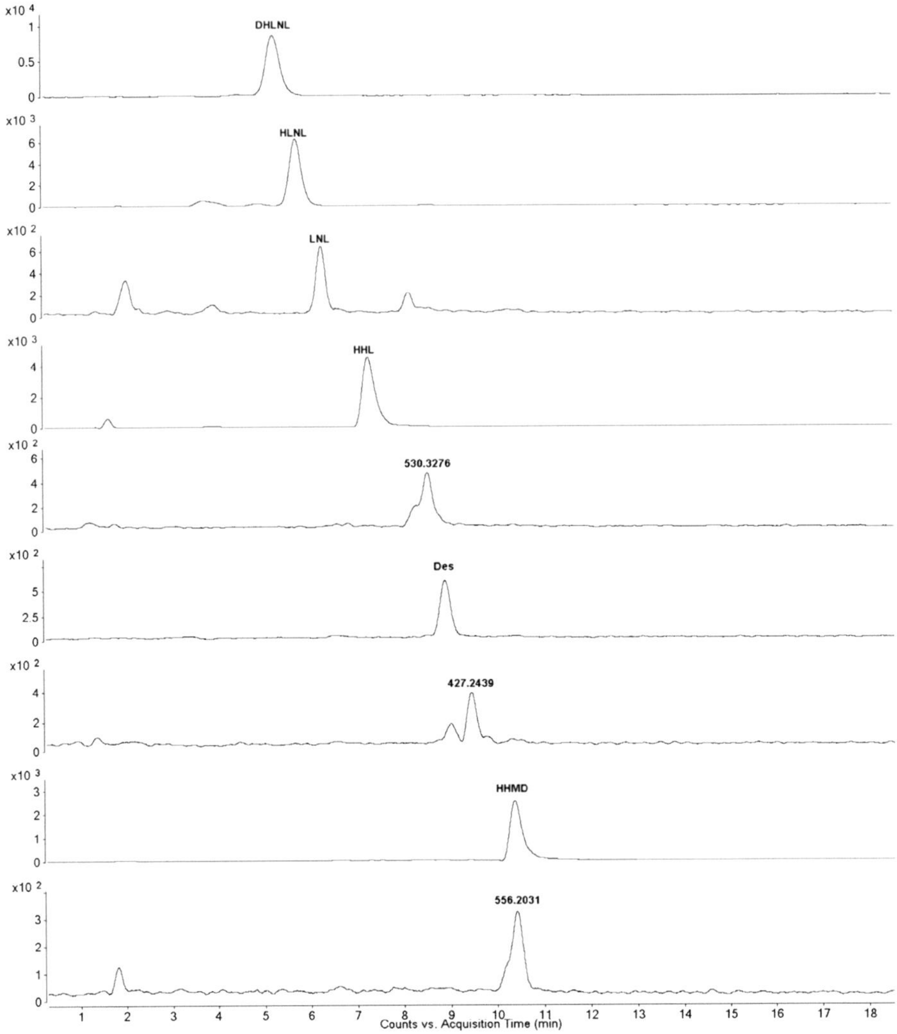

Figure 8.10 Extracted ion chromatograms of crosslinks from deer skin analyzed on the Diamond Hydride column under aqueous normal phase conditions. Solutes: DHNLH = dihydroxylsinonorleucine; HLNL = hydroxylsinonorleucine; LNL = lysinonorleucine; HHL = histidinohydroxylsinonorleucine; 530.3276 = unidentified crosslink; Des = desmosine; 427.2439 = unidentified crosslink; HHMD = histidinohydroxymerodesmosine; 556.2031 = unidentified crosslink. Reprinted with permission from ref. 61.

is so hydroscopic, it readily absorbs water from any source. The most common source is small traces of water in the mobile phase. With the amount of water on the surface of silica being variable due to the exact composition of the mobile phase, there often are observed variable retention times in both inter- and intra-day measurements. To somewhat overcome this problem, mobile phase solvents are generally rigorously dried to try to maintain a constant water content on the stationary phase. Because silica hydride is mildly hydroscopic, it has low affinity for water as demonstrated in the water layer determinations cited above. Therefore, the limited studies available do not utilize any drying of the mobile phase solvents and no significant variations in retention time on either a short- or long-term basis were observed.

There are two studies on the use of silica hydride-based stationary phases in organic normal phase. The significant feature in both of these studies is that excellent results were obtained with modified silica hydride surfaces. In one report, the pharmaceuticals loratadine and carvone are separated on a C18 silica hydride column using a gradient using hexane and dichloromethane as the organic solvents [62]. The gradient goes from 95% to 50% hexane in 6.5 min. A more extensive study was published that documented the retention of various substituted phenols on several silica hydride columns [10]. In this study, several different types of organic solvents were tested as well as modified and unmodified silica hydride materials. In all cases, it was possible to obtain organic normal-phase retention and none of the solvents were dried before use. Based on this data, it seems possible that this mode could be used for both analytical and preparative applications.

8.4 Method Development Strategies on Silica Hydride Phases

The utility of these columns is found in the wide range of solvents that can be used with them and the unique and strong retention they can have for polar compounds while at the same time retaining nonpolars. Selection of the column (C18, C8, cholesterol, or diamond hydride) will be determined by the compounds to be analyzed. Mixtures that will be highly polar and do not contain nonpolar compounds might be better suited to be analyzed by the diamond hydride, diol, or amide where mixtures with both polar and nonpolar compounds might be better suited to the C18, C8, cholesterol, or phenyl columns. The following sequence of steps can be used as a generic method development strategy to provide a viable analytical protocol for silica hydride stationary phases. Further modification of results of these steps can be done to formulate a validated method.

8.4.1 Mobile Phase Selection

The first step is to properly condition the column. This often depends on the type of sample that will be analyzed. For urine, plasma, or other biological samples, conditioning with 80% acetonitrile/DI water/0.2% acetic acid (or 0.1% formic acid) is recommended for 20 min. Cleaning with 50% MeOH/50% water is recommended for every 10 injections for 5 min. If standard samples are used, conditioning with 80% acetonitrile/DI water/0.2% acetic acid (or 0.1% formic acid) is recommended for 20 min. Subsequently cleaning with 50% MeOH/50% water is recommended every 100 injections or sooner for 5 min if a change in performance is noticed. After the column has been installed and conditioned, it is a good idea to start with a typical gradient run. This usually would involve an acidified mobile phase of water and acetonitrile. Acidify the mobile phase with up to 0.5% of an acid such as formic or acetic acid. If MS is not used as a detector, trifluoroacetic acid (TFA) is another good candidate as an additive.

8.4.2 Equilibration for Reversed-Phase

For columns that can operate in both modes or if only a reversed-phase method is being considered, then equilibration with a suitable mobile phase is necessary before beginning the method development process. Typically this would be a mobile phase of 95:5 water/organic solvent containing any additives (formic acid, acetic acid, TFA, etc.) to be used in the final method. The equilibration should utilize approximately six column volumes of mobile phase before the first sample injection.

8.4.3 Reversed-Phase Gradient

In this step, a shallow gradient is run for 95% water to 40% water over 20 min on a 75 mm or longer column. For longer columns, the gradient should be increased proportionately. This shallow gradient will provide an overall assessment of the type of compounds present in a mixture and can be used to optimize gradient or isocratic conditions for finalizing and analytical method or protocol. For example, if less retention or sharper peaks are required, then a steeper gradient from the same starting and final mobile phase compositions can be created.

8.4.4 Equilibrate for Aqueous Normal Phase

Generally, at the end of a reversed-phase run the mobile phase will already contain a high amount of acetonitrile. At this point, the acetonitrile content in the mobile phase should be increased to 100%. The pure acetonitrile should then be passed through the column for at least 2 min with a flow rate of between 0.5 and 1.0 mL/min.

8.4.5 Aqueous Normal-Phase Gradient

Using the same mobile phase utilized for the reversed-phase gradient, set up a gradient that runs from 90% to 40% acetonitrile over 20 min for a 75 mm column. For a longer column, adjust the gradient appropriately. As in the reversed-phase protocol, this type of gradient allows for the determination of either an optimum gradient or isocratic mobile phase composition. For sharper peaks and less retention, a steeper gradient encompassing the same starting and end points in mobile phase composition should be used.

8.4.6 Compare Reversed-phase and Aqueous Normal-Phase Data

The results from the two gradient runs should be compared to determine the appropriate approach for analysis of the sample tested. From these results it can concluded that either the reversed-phase or aqueous normal-phase approach will provide the desired separation required. In some cases, the comparison will demonstrate that an isocratic mobile phase can retain and separate both polar and nonpolar compounds.

8.5 Conclusions

Silica hydride-based separation materials are being recognized as an alternative to traditional silica stationary phases. Its unique properties provide chromatographic advantages that can be exploited for improved and more efficient separations. The low water content is one feature that has the broadest impact on differentiating silica hydride from ordinary silica. The resulting mildly hydrophobic surface surprisingly retains both polar and nonpolar compounds. The ability to retain polar compounds has been attributed to the strongly negative surface as determined by zeta potential measurements. This negative charge is thought to be similar to oil droplets in organo/aqueous solvents. In this case, hydroxides from the auto dissociation of water are responsible for the high negative zeta potential. The excellent analytical reproducibility of the silica hydride materials is also another distinguishing feature. The ANP capabilities of silica hydride phases have produced the majority of applications to date since they often have provided superior performance to HILIC. However, rugged and reproducible reversed-phase and organic normal phase analyses can also be done on silica hydride columns. It is expected that use of these materials will continue to expand as more scientists become aware of their unique properties.

References

1. MacNeill, R., and Taylor, P. (2015). The versatility in LC selectivity attainable with the silica base and associated bonded phases, *Bioanalysis*, **7**, pp. 637–642.
2. Borges, E. M. (2015). Silica, hybrid silica, hydride silica and non-silica stationary phases for liquid chromatography, *J. Chromatogr. Sci.*, **53**, pp. 580–597.
3. Sandoval, J. E., and Pesek, J. J. (1989). Synthesis and characterization of hydride modified porous silica material as an intermediate in

the preparation of chemically bonded chromatographic stationary phases, *Anal. Chem.*, **61**, pp. 2067–2975.

4. Sandoval, J. E., and Pesek, J. J. (1991). Hydrolytically stable bonded chromatographic stationary phases prepared through hydrosilation of olefins on a hydride-modified silica intermediate, *Anal. Chem.*, **63**, pp. 2634–2641.
5. Montes, M. C., van Amen, C., Pesek, J. J., and Sandoval, J. E. (1994). Chromatographic evaluation of alkyl-bonded phases through olefin hydrosilation on a hydride-silica intermediate, *J. Chromatogr.*, **688**, pp. 31–45.
6. Pesek, J. J., Matyska, M. T., Hemphala, H., Christensen, P. (1995). High performance liquid chromatographic characterization of diol bonded phases synthesized via a hydride intermediate, *J. Liq. Chromatogr.*, **18**, pp. 2507–2525.
7. Pesek, J. J., Boysen, R. I., Hearn, M. T. W., Matyska, M. T. (2014). Hydride-based stationary phases: A rapidly evolving technology for the development of new bio-analytical methods, *Anal. Methods*, **6**, pp. 4496–4503.
8. Brown, L., Ciccone, B., Pesek, J. J., Matyska, M. T. (2003). An evolution in separation media for HPLC, *Am. Lab.*, **35**, pp. 23–29.
9. Pesek, J. J., and Matyska, M. T. (2006). How to retain polar and nonpolar on the same column with an isocratic mobile phase, *LCGC*, **24**, pp. 296–303.
10. Pesek, J. J., Matyska, M. T., and Sharma, A. (2008). Use of hydride-based separation materials in organic normal phase chromatography, *J. Liq. Chromatogr. Rel. Technol.*, **31**, pp. 134–147.
11. Soukup, J., and Jandera, P. (2014). Adsorption of water from aqueous acetonitrile on silica-based stationary phases in aqueous normal phase liquid chromatography, *J. Chromatogr. A*, **1374**, pp. 102–111.
12. Kulsing, C., Yang, Y., Matyska, M. T., Pesek, J. J., Boysen, R. I., and Hearn, M. T. (2015). Prediction of the zeta potentials and ionic descriptors of a silica hydride stationary phase with mobile phases of different pH and ionic strength, *Anal. Chim. Acta*, **859**, pp. 79–86.
13. Beattie, J. K., and Djerdjev, A. M. (2004). The pristine oil/water interface: Surfactant-free hydroxide-charged emulsions, *Agnew. Chem., Int. Ed.*, **43**, pp. 3568–3571.
14. Zangi, R., and Engberts, J. B. (2005). Physisorption of hydroxide ions from aqueous solution to a hydrophobic surface, *J. Am. Chem. Soc.*, **127**, pp. 2272–2276.

15. Kudin, K. N., and Car, R. (2008). Why are water-hydrophobic surfaces charged?, *J. Am. Chem. Soc.*, **130**, pp. 3915–3919.

16. Kulsing, C., Nolvachi, Y., Marriott, P., Boysen, R., Matyska, M., Pesek, J. and Hearn, M. (2015). Insights into the origin of the separation selectivity with silica hydride adsorbents, *J. Phys. Chem., Part B*, **119**, pp. 3063–3069.

17. McCalley, D. V. (2010). Study of the selectivity, retention mechanisms and performance of alternative silica-based stationary phases for separation of ionised solutes in hydrophilic interaction chromatography, *J. Chromatogr. A*, **1217**, pp. 3408–3417.

18. Pesek, J. J., Matyska, M. T., and Kim, A (2013). Evaluation of silica hydride-based stationary phases for the analysis of drugs of abuse, *J. Sep. Sci.*, **36**, pp. 2760–2766.

19. Nguyen, T., Felker, P., Hurst, J. W., Ly, C., Jarman, S., Diep, D., Pham, C., Pesek, J. J., Matyska, M. T., Young, J, E., and Takeoka, G. (2016). LC-MS characterization of phenolics in mesquite flour, *LCGC North Am.*, **34**, pp. 28–31.

20. Pesek, J., Matyska, M., Hoffman, J. F., Madruga, N., Crizel, R. I., Elias, M. C., Vanier, N. L., and Chaves, F. C. (2017). Analysis of mycotoxins in grains using silica hydride-based stationary phases, *J. Sep. Sci.*, **40**, pp. 1953–1959.

21. Young, J. E., Pan, Z., Teh, H. E., Menon, V., Modegerer, B., Pesek, J. J., Matyska, M. T., and Takeoka, G. (2017). Phenolic composition of phenolic extracts using an LCMS approach with silica hydride columns, *J. Sep. Sci.*, **40**, pp. 1449–1456.

22. http://kb.mtc-usa.com/article/AA-00789/0/.

23. Pesek, J. J., Matyska, M. T., Sieng, M., and Doan, L. (2016). Analysis of capsaicinoids in hot sauces using a silica hydride-based stationary phase, *Curr. Chromatogr.*, **3**, pp. 12–16.

24. Young, J. E., Lim, M. V., Topete, J., Hang, H., Gahol, M., Pesek, J. J., and Matyska, M. T. (2016). Improved sensitivity and specificity of *trans*-resveratrol in red wine analysis with HPLC-UV and LC-MS, *LCGC North Am.*, **34**, pp. 205–213.

25. Dang, A., Sieng, M., Pesek, J. J., and Matyska, M. T. (2015). Determination of bisphenol A in receipts and carbon paper by HPLC-UV, *J. Liq. Chromatogr. Rel Technol.*, **38**, pp. 438–442.

26. D'Orazio, G., Rocco, A., and Fanali, S. (2012). Fast liquid chromatography using columns of different internal diameters packed with sub-2 μm silica particles, *J. Chromatogr. A*, **1228**, pp. 213–220.

27. Pesek, J. J., Matyska, M. T., and Lee, P. (2011). Synthesis of a preparative C30 stationary phase on a silica hydride surface and its application to carotenoid separation, *J. Liq. Chromatogr. Rel. Technol.*, **34**, pp. 231–240.

28. Rocchi, S., Rocco, A., Pesek, J. J., Matyska, M. T., Capitani, M., and Fanali, S. (2015). Use of a novel sub-2 μm vancomycin silica hydride stationary phase, *J. Chromatogr. A*, **1381**, pp. 149–159.

29. Pesek, J. J., Boysen, R. I., Hearn, M. T. W., and Matyksa, M. T. (2014). Hydride- based stationary phases: A rapidly evolving technology for the development of new bioanalytical methods, *Anal. Methods*, **6**, pp. 4496–4503.

30. Pesek, J. J., Matyska, M. T., Fischer, S. M., and Sana, T. R. (2008). Analysis of hydrophilic metabolites by high-performance liquid chromatography-mass spectrometry using a silica hydride-based stationary phase, *J. Chromatogr. A*, **1204**, pp. 48–55.

31. Chalkraft, K. R., and McCarry, B. E. (2013). Tandem LC columns for the simultaneous retention of polar and nonpolar molecules in comprehensive metabolomic analysis, *J. Sep. Sci.*, **36**, pp. 3478–3485.

32. Gouzy, A., and Larrouy-Maumus, G. (2014). *Mycobacterium tuberculosis* exploits asparagine to assimilate nitrogen and resist acid stress during infection, *PLOS Pathogens*, **10**, e1003928, pp. 1–14.

33. Jenkins, S., Fischer, S. M., Chen, L., and Sana, T. R. (2013). Global LC/MS metabolomics profiling of calcium stressed and immunosuppressant treated Saccharomyces cerevisiae, *Metabolites*, **3**, pp. 1102–1117.

34. Pesek, J. J., Matyska, M. T., Loo, J. A., Fischer, S. M., and Sana, T. R. (2009). Analysis of hydrophilic metabolites in physiological fluids by HPLC-MS using a silica hydride-based stationary phase, *J. Sep. Sci.*, **32**, pp. 2200–2208.

35. Putluria, N., Shojaiegn, A., and Vasu, V. T. (2011). Metabolic profiling reveals potential markers and bioprocesses altered in bladder cancer progression, *Cancer Res.*, **15**, pp. 7376–7386.

36. Hellmuth, C., Koletzko, B., and Peissner, W. (2011). Aqueous normal phase cysteine and methionine by liquid chromatography-tandem mass spectrometry, *J. Chromatogr. B*, **879**, pp. 83–89.

37. Gouzy, A., and Larrouy-Maumus, G. (2013). *Mycobacterium tuberculosis* nitrogen assimilation and host colonization require aspartate, *Nat. Chem. Biol.*, **9**, pp. 674–676.

38. Sana, T. R., Gordon, D. B., Fischer, S. M., Tichy, S. E., Kitagawa, N., Cai, C., Gosnell, W. L., and Chang, S. P. (2013). Global mass spectrometry based

metabolomics profiling of erythrocytes infected with *Plasmodium falciparum*, *PLUS ONE*, **8**, p. e60840.

39. Marcobal, A., Kashyap, P. C., Nelson, T. A., Aronov, P. A., Donia, M. S., Spormann, A., Fischbach, M. A., and Sonnenburg, J. L. (2013). A metabolomic view of how the human gut microbiota impacts the host metabolome using humanized and gnotobiotic mice, *ISME J.*, **7**, pp. 1933–1943.
40. Eoh, H., and Rhee, K. Y. (2013). Multifunctional essentiality of succinate metabolism in adaptation to hypoxia in *mycobacterium tuberculosis*, *Proc. Natl. Acad. Sci. U. S. A.*, **110**, pp. 6454–6459.
41. Chakraborty, S., Gruber, T., Barry III, C. E., Boshoff, H. I., and Rhee, K. Y. (2013). Para-aminosalicylic acid acts as an alternate substrate of folate metabolism in mycobacterium tuberculosis, *Science*, **339**, pp. 88–91.
42. Pesek, J. J., Matyska, M. T., Duley, J., Zamzami, M., and Fischer, S. M. (2010). Aqueous normal phase (ANP) retention of nucleotides on silica hydride-based columns. Method development strategies for analytes relevant to clinical analysis, *J. Sep. Sci.*, **33**, pp. 930–938.
43. Patti, G. (2011). Separation strategies for untargeted metabolomics, *J. Sep. Sci.*, **34**, pp. 3460–3469.
44. Pesek, J. J., Matyska, M. T., and Young, J. E. (2014). Analysis of thiopurines using aqueous normal phase chromatography, *J. Pharm. Biomed. Anal.*, **95**, pp. 102–106.
45. Pesek, J. J., Matyska, M. T., Modereger, B., Hasbun, A., Phan, V. T., Mehr, Z., Guzman, M., and Watanable, S. (2016). The separation and analysis of symmetric and asymmetric dimethylarginine and other isobaric compounds using aqueous normal phase chromatography, *J. Chromatogr. A*, **1441**, pp. 52–59.
46. Pesek, J. J., Matyska, M. T., and Dang, A. (2015). Analysis of ethyl glucuronide and ethyl sulfate using aqueous normal phase chromatography/mass spectrometry, *J. Sep. Sci.*, **38**, pp. 1515–1520.
47. Cifkova, E., Hajek, R., Lisa, M., and Holcapek, M. (2016). Hydrophilic liquid interaction chromatography-mass spectrometry of (lyso)-phosphaditic acids, (lyso)phophaditylserines and other lipid classes, *J. Chromatogr. A*, **1450**, pp. 65–73.
48. Webster, G. K., Elliott, A., Dahan, A., and Miller, J. M. (2011). Analysis of PEG 400 in perfusate samples aqueous normal phase (ANP) chromatography with Evaporative light scattering detection, *Anal. Methods*, **1218**, pp. 742–744.

49. Young, J. E., Matyska, M. T., and Pesek, J. J. (2011). Liquid chromatography/ mass spectrometry compatible approaches for the quantitation of folic acid in fortified juices and cereals using aqueous normal phase conditions, *J. Chromatogr. A*, **1218**, pp. 2121–2126.

50. Dang, A., Matyska, M. T., and Pesek, J. J. (2013). The use of aqueous normal phase chromatography as an analytical tool for food analysis. Determination of histamine as a model system, *Food Chem.*, **141**, pp. 4226–4230.

51. Young, J., Matyska, M. T., and Pesek, J. J. (2016). Robust HPLC-refractive index analysis of simple sugars in beverages using silica hydride columns, *Curr. Nutr. Food Sci.*, **12**, pp. 125–131.

52. Le, R., Young, J. E., Pesek, J. J., and Matyska, M. T. (2013). Separation of 1,3- Dimethylamine and other polar compounds in dietary supplement formulations using aqueous normal phase chromatography with mass spectrometry, *J. Sep. Sci.*, **36**, pp. 2578–2583.

53. Young, J. (2016). Advances in chromatographic analysis of foods and beverages: Modern stationary phases for challenging compounds, *Agro FOOD Ind. Hi Tech.*, **27**, pp. 14–17.

54. Nolvachai, Y., Kulsing, C., Boysen, R. I., Matyska, M. T., Pesek, J. J., Marriott, P. J., and Hearn, M. T. W. (2015). Comparison of the performance of different silica hydride particles for the solid-phase extraction of non-volatile analytes from dark chocolate with analysis by gas chromatography-quadrupole mass spectrometry, *Food Chem.*, **174**, pp. 434–439.

55. Pesek, J. J., Matyska, M. T., and Young, J. E. (2012). How silica hydride HPLC columns can benefit the modern pharmaceutical laboratory, *Sep. Sci.*, **4**, pp. 11–15.

56. Pesek, J. J., Matyska, M. T., and Dang, A. (2012). Analysis of cycloserine and related compounds using aqueous normal phase chromatography/mass spectrometry, *J. Pharm. Biomed. Anal.*, **64–65**, pp. 72–76.

57. Gea, J., Liu, F., Holmes, E. H., Ostrander, G. K., and Li, Q. X. (2012). Aqueous normal phase liquid chromatography coupled with tandem time-of-flight quadrupole mass spectrometry for determination of zanamivir in human serum, *J. Chromatogr. B*, **906**, pp. 58–62.

58. http://kb.mtc-usa.com/article/AA-01460/90/.

59. Boysen, R. I., Yang, Y., Chowdhury, J., Matyska, M. T., Pesek, J. J., and Hearn, M. T. W. (2011). Simultaneous separation of hydrophobic

and hydrophilic peptides with a silica hydride stationary phase using aqueous normal phase conditions, *J. Chromatogr. A*, **1218**, pp. 8021–8026.

60. http://kb.mtc-usa.com/article/AA-01232/0/.
61. Naffa, R., Holmes, G., Meekyung, A., Harding, D., and Norris, G. (2016). Liquid chromatography electrospray ionization mass spectrometry for the simultaneous quantitation of collagen and elastin crosslinks, *J. Chromatogr. A*, **1478**, pp. 60–67.
62. http://kb.mtc-usa.com/article/AA-00800/0/.

Chapter 9

Ion Chromatography Columns

Christopher Pohl
Chemistry R&D, Thermo Fisher Scientific,
1228 Titan Way, Sunnyvale, California 94085, USA
Chris.Pohl@thermofisher.com

This chapter provides an overview of ion chromatography columns including the early history of ion chromatography columns and the state-of-the-art concerning the architecture of ion chromatography columns.

9.1 Introduction

Ion chromatography is a highly specialized branch of liquid chromatography, generally making use of columns specifically designed for ion chromatography and intended to be used with ion chromatography instrumentation. In general, ion chromatography is performed with a specialized pumping system designed for the

Chromatographic Method Development
Edited by Gregory K. Webster and Laila Kott

ISBN 978-981-4800-53-2 (Hardcover), 978-0-429-20172-1 (eBook)
www.jennystanford.com

delivery of highly acidic and highly basic mobile phases without risk of corrosion as a consequence of the extreme pH associated with such mobile phases. Also, although there are notable exceptions, ion chromatography is commonly practiced while making use of a conductivity detector. Because a conductivity detector is a bulk property detector [1], the conductivity signal arising from conductive mobile phases is highly susceptible to environmental effects. For that reason, ion chromatography is usually practiced in a temperature-controlled environment to minimize thermal effects on the conductivity signal. In addition, the signal-to-noise ratio associated with detection of ionic species against a background conductivity of the mobile phase is relatively poor given the fact that the concentration of the mobile phase is typically much greater than the concentration of analytes being detected in the presence of the mobile phase conductivity. For that reason, most ion chromatography is practiced making use of a suppressor device [2, 3]. The suppressor device is unique among analytical techniques in that it reduces or eliminates the mobile phase conductivity while simultaneously enhancing the conductivity of the analyte ion. While the operation of the suppressor device is beyond the scope of this chapter, it is relevant to the design and architecture of ion chromatography columns. Because the best conductivity detector performance is achieved when a strong acid mobile phase is used for the analysis of cations and an alkali metal hydroxide mobile phase is used for the analysis of anions, this necessitates the use of stationary phase materials compatible with long-term exposure to mobile phases at extreme pH. Thus, while silica-based materials are the mainstay of HPLC, the poor hydrolytic stability of silica at high pH and the poor ligand stability at low pH render silica-based stationary phases a poor choice for ion chromatography. In contrast, polymeric stationary phases, especially polymers based on styrenic monomers are exceptionally stable at high pH and low pH, making columns derived from such monomers an excellent choice for use in ion chromatography. For that reason, the bulk of the discussion in this chapter will focus on polymeric stationary phases suitable for use in ion chromatography.

9.2 Cation-Exchange Phases

As previously mentioned, the earliest phases in ion chromatography were based on cross-linked polystyrene since they provided the necessary chemical stability to withstand long-term exposure to mineral acids. While there were examples of early silica-based cation-exchange materials, these tended to be used with non-suppressed ion chromatography given the relatively poor chemical stability of silica-based materials when exposed to extremes of pH. Although detection limits are 10–50 times worse when operating in non-suppressed mode [3] than what can be achieved using a suppressor, operation in non-suppressed mode enables the use of mobile phases with more moderate pH conditions that are not as corrosive to silica. Even so, silica-based phases tend to be so unstable that even acidic samples can damage the stationary phase. As a consequence, use of silica-based stationary phases became increasingly uncommon as newer high-performance polymeric phases came on the scene.

9.2.1 Surface-Sulfonated Cation-Exchange Phases

The earliest cation-exchange phases in ion chromatography [2] were composed of 2% cross-link polystyrene particles that were lightly surface sulfonated to introduce an extremely thin layer of sulfonated sites on the exterior surface. Use of sulfuric acid as the sulfonation reagent allows for a shell progressive reaction as the reagent proceeds to sulfonate the particle in a highly controllable manner. By adjusting the reaction time and reaction temperature, a wide variety of strongly acidic cation-exchange materials with a wide range of cation-exchange capacities suitable for use in ion chromatography can be prepared.

The 2% cross-link used in early materials was optimized for monovalent cation selectivity. Higher cross-link levels reduced the resolution of ammonia from potassium. However, such low cross-link levels rendered the substrate particle highly susceptible to excessive swelling in the presence of organic solvent. Because excessive swelling can cause column bed disturbances during swelling and subsequent shrinking back to the original dimensions when the solvent is removed from the mobile phase, such materials were restricted to use with largely aqueous mobile phases. At first glance, this might not seem like much of a problem, but the addition of solvent is often required to remove highly retained analytes from the stationary phase. Restrictions on the amount of solvent made it difficult to remove such contaminants from the stationary phase and reduced the long-term lifetime of such columns.

One disadvantage of using sulfonated cation-exchange materials for ion chromatography arises from the high-affinity such materials have for divalent cations. The concentration of hydronium necessary for elution of calcium or magnesium exceeded the practical mobile phase concentration range suitable for use with suppressor devices. As a result, the earliest suppressed ion chromatography methods utilized different eluent systems for monovalent cations than for divalent cations. Hydronium-based eluent systems were used for the elution of monovalent cations while acidic solutions of meta-phenylenediamine were used for the elution of divalent cations.

While hydronium-based mobile phases were not problematic for long-term use, meta-phenylenediamine-based eluent systems were problematic both because they tended to degrade as used in most common applications and because this eluent species was difficult to remove from the cation-exchange phase after initial use. Meta-phenylenediamine tends to undergo oxidation reactions, producing polymeric byproducts difficult to remove from the stationary phase.

9.2.2 Sulfonated Latex-Based Cation-Exchange Phases

The first commercial introduction of latex-based cation-exchange phases took place in 1986 more than 10 years after the first commercial introduction of ion chromatography. While anion-exchange phases making use of an anion-exchange latex coating on nonporous surface-sulfonated substrate particles had been in use for more than a decade (see Section 9.3.1), preparation of the cation-exchange variant arrived much later due to the difficulty of synthesizing colloidally stable, fully sulfonated cation-exchange latex particles. The first column of this type, the

IonPac CS3 column was based on a three-layer electrostatically bonded pellicular architecture (see Fig. 9.1 below).

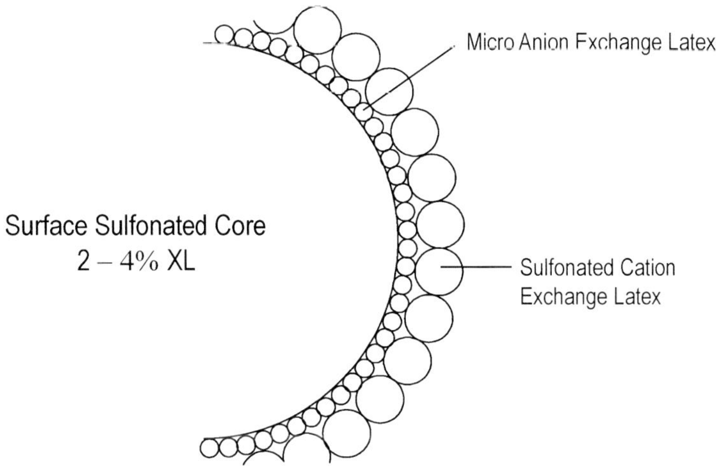

Figure 9.1 Schematic representation of the CS3 stationary phase.

The construction of CS3 chromatographic material starts with 2% cross-link polystyrene particles; the substrate is surface sulfonated as described in Section 9.2.1. Sulfonation conditions are adjusted to produce a very thin sulfonation layer to minimize any potential cation-exchange retention after assembly of the pellicular structure. Then, a layer of colloidal anion-exchange particles was deposited on the surface-sulfonated substrate. As a consequence of the minimal sulfonation of the substrate, application of the anion-exchange particles eliminated all traces of cation-exchange properties from the surface-sulfonated layer as a consequence of the intercalation of the anion-exchange particles into the surface-sulfonated cation-exchange layer at the particle surface. Finally, a layer of fully sulfonated cation-exchange latex was attached to the anion-exchange latex coating to produce the final product. As with the previous coating step, only traces of anion-exchange retention were detectable after application of the cation-exchange latex coating.

Cation-exchange stationary phases based on sulfonated latex provided significant improvements in chromatographic performance compared to the surface-sulfonated materials. Surface-sulfonated materials result in a cross-link gradient from the outermost surface of the particle to the sulfonation boundary adjacent to the edge of the sulfonated substrate particle which comprises the majority the particle volume. Because swelling of the sulfonated particle is constrained by the rigid core at the boundary between the sulfonated zone in the un-sulfonated zone of the particle, this effectively increases the cross-link at the boundary between the two phases. In contrast, at the exterior surface where swelling is relatively unconstrained by the influence of the un-sulfonated substrate, the effective cross-link is close to what one would expect for the homogeneously swollen sulfonated polymer. Evidence of this cross-link gradient can be seen in two different ways. First, when using latex particles which feature cross-link homogeneously distributed throughout the particle, 5% cross-link is necessary to center ammonia between sodium and potassium while in the case of surface-sulfonated substrate, 2% cross-link is optimal for providing the same selectivity. Second, chromatographic performance for relatively poorly hydrated cations is considerably better with the latex-based stationary phases. Because the boundary between the sulfonated zone on the un-sulfonated zone is relatively poorly hydrated, this tends to be an area where weakly hydrated cations exhibit the highest affinity. However, because this poorly hydrated boundary zone is the farthest away from the particle surface, mass transport to and from this region is relatively slow, resulting in relatively poor peak shape for poorly hydrated cations.

A second innovation coincided with the commercial introduction of cation-exchange materials utilizing cation-exchange latex coatings: an improved eluent system, 2,3-diaminopropionic acid. In contrast to the earlier HCl/HCl-meta-phenylenediamine eluent systems which required separate dedicated columns, one column for monovalent cations and a second column for divalent cations, 2,3-diaminopropionic acid is relatively easy to elute from a cation-exchange latex making it straightforward to switch from the analysis of monovalent cations to the analysis of divalent cations. The selectivity of this new eluent system was not optimal for separation of common monovalent cations (lithium, sodium, ammonium, and potassium) under conditions suitable for elution of common divalent cations (magnesium and calcium), 2,3-diaminopropionic acid. The best chromatography for monovalent cations was still seen with simple mineral acid eluent system.

The chief advantage of this new eluent system is that it made it easy to switch back and forth from a mineral acid eluent optimal for monovalent cations to the HCl and 2,3-diaminopropionic acid mixture suitable for elution of divalent cations.

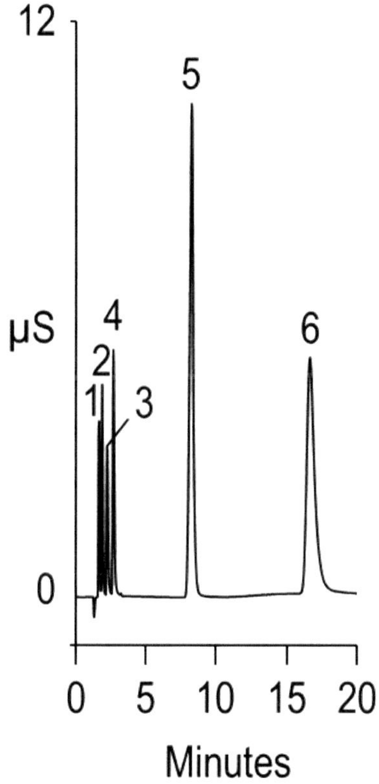

Figure 9.2 Separation of cations on an IonPac CS11 column. Flow rate 0.25 mL per min, 40 mM hydrochloric acid with 4 mM 2,3-diaminopropionic acid, ambient temperature, suppressed conductivity detection, 2.5 µL injection volume. Peaks: (1) 0.5 ppm lithium; (2) 2 ppm sodium; (3) 2.5 ppm ammonium; (4) 5 ppm potassium; (5) 15 ppm magnesium; (6) 25 ppm calcium.

The third innovation related to this class of stationary phases was the introduction of latex particles composed the styrenic monomers with dramatically different levels of reactivity activation concerning sulfonation. By combining divinylbenzene with a styrenic monomer with reactivity activated to sulfonation and a second styrenic monomer with reactivity deactivated to sulfonation, it was possible to achieve greater spacing between the sulfonated groups within the sulfonated latex particle. Columns based on this chemistry include the IonPac CS10 and the IonPac CS11. Because retention of divalent ions is much more sensitive to ion-exchange site concentration than the retention of monovalent ions, this allows the manipulation of the selectivity of divalent species relative to monovalent species. As a consequence, for the first time it was possible to separate the common cations (lithium, sodium, ammonium, potassium, magnesium, and calcium) under isocratic conditions using an HCl–2,3-diaminopropionic acid eluent system (see Fig. 9.2 above).

In addition to these phases, there are also several companies (Wescan, Alltech, and Hamilton) making surface-sulfonated macroporous cation-exchange resins. For the most part, these columns were operated in nonsuppressed mode utilizing eluent systems such as HCl-ethylenediamine. Such materials provided relatively poor chromatographic performance due to the relatively high cross-link of macroporous substrate particles.

9.2.3 Polymer-Encapsulated Weak Acid Cation-Exchange Phases

In 1987, Prof. Schomburg and coworkers [4] developed an entirely new method of stationary phase attachment. Their work originally focused on developing new methods of attaching stationary phases to alumina. Given the fact that alumina is much more pH stable than silica, Prof. Schomburg and the team was working on the development of methods for attaching a stationary phase to alumina. Because alumina is not amenable to direct silane bonding, which enables facile modification of the surface chemistry of silica substrate, they devised a novel encapsulation methodology involving deposition of a thin polymer film which was subsequently cross-linked in situ. They chose copolymers based on butadiene since such polymers contain residual double bonds along the polymer backbone. Among the polymers they investigated was butadiene-maleic anhydride copolymer. They found that this polymer when fully hydrolyzed and subsequently cross-linked in situ was useful for the separation of both monovalent and divalent cations under isocratic conditions using a relatively simple mobile phase. By using a weak acid cation-exchange site rather than a strong acid cation-exchange site, the concentration of cation-exchange sites and the corresponding monovalent-divalent selectivity can be manipulated by adjusting the mobile phase pH. As they discovered, alumina was not a very good substrate for the separation of inorganic cations due to the underlying selectivity of alumina for inorganic cations. For that reason, they ultimately ended up using silica as the substrate for the butadiene-maleic anhydride copolymer. By combining a silica substrate with suitable mobile phase pH, they were able to achieve the best separations ever reported for the common inorganic cations under isocratic conditions.

This development led to the wholesale shift of stationary phase design to the use of weak acid cation-exchange phases for ion chromatography. Several manufacturers, including Metrohm, Shodex, Tosoh, and Dionex, developed silica-based analogs of the phase described by Prof. Schomburg. However, there are significant disadvantages of this synthetic approach when combined with a suppressor. First, the underlying silica has a limited pH stability which constrained the pH of samples that can be utilized with such materials. Second, the in situ surface cross-linking methodology developed by Prof. Schomburg exhibits a significant amount of "bleed" arising from non-covalently attached polymer fragments trapped within the cross-linked matrix. These fragments are captured by the suppressor, slowly converting the suppressor to a cation-exchange column, causing poor recovery and bad peak shapes as polymer fragments accumulate on the suppressor. For this reason, such materials are only recommended for use in non-suppressed mode.

9.2.4 Polymer-Grafted Weak Acid Cation-Exchange Phases

As a consequence of the polymer bleed problem associated with encapsulated weak acid cation-exchange phases, Dionex embarked upon a development process with the aim of replicating the selectivity of Prof. Schomburg's column while circumventing the problems associated with the substrate and the attachment method. Ultimately, they developed the first polymer-based weak acid cation-exchange resin (the IonPac CS12 column) [5]. The CS12 column utilizes radical polymerization to covalently graft weak acid cation-exchange polymers to the surface of a macroporous substrate particle. Only a small fraction of the cation-exchange sites in such a material is ionized at any one time, necessitating a relatively high capacity to achieve sufficient retention under suitable eluent conditions. A high-surface-area macroporous polymer allows the preparation of material of high capacity while still providing relatively good mass transport kinetics. Later, improvements in the grafting chemistry and inclusion of a vinyl phosphonate comonomer were used to prepare the IonPac CS12A column exhibiting both improved peak shape for common inorganic cations as well as better selectivity for magnesium, manganese, and calcium (Fig. 9.3). The IonPac CS12A column is still the world's best selling cation-exchange ion chromatography column, more than 20 years since its initial introduction in 1995. It is cited in a wide variety of publications [6–8] and used in many proprietary analytical methods throughout the world.

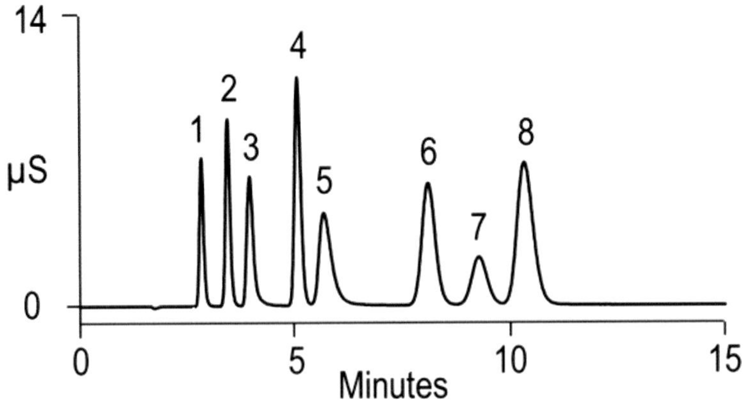

Figure 9.3 Separation of cations on a 4 mm ID IonPac CS12A column. Flow rate 1 mL per min, 20 mM methanesulfonic acid, ambient temperature, suppressed conductivity detection, 25 µL injection volume. Peaks: (1) 0.5 ppm lithium; (2) 2 ppm sodium; (3) 2.5 ppm ammonium; (4) 5 ppm potassium; (5) 10 ppm diethylamine; (6) 2.5 ppm magnesium; (7) 2.5 ppm manganese, (8) 10 ppm calcium.

With the initial success of such polymer-based weak acid cation-exchange phases, leading manufacturers throughout the world have introduced a variety analogous weak acid cation-exchange materials. A wide variety of such materials are commercially available today [9]. Many of these materials were specifically designed for use in nonsuppressed ion chromatography, but most of these materials can be used in suppressed ion chromatography with the exception those using the polymer encapsulation methodology of Prof. Schomburg.

9.2.5 Polymer-Grafted Strong and Weak Acid Cation-Exchange Phases

One disadvantage of using a weak acid cation-exchange phase under acidic conditions is that mass transport is adversely affected by the relatively poor hydration of carboxylic acids when largely in a protonated state. Poor hydration is due in part to the fact that the protonated carboxylic acid is significantly less hydrophilic than the salt form of the corresponding acid. The second factor that adversely affects the chromatographic performance of carboxylic acid phases under conditions where the phase is largely in a protonated state is the tendency of protonated carboxylic acid groups to form stable hydrogen bonded dimers with adjacent carboxylic acid groups. These hydrogen bonded dimers form in situ cross-links which further dehydrate the polymer network. The combination of these two factors results in relatively slow mass transport within the stationary phase under acidic conditions. One way to mitigate this problem is to copolymerize a carboxylic acid monomer with more acidic monomers under acidic conditions. An example of such a column is the IonPac CS20 column, shown in Fig. 9.4. It uses a terpolymer composition consisting of styrene sulfonate, vinyl phosphonate, and a carboxylic acid monomer. By polymerizing the three monomers under acidic conditions such that the carboxylic acid monomer is fully protonated, electrostatic forces substantially reduce the probability of two of the more acidic monomers ending up as adjacent monomers in the polymer network. Such an approach greatly reduces the affinity of such polymers for divalent species, eliminating the need for a divalent eluent cation to elute alkaline earth cations. Such a stationary phase essentially bridges the gap between older style sulfonated cation-exchange materials which have excellent monovalent cation selectivity but exhibit excessive retention for divalent cations and weak acid cation-exchange phases which have relatively poor monovalent cation selectivity but are easily eluted with hydronium.

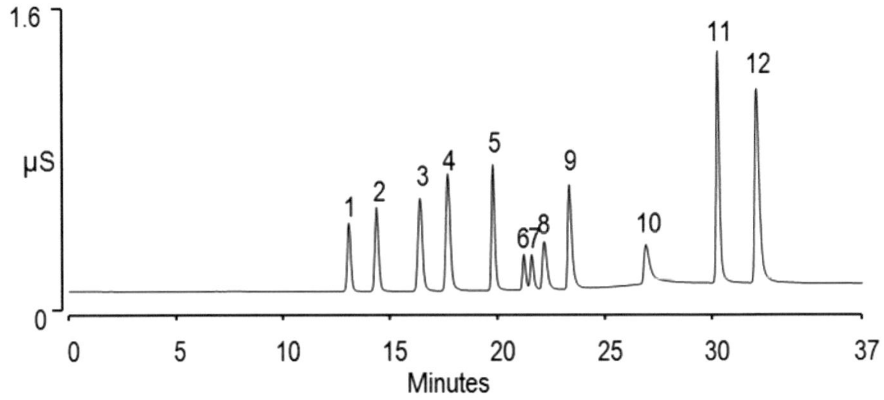

Figure 9.4 Separation of cations on 4 mm ID IonPac CS20 and CG20 columns. Flow rate 1.2 mL per min, 2–4 mM methanesulfonic acid (MSA) from 0 to 15 min, 4–15 mM methanesulfonic acid (MSA) from 15 to 20 min, 15–50 mM methanesulfonic acid (MSA) from 20 to 25 min, 50 mM methanesulfonic acid (MSA) from 25 to 37 min, 50°C, suppressed conductivity detection, 10 µL injection volume. Peaks: (1) 0.25 ppm lithium; (2) 1 ppm sodium; (3) 1.25 ppm ammonium; (4) 2.5 ppm potassium; (5) 5 ppm dimethylamine; (6) 5 ppm trimethylamine; (7) ethylmethylamine; (8) 5 ppm morpholine; (9) 5 ppm diethylamine; (10) 5 ppm piperidine; (11) 1.25 ppm magnesium; (12) 2.5 ppm calcium.

9.2.6 Cation-Exchange Phases for Transition Metal Separations

Although suppressed ion chromatography is commonly applied to the analysis of alkali metals, alkaline earth metals, ammonia and amines, suppressed ion chromatography is not compatible with transition metal detection. Application of suppressed ion chromatography to cations involves the use of a suppressor in the hydroxide form. Hydroxide from the suppressor replaces the strongly acidic anionic component of the mobile phase and subsequently reacts with the hydronium counterion to produce water. Also, the suppression reaction increases the conductivity of the cation by converting it to its highly conductive hydroxide form. Although this approach works for alkali metals and alkaline earth metals, the same approach does not work well for transition metals because the hydroxides of transition metals are only sparingly soluble in water. As a consequence, nearly all transition metal detection is accomplished either by post-column reaction with a chromophoric chelating agent or by use of a conventional spectroscopic detector such as ICP or ICP-MS. Given the fact that transition metal detection systems do not usually make use of a suppressor, stationary phase design doesn't focus on simple cation-exchange retention mechanisms. Selectivity of both weak acid cation-exchange materials and strong acid cation-exchange materials is poor for transition metals due to the relatively similar hydration properties of this class of metals.

Several approaches have been utilized to improve selectivity transition metals. One approach is to add a chelating species to the mobile phase that converts the metal to an anionic species. Cation-exchange retention will be reduced in proportion to the extent that the transition metal is in equilibrium with the anionic form of the metal complex. By choosing an appropriate anionic complexing agent, it is possible to manipulate the selectivity of the separation if the retention time for a specific pair of transition metals is identical when using a non-complexing mobile phase. The best results are achieved when using a complexing agent with widely differing formation constants with the transition metals of interest as this will produce the highest selectivity. If conditions are chosen such that all transition metals are 100% in the form of the anionic complex, there will be no retention via cation-exchange. Thus, such conditions should be avoided when operating in cation-exchange mode. An alternative is to use anion-exchange to separate the anionic complexes of transition metals. In this case, the extent to which the transition metal forms an anionic complex will enhance retention with maximum retention being achieved under conditions where the complex is essentially 100% complex with the anionic chelating agent. If multiple transition metals have similar formation constants with the anionic complexing agent, it is likely that the complexes will co-elute via anion-exchange although some transition metals will be separated under these conditions because different transition metals differ in their coordination numbers. However, the best option is to avoid conditions where the transition metals are 100% in the complexed form when using anion-exchange.

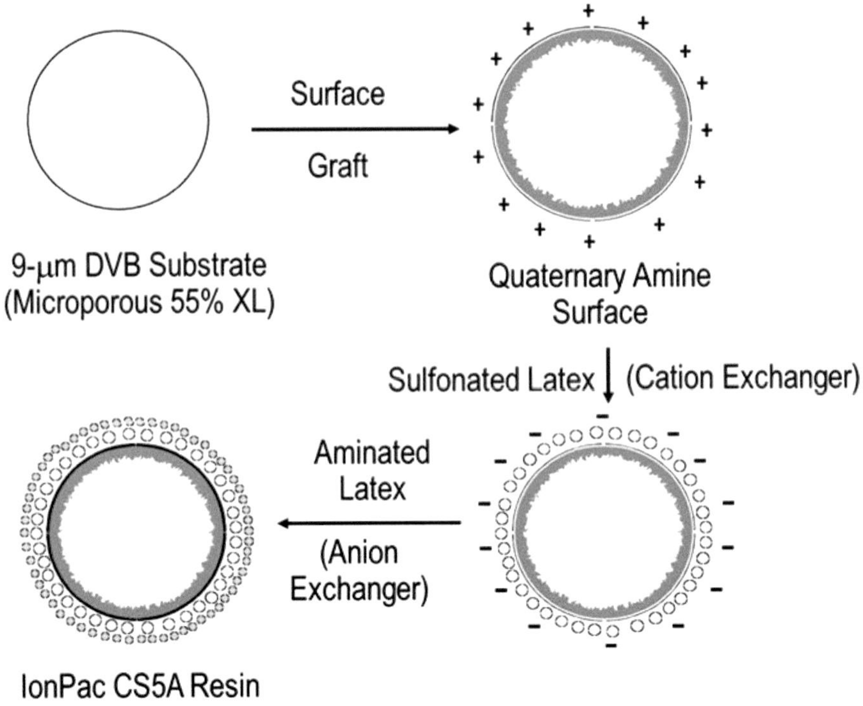

Figure 9.5 Schematic of CS5A resin synthesis and structure.

A more powerful option is to use a material exhibiting both anion-exchange and cation-exchange properties. Such a stationary phase allows one to use either cation-exchange mode, anion-exchange mode or a combination of the two modes. Operating with a dual-mode stationary phase allows one to avoid poor retention and poor selectivity associated with using anion-exchange or cation-exchange alone. The Dionex IonPac CS5A column was designed specifically to allow operation in dual ion-exchange mode. The phase architecture is illustrated in Fig. 9.5. The substrate particle is composed of 55% cross-link DVB-ethylvinylbenzene nonporous particles. An initial, extremely thin anion-exchange layer is produced by grafting the surface with a water-soluble cationic monomer. A 140 nm 10% cross-link strong acid cation-exchange latex coating is applied directly to the surface to introduce cation-exchange sites. Due to the minimal nature of the underlying anion-exchange layer, this retention layer is quantitatively negated by the cation-exchange nanoparticle coating. A 76 nm 2% cross-link strong base anion-exchange latex coating is applied to the previous layer to make a cation-exchange/anion-exchange composite. The combination of the two outermost ion-exchange latex layers results in useful retention zones for transition metals in either cation-exchange or anion-exchange mode or a combination of the two modes. An example separation of the CS5A column operating in the anion-exchange mode is shown in Fig. 9.6. In this case, the ligand pyridine-2,6-dicarboxylate forms a relatively stable complex with transition metals and the transition metals are separated as the anionic complex.

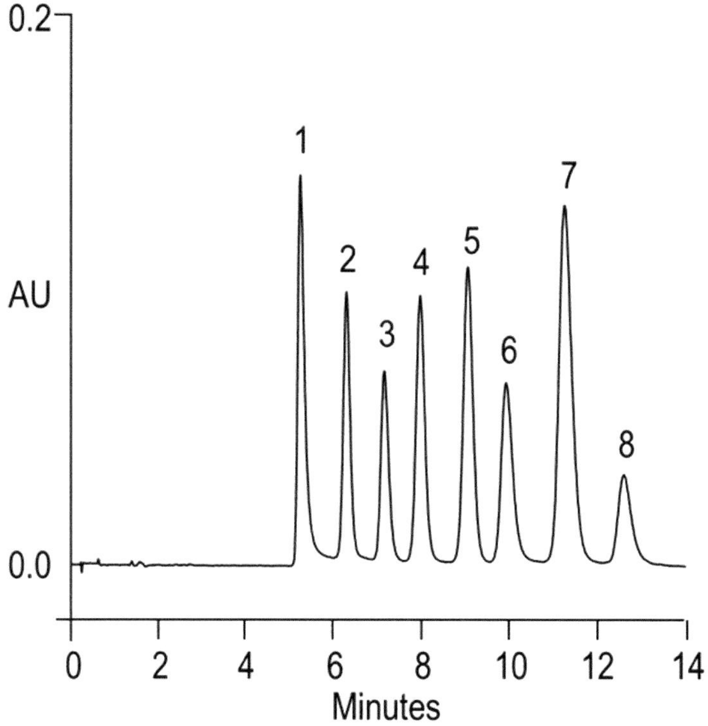

Figure 9.6 Separation of transition metal cations on 4 mm ID IonPac CS5A and CG5A columns. Flow rate 1.2 mL per minute, MetPac PDCA eluent, ambient temperature, absorbance detection at 530 nm with PAR in MetPac post-column reagent diluent, 50 µL injection volume. Peaks: (1) 1.3 ppm iron (III); (2) 1.3 ppm copper; (3) 2.6 ppm nickel; (4) 1.3 ppm zinc; (5) 1.3 ppm cobalt; (6) 6 ppm cadmium; (7) 2.6 ppm manganese; (8) 1.3 ppm iron (II).

9.3 Anion-Exchange Phases

Anion-exchange phases in ion chromatography are by far the most diverse category of stationary phases. In the case of cations, elemental spectroscopy instrumentation provides an alternative to ion chromatography, diminishing the popularity of this mode of ion chromatography. However, in the case of anions, elemental spectroscopy is not capable of distinguishing between the various anions commonly found in samples of analytical interest. Considering the wide variety of anions found in nature and industrial and pharmaceutical samples, a wide variety of different phases have been developed to provide optimal selectivity for critical separation needs. As in the case of cation-exchange phases, the vast majority of all anion-exchange phases are based on polymeric substrates. The extreme pH is associated with anion-exchange separations when operated in conjunction with the suppressor are incompatible with inorganic substrates. Although there were a few silica-based phases developed in the early days of ion chromatography, such phases have largely disappeared due to the poor stability of such materials under typical high pH operating conditions.

9.3.1 Quaternary Latex-Based Anion-Exchange Phases on Nonporous Substrates

The earliest anion-exchange phases used in ion chromatography [2] made use of surface-sulfonated 2% cross-link polystyrene beads as a substrate. This was a natural starting point given the fact that such materials had already been developed for separation of cations (see Section 9.2.1). Preparation of the anion-exchange phase was achieved by coating the surface-sulfonated substrate with a layer of anion-exchange nanoparticles. Although Hamish Small initially prepared the particle coating materials by grinding macroporous anion-exchange materials, the chemists at Dow later applied their considerable expertise in making emulsion polymers to prepare anion-exchange materials via emulsion polymerization. Preparation of the anion-exchange nanoparticles involved the co-polymerization of vinylbenzyl chloride (VBC) with divinylbenzene. Subsequently, the active anion-exchange nanoparticles were produced by reacting the cross-linked vinylbenzyl chloride nanoparticles with a suitable tertiary amine, forming particles with quaternary anion-exchange sites distributed homogeneously throughout the nanoparticles. Finally, the anion-exchange composite was prepared by mixing a suitable quantity of anion-exchange nanoparticles with separately prepared surface-sulfonated cross-linked polystyrene beads. Such composites form essentially instantaneously upon mixing the oppositely charged materials in an aqueous slurry medium. Although the attachment mechanism is electrostatic, the large number of simultaneous electrostatic interactions associated with the attachment of the nanoparticles to the substrate surface preclude elution of the electrostatically attached nanoparticles under any known conditions producing an exceptionally stable composite.

The earliest phases using this synthesis approach used relatively large particle size substrate particles (20–30 μm) and relatively large diameter nanoparticles (200–300 nm) to prepare the stationary phase coating. Such materials were of relatively low capacity and low efficiency. The low capacity of these materials was dictated by the relatively limited suppression capacity of packed bed suppressors in use at the time. These low capacity stationary phases allowed the use of relatively dilute mobile phases, thus prolonging the life of the suppressor. However, the relatively low efficiency of such materials necessitated relatively long analysis times, generally greater than 30 min for a typical analysis. Later developments in anion-exchange phases made use of much smaller substrate particles and smaller anion-exchange nanoparticles. By reducing the diameter of substrate particles and anion-exchange nanoparticles while maintaining the same diameter ratio for the particles, column capacities are held constant while chromatographic efficiency is improved. Progressive reduction in both particle diameters resulted in significant improvements in chromatographic performance and speed of analysis. Figure 9.7 shows an example chromatogram of one of the more modern variants of this stationary phase architecture.

The final development for this class of ion-exchange materials was the shift to the use of 55% divinylbenzene for the substrate particles. As in the case of 2% cross-link polystyrene particles, these particles are surface sulfonated before being coated with anion-exchange nanoparticles. The chief advantage of such high cross-link materials is that these materials are immune to swelling in organic solvents due to the high cross-link. Use of 2% cross-link polystyrene particles limited the use of organic solvent to no more than 5% to avoid excessive swelling and column damage. In contrast, these new highly cross-linked substrate particles allowed the use of high concentrations of organic solvents to clean hydrophobic contaminants from the active stationary phase and the use of organic solvents to manipulate ion-exchange selectivity. Figure 9.8 shows the influence of organic solvent on selectivity with the AS4A-SC column which utilizes 55% cross-link substrate particles.

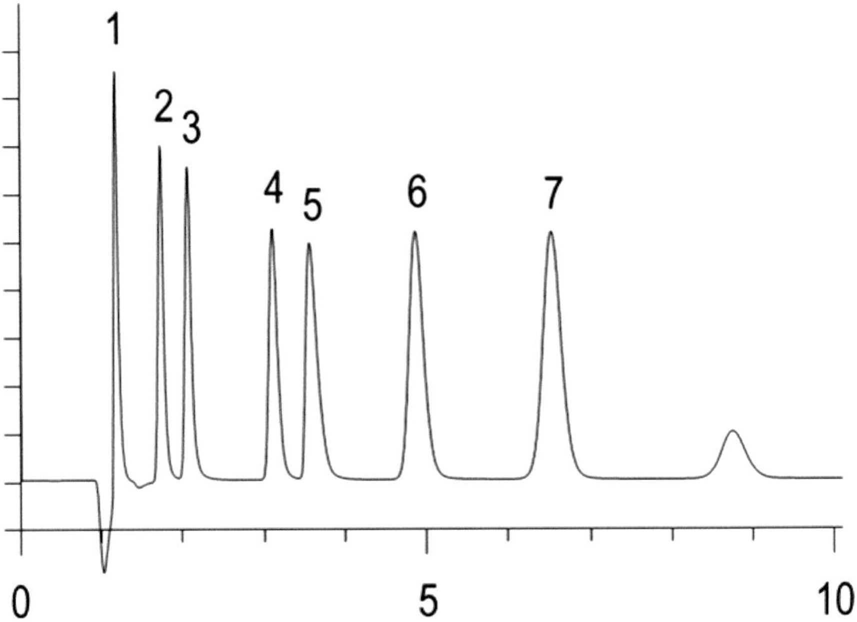

Figure 9.7 Separation of anions on a 4 mm ID IonPac AS4A column. Flow rate 2 mL per minute, 1.7 mM sodium bicarbonate and 1.8 mM sodium carbonate, ambient temperature, suppressed conductivity detection, 25 μL injection volume. Peaks: (1) 1 ppm fluoride; (2) 1.5 ppm chloride; (3) 2.5 ppm nitrite; (4) 4 ppm bromide; (5) 4 ppm nitrate; (6) 10 ppm phosphate; (7) 5 ppm sulfate.

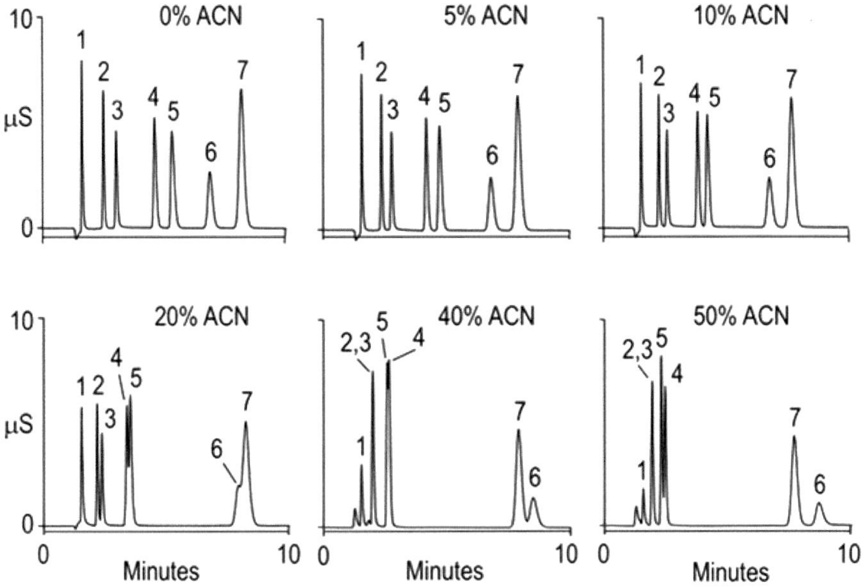

Figure 9.8 Separation of anions on a 4 mm ID IonPac AS4A-SC column. Flow rate 2 mL per minute, 1.7 mM sodium bicarbonate and 1.8 mM sodium carbonate, ambient temperature, suppressed conductivity detection, 25 μL injection volume. Peaks: (1) 2 ppm fluoride; (2) 3 ppm chloride; (3) 5 ppm nitrite; (4) 10 ppm bromide; (5) 10 ppm nitrate; (6) 15 ppm phosphate; (7) 15 ppm sulfate.

9.3.2 Quaternary Latex-Based Anion-Exchange Phases on Wide-Pore Substrates

From the time of initial commercialization of ion chromatography, suppressor capacity had been somewhat limited. Initially, packed bed suppressors needed to be periodically regenerated, so it was desirable to reduce the frequency of regeneration by using relatively low capacity columns. Subsequently, packed bed suppressors were supplanted by fiber suppressors which allowed continuous regeneration, but these too had limited dynamic capacity. As a result, the maximum eluent concentration was constrained to low millimolar concentrations. However, by the mid-1980s, the first high-capacity suppressors were introduced. These suppressors were capable of suppressing up to 100 mM eluents, greatly expanding the range of eluent concentrations that could be utilized

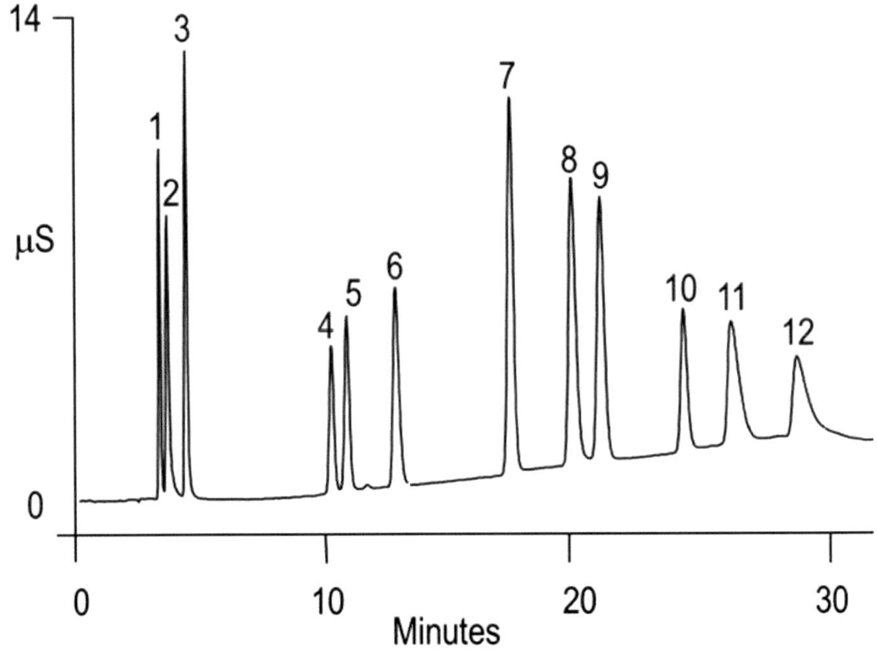

Figure 9.9 Separation of anions on a 4 mm ID IonPac AS10 column. Flow rate 1 mL per minute, 50–124 mM sodium hydroxide in 30 min, ambient temperature, suppressed conductivity detection, 20 μL injection volume. Peaks: (1) 1.2 ppm fluoride; (2) 30 ppm acetate; (3) 5 ppm formate; (4) 10 ppm selenite; (5) 1.8 ppm chloride; (6) 9 ppm nitrite; (7) 10 ppm sulfate; (8) 10 ppm oxalate; (9) 10 ppm selenate; (10) 9 ppm phosphate; (11) 10 ppm bromide; (12) 6 ppm nitrate.

in ion chromatography. The availability of higher dynamic capacity suppressors led to an interest in the development of new anion-exchange phases with higher capacity. By using higher capacity columns, a wider range of sample types could be directly injected into the column without peak shape distortion associated with high ionic strength samples. One of the earliest materials with high capacity made use of quaternary anion-exchange latex particles coating porous substrates rather than nonporous substrates. Such an architecture enables the production of materials with significantly higher capacities while using the same stationary phase synthesis methodologies developed for use with nonporous substrates. One challenge with such materials is that the pore size needs to be large enough to allow for anion-exchange nanoparticles to coat both the interior and the exterior surface of the particles. Preparation of a suitable material requires very large pore size substrate particles with nanoparticles significantly smaller than the nanoparticles used with nonporous substrates. The first column made using this approach was the IonPac AS10 column (see Fig. 9.9). It utilized particles with an average pore size of around 200 nm with 65 nm anion-exchange nanoparticles. The wide-pore substrate particles used in the AS10 column were prepared from 55% divinylbenzene which imparted resistance to swelling under the influence of organic solvents. As mentioned previously, using 2% cross-link polystyrene limited the use of an organic solvent to less than 5%. Subsequently, this stationary phase morphology has been used in a wide variety of different columns including the IonPac AS9-HC, the IonPac AS11-HC, the IonPac AS12A, and the IonPac AS16 targeting oxyhalides in drinking water, organic acids in foods, analysis of common anions in drinking water and analysis of perchlorate in drinking water, respectively.

9.3.3 Adsorbed Coating Anion-Exchange Phases

The late 1980s saw the introduction of a new class of anion-exchange phases, anion-exchange phases based on an adsorbed coating. Examples of such materials include the Sarasep AN1 and AN300. The main advantage of these materials is their relatively high retention for fluoride under conditions that allow elution of more strongly retained anions in under 12 min. Fluoride tends to be very weakly retained by anion-exchange materials, eluting near the void volume on most columns. Because of the proximity of fluoride to the void volume, trace analysis of fluoride in real samples is often difficult. The new columns from Sarasep somehow circumvented this problem, greatly facilitating the quantitation of fluoride in real drinking water samples. Although never presented to the market as an adsorbed coating, treatment of these columns with organic solvent removes the vast majority of the stationary phase, demonstrating that the stationary phase is not permanently attached to the substrate surface. Although the exact nature of the stationary phase has not been revealed, it is speculated that the stationary phase is prepared via deposition of a low-molecular-weight cationic polymer onto the surface of a high-surface-area hydrophobic substrate. Presumably, sufficient retention in a purely aqueous environment provides stable retention via a hydrophobic interaction with the substrate. While such materials represented a major step forward anion-exchange selectivity of columns designed for use with carbonate eluents, the poor compatibility of such phases with organic solvent containing mobile phases limited the utility of such materials. As such, although these columns are still commercially available, they have been for the most part supplanted by later materials with similar fluoride selectivity by providing much better solvent compatibility.

9.3.4 Polymer-Grafted Strong Base Anion-Exchange Phases

Development of the Sarasep AN1 stationary phase demonstrated the feasibility of providing improved resolution of fluoride from the system void without sacrificing analysis time. However, given the fact that solvent compatibility of the Sarasep AN1 was poor, this drove research into the development of new classes of anion-exchange materials with improved fluoride resolution from the void while still maintaining solvent compatibility common to other modern ion chromatography stationary phases. Subsequent research showed that achieving this selectivity required a stationary phase architecture enabling high levels of hydration of the stationary phase. All of the previous anion-exchange phases had been based on moderate levels of cross-link which restricts hydration of the stationary phase. One way around this problem is to graft enormous numbers of linear polymer strands to the surface of a high-surface-area polymeric substrate. Such a material provided the necessary retention even in the absence of any cross-linking. The IonPac AS14 exemplifies grafted materials of this architecture based on a pH stable methacrylate polymer, while the IonPac AS14A exemplifies grafted materials of this architecture based on an, even more, pH stable styrenic polymer. This architecture provides greatly improved fluoride resolution without sacrificing solvent compatibility (see Fig. 9.10).

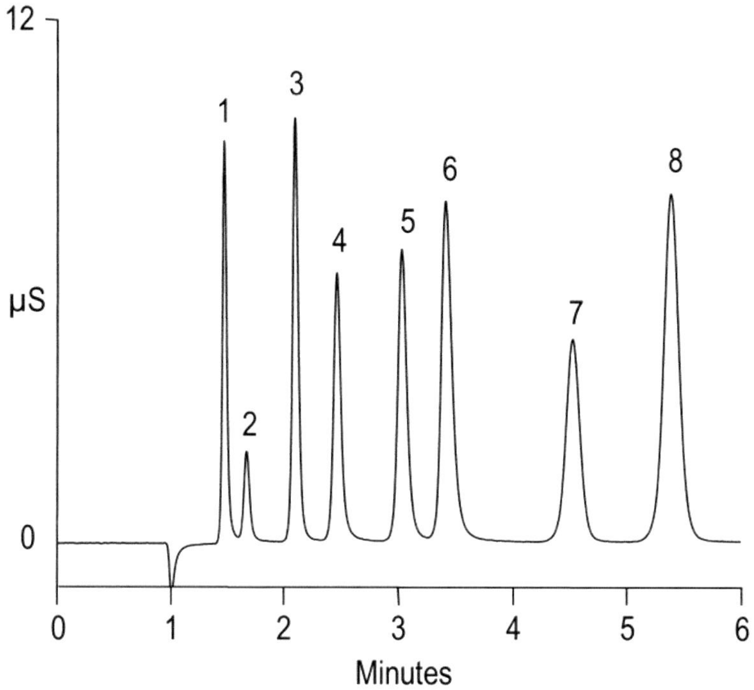

Figure 9.10 Separation of anions on a 3 mm ID by 150 mm IonPac AS14A column with 3 mm ID AG14A column. Flow rate 0.8 mL per min, 1 mM sodium bicarbonate and 8 mM sodium carbonate, 30°C, suppressed conductivity detection, 25 µL injection volume. Peaks: (1) 5 ppm fluoride; (2) 20 ppm acetate; (3) 10 ppm chloride; (4) 15 ppm nitrite; (5) 25 ppm bromide; (6) 25 ppm nitrate; (7) 40 ppm phosphate; (8) 30 ppm sulfate.

9.3.5 Surface-Chloromethylated High Cross-Link Macroporous Anion-Exchange Phases

Early in the development of ion chromatography phases, Barron and Fritz [10] developed an aqueous-based chloromethylation methodology using paraformaldehyde in HCl. This approach offered two potential benefits over conventional chloromethylation using chloromethyl ether. First, it avoided the use of chloromethyl ether which invariably contains low levels of bis-chloromethyl ether, a potent carcinogen. Second, because the reaction takes place in an aqueous environment, the reaction is largely restricted to the surface of the substrate particle, improving chromatographic performance. The first benefit (avoidance of bis-chloromethyl ether) is highly suspect as it is likely that the paraformaldehyde-HCl synthesis methodology does create bis-chloromethyl ether in situ. The benefit of a surface constrained reaction because the reaction was aqueous based also proved to be highly suspect as the improved chromatographic performance was only observed with aromatic eluents commonly employed in nonsuppressed ion chromatography. Aromatic eluents tend to mask the high-energy sites associated with poorly hydrated subsurface ion-exchange sites, making it difficult to assess whether or not chloromethylation sites are truly constrained to the surface.

While the details of the synthesis have not been published, both the chromatographic performance and the resin specifications suggest that Hamilton PRP-X100 is synthesized using the above paraformaldehyde-HCl functionalization methodology. The chromatographic performance with hydrophilic eluent species such as carbonate-bicarbonate suggests that a significant fraction of the ion-exchange sites are poorly hydrated subsurface ion-exchange sites. In fact, Hamilton PRP-X110S is supplied in a carbonate-bicarbonate storage solution pre-equilibrated with thiocyanate, added to the solution to mask any poorly hydrated ion-exchange sites. Using thiocyanate to mask poorly hydrated sites presents several problems for the user. First, any thiocyanate in the mobile phase adds to the background conductivity, raising detection limits. Second, because the affinity of thiocyanate for the stationary phase is much higher than the affinity of either carbonate or bicarbonate for the stationary phase, long equilibration times are associated with each new batch of thiocyanate containing eluent which of necessity will be slightly different regarding relative concentrations. Production of stationary phase is free from such poorly hydrated ion-exchange sites is far preferable to adding masking reagents to the mobile phase.

9.3.6 Polyvinyl Alcohol Backbone Anion-Exchange Phases

Anion-exchange materials based on polyvinyl alcohol substrates represents another new class of anion-exchange materials widely used today in ion chromatography. Such materials were initially used for the preparation of reversed-phase polymers after extensively esterifying surface alcohols with activated long chain carboxylic acids. Such materials offer many advantages including good peak shapes for relatively poorly hydrated anions or relatively hydrophobic anions. These materials also provide excellent resolution of fluoride from the system void. Due to the relatively unrestricted hydration of the ion-exchange sites on the surface and the very hydrophilic polymer substrate, favorable conditions are achieved for retention of fluoride. Shodex produces columns using this architecture [9]. The most notable of these is the SI-50-4E column which provides excellent selectivity and peak shapes for the common anions. Metrohm also sells a number of these phases as OEM products [9]. They are easily identified in that their specifications exactly match those from the Shodex website for the corresponding column.

One disadvantage of using polyvinyl alcohol as the basis for constructing the substrate particle is the limited physical strength of such materials. While highly cross-linked substrate particles composed of styrenic monomers tend to be a very high strength, with a typical operating flow rate in the 1–2 mL per minute range when using 4 mm column hardware, analogous polyvinylalcohol materials typically have a maximum flow rate of 0.8 mL per minute. This lower flow rate limit is due to the lower physical strength of polyvinyl alcohol-based materials. Such a flow rate constraint limits the overall speed of analysis when using smaller particle size media.

9.3.7 Hyperbranched Anion-Exchange Phases on Wide-Pore Substrates

The most recently developed stationary phase synthesis architecture utilizes synthesis of hyperbranched anion-exchange materials that are electrostatically attached to wide pore substrates. Such materials are synthesized using a novel synthesis methodology [11] utilizing two main ingredients: a primary amine and a diepoxide monomer. In the first part of the synthesis, the primary amine and the diepoxide monomer are allowed to react in the presence of a wide-pore surface-sulfonated substrate at elevated temperature.

A low-molecular-weight polymer with tertiary amine functionality at regular intervals along the polymer backbone is formed under these conditions. A fraction of this polymer is electrostatically attached to the surface of the wide-pore substrate. The excess free polymer is then removed. Subsequently, the polymer-coated substrate is allowed to react with the diepoxide. The reaction of the diepoxide with the polymer quaternizes the tertiary amine groups distributed along the polymer backbone while generating pendant epoxy groups at each reaction site. All unreacted diepoxide is then removed and replaced with a solution containing primary amine. The primary amine reacts with all of the pendant epoxide moieties to produce pendant secondary amine groups. The sequence of reacting the polymer first with the diepoxide monomer and then with the primary amine is termed "a reaction cycle." Subsequently, the excess primary amine is removed and replaced again with diepoxide monomer. This time the diepoxide monomer reacts twice with the pendant secondary amine groups to produce two pendant epoxide moieties for each secondary amine group produced by the previous reaction. After removing any unreacted diepoxide, the resulting polymer is again exposed to the primary amine to produce secondary amine groups, completing the second reaction cycle. Because each secondary amine group can react with two epoxy monomers, this results in a doubling of the number of ion-exchange sites at the end of the second reaction cycle. Reaction cycles can be repeated multiple times until the desired selectivity and capacity is achieved.

Of course, the above is only a general description of what is possible using reactions between amines and epoxides. A variety of different materials are readily produced using a variety of reagents. In total, more than a dozen phases are commercially available utilizing some version of this chemistry. An example chromatogram from a column synthesized using this chemistry platform is shown in Fig. 9.11. The combination of an extremely hydrophilic stationary phase backbone with a relatively unrestricted hydration of the stationary phase allows for the application of such materials to an extremely wide variety of analytes ranging from highly hydrated anions such as fluoride to poorly hydrated polarizable anions such as hexafluorophosphate with little if any, compromise in peak shape.

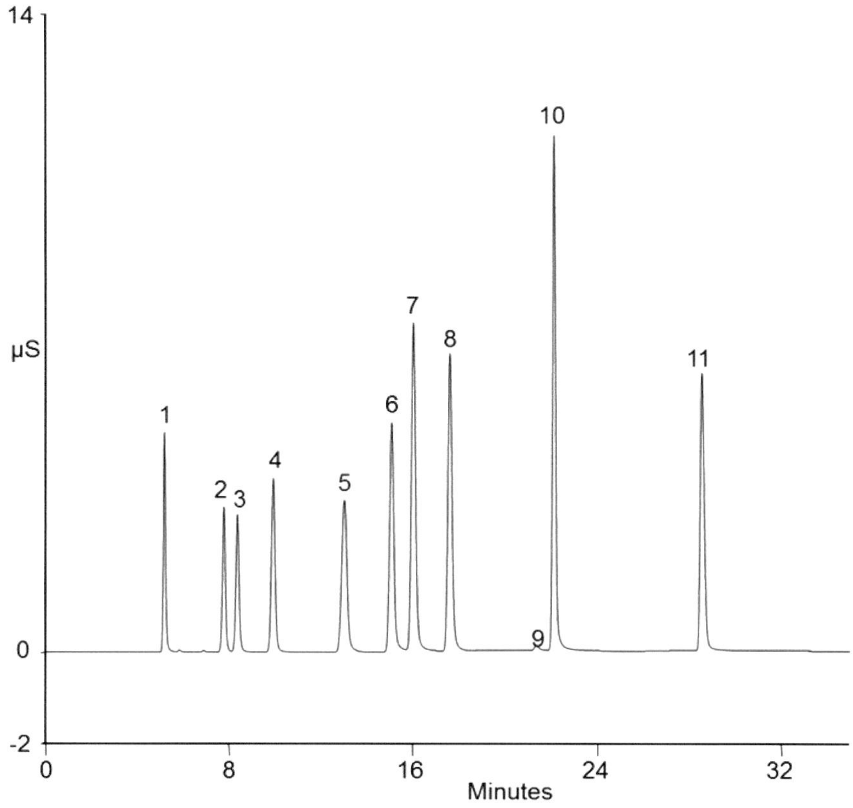

Figure 9.11 Separation of anions on 4 mm ID IonPac AS19-4 μm and AG19 columns. Flow rate 1 mL per minute, 10 mM potassium hydroxide 0–10 min, 10–45 mM potassium hydroxide 10–25 min, 30°C, suppressed conductivity detection, 10 μL injection volume. Peaks: (1) 3 ppm fluoride; (2) 10 ppm chlorite; (3) 20 ppm bromate; (4) 6 ppm chloride; (5) 15 ppm nitrite; (6) 25 ppm bromide; (7) 25 ppm chlorate; (8) 25 ppm nitrate; (9) carbonate; (10) 25 ppm sulfate; (11) 40 ppm phosphate.

References

1. Scott, R. P. W. (1986) Bulk Property Detectors, in *Journal of Chromatography Library* **33**, ed. Scott, R. P. W., Chapter 3 (Elsevier, Amsterdam) pp. 49–87.
2. Small, H., Stevens, T. S., and Bauman, W. C. (1975) Novel ion exchange chromatographic method using conductimetric detection, *Anal. Chem.*, **47**, pp. 1801–1809.

3. Small, H. (1989) *Ion Chromatography*, Plenum, New York.
4. Kolla, P., Köhler, J., and Schomburg, G. (1987) Polymer-coated cation-exchange stationary phases on the basis of silica, *Chromatographia*, **23**, pp. 465–472.
5. Jensen, D., Weiss, J., Rey, M., and Pohl, C. (1993) Novel weak acid cation-exchange column, *J. Chromatogr. A*, **640**, pp. 65–71.
6. Rey, M., Pohl, C. (1996) Novel cation-exchange stationary phase for the separation of amines and of six common inorganic cations, *J. Chromatogr. A*, **739**, pp. 87–97.
7. Richens, D., Simpson, D., Peterson, S., McGinn, A., Lamb, J. (2003) Use of mobile phase 18-crown-6 to improve peak resolution between mono- and divalent metal and amine cations in ion chromatography, *J. Chromatogr. A*, **1016**, pp. 155–164.
8. Kadnar, R. (1999) Determination of amines used in the oil and gas industry (upstream section) by ion chromatography, *J. Chromatogr. A*, **850**, pp. 289–295.
9. Pohl, C. (2005) Ion chromatography, in *Separation Science and Technology; Handbook of Pharmaceutical Analysis by HPLC* **6**, eds. Ahuja, S., and Dong, M., Chapter 8 (Academic Press, Amsterdam) pp. 219–254.
10. Barron, R., Fritz, J. (1984) Effect of functional group structure on the selectivity of low-capacity anion exchangers for monovalent anions, *J. Chromatogr. A*, **284**, pp. 13–25.
11. Pohl, C., Saini, C., (2008) New developments in the preparation of anion exchange media based on hyperbranched condensation polymers, *J. Chromatogr. A*, **1213**, pp. 37–44.

Chapter 10

Ion Chromatography: Method Development

Laila Kott

Xenon, Burnaby, B.C., Canada

lkott@lu.lv

This chapter provides the methodology for ion chromatographic method development, which includes selection of eluant systems and detectors, as well as sample preparation. Discussion also includes the choice of column chemistries and gradient separations.

10.1 Introduction

The use of ion exchangers has been known throughout history. Aristotle reported the desalination of seawater using sands and soils, but the widespread use of synthetic ion exchangers did not come about until scientists discovered the ion-exchange capabilities

Chromatographic Method Development
Edited by Gregory K. Webster and Laila Kott

ISBN 978-981-4800-53-2 (Hardcover), 978-0-429-20172-1 (eBook)
www.jennystanford.com

of crushed gramophone records [1]. Since then the uses of ion-exchange resins have ranged from the desalination of water, to recovery of toxic or low abundant materials from industrial waste and mine tailings to the current use as columns used in ion chromatography, as described in this chapter.

Ion chromatography as a commercially available system has been around since the 1970s. In the pharmaceutical industry, it is largely used for counterion analysis, but has also been widely used with amino acids, peptides and other biological-based components (i.e., proteins, carbohydrates). It has use outside of the pharmaceutical industry for drinking and wastewater analysis and sulfite analysis in the wine and the food industry [2–5]. Many methods are developed to analyze or detect one or two components, often making the overall method development uncomplicated. However, in the presence of complicated matrices, ion chromatographic method development can be challenging. This chapter will cover the components of the ion chromatograph, how to choose the components for a given analysis, method development, and general system troubleshooting.

10.2 Components of the System

The setup of the ion chromatographic (IC) system is similar to the liquid chromatographic (LC) system with some key differences. First and foremost, an IC system is designed to detect ions, much like mass spectrometry, such that the column, the suppressor, and the eluant are all chosen depending on whether the analysis is looking for anions or cations. As described in the previous chapter (Chapter 9), there are many choices of columns, depending on the target analyte, but unlike like LC systems the IC system needs to be considered as a whole, not just what column to use.

An IC system consists of an injector, a pump, an eluant (which can now be generated in situ by an eluant generator), a column and usually a guard column, a suppressor (not 100% ubiquitous but becoming more and more common) and a detector, all connected to a chromatography data system (see Fig. 10.1). The components that differ from liquid chromatography: column, suppressor, eluant, and detector will be discussed. For the components that are similar to the LC, they are programmed and maintained, as they would be on an LC system.

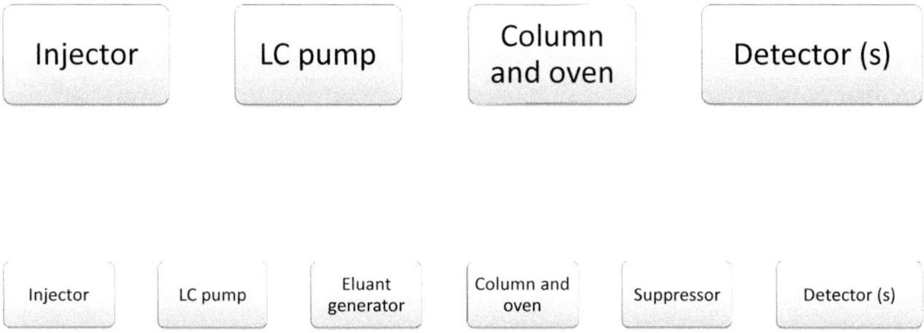

Figure 10.1 A comparison of the main components in an ion chromatography system, compared to those in a liquid chromatography system. Note that the eluant can be routed back to the suppressor for regeneration.

10.2.1 Columns

Unlike liquid chromatography columns, ion-exchange columns are predominantly polymer based and due to the nature of the analyte (ions) are designed to be stable under more extreme pH conditions than typical (U)HPLC columns. This is in part due to the elution of analytes in IC being accomplished by acids or bases, typically methanesulfonic acid, potassium hydroxide or potassium carbonate. Since the predominant mode of detection for IC is conductivity, it is important to keep the overall ionic strength of the eluant as low as possible, which in turn requires less suppression of the signal. As such, columns with low ion-exchange capacities are preferred for standard pharmaceutical analyses, as they require less eluant and therefore, also less suppression [6]. In IC the type and design of the stationary phase in particular plays a significant role in controlling selectivity due to the range of stationary phase chemistries available; this being in contrast to reversed-phase liquid chromatography, which is dominated by similar octadecylsilica-based stationary phases [7].

Column selection begins first and foremost with type of analysis, anionic or cationic. Figure 10.2 shows a representation of the functional groups found on both the cationic and anionic column types.

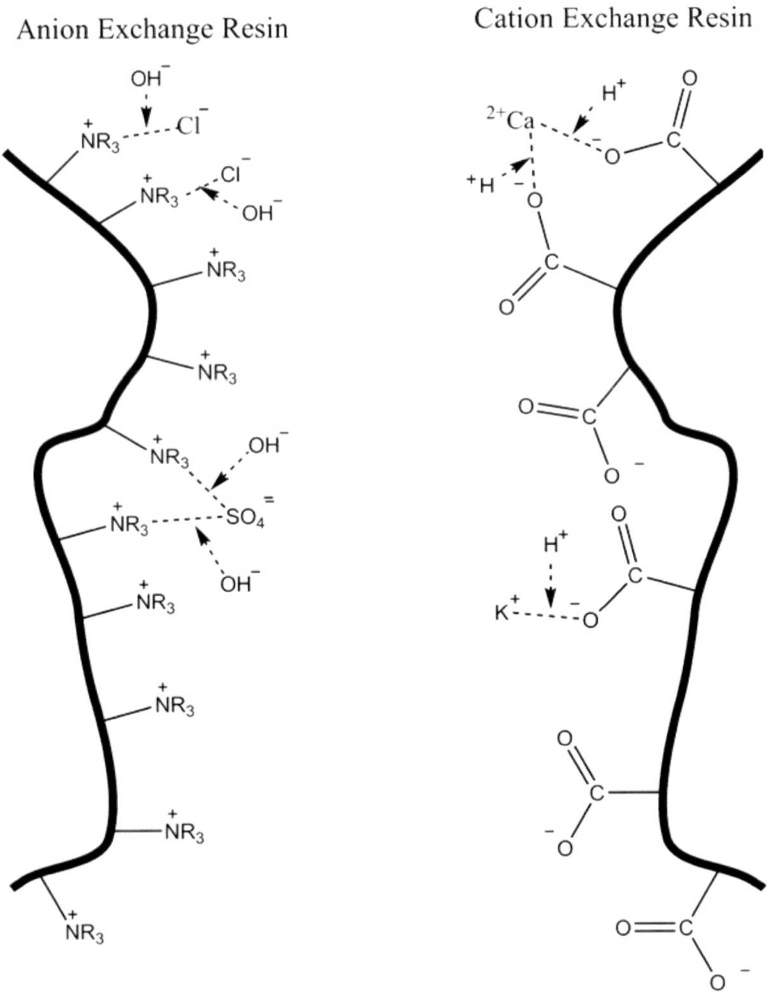

Figure 10.2 Schematic showing the monovalent and divalent ionic interactions on anion and cation resins. Divalent ions require two hydroxyl or hydronium ions to be eluted off the column, whereas monovalent ions require only one.

10.2.1.1 Anionic columns

There are a wide variety of IC columns available for anion analysis. A typical starting point for anion counterion analyses would be to consider the type of anion and the capacity of the column. Lower capacity is better for suppressed systems as mentioned in Section 10.2.1 earlier. The functional group for anion analysis is usually quaternary amine based, with different groups attached to the amine, giving different selectivities and capacities of exchanger (weak, medium or strong).

Some suggested columns to try first are *Dionex IonPac AS analytical columns manufactured by Thermo Scientific, Metrosep A Supp 17 by Metrohm, Agilent-AX* and *Hamilton PRP-X100 columns*. These columns and their equivalent guard columns come in both 2 and 4 mm diameters. Choice of column can also depend on detector type (i.e., conductivity vs. UV) and if the system to be used is equipped with suppressors or not. On systems using suppressors, one must always remember that the column diameter must match the suppressor diameter. The above-mentioned IC columns are designed to separate and analyze most inorganic anions and organic acids using hydroxide or carbonate as the eluant and conductivity detection. It is of note that some columns are designed to work with a specific eluant. Both weakly retained monovalent anions such as fluoride, acetate, formate, chloride and strongly retained multivalent ions, such as phosphate and citrate can be efficiently eluted in the same run by using isocratic or gradient elution typically within 10 to 20 min. If the ion chromatographic analysis has a more complicated matrix, is to be used for analyzing waste water, drinking water, specific species defined by the EPA or for larger biologic molecules like polypeptides, oilgonucleotides or proteins, then a little research and comparison of available column technologies will help in choice of column.

10.2.1.2 Cationic columns

There are a wide variety of IC columns available for cation analysis. A typical starting point for cation counterion analyses would be to consider the capacity of the column. As mentioned before, lower capacity is better for suppressed systems. The functional group for cation analysis is usually carboxylic (weak exchanger) or sulfonate (strong) based, giving different selectivities and capacities of exchanger.

Some suggested columns to try first are: *Dionex IonPac CS columns manufactured by Thermo Scientific Dionex, Metrosep C Supp 1 by Metrohm Agilent-CX and Hamilton PRP-X200 columns.* As with the anionic columns, these columns and their equivalent guard columns come in both 2 and 4 mm diameters, but one must always remember that the column diameter must match the suppressor diameter, if one is being used. All the considerations above apply to choice of cationic column: use of suppressor, type of detection, choice of eluant. These IC columns are designed for the separation and analysis of alkali, alkaline earth metals and ammonium with conductivity detection. Both monovalent and mulitivalent cations can be efficiently eluted in the same run by using isocratic or gradient elution typically within 10 to 20 min. Similarly as for anions, if the cationic matrix is more complex, is to be used for analyzing wastewater, drinking water, specific species defined by the EPA, transition metals or alkylamines and alkanolamines, then, again, a little research and comparison of available column technologies will help in choice of column.

10.2.2 Eluants

In ion chromatography, once the ion is retained by the stationary phase, it falls to the eluant to disrupt the interaction of the sample ion and the stationary phase to elute the ion. Some ion chromatographs come with the ability to generate the eluants in situ/inline, whereas other systems require that eluants are prepared as they would be for liquid chromatographic separations, at fixed concentrations. Since pH and ion concentration play such an important role in the IC elution process, special care must be taken when preparing IC eluants.

For eluants prepared offline (not using the eluant generators), they must be prepared with fresh de-ionized water, with sufficiently low conductivity (18.2 MΩ/cm at 25°C is typical) and should be free of organic compounds as not to compromise IC separation sensitivity. If unsure of the water quality or conductivity, boil the water to remove any dissolved CO_2, which can convert to carbonate and change the background conductivity of the eluant. Other considerations when preparing eluants are as follows:

- Eluants should be prepared and stored in plastic bottles to avoid contamination of leached ions from glass bottles.
- The eluant should be degassed online or helium purged before use.
- To avoid clogging of the IC system and to extend the life of the column, the eluant should be filtered using a chemically inert filter with 0.45 μm pores or smaller. Cellulose filters are not stable at high pHs, so filters such as the "hydrophillic" versions of polyvinylidenedifluoride (PDVF) or polytetrafluoroethylene (PTFE) are recommended. Note if they are not the hydrophilic version of these filters, they cannot be used for filtering aqueous solutions.
- The concentration of the eluants should be at a high enough concentration to cover the elution of the most tightly bound ion. Typical concentrations are
    - carbonate/Bicarbonate up to 25 mM;
    - hydroxide (sodium or potassium) up to 130 mM;
    - methanesulfonic acid (MSA) up to 65 mM.

For in situ eluant generation, a generator cartridge needs to be installed on the IC system and a source of fresh de-ionized water is required, with sufficiently low conductivity (18.2 MΩ/cm at 25°C is typical), free of organic compounds and stored in plastic mobile phase bottles, as for offline eluant preparation above. For properly calibrated systems, the in situ eluant generation uses the flow rate (inverse relationship) and applied current to deliver the correct concentration, following Faraday's laws of electrolysis and using the proportionality:

$$[\text{Eluent charge concentration(mM)}] \propto \frac{\text{Applied Current(mA)}}{\text{Flow Rate(mL/min)}} \quad (10.1)$$

10.2.2.1 Anionic eluants

As introduced above, eluants elute solutes based on their ionic strength. The common anion eluants in order of their ionic strength are:

Borate < Hydroxide < Bicarbonate < Carbonate.

In general, the predominant method of detection for inorganic and organic anions is by conductivity and therefore, hydroxide tends to be the preferred anion eluant due to its low background conductivity.

Hydroxide is a monovalent ion therefore, higher concentrations of the eluant would be required for it to successfully elute multivalent ions. Most commercial suppressors can handle concentrations above 130 mM, allowing for use of hydroxide as an eluant for more highly charged anions. However, as a good general rule, the valencies of the anions to be eluted and the eluant itself should be comparable. For example, if a sample mixture contains a many divalent anions, it would be wise to choose an eluant that also contains bivalent anions, such as sodium carbonate.

10.2.2.2 Cationic eluants

The typical cation eluants in order of ionic strength are

Hydrochloric acid < Nitric acid < Methanesulfonic acid (MSA) < Sulfuric acid.

In general, MSA is the first choice due to its low background of conductivity. Divalent cations show a high affinity for strong ion-exchange columns and cannot be eluted with dilute mineral acids, which would require higher acid concentrations, leading to a higher background conductivity, thereby making MSA the best starting point for method development. It is important to note that organic additives in the eluant such as trifluroacetic acid (TFA) that can act as ion pairing agents are not recommended, as they will shield the charge required for separation.

10.2.3 Suppressor Theory, Operation, and Troubleshooting

Not all systems running ion chromatography have suppressors. For those that do, the suppressor plays a central role as it is used to reduce the conductivity of the eluant to as near to zero as it can eletrolytically, which enhances the conductivity of sample ions and thereby increases the sensitivity of detector. Due to this signal enhancement by the suppressors, they are often considered to be part of the detection system. Since the incoming sample, column, and eluant are anion/cation specific, so must be the suppressors.

Suppressors work by converting the bulk of the effluent, which is in the form of the eluant (i.e., MSA or hydroxide) and the counterions of the analytes, to either hydroxide ions for cations or hydronium ions for anions. The result of the conversions means that only the analyte ions of interest are detected by the conductivity detector. The conversions are effected by having hydroxide ions forming at the cathode of a cation suppressor, which then migrate through an anion-exchange membrane to neutralize the MSA eluant. The MSA^- ions migrate to the anode, again through an anion-exchange membrane, where they combine with locally generated hydronium ions, maintaining neutrality. A bit of gas is generated at both electrodes, which goes to waste. This is the cause of the apparent bubbles eluting out of the suppressor from the regenerant out port and is an indication of proper suppressor function. The same is true for the anion suppressor with hydronium ions being produced at the anode which migrate through a cation-exchange membrane to neutralize the eluant (hydroxide or carbonate). The remaining counter ion (typically sodium or potassium) migrates to the cathode where it combines with some locally generated hydroxide and flows to waste.

Currently the most common type of suppressors are the membrane-based suppressors, as they have a high suppression capacity and can be continuously regenerated. The electrolytic self-regenerating suppressor uses the effluent from the conductivity detector, which as described above contains only the analyte of interest and hydronium or hydroxide ion, as the source of regenerant. Some suppressors use an external chemical regenerant or an external source of de-ionized water with a pump as a regenerant. An external regenerant would be required if the effluent is being diverted to a second detector, i.e., to a mass spectrometer, if organic solvents are used in the eluant or if the flow rate is so low (as can be the case with microbore columns) that a higher flow is required for regeneration. When developing a method, the column diameter should be matched with a suppressor with a similar diameter.

Figures 10.3 and 10.4 show a schematic of how the cation and anion suppressors exchange out the highly charged eluant and use the detector effluent to self-regenerate. The detector effluent, which has only a few anions in it, is directed back through the suppressor as a regenerant. The water in the regenerant aides in the suppression and regeneration of the suppressor.

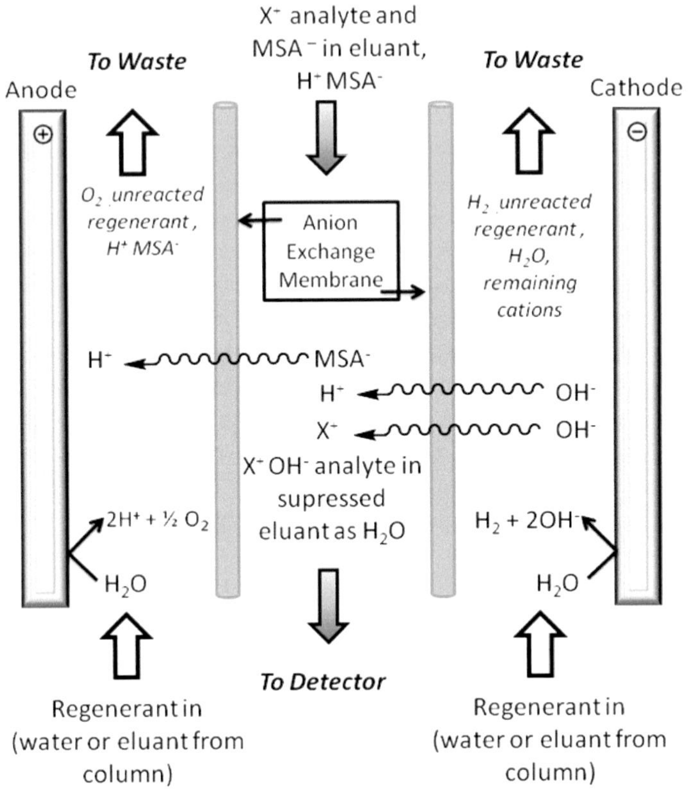

Figure 10.3 Schematic showing the ionic flows through the suppressor to the detector and then from the detector back through the suppressor for regeneration: Cation.

10.2.4 Detectors

10.2.4.1 Conductivity detector (CD)

The conductivity detector (CD) is the most common detector for IC analysis, as all ions are electrically conducting, making this a universal detector for this type of chromatography. The CD is a modular detector with an integrated cell and has a signal range up to 15000 μS/cm. The detector is not ion specific *per se*, however, since the columns and the suppressors are ion specific, the counter ion to the analyte ion will be unretained on the column and will only be part of the background conductance.

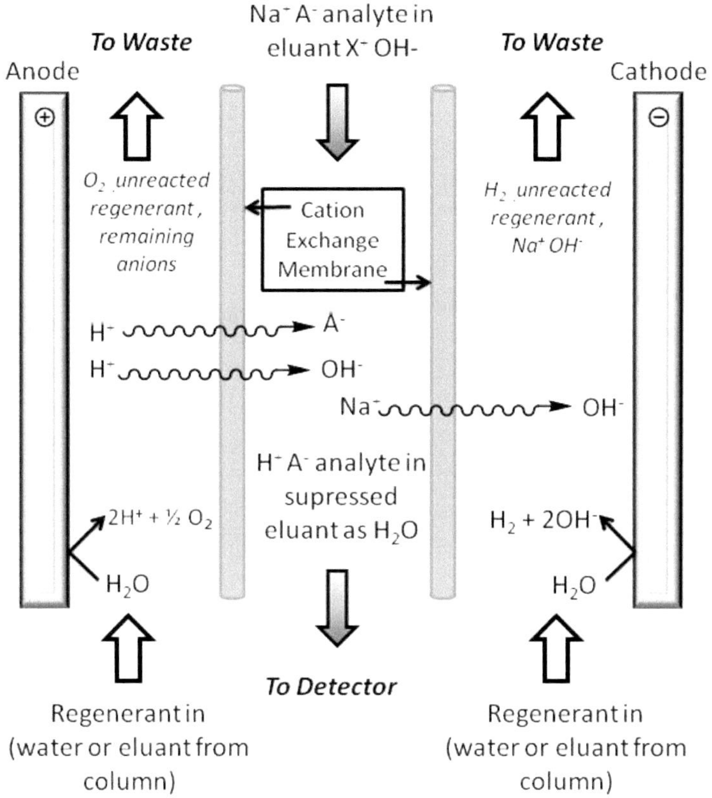

Figure 10.4 Schematic showing the ionic flows through the suppressor to the detector and then from the detector back through the suppressor for regeneration: Anion.

When a sample is eluted from the suppressor, the sample ions enter the detector where, since they are already naturally charged, they increase the conductivity in the detector. The response above baseline will then be the result of the analyte(s) of interest and their response will be proportional to their concentration.

The signal from the detector will be positive or negative, but the chromatography software will ensure that the data signals are displayed as positive peaks. Table 10.1 shows some typical specifications of a conductivity detector, based on sodium ion measurements.

Table 10.1 Typical specifications of conductivity detectors using sodium ion [8]

Sensitivity	5×10^{-9} g/mL
Linear dynamic range	5×10^{-9} to 1×10^{-6} g/mL
Response index[a]	0.97 to 1.03

[a]Which indicates that all monovalent ions have the same response, all divalent ions have similar response, etc.

The conductivity of a solution is measured by applying an alternating voltage between electrodes in a conductivity cell. The more ions pass through the cell, the more current is conducted through the solution, the greater the signal that is measured as there is a higher electrical conductivity. The electrical conductivity can be described using the following equations. Ohm's Law states that an applied potential (V in volts) is directly related to the current (I in amps) and the resistance (R in ohms):

$$V = R \tag{10.2}$$

The resistance of a solution depends on the type and concentration of ionic species in the solution and temperature. The conductance of a solution (G, units of Siemens, S) is the reciprocal of resistance:

$$G = \frac{1}{R} \tag{10.3}$$

The conductivity (k) which has units of S/cm is also related to the reciprocal of resistance and is also dependent on the cross-sectional area of the electrode (A) and the distance between the electrodes (d):

$$k = \frac{d}{AR} = \frac{d}{A}G = KG, \tag{10.4}$$

where K is the cell constant for the detector, with units of cm^{-1}.

Since conductance of electrolytes will vary with concentration (C), the equivalent or molar conductance (Λ) is defined per 1000 cm^3 as

$$\Lambda = \frac{1000k}{C} \tag{10.5}$$

Combining Eqs. 10.4 and 10.5, we get conductance defined as:

$$G = \frac{\Lambda C}{1000K} \tag{10.6}$$

Defining λ_+ and λ_- as the ionic conductance from both the positive and negative ions, or in the case of an infinitely dilute solution, the *limiting* equivalent ionic conductance for the cationic and anionic species in solution, Eq. 10.6 becomes:

$$G = \frac{(\lambda_+ + \lambda_-)C}{1000K} \tag{10.7}$$

Using the values for the limiting equivalent ionic conductance for common eluant and solute species typically used in ion chromatography (listed in Table 10.2), the background conductance of eluants can be calculated. For example, for a 5 mM KOH eluant in a cell with a cell constant K of 10 and using Eq. 10.7 and the values in Table 10.2, the background conductance due to the eluant would be:

$$G = \frac{(73.5 + 198.6)\ 0.005}{1000 \times 10}$$

$$G = \frac{(272.1)\ 0.005}{10000}$$

$$G = 0.00013605\ \ S = 136.05\ \text{mS}$$

Therefore, without suppression, the background conductance of an isocratic method using 5 mM KOH would be 136.05 μS. This calculation can also give you an idea on the effectiveness of your suppressor or if the magnitude of the (counter) current being applied to the suppressor is sufficient to negate the background conductance from the eluant.

Table 10.2 Limiting equivalent ionic conductances of commonly used ions in IC chromatography at 25°C [9, 10]

Ion		λ_n (S·cm^2/eq) [per mol]
Hydroxide	OH^-	198.6
Fluoride	F^-	55.4
Chloride	Cl^-	76.3
Bromide	Br^-	78
Iodide	I^-	77
Nitrate	NO_3^-	71.4
Sulfate	SO_4^{2-}	80 [160]
Phosphate	PO_4^{3-}	69 [207]
Chlorite	ClO_2^-	52
Chlorate	ClO_3^-	64.6
Hydrogen carbonate	HCO_3^-	44.5
Carbonate	CO_3^{2-}	72 [144]
Thiocyanate	SCN^-	66
Formate	$HCOO^-$	55
Acetate	CH_3COO^-	41
Ethanolate	$CH_3CH_2COO^-$	36
Benzoate	$C_6H_5COO^-$	32
Oxalate	$C_2O_4^{2-}$	74 [148]
Hydronium	H^+	349.8
Lithium	Li^+	38.7
Sodium	Na^+	50.1
Potassium	K^+	73.5
Ammonium	NH_4^+	73.5
Magnesium	Mg^{2+}	53.05 [106.1]
Tetraethylammonium	$N(C_2H_5)^+$	33
Calcium	Ca^{2+}	59.5 [119.0]
Zinc	Zn^{2+}	52.8 [105.6]

As long as the dissociation of the ions is complete (which is usually true for low concentrations typically used in ion chromatography), conductivity is linear with ion concentration.

Temperature also affects the conductivity of a solution. Equation 10.8 best describes the effect, based on a comparison of several equations to describe the temperature effects on conductivity detection [11]:

$$G_{25} = G_T \left(\frac{\eta_T}{\eta_{25}} \right)^K, \tag{10.8}$$

where G_{25} and G_T are the conductances at 25°C and T°C and η_{25} and η_T are the viscosities of water at 25°C and T°C. K is a constant.

To ensure reproducible results, it becomes important to control the temperature of the detector.

10.2.4.2 Other detectors and their uses

UV detectors can be used in ion chromatography (more commonly on non-suppressed systems), however, since most standard counter ions do not have any UV activity, it is not one of the more standard detectors. It can be useful when analyzing highly aromatic ions, amino acids or ions that can become detectible through derivatization. These detectors have a large analytical range, provided that signal absorption remains below 1 AU, adhering to Beer's law.

Amperometric detection is a type of electrochemical detection which uses a working electrode, a counter electrode and a reference electrode to measure either the oxidation or reduction of compounds. This detector is useful for compounds with high pKs, yielding low dissociation which would make them difficult to detect by conductivity. It is also useful for the detection of nonconductive compounds and increases the detection limits for ionic compounds. Amperometric detection has been applied to analysis of cyanamide [12], sulfite ions [13] and carbohydrates [2].

10.3 Method Development

Method development can start in a similar way as with developing a method for liquid chromatography. The first step in both method development plans is to search the literature for a good starting point and to determine the theoretical pKa of the molecule(s) of interest.

From the literature, there are a few review articles for novel ionic compounds, and data has been compiled using the "Virtual Column" software to optimize and compile IC retention data for different IC columns [14] however the current predominant use of ion chromatography is mostly for methods that involve only a few compounds to separate, so often a thorough literature search is not required. Determining the theoretical pKa, however, is very useful. In liquid chromatography, one is concerned about keeping the analyte neutral, so that it interacts as predicted with the stationary phase of the column. The opposite is true for ion chromatography. The analyte needs to be charged so as to interact with the stationary phase of the column, and only when the ion concentration of the mobile phase reaches a point where the analyte cannot compete and is less likely to interact with the column than the eluant itself, will the analyte of interest elute. As one might expect, monovalent compounds are more easily displaced than multivalent ions, which would, in most cases, require a higher concentration eluant.

Looking at the pKa curves also helps to highlight whether the analyte should be analyzed in anion or cation mode and whether a concentration or a pH gradient or both might be needed. It is essential to understand this as anions and cations require different columns, suppressors, and eluants as discussed above. The first requirement of ion chromatography method development is to determine what ion type (anion or cation) will be analyzed.

10.3.1 Sample Preparation and Diluents

In general, de-ionized water is the best solvent for IC samples; organic solvent use in sample diluents is not preferred for IC analysis and can cause issues with some of the columns and suppressors. When preparing samples in de-ionized water, use of sonication, stirring or shaking can improve the solubility of sample.

Plasticware, plastic pipettes, or pipette tips and plastic vials should be used exclusively when making standards, samples, and solutions in ion chromatography as ions such as sodium can be leached from glass.

If the sample does not readily dissolve in water but the ion of interest can be extracted into the aqueous phase, filtering off the insoluble part of the sample and analyzing only the supernatant is a sample preparation approach. This approach can also avoid any interference from high concentrations of the parent and reduce wear and tear on the guard and analytical columns. There are a variety of plastic filter vials available, where filtration can be performed directly in an IC vial which may then be placed directly into the autosampler. Spiked recovery studies should be performed into the appropriate matrix (e.g., drug substance for counterion analysis) to verify recovery using this in-vial filtration sample preparation approach.

If use of some organic diluent is unavoidable, this may require some replumbing of the IC system to avoid contact of the organic solvents with components susceptible to such solvents. Even if only a small amount of organic solvent is required, the guard column may need to be replaced more frequently. Organic solvents in the sample preparation may also impact the suppressor, column, and baseline. Additionally, sample precipitation from contact with the 100% aqueous mobile phase could be an issue depending on the solubility of sample. Consider a small amount of organic solvents such as methanol, ethanol, isopropanol, acetonitrile, and dimethylsulfoxide (DMSO) to improve the solubility of the sample only if the above recovery experiments fail and it is deemed necessary.

A final sample preparation possibility is to do an aqueous extraction by an organic solvent that is not miscible with water such as ethyl acetate, heptanes, and methylene chloride. Once dissolved, the extraction can be performed to get the ion of interest into an ion chromatographic compatible diluent, namely water.

10.3.2 Isocratic and Gradient Elution

The back pressures in ion-exchange columns are similar to core shell columns, which, in general, means that flow rates for ion chromatography can be faster than those in reverse-phase chromatography (with non-core shell columns). One must still keep below the pressure limits of the column and suppressor being used, but flow rates up to 2.5 mL/min can be employed.

For an isocratic method, the retention of monovalent and divalent ions shift to earlier retention times with increasing concentration of eluant. Consequently, if a shorter method is desired, the eluant concentration can be increased, but one must keep in mind the changes required for the suppressor (an increase in suppression current is required) and depending on how high the concentration of eluant used is, there may be more noise in the baseline, resulting in higher limits of detection.

Gradients can also decrease the run time, and have the advantage that they do not require the higher eluant concentrations for the entire run. Gradient elutions can be adjusted for the elution of monovalent species early on and then an increase in eluant concentration can have the divalent ions elute quickly after. However, for multivalent ions, elution depends more on pH as the valency of these ions is pH dependent. Phosphate is a good example of this.

Effect of pH on the speciation of phosphoric acid:

$$H_3PO_4 \leftrightarrow H^+ + H_2PO_4^- \quad pK_1 = 2.16$$

$$H_2PO_4^- \leftrightarrow H^+ + HPO_4^{2-} \quad pK_2 = 7.21$$

$$HPO_4^{2-} \leftrightarrow H^+ + PO_4^{3-} \quad pK_3 = 12.33$$

For these multivalent ions and also for biologic samples such as amino acids, a pH gradient instead of a concentration gradient (also known as a salt gradient) needs to be employed. In these cases two (or more) mobile phases are prepared at varying pHs and elution occurs as the pump gradient changes from one to the other thereby increasing or decreasing the pH. The gradients can be simple, changing gradually only one or two pH units or can cover a larger range. Depending on the complexity of the solution being analyzed, (i.e., two amino acids vs. a protein digest) several pH buffers may need to be employed. One must also consider ionic strength, as pH gradients can result in both higher pH and ionic strength, essentially having two modes of elution working together [15]. The choice of which gradient type to use depends on how close the other components in the sample elute. If there are only two or three components then a combination gradient can be used for sharp peaks and a quick method. If, however, the other components are all closely eluting then a more controlled pH gradient, maintaining the ionic strength of the eluant throughout the change in pH may be required.

Gradients can be further sped up by starting at higher initial isocratic eluant concentrations. One, however, is not limited to just one gradient. If the sample has a complex matrix and/or ions of different valencies, a method may require two or more different gradients combined with isocratic sections.

It is recommended to always have some length of isocratic portion as the first step. This is because IC systems can vary in dead volume (dwell time) and set up, which can make method transfers more challenging as the gradients do not always match. If there is an isocratic portion at the beginning of the method, then the gradient start time can be used to compensate for the difference in dead volume/dwell time. This is accomplished by starting the gradient sooner or later in the run, depending on whether the instrument transferring in has a larger or smaller dead volume/dwell time.

10.3.3 Suppressor Current Settings

For electrolytic self-regenerating suppressors, it is very important to select the proper operating current. A current setting set too high may permanently damage the suppressor. The optimum current setting produces just enough hydronium or hydroxide ions to displace the eluant counterions, thereby neutralizing the eluant.

The optimum current for the suppressor can be calculated by the following equation, which is based on Eq. 10.1:

$$\text{Current(mA)} = [\text{Flow Rate(mL/min)}] \times [\text{Eluent Concentration(mN)}] \tag{10.9}$$

Some software packages will calculate it or give a range of applicable currents for an isocratic separation based on the choice of eluant, its concentration, and flow rate.

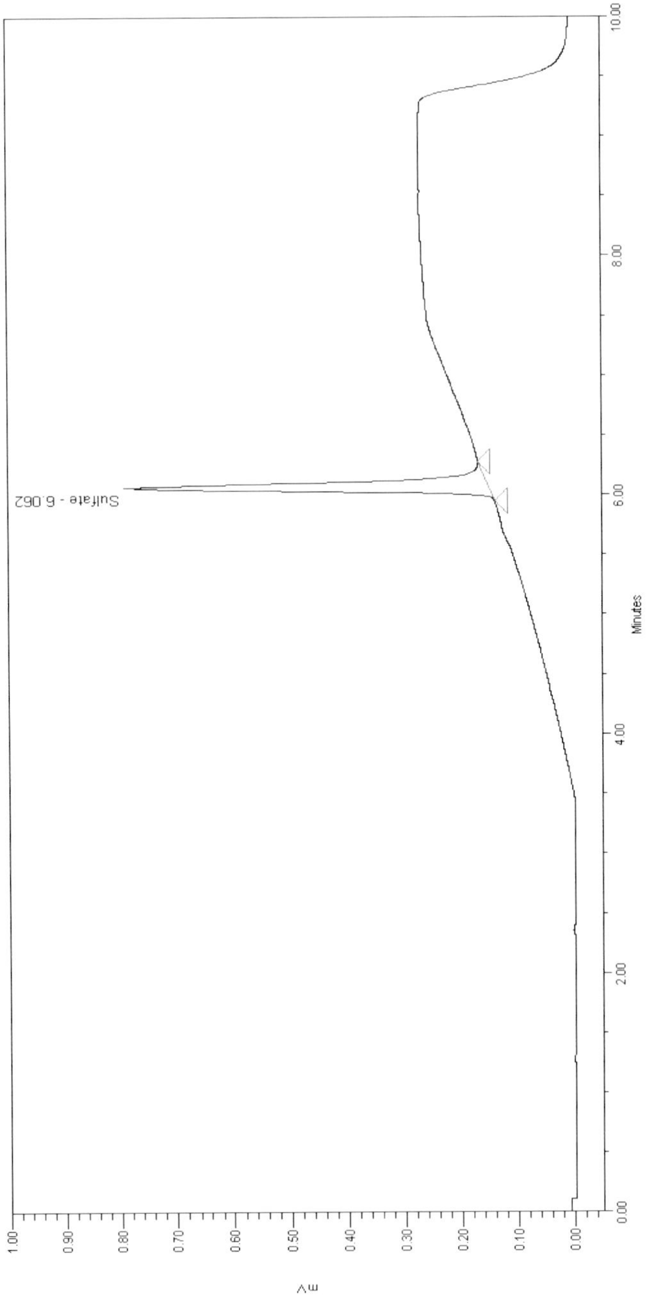

Figure 10.5 Chromatogram showing the change in the baseline as the eluant gradient changes the effluent conductivity above the suppressor setting for 1 mM hydroxide, which was set 7 mA.

Using the relationship in Eq. 10.9, the current setting on the suppressor can be adjusted during method development to compensate for the background conductance of the eluant in order to maximize your analyte signal. During method development, this parameter should be adjusted to give a zero baseline (or as closc to it as is reasonable).

When running a concentration gradient elution, one sees the baseline rise as the eluant concentration increases. A gradient can improve the separation and give sharper peaks, while running in a shorter time. However, with a fixed suppressor value, as the gradient background conductivity increases past the set suppressor current one sees a rise in the baseline. This can result in a lower detection limit as the peak of interest may now have to be integrated over a sloped baseline. Figure 10.5 shows the effect of a 1 mM to 15 mM gradient over 4 min and the resulting rise and then subsequent fall of the baseline as the gradient is returned to its starting concentration.

For gradient separations, a current "ramp" can be programmed similar to the eluant concentration ramp by calculating the appropriate current at each programmed gradient concentration adding the current change at the same time or similar times as the gradient is changed. There might be a need for a slight delay in the time program needed to change the suppressor current as the change in suppressor current will be immediate, but the gradient concentration change will need time to migrate through the system. The delay in the suppressor current change will be related to the system volume and flow rate.

Suppressors, just like columns, will degrade over time and are considered consumables. The question then arises: How do we know when to change the suppressor? High backpressure, decreased resolution, and non-Gaussian peaks are indicative of an aging column, however, some of these symptoms may also be due to the suppressor and it may be difficult to tell which part of the system is at fault. Increased background, therefore having to increase the current setting and loss of linearity or 100% noisy baseline with no peaks are also indicative of a deteriorating suppressor.

10.3.4 Temperature Effects

Temperature can affect chromatographic separations on an ion chromatography column in a similar manner as it does in liquid chromatography. Elution will be quicker at higher temperatures and compounds will elute later at lower temperatures.

Increasing the temperature decreases the viscosity of solutions and increases ion mobility. This directly affects the conductivity of a solution, and as such, temperature control of the CD is also important. A general rule for setting the temperature on the CD is that it should be at or above the column temperature, such that there is not a cooling gradient across the solution when entering the detector. Typical column temperatures for analysis in ion chromatography are 25 to 35°C. A general starting detector temperature for method development for that range of column temperatures would be 35°C. The maximum temperature for these detectors is generally around 60°C.

10.4 General System Troubleshooting

Troubleshooting in ion chromatography is similar to that in other types of chromatography, and a lot of the first pass troubleshooting one does applies to these systems as well. Some basic symptom/errors and how to correct them are as follows:

- High back pressure
  - Reduce flow rate
  - Check for blockage in system and/or column
  - Sample crashing out of solution along the flow path
- Baseline drift
  - Wait for system to stabilize (this can take a while)
  - Check for small leaks
  - If running a gradient, try a suppressor gradient
- Large sodium peak
  - Use plasticware and plastic pipettes
- Sensitivity has decreased
  - Column, suppressor or both are spent

- High background conductivity
  - Eluant is contaminated
  - Wrong suppressor setting; check to make sure that the current setting is appropriate for the flow rate and eluant concentration used
- High noise spikes with normal conductivity
  - Air in the system; check valves and connections for leaks
- High background conductivity with high noise
  - Replace suppressor

10.5 Applications

Most ion chromatographic applications in the pharmaceutical industry are currently counterion analyses, analyses of ionic impurities or detection of certain genotoxic impurities, which often require the detection and separation of one or two ions only, making method development fairly simple. However, as drug development expands to new modalities, including compounds such as nucleotides, amino acids, peptides, polypeptides, and proteins, all of which are now becoming part of or in fact, the active pharmaceutical ingredient, ion chromatography will be further challenged with their analyses. Chapter 9 and Paull and Nesterenko [7] have compiled lists of commercially available columns for amino acid and analysis of other biologics, showing that column developers are expanding on their stationary phase chemistries, to keep up with the trend towards analyzing more biologics.

Along with the newer horizons in drug development, there is continued scrutiny on the environment. There are many EPA and USP methods currently using ion chromatography and that number is likely to rise. The monitoring of wastewaters from sewage plants, mining operations, water runoff from farming, air quality monitoring are all areas where ion chromatography is being used. As acceptable limits for human exposure of metals and other ionic toxins in foodstuffs and drinking water are decreasing so must the limits of detection in the analytical methodology. This may lead to more interest in IC-MS methodology, which is available commercially but has yet to be employed to the same degree as other mass spectrometric combinations.

Acknowledgments

The author would like to thank Anthony Landry, Jelena Petrovic, and Amy McCusker for their help.

References

1. Eccles, H., and Greenwood, H. (1992) Chelate ion-exchangers: The past and future applications, A users view. *Solvent Extraction and Ion Exchange*, **10**, pp. 713–722.
2. Bhattacharyya, L., and Rohrer, J. S. (eds.) (2012) *Applications of Ion Chromatography for Pharmaceutical and Biological Products* (John Wiley & Sons, New Jersey, USA).
3. Herbranson, D. E., Eliason, M. E., and Karnatz, N. N. (1987) Development of a high performance ion chromatographic (HPIC) method for the determination of sodium metabisulfite in parenteral formulations. *Journal of Liquid Chromatography,* **10(5)**, pp. 3441–3450.
4. Koch, M., Köppen, R., Siegel, D., Witt, A., and Nehls, I. (2010) Determination of total sulfite in wine by ion chromatography after in-sample oxidation. *Journal of Agricultural and Food Chemistry,* **58**, pp. 9463–9467.
5. Iammarino, M., and Di Taranto, A. (2013) Determination and validation of an ion chromatography method for the simultaneous determination of seven food additives in cheeses. *Journal of Analytical Sciences, Methods and Instrumentation,* **3**, pp. 30–37.
6. Weiss, J. (2004) *Handbook of Ion Chromatography, Volume 1*, 3rd ed. (Wiley-VCH Verlag GmbH & Co., Germany).
7. Paull, B., and Nesterenko, P. N. (2005) Novel ion chromatographic stationary phases for the analysis of complex matrices. *The Analyst,* **130**, pp. 134–146.
8. Scott, P. W. R. (2013) *Ion Chromatography: Theory and Practice in Chromatographic Analysis of Cationic and Anionic Species*, Chrom-Ed Series Book 5 (Integritext, UK).
9. Shpigun, O. A., and Zoltov, Y. U. (1988) *Ion Chromatography in Water Analyses* (Ellis Horwood, Chichester, UK) p. 85.

10. *A Practical Guide to Ion Chromatography*. (2007) Sequant AB, Sweden. ISBN 978-91-631-8056-9.

11. Haddad, P. R., and Jackson, P. E. (1990) Conductivity detection, in *Ion Chromatography (Journal of Chromatography Library Volume 46)*, Chapter 9 (Elsevier Science, Sydney, Australia).

12. Nair, J. B. (1994) Determination of trace levels of cyanamide in a novel potassium channel activator bulk drug by pulsed electrochemical detection. *Journal of Chromatography A,* **671**, pp. 367–374.

13. Casella, I. G., and Marchese, R. (1995) Sulfite oxidation at a platinum glassy carbon electrode. Determination of sulfite by ion exclusion chromatography with amperometric detection. *Analytica Chimica Acta,* **311**(2), 199–210.

14. Park, S. H., Shellie, R. A., Dicinoski, G. W., Schuster, M. T., Haddadd, P. R., Szucs, R., Dolan, J. W., and Pohl, C. A. (2016) Enhanced methodology for porting ion chromatography retention data. *Journal of Chromatography A,* **1436**, pp. 59–63.

15. Wang, Y., Moreno, G. T., Zhang, B., Zhang, L., Farnan, D., Patapoff, T. Ionic strength-mediated pH gradient ion exchange chromatography. WO 2014/078729 A l, 22 May (2014).

Chapter 11

Fundamentals, Properties, and Applications of Stationary Phases for Gas Chromatography Method Development

Omprakash Nacham and Jared L. Anderson

Department of Chemistry, Iowa State University,
1605 Gilman Hall, Ames, Iowa, USA

onacham@iastate.edu, andersoj@iastate.edu

Gas chromatography (GC) is an important separation technique that has experienced great popularity within numerous disciplines of the chemical sciences. Technological advancements in column fabrication as well as the development of new stationary phase chemistries have played a significant role in increasing the applicability of GC. To address the challenges associated with complex samples, a wide range of stationary phases with different chemical selectivities and thermal properties have been developed.

Chromatographic Method Development
Edited by Gregory K. Webster and Laila Kott

ISBN 978-981-4800-53-2 (Hardcover), 978-0-429-20172-1 (eBook)
www.jennystanford.com

The introduction of poly(siloxane) and polyethylene glycol stationary phases has greatly improved the separation power of GC. Importantly, the development of new stationary phase bonding techniques to immobilize these stationary phase materials has significantly increased the thermal stability, durability, and lifetime of GC columns. Ionic liquids (ILs) and polymeric ionic liquids (PILs) have emerged as promising stationary phase materials. They demonstrate the potential to complement nonionic stationary phases because of their interesting and useful physicochemical properties. The structural tunability and high thermal stability of ILs have provided potential alternatives to conventional stationary phases, particularly in the analysis of polar compounds (e.g., hydrogen bond donors) as well as for high-temperature GC separations. Techniques that are employed during stationary phase immobilization/deposition greatly influence the separation efficiency of the GC column. As a result, a fundamental understanding of stationary phase properties, column evaluation, and principles involved in column fabrication methods is useful in GC analysis. This chapter provides an overview on the properties, applications, and procedures used in the fabrication of GC columns that are essential in aiding GC method development.

11.1 Introduction

Gas chromatography (GC) is a separation technique that has been widely used for the analysis of volatile and semi-volatile chemical constituents. GC has become a workhorse separation method in a variety of applications, including petroleum analysis [1], food sample testing [2], environmental sample screening [3], and drug analysis [4]. The remarkable work on GC in the 1940s by A. J. P. Martin and R. L. M. Synge is considered a milestone in the evolution of modern GC techniques.

Advances in column technology have played a significant role in transforming the face of GC. Before optimizing any GC analysis, the selection of an appropriate column type as well as stationary phase is an important prerequisite. GC columns are broadly divided into two classes, namely, packed columns and open tubular columns. In packed columns, a solid material (possessing high surface area) or a liquid coated on the solid support typically functions as the stationary phase. High sample capacity, low phase ratio (ratio of the volume of gas to liquid phase), and a plethora of stationary phase materials that are compatible with column fabrication procedures are characteristics of the classical packed columns. However, lengthy analysis times, low column efficiency, poor resolution, and band broadening are some drawbacks of packed columns [5, 6]. Problems associated with classical packed columns are partly alleviated by the introduction of micropacked columns. They possess inner diameters lower than 1 mm with particle sizes ranging from 40–100 μm. This type of column has been successfully employed for trace level analysis and exhibit potential for coupling with mass spectrometry (MS) [6]. Yet, relatively low column lengths, demand for modified injection devices to withstand high inlet pressures, and reduced throughput has significantly limited practical applications of the micropacked columns [7]. Despite their limitations, classical packed columns and micro-packed are commonly utilized for preparative-scale separations.

The growth and developments in open tubular column technology have largely increased the applicability of gas-liquid chromatography (GLC) to many areas of research within the chemical sciences. The concept of open tubular columns and its potential advantages over packed columns were first explained by Golay in the late 1950s [8]. When a stationary phase (liquid or adsorbent) is either coated on an inert support or directly applied on the walls of packing material with capillary dimensions, the resulting columns are called open tubular (capillary) GC columns. High peak capacity, low peak dispersion, high plate numbers, good reproducibility, and relatively low analysis times are examples of attractive features for open tubular columns [9]. Among the different open tubular column configurations, wall-coated open tubular columns (WCOTs) have enjoyed much great success and exhibit compatibility with a broad range of GC instrumentation. Owing to their versatile nature and favorable stationary phase properties, WCOTs are considered as a first choice of columns for most analytical GC separations.

The selectivity of a stationary phase plays a crucial role in separating multi-component samples. In parallel to advancements in column technology, a significant amount of effort has been invested towards the development of innovative new stationary phase materials that can impart the desired selectivity and high thermal stability to satisfy the increasing demands of complex samples. Today, a wide range of GC columns are available with different chemical selectivities and thermal properties. As a result, selection of an appropriate stationary phase that can provide high selectivity as well as good separation efficiency is an important task and holds significant impact in GC analysis.

This book chapter provides an overview on the various column fabrication procedures and different types of stationary phases utilized in GC analysis. The first section of the chapter focuses on fundamentals involved in column preparation, evaluation, and their influence on gas chromatographic separation. The second section of this chapter encapsulates the properties, applications, and limitations of current stationary phases such as poly(siloxanes) and polyethylene glycols. Finally, emerging stationary phases including ionic liquids (ILs) and PILs will be discussed along with their advancements with respect to improved chemical selectivities and thermal stabilities for providing solutions to separating complex sample.

11.2 Column Preparation and Evaluation

11.2.1 Materials and Methods Involved in Preparation of WCOT Capillary Columns

The tubing materials used in the fabrication of capillary GC columns include stainless steel, nickel, glass, and fused silica [6]. Capillary GC columns manufactured with fused silica and glass are more common than other tubing materials. The introduction of the glass capillary drawing machine in the early 1970s by Desty revolutionized capillary column technology [10]. In comparison to other tubing materials, glass is cheap, easily deformable, and has a homogenous surface. For nearly a decade, glass capillary columns have remained as a mainstay in GC separations. However, the poor wettability, presence of high amounts of residual metallic impurities (sodium, calcium, barium), and chemical reactivity with analytes present serious challenges in the use of glass capillary columns [9].

Fused silica capillary columns were first introduced by Dandeneau and Zerenner and co-workers in the 1970s [11]. Fused silica capillary columns offer greater flexibility and chemical inertness when compared to glass columns. They are mostly composed of primary silanol groups and contain less than 1 part per million (ppm) of metallic impurities and are supported externally by a polymeric coating (e.g., polyimide). Today, most capillary GC columns are prepared using fused silica tubing.

To coat a thin layer of liquid stationary phase on the surface of tubing material, the surface tension of the stationary phase should be lower than the critical surface tension of the coating surface. Surface modification reactions are often employed to increase the free surface energy (reduce the surface tension) as well as to deactivate the free silanol groups. There are different types of surface modification reactions available depending on the column tubing material and stationary phase chemical composition. Acid-leaching, etching with reactive gases [12], exhaustive silylation, and particle-deposition [13] are some common methods to increase the free surface energy of the glass and fused silica capillaries [14, 15]. For example, acid-leaching using 20% (v/v) hydrochloric acid at 100–170°C followed by silianization has greatly improved the surface wettability of fused silica columns for coating non-polar and moderately polar stationary phases. Most particle deposition methods use a suspension of sodium chloride or barium carbonate and are utilized for roughening glass surfaces. The surface activity of fused silica or glass capillary columns can be reduced by either high-temperature silylation reactions or condensation with poly(alkylhydrosiloxanes) [16].

WCOT columns (glass or fused silica) are coated using either the dynamic or static coating method. In the dynamic coating method, a solution of the stationary phase (5–15% w/v) is typically prepared in a suitable coating solvent and forced through the column by an inert gas. Subsequent to coating, the temperature of the gas is increased to evaporate the coating solvent. Various modifications to this method were made to provide uniform and reproducible coatings. Importantly, a short plug of mercury is introduced between the coating solution and a driving gas. As the mercury plug is moving through the column, a uniform and thin layer of coating solution develops on the wall of the capillary column [17]. This method has largely improved the reproducibility of the coating. However, an important weakness associated with the dynamic coating method is the difficulty in predicting the thickness of the stationary phase. Alternatively, static coating methods enable the preparation of columns with a predictable thickness. In static coating methods, the desired coating solution (0.02–4.0% v/v) is prepared in a suitable volatile organic solvent (e.g., dichloromethane, pentanes, etc.). After completely filling the capillary, the coating solvent is evaporated by sealing one end of the column and attaching another end to the vacuum source. As the solvent is evaporated under controlled vacuum conditions, a thin layer of stationary phase is deposited on the column surface. The thickness of the stationary phase is related to concentration of coating solution as well as inner diameter of the capillary column [9, 18]. Although static coating methods are superior for producing reproducible and thin stationary phases, these methods can be time-consuming and laborious.

The demand for new column fabrication procedures requiring less time and yielding more reproducible coatings is rapidly increasing. Sol-gel coating techniques minimize the total number of steps involved in the column fabrication procedure in which surface roughening, deactivation, coating and/or immobilization are performed in a single step. Wang et al. employed the sol-gel coating technique for the fabrication of a PDMS column (efficiency of 3400 plates/meter; 10 m × 250 μm i.d.) and demonstrated its utility for separating a mixture of saturated and unsaturated fatty acid methyl esters [19]. Using this technology, two columns, SolGel-ms™ and SolGel-WAX™, have been recently commercialized. The applicability of sol-gel techniques for different classes of stationary phase materials is still under constant development.

11.2.2 Evaluation of Gas Chromatographic Column Performance

The quality of open tubular GC columns is gauged based on three important parameters, including efficiency, chemical inertness, and thermal stability. Numerical values such as plate number (N), effective plate number (N_{eff}), and plate height (H) are commonly used to express column efficiency. These values are determined by measuring the retention time of an inert chemical substance (alkanes, aromatics, etc.) under isothermal conditions [20]. The selected probe should exhibit reasonable retention time and enough solubility in the stationary phase. The stationary phase film thickness, column dimensions, carrier gas type, and velocity largely influence the column efficiency [21]. For example, as the inner diameter of the column decreases, the efficiency of the column increases. For a commercial WCOT column (30 m × 0.25 m i.d.) with a film thickness of 0.25 μm, *N* values vary typically from 5000–6000 plates/meter (non-polar stationary phase) [9]. However, depending on the physicochemical properties (i.e., viscosity, polarity, surface tension) of the stationary phase and applied coating method, efficiencies can be as low as 1000 plates/meter. Often low column efficiency results from improper coating techniques that can cause uneven film formation or a continuous film with variable thickness resulting in low efficiency [21]. An efficiency test is a quick and easy diagnostic tool to assess the column quality, but it should be complemented with other tests to obtain comprehensive information on column performance.

Subjecting the coated capillary columns to standardized test mixtures enables access to information about the chemical inertness and separation efficiency. The Grob test and Luong test are commonly used to examine the secondary interactions of the stationary phase [22, 23]. These interactions result in peak broadening, peak tailing, irreversible adsorption, and degradation. The Grob test mixture is a solution composed of twelve different analytes with a wide range of polarities. On the basis of peak shape and peak resolution of the test analytes, the column performance is estimated. For example, based on the peak shapes of alcohols (1-octanol and nonanol), column activity due to residual silanol groups on non-polar stationary phase can be determined.

Film instability and decomposition and/or volatilization of the stationary phase at high column temperatures play a crucial role in diminishing the column performance. Therefore, the identification of maximum allowable operating temperature (MAOT) of a stationary phase prior to GC separation is an important prerequisite. To determine the MAOT, a thin layer of the stationary phase is deposited on the capillary surface and subjected to a programmed temperature run ranging from 40 to 400°C using a low-temperature ramp. During these measurements and at a specific temperature or range of temperatures, chemical constituents (intact stationary phase or decomposed products) that evolve from coating materials contribute to a significant rise in the baseline which is considered as the MAOT of the stationary phase [24–26]. Alternatively, the determination of a column's efficiency before and after holding the column at high temperatures (approximately 12 h) provides valuable information on the thermal stability and chemical inertness of the stationary phase (isothermal GC method) [21]. Coupling any of these two techniques to sensitive detectors such as mass spectrometry (MS) enables the structural information of any decomposed products. In research laboratories as well as in the production environment, this data can be used to investigate decomposition mechanisms and aids in tailoring the chemical structure to reduce overall thermal decomposition. In some cases, data obtained from thermogravimetric analysis (TGA) is used for establishing the thermal stability of new stationary phase materials such as ionic liquids. Due to lower sensitivity of the TGA method when compared to GC-based methods, thermal decomposition temperatures obtained from TGA are not considered a true estimation of MAOT for a GC stationary phase.

11.3 Selectivity of Gas Chromatographic Stationary Phases

Various models have emerged for the classification of stationary phases and to facilitate the column selection process for GC method development. Solvent strength (polarity) and solvent selectivity are the primary basis for earlier models to classify stationary phases. Since the solvation process is the result of multiple and independent interactions between a solute and stationary phase, the use of single value polarity scales does not always provide reliable information on stationary phase properties. On the other hand, solvent selectivity models are formulated by considering various solute-solvent intermolecular interactions including dispersion, induction, orientation, and hydrogen bonding. The system of phase constants introduced by Rohrschneider (further developed by McReynolds) and the Abraham solvation parameter model are the two most common models used to classify stationary phases based on selectivity [27–29].

The Rohrschneider–McReynolds system assumes that intermolecular interactions are additive, and their individual contribution to retention can be evaluated from the difference in retention index values for a series of prototypical compounds on the stationary phase to be classified. Within this model, squalane is used a reference nonpolar phase and retention index values are determined with respect to squalane at 120°C. Although McReynolds phase constants are successfully used to characterize various stationary phases, this method has some limitations. The measurement of retention indices at 120°C results in minimum interactions with some stationary phases and complicates phase constant determination. In addition, a fewer number of test solutes can be used to derive the phase constants and it is assumed that a single dominant intermolecular interaction of test solute is responsible for the solvation process. Selectivity scales based on Gibbs free energy of solution have been employed to understand the retention characteristics of the stationary phase. However, these methods turned out to be unreliable and provided inexplicable results for various stationary phases chemistries [30].

11.3.1 Solvation Parameter Model

Today, a vast majority of new stationary phases are classified using system constants generated from the Abraham solvation parameter model. This model assumes that the solvation process occurs in three steps. Initially, a cavity similar to a solute's volume is formed in the stationary phase. Subsequent to reorganization of solvent molecules around the cavity, the solute is transferred into the cavity and interacts with solvent molecules. It can be assumed that the free energy associated with cavity reorganization is negligible compared to other processes, and gas-solute and solute-solute interactions in the gas phase are negligible. The retention equation describing the solvent parameter model describing all possible interactions between the solute and stationary phase is given in Eq. 11.1.

$$\log k = c + e\mathrm{E} + s\mathrm{S} + a\mathrm{A} + b\mathrm{B} + l\mathrm{L} \quad (11.1)$$

In this method, a set of probe molecules with various functional groups is applied on stationary phase coatings, and the dependent variable (log k) chromatographic retention factor for each probe molecule is measured. The retention factor is a function of interactions with the stationary phase and functional groups on the probe molecule. As shown in Eq. 11.1, each probe molecule has a unique set of solute descriptors defined as follows: E is excess molar refraction, S is the solute dipolarity/polarizability index, L is the solute gas-liquid distribution coefficient at 298 K, A and B are solute hydrogen bond acidity and basicity, respectively. The values for the solute descriptors can be obtained from literature. The coefficients of solute descriptors are called the system constants (e, s, a, l, b) which are temperature dependent and provide a measure of interactions between the stationary phase and solute molecules. The system constant e provides the information of solute-stationary phase interaction via π–π and n–π interactions, s is a measure of stationary phase dipolarity, l defines the ability of the stationary phase to undergo dispersive interactions with solute molecules, a and b are measure of stationary phase hydrogen bond basicity and acidity, respectively. These system constants are calculated by subjecting the retention factors and solute descriptors to multiple linear regression analysis [24, 31–34]. Tables 11.1 and 11.3 represent the system constants of selected GC stationary phases. More information regarding the system constants of various stationary phases at different temperatures can be obtained from several sources [35–38].

11.4 Stationary Phase Materials

Since the introduction of GC, numerous liquid materials have been used as stationary phases. Unfortunately, most of the early stationary phase materials were abandoned due to lack of reproducibility in their production, undefined chemical composition, poor retention reproducibility, and low thermal stability [9]. However, with greater advancements in chemical synthesis and column fabrication techniques, liquid stationary phases that can circumvent these challenges while providing improved selectivity and chromatographic performance have been produced.

Table 11.1 System constants at 120°C for representative poly(siloxane)-based stationary phases from different column manufacturers

Stationary phase	Solvation parameter model System constants *e*	*s*	*a*	*b*	*l*	Ref.
DB-1	0	0.215	0.175	0	0.507	[35]
HP-5[a]	0.004	0.306	0.199	0	0.585	[35]
Rtx-20[a]	0.030	0.535	0.245	0	0.545	[35]
DB-35[a]	0.074	0.640	0.277	0	0.531	[35]
Rtx-65[a]	0.120	0.828	0.330	0	0.528	[35]
DB-225[b]	−0.030	1.232	1.176	0	0.457	[35]
DB-200[c]	−0.0335	1.041	0.147	0	0.465	[35]

DB-1: 100% PDMS.

[a]Number indicates the percentage of diphenylsiloxane monomer.

[b]50%cyanopropylphenylsiloxane monomer.

[c]Poly(dimethyltrifluoropropylsiloxane).

11.4.1 Aliphatic and Aromatic Hydrocarbon-Based Stationary Phases

Separations on hydrocarbon-based stationary phases are mostly due to dispersive-type (cohesive) interactions. High-molecular-weight hydrocarbons such as hexadecane, squalane, Apolane-87 and different grades of Apiezons are common hydrocarbon-based stationary phases [39]. Attempts to prepare hydrocarbon phases with some degree of branching as well as incorporation of halogens as substituents has resulted in diminished selectivity [6]. This is likely due to orientation effects and reduced dispersion interactions with the solute molecules. Squalane and Apiezon require extensive chemical purification as their precursor materials are often obtained from natural sources. There are several types of Apiezon, including Apiezon A, B, C, L, M, MH, and W. Of all these grades, Apiezon MH and L are commonly used because of their better quality and reproducibility in production [9]. Apolane-87 is a well-characterized synthetic hydrocarbon phase possessing higher thermal stability than other hydrocarbon phases. Sojak and co-workers developed WCOT columns with Apolane-87 and subsequently used them for the separation of linear pentadecene isomers. It has been claimed that Apolane-87 exhibited superior selectivity and higher thermal stability than squalene [40].

Hydrocarbon phases are used for the analysis of petrochemical samples and for the separation of perfluorocarbons from partial or non-fluorinated hydrocarbon samples [6]. They are mostly used as non-polar reference phases in various stationary phase classification schemes. Important shortcomings of hydrocarbon stationary phases include susceptibility to oxidation, mixed retention mechanisms (providing active sites on the column walls), poor wettability, volatility, and low MAOT. The chemical stability of aliphatic phases was significantly improved by the introduction of perfluorocarbon-based stationary phases [41]. These are mostly used for the separation of reactive chemicals such as metal fluorides, halogens, and halide compounds of hydrogen, sulfur, and phosphorus where the conventional phases are prone to degradation. In some cases, highly-fluorinated stationary phases are used for the separation of isomeric perfluorocarbons and Freons [42, 43].

To overcome problems related to volatility and thermal stability of aliphatic hydrocarbons, polyaromatic compounds were investigated as liquid stationary phases. Compounds such as benzyldiphenyl, alkylnapthalenes, polyphenyl tar, and linear polystyrene are among some earlier liquid stationary phases used for the separation of aromatics and heterocyclic compounds. However, the minimum column operating temperature required for these stationary phases range from 60 to 230°C. This shortened range has reduced their suitability as stationary phases towards a wide variety of samples [6]. Most of the aliphatic and aromatic hydrocarbon phases have been replaced by robust and thermally stable poly(organosiloxane)-based liquid stationary phases.

11.4.2 Poly(Siloxane)-Based Stationary Phases

Poly(siloxanes) represent one of the most popular stationary phase materials in capillary GC. The advances in silicone chemistry have provided the opportunity to exploit poly(siloxanes) as indispensable stationary phases for many complicated separations. Relative to hydrocarbons, the high bond energy of Si-O and Si-C are responsible for high thermal stability and low vapor pressure of poly(siloxanes) [44, 45]. Depending on their chemical composition, the viscosities of poly(siloxane) phases are slightly affected by temperature. This represents an important advantage over aliphatic and aromatic hydrocarbon stationary phases. In addition, good wettability, low glass transition temperature, and stable film formation capabilities have made these materials suitable GC stationary phase candidates.

Figure 11.1 depicts representative chemical structures of poly(siloxanes). Substituents groups such as methyl, vinyl, phenyl, and 3-cyanopropyl are attached as R groups to tailor the chemical selectivity of the resulting poly(siloxanes). Poly(dimethylsiloxane) (PDMS; R = methyl) is the most commonly used non-polar stationary phase. Because of its wide operating temperature range, PDMS (DB-1; commercial identity) is capable of resolving samples with a broad range of boiling points. In most cases, this is the first liquid stationary phase of choice when there is little information

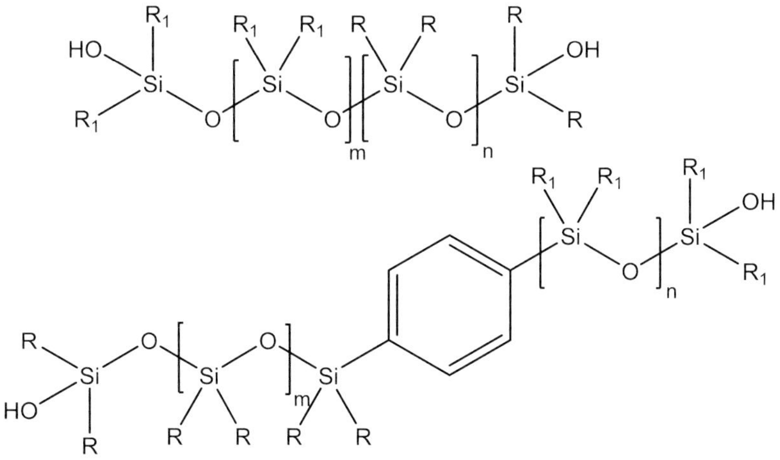

R_1= Cyanopropyl, phenyl, cyclodextrin, trifluoropropyl etc.,
R = Methyl

Figure 11.1 Common structural motifs of poly(siloxane)-based stationary phases containing varying functional groups that result in different chromatographic selectivity.

on a sample's composition. By increasing the percentage of diphenyl or cyanopropyl monomer content in the PDMS phase, poly(siloxanes) with a wide range of polarities are generated. This effect can be clearly observed from the change in *s* (system constant used to measure dipolarity of the stationary phase) value of the stationary phases. For example, by introducing 65% diphenylsiloxane monomer (Rtx-65) in PDMS, the *s* value increases from 0.215 to 0.828, as shown in Table 11.1. The high dipolar-type interactions in Rtx-65 are used for the separation of phenols and fatty acid samples. Table 11.2 presents the commercial identities and selected applications of various poly(siloxane) phases. For Table 11.2, a representative column is chosen from each class of poly(siloxanes) and their recent gas chromatographic separations are reported.

11.4.2.1 Preparation of poly(siloxane)-based stationary phases for open tubular columns

For the preparation of alkyl or phenyl substituted poly(siloxanes), dichlorosilanes or dialkoxysilanes are usually subjected to acid/base hydrolysis [44, 46]. In the presence of functional groups such as esters and cyclodextrin (which are sensitive to hydrolysis conditions), poly(siloxanes) phases are often prepared via hydrosilylation techniques [47, 48]. Owing to the depolymerization effects of acid/base catalysts on poly(siloxanes) via cyclosiloxane formation at high temperatures, several purification methodologies are employed following the synthesis of poly(siloxanes) [9].

The thermal stability and viscosity of poly(siloxane) phases are highly dependent on the attached functional group and its position relative to the siloxane bond. The presence of electron withdrawing groups, such as cyano and fluoro, adjacent to the siloxane bond greatly enhance the decomposition at lower temperatures. To maintain the desired selectivity and thermal stability, a minimum distance of three carbon atoms from the siloxane bond is usually maintained. Furthermore, it was observed that the incorporation of bulky groups such as phenyl (Fig. 11.1) and corborane in the polymer backbone resulted in improved thermal stability with minimal cyclosiloxane formation.

Table 11.2 Applications of poly(siloxane)-based stationary phases

Stationary phase	Commercial identity	Representative column	Sample types	Sample matrix	Ref.
100% PDMS (non-polar)	AT-1, BP-1, CB, CP-sil-5, DB-1, Equility-1, HP-1 ms, Optima-1 ms, DB-1 ms, ZB-, AT-1 ms, Sol-gel-ms, HP-1, MDN-1, Optima-1, OV-1, OV-101, SE-30, Rtx-1, SPB-1, VF-1 ms	DB-1	Steroids, organochlorine pesticides, mixture of essential oils, aromatic and saturated hydrocarbons, pyrethroid pesticides, chemical warfare agents	Air; ginger extract; crude oil; soil; shale gas; tea.	[81–87]
5% phenyl, 95% dimethysiloxane (non-polar)	AT-5, AT-5 ms, BP-5, BPX-5 ms, CP-Sil-8, CP-Sil-8, CB-ms, DB-5, Equity-5, HP-5, HP-5 ms, MDN-5, MDN-5 s, MTX-5, Optima-5, Optima-5 ms, OV-23, Permabond SE-52, PB-5, RTX-5, RTX-5 ms, SPB-5, Ultra-2, ZB-5, 007-5	DB-5	Alkylphenol, parabens, aroma compounds, polychlorinated dibenzofurans, polychlorinated dibenzo-p-dioxins, polybrominated dibenzo-p-dioxins, essential oils, Polychlorinated biphenyls, unsaturated isoprenoids	Urine, blood; breast milk; particulate matter, pomegranate; agricultural soils; fresh water.	[76–80]
6% Cyanopropyl phenyl, 94% dimethylsiloxane (mid-polar)	AT-624, BP-624, CP-624, CP-30, DB-624, DB-30, DB-VRX, HP-624, HP-30, Optima-624, Optima-30, PE-624, PE-30, RTX-624, RTX-30, RTX-Volatiles, VF-624, ZB-5, ZB-624,007-624, 007-502	DB-624	Low-molecular-weight silanol and siloxanes; volatile organic compounds (VOCs); residual solvents; odor-active sulfur and nitrogen compounds; flavoring excipients	Biogas; gas; municipal sludge sample; pharmaceutical drugs and intermediates; fermented sausage; pork liver powder.	[92–97]

(Continued)

Table 11.2 (*Continued*)

Stationary phase	Commercial identity	Representative column	Sample types	Sample matrix	Ref.
35% phenyl 65% dimethylsiloxane (mid-polar)	AT-35, BPX-35, DB-35, HP-35, MDN-35, RTX-35, SPB-35	DB-35	Pesticide residues; polycyclic aromatic compounds and tobacco specific N-nitrosamines; methylmalonic acid (biomarker for organic aciduria); plant primary metabolite profiling	Tobacco; fruit samples; cigarette smoke; urine; grape leaves.	[88–91]
14% cyanopropyl phenyl 86% dimethylsiloxane (mid-polar)	BPX-10, CB-1701, CP-Sil-19CB, DB-1701, DB-1701P, HP-1701, Optima-1701, Rtx-1701, SPB-1701, ZB-1701, 007-1701	DB-1701	Organophosphorus pesticides; organochlorine pesticides; trihalomethanes; trimethyl silyl derived monosaccharides	Plant; tobacco products; skimmed milk; drinking water; plant.	[101–103]
50% Cyanopropyl phenyl 50% dimethylsiloxane (polar)	BP-225, BPx-50, HP-225, DB-225, Optima-225, OV-225, RTX-50, RTX-225, SP-2250, ZB-1701, 007-225	DB-225	Pectic polysaccharides; mono unsaturated fatty acid profiling	Fruit samples; plasma; serum.	[98–100]

Earlier approaches used in the preparation of poly(siloxanes) relied on the direct application of coating solutions on the wall of capillary materials. These columns are unsuitable for long-term use and their low thermal stabilities precluded their use in high-temperature GC separations. The introduction of simultaneous stationary phase immobilization and cross-linking techniques has greatly improved film stability, durability, and thermal stability of poly(siloxanes). Techniques such as thermal condensation and radical initiated cross-linking methods are used to immobilize poly(siloxanes) on the walls of open tubular columns. Most of the modern GC capillary columns are produced using one of these two techniques. In thermal immobilization techniques, non-polar or moderately polar poly(siloxanes) with terminal silanol groups are used as starting materials. A high column temperature is used to cross-link the linear polymer chains and to react silanol-terminated poly(siloxanes) with the walls of capillary materials (fused silica/glass). Occasionally, a small of percentage of cross-linking agent such as alkyltrimethoxysilanes is also included in the coating mixture to enhance the reaction efficiency as well as improve the thermal stability [49, 50].

In free radical immobilization methods, endcapped poly(siloxanes) are cross-linked onto the deactivated column using a variety of radical initiators. Dicumyl peroxide [51], azo-compounds [52], γ-radiations [53] are some examples of radical generators. Unlike thermal techniques, organic substituents attached to the poly(siloxanes) participate in cross-linking reactions and no new siloxane bonds are formed with adjacent polymer chains. However, the success of cross-linking largely depends on the chemical nature of substituents. In general, neat PDMS is a good substrate for cross-linking whereas the presence of cyano, phenyl, and long alkyl chain lengths reduces its reaction efficiency. In such cases, a small percentage of poly(siloxane) monomers endcapped with functional groups such as tolyl and vinyl are added to enhance the reaction efficiency. These immobilization techniques not only help in obtaining columns with high thermal stability, but also enable thicker coatings (1.0–8.0 μm) and provide stationary phases that can tolerate solvent rinsing procedures. The latter has a significant advantage in extending the column lifetime and is especially useful for removing non-volatile components that can originate from complex sample matrices [44, 54].

Immobilization approaches involving sol-gel technology has reduced the total number of steps involved in conventional immobilization methods. However, only a few column types have been commercialized using this method. Nevertheless, technological advancements made in stationary phase bonding chemistry have enormously increased the types of applications and samples that can be analyzed by GC.

11.4.2.2 Modified poly(siloxane) stationary phases for chiral analysis

Since the separation of isomeric molecules with similar vapor pressure is a formidable task using conventional poly(siloxanes), chiral selectors are often either tethered or dissolved in poly(siloxane) phases for chiral GC applications. Crown ethers [55], chiral amino acids [56], calixarenes [57], resorcinarenes [58], and cyclodextrins [59, 60] are some common chiral selectors often used in chiral GC analysis. In chiral separations, hydrogen bonding, coordination, steric interactions, and inclusion formation are four important types of interactions responsible for enantioselectivity [9]. The chemical cross-linking of chiral selectors (amino acids, cyclodextrin, and metal bound ligands) to poly(siloxanes) and subsequent immobilization and/or coating on GC capillary has increased the applicability of chiral GC analysis [61].

11.4.2.3 Chiral separations using α-amino acid-based poly(siloxanes)

The inspiration for the design of amino acid-based chiral stationary phases (CSPs) was conceived from the stereoselectivity of enzymes which catalyze biochemical reactions to produce a stereospecific product. The chiral poly(siloxane) stationary phase containing valine diamide was termed as Chirasil-Val. Commercially, this product is available in both enantiomeric L-Val and D-Val forms. On D-Val column, L-amino acids elute earlier than D-amino acids and vice versa. This is highly useful in performing method validation and enantiomer labeling experiments [61]. These columns possess thermal stability from 70–250°C and exhibit low tendency for column bleeding. Chiral separations on diamide-based stationary phases predominantly occur via hydrogen bonding interactions. A variety of synthetic approaches have been investigated in the preparation of different Chirasil-Val derivatives to improve their enantioselectivty and thermal stability. Recently, Levkin et al. prepared a new binary CSP comprising valine diamide and β-cyclodextrin on poly(siloxane) backbone using platinum-catalyzed hydrosilylation (Chirasil-DexVal-C11). Interestingly, Chirasil-DexVal-C11 exhibited superior performance in the enantiomeric separation of proline amino acid, which is problematic using only neat valine-diamide-based CSPs [62].

11.4.2.4 Chiral separations using cyclodextrin-based poly(siloxanes)

Enantiomeric separations using cyclodextrin phases is possible due to inclusion complexation between the guest (analyte) and the host (cyclodextrin) molecules. Cyclodextrins (CDs) are oligosaccharides containing 6, 7, or 8 glucose units, designated as α, β, γ cyclodextrins, respectively [9, 63]. The melting point and solubility of CDs are typically varied by either O-alkylation or O-acylation reactions. CDs are coated as a stationary phase in either diluted or undiluted form. In the diluted approach, the polarity of CD is adjusted according to dissolving solvent such as poly(siloxanes) and subsequently the CD solution is coated using static coating method. In the undiluted approach, low-melting CD derivatives (per-O-pentylated and 3-O-acylated-2,6-di-O-pentylated) are directly coated on the walls of open tubular columns [64]. In the diluted methods, modified CDs such as permethylated β-CD and 2,6-di-O-methyl-3-O-trifluoro-β-CD are often dissolved in semipolar poly(siloxane) (OV-1701) for the preparation of CD columns [65, 66].

Blum and co-workers developed second-generation CDs that contain bulky tert-butyldimethylsilyl (TMDS) residues which are dissolved in poly(siloxane) PS-086 and are used for chiral GC up to 250°C [67]. Compared to permethyl-β-CD, the TBDMS-derivatives showed improved solubility in poly(siloxanes) and exhibited higher enantioseparation factors. An important advancement in CD-based CSP preparation is chemical linking of CDs to poly(dimethysiloxanes) (Chirasil-Dex). This eliminates the requirement for a semi-polar solvent matrix. Using either thermal or radical initiated immobilization techniques, CD linked poly(siloxanes) are immobilized on fused silica and glass capillaries [59, 68]. Commercial products such as Chirasil-Dex-CB involve a bonded chiral stationary phase which has an operating temperature range from –25 to 200°C. This type of CSP are successfully used in the separation of a wide variety of racemic compounds. New developments in CD-based CSPs include the emergence of linear dextrins based on maltohepatose. These phases comprise linear oligosaccharides (maltohepatose) with tert-butyl dimethylsilyl and acetyl groups on primary and secondary hydroxyl sites. The chiral selectivity of linear dextrin was demonstrated by separating α-amino acid derivatives and halogenated compounds on fused capillary columns [69]. Alternative solvents for dissolving the CD chiral selector have also been explored. It was shown that ionic liquid-based CDs exhibited broad enantioselectivities and greater thermal stabilities [70]. Poly(siloxanes) modified with calixarenes, benzo-crown ethers, and resorcinarene represent a new class of CSPs. These are mainly limited to research laboratories and require numerous developmental studies prior to commercialization.

11.4.3 Poly(Alkylene Oxide)-Based Stationary Phases

In addition to poly(siloxanes), poly(alkylene oxides) represent the most popular contemporary GC stationary phases used for the separation of polar compounds. These are polymers of ethylene/propylene oxide units with terminal hydroxy groups. The hydrogen bonding capabilities of these stationary phases are extensively exploited for separating different classes of polar compounds. By varying the molecular weight and chemical composition, the selectivity and thermal stability of the resulting polymer can usually be controlled [6]. Poly(ethylene oxide) (PEG) materials are prepared by polymerizing ethylene oxide using metal hydroxide as catalysts. The polydispersity (molecular weight distribution index) of PEG has a significant impact on the viscosity, thermal stability, and selectivity of the stationary phase. The low-molecular-weight PEGs significantly interfere with analyte retention. Previous reports suggest that conditioning the column at high temperatures relieves the effects of low-molecular-weight PEGs.

Often PEG phases are prone to auto oxidation at high temperature due to the presence of ether linkages. This situation is exacerbated in the presence of oxygen and leads to shorter column lifetimes and low thermal stabilities (<200°C). The use of oxygen scavengers in the flowing carrier gas has significantly improved the column performance [71]. Since water molecules can hydrogen bond with alkylene oxide chains, the use of moisture free carrier gas is warranted to ensure reproducible analysis. Carbowax 20M is the most commonly used PEG stationary phase in both packed and open tubular columns. The column operating temperature of Carbowax 20M (bonded to fused silica capillary) ranges from 60°C to 250°C. For applications requiring high column temperatures, more pure and high-molecular-weight Superox-20M can be used [9, 72]. PEG phases are mostly used for separating mixture of amines (primary, secondary, and tertiary) [73], alcohols from hydrocarbons, and ketones, aldehydes from aromatics [6]. The polarity of the PEG phase can be modified by chemical linking to PDMS. DX-1 and DX-4 are PEG phases that contain 90% and 15% poly(dimethylsiloxne), respectively. The condensation of terminal hydroxy groups of PEG with 2-nitroterephthalic yielded a new stationary phase, FFAP (free fatty acid phase; commercial identity: FFAP; SP-1000; Carbowax 20M-dinitroterephthalate) mostly used for the separation of organic acids, alcohols, amides, ketones, and lactones. Due to the presence of free carboxylic acid groups within the substituent, the FFAP phase is not used for separation of basic compounds, aldehydes, and other compounds that can react with acidic groups.

11.5 Ionic Liquids and Polymeric Ionic Liquids as Emerging Stationary Phase Materials

Although PEG and poly(siloxane)-based stationary phases have been widely used in GC, the quest for new stationary phase materials that can complement them as well as increase the chemical separation space is highly desirable. ILs represent a new class of non-molecular solvents with interesting physicochemical and solvation properties. These are molten organic salts comprising bulky organic cations and inorganic/organic anions that exhibit melting points at or below 100°C [74]. Since the introduction of air-stable 1,3-dialkylimdazolium-based ILs, the applicability of ILs has been greatly expanded to various areas in the chemical sciences [75]. The structural tunability of ILs to tailor their physicochemical properties is highly advantageous for designing optimal GC stationary phases. In addition, the unique solvation properties (e.g., hydrogen bond acidity) as well as high thermal stability enable them to be used as potential alternatives to conventional stationary phases for the analysis of samples that require high temperatures.

11.5.1 Influence of Physicochemical Properties of ILs on Column Performance

Important physicochemical properties of ILs including melting point, viscosity, thermal stability, surface tension, and chemical inertness play crucial roles in developing new IL-based stationary phases. Consequently, these properties greatly influence the separation performance of the column.

11.5.1.1 Melting point

The melting point of the IL determines the minimum operating temperature of the column. When an IL is used as a GC stationary phase at a temperature below its melting point, low column efficiency usually results due to mixed retention mechanisms (adsorption and partition) [32]. Therefore, room-temperature ILs (RTILs; melting point/glass transition below room temperature) are of significant interest since they extend the operating temperature range of the stationary phase. Based on previous studies, structural features such as low ion symmetry, charge delocalization, and high conformational freedom yield ILs with low melting points [104, 105]. Figure 11.2 represents typical structures of cations and anions of low melting point ILs. For most of the ILs containing the same cation composition, halide-based anions usually exhibit higher melting points than the corresponding bis[(trifluoromethyl)sulfonyl]imide [NTf_2^-], trifluoromethanesulfonate [TfO^-], tetrafluoroborate [BF_4^-], and hexafluorophosphate [PF_6^-] anions [32].

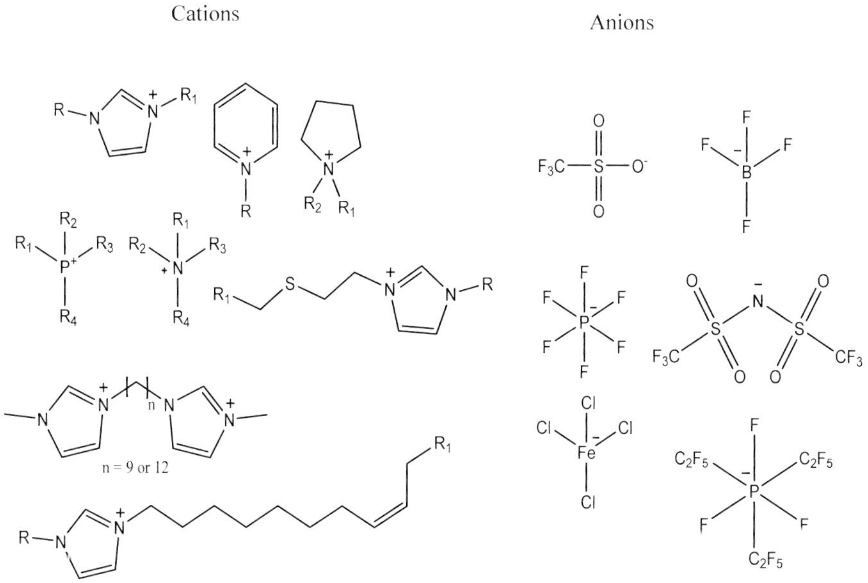

Figure 11.2 Representative structures of cations and anions often used in preparation of room-temperature ionic liquids.

11.5.1.2 Viscosity

The viscosity of the IL is an important property that influences the film stability of the stationary phase, particularly at high temperatures. Owing to long-range columbic forces and intermolecular interactions (e.g., hydrogen bonding), ILs exhibit at least two to three orders of magnitude higher viscosities than organic liquids at room temperature. Within the class of 1-alkyl-3-methylimidazolium-based ILs, increasing the length of the alkyl side chain causes a significant increase in viscosity [21, 106]. Effective charge delocalization tends to reduce the viscosity of ILs, whereas substituents that reduce ion-symmetry often result in increased viscosity. Halide-based anions enhance the viscosity of the IL by participating in hydrogen bonding interactions while [NTf_2] anions tend to reduce the viscosity due to charge delocalization and poor hydrogen bonding capabilities [32, 107]. Other approaches to increase the viscosity involve incorporation of functional groups that can result in increased intermolecular interactions (e.g., π–π and hydrogen bonding). However, these approaches are only applicable if the chemical selectivity of stationary phases is not compromised.

11.5.1.3 Thermal stability

In contrast to molecular liquids, the columbic interactions within ILs prevent the escape of ions into the gas phase and subsequently reduce their vapor pressure over a wide range of temperatures. For this reason, the MAOT of an IL is mostly based on the film instability and onset of thermal decomposition rather than its vapor pressure. The examination of bleed profiles of IL-based stationary phases at high temperatures using MS suggested that thermal decomposition of the cation/anion is greatly responsible for origin of the bleed components [108].

Apart from low vapor pressures, high thermal stability is another important feature of ILs that has increased their popularity in GC. The chemical properties of both the cation and anion are critical in determining the thermal stability of the IL. ILs comprising less coordinating or weak nucleophilic anions (e.g., $[NTf_2^-]$ and $[TfO^-]$) exhibit higher thermal stability than halide-based anions. In comparison to imidazolium cations, quaternary ammonium cations possess low thermal stability due to their high propensity to participate in Hofmann elimination and reverse Menshutkin reactions at high temperatures [109, 110]. ILs derived from phosphonium cations usually exhibit higher thermal stability than their imidazolium and other cyclic quaternary ammonium counterparts. It was shown that thermal decomposition of imidazolium-based ILs progress via N-dealkylation by attack of the anion to generate the amine [111, 112]. When compared to monocationic systems, ILs with multiple cations (dication and trication) demonstrated high thermal stabilities. For example, the thermal stability of tricationic imidazolium-based ILs containing $[NTf_2^-]$ anions are approximately 100°C higher than the corresponding monocationic analogs. Payagala and co workers prepared various tricationic ILs and investigated their physicochemical properties and employed them as GC stationary phases. It was observed that tripropylphosphonium-based ILs containing $[NTf_2^-]$ anions exhibited thermal stability as high as

330°C (based on isothermal GC method, refer to Section 11.2.2) [113].

11.5.1.4 Surface tension

Surface tension and wetting ability are important properties that determine the film forming capacity of ILs. For 1-alkyl-3-methyl-imidazolium-based ILs, an increase in the alkyl chain length results in a gradual decrease in the surface tension [114]. Larger anions such as [NTf_2^-] and [PF_6^-] usually yield ILs with low surface tension [115]. In general, ILs with high surface tension often pose challenges during the static coating process. In such cases, the structure of cation and/or anion can be modified to produce a lower surface tension. On the other hand, surfactant doping can also be used to reduce the surface tension of the IL. However, it should be emphasized that these approaches may alter the chemical selectivity of the resulting IL stationary phase [116].

The purity of ILs largely influences its physicochemical properties as well as separation efficiency when used as a stationary phase. The presence of halide impurities in ILs often results in a stationary phase with low thermal stability and tailing peaks (especially for alcohols and acidic compounds). In addition, water contamination often leads to lower viscosity and surface tension, and presents severe challenges in the coating process. To circumvent the coating challenges, ILs are often dried under reduced pressure for longer times to remove volatile organic impurities and water prior to coating.

11.5.2 Imidazolium and Phosphonium-Based IL Stationary Phases

Owing to their relatively high thermal stability and stable film forming capabilities, imidazolium and phosphonium cation motifs are extensively exploited for generating IL-based stationary phases. Indeed, most of the commercial IL columns are also manufactured using these cations. For more details on other cation systems (ammonium, pyridinium, piperidinium, etc.) and their relevant GC applications, several publications have extensively discussed this aspect [21, 32].

The 1,3-dialkylimidazolium cations constitute the most studied single cation IL-based GC stationary phases. The anionic component of 1,3-dialkylimidazolium ILs largely influences the thermal stability as well as melting point of the resulting ILs. In most cases, ILs containing $[NTf_2^-]$ anions yielded low melting points and high thermal stability compared to structural counterparts that contain halide-based anions. The applicability of 1,3-dialkylimidazolium-based ILs as GC stationary phases was first reported by Armstrong and co-workers in the late 1990s [117]. The separation behavior of ILs for various organic molecules revealed that ILs exhibit dual nature selectivity where they can function as nonpolar stationary for separating nonpolar molecules and as a polar stationary phase for separating polar molecules.

The chemical selectivity of IL stationary phases can be readily varied by simple anion/cation exchange reactions. For example, the 1-butyl-3-methylimidazolium (BMIM) IL containing chloride anion showed a selectivity for polar organic molecules via hydrogen bond donor/acceptor interactions whereas analogous the $[PF_6^-]$ anion-based IL demonstrated higher selectivity for non-polar compounds [117]. Within the 1,3-dialkylimidazolium-based ILs, the hydrogen bond basicity of the stationary phases is largely determined by the anion. Based on the Abraham solvation parameter model, the hydrogen basicity of various imidazolium-based ILs comprising different anions follows: chloride > octylsulfate > hexafluoroantimonate ≈ $[TfO^-]$ > $[NTf_2^-]$ > $[PF_6^-]$. Polar molecules such as alcohols and acids (proton donors) exhibited severe peak tailing when separated using ILs containing $[NTf_2^-]$ anions. In these situations, ILs containing $[TfO^-]$ anions may be potential alternatives to improve the chromatographic performance [24].

In comparison to most of the nonionic stationary phases, the cation functionalized imidazolium-based ILs yielded high hydrogen bond acidic stationary phases. Anderson and co-workers examined the solvation properties of various imidazolium and pyrrolidinium ILs containing the tris(pentafluoroethyl)trifluo rophosphate (FAP^-) anion using solvation parameter model. It was observed that in comparison to unfunctionalized cations, ILs with 2-hydroxyethyl substituents possessed high hydrogen bond acidities and demonstrated superior selectivity for compounds that contain proton acceptor groups. The tunability of imidazolium-based ILs for structure modification in pursuit of desired chemical selectivity was greatly exploited by incorporating different functional groups [118].

Phosphonium-based ILs represent higher thermal stability stationary phases relative to other cation systems. However, structural tuning of phosphonium-based cations is not straightforward and often involves multiple synthesis procedures. As a result, fewer of these stationary phases are reported compared to imidazolium-based ILs. Breitbach and co-workers reported the synthesis of eight monocationic phosphonium ILs with different anions and investigated their solvation properties and thermal stabilities. It was shown that with same anion composition, the hydrogen bond basicity of phosphonium-based ILs is greater than imidazolium ILs. Due to lack of acidic hydrogen atoms, most of these ILs exhibited low or negative hydrogen bond acidity system constants. In addition, phosphonium-based ILs demonstrated superior capabilities in separating a mixture of 24 flavor and fragrance homologous ester compounds. For the trihexyltetradecylphosphonium triflate [P_{66614}^{+}] [TfO^{-}] IL a thermal stability as high as 405°C was observed. Phosphonium-based ILs tend to interact less with solutes through π–π or n–π interactions [119]. Table 11.3 represents the properties, applications, and system constants (determined at 100°C) of selected IL-based columns.

Multi-cationic systems using either imidazolium or phosphonium cations yielded ILs with high thermal stability and stable films at high temperatures. Multi-cationic platforms provide additional opportunities to impart desired chemical selectivity and the ability to tune the physicochemical properties including surface tension, viscosity, and thermal stability of the resulting IL. Most of the commercial ILs columns are manufactured using either dicationic imidazolium or phosphonium cation motifs with [NTf_2^{-}] anions, as shown in Table 11.3. Many of the commercial IL columns are non-bonded stationary phases. SLB-IL100 and SLB-IL111 are extensively used for the separation of fatty acid methyl esters (FAMES) from various sample matrices [35, 120, 121].

Table 11.3 Chemical composition, MAOT, and system constants of selected ionic liquid-based stationary phases

Ionic liquid and maximum operating temperature (°C)	Solvation parameter model System constants					Application	Ref.
	e	*s*	*a*	*b*	*l*		
1-benzyl-3-methylimidazolium [TfO]; 260	0	1.70	2.41	0	0.47	Selectivity and thermal stability compared to different stationary phases	[24]
1-(4-methoxyphenyl)-3-methylimidazolium [TfO]; 260	0.28	2.05	2.03	0.16	0.41		
1,2-dimethyl-3-(3-cyanopropyl)imidazolium [NTf_2]; 250	0.25	1.81	1.25	0.30	0.35	Analysis of n-alkanes, esters, ketones, and alcohols	[127, 128]
1,2-dimethyl-3-(3-hydroxypropy)imidazolium [NTf_2]; 250	0.21	1.81	1.25	0.30	0.35		
1-butyl-1-methylpyrrolidinium [NTf_2]; 300	0	1.44	1.25	0	0.48		
1-butyl-3-(2-hydroxycyclohexyl) imidazolium [BF4]; 250	−0.29	0.67	3.11	0.31	0.49	Analysis of FAMES, alkanes, and esters	[130]
1-(1,2-dimethyl)-3-propylimidazolium [NTf_2]; 320	—	—	—	—	—	Analysis of n-alkanes, esters, and alcohols	[127]
2-methyl-1-propylpyridinium [NTf_2^-]; 250	0.06	1.62	1.19	0.17	0.31	Analysis of n-alkanes, alcohols, esters, phenols, and organic acids	[128]
3-methyl-1-propylpyridinium [NTf_2]; 250	0.07	1.60	1.15	0.15	0.31		
4-methyl-1-propylpyridinium [NTf_2^-]; 250	0.10	1.57	1.13	0.10	0.32		

Ionic liquid and maximum operating temperature (°C)	Solvation parameter model: System constants *e*	*s*	*a*	*b*	*l*	Application	Ref.
1,9-bis(2,3-dimethylimidazolium)nonane [NTf_2]; 250	0.37	1.62	1.13	0.00	0.14	Analysis of alcohols, esters, phenols, and organic acids	[128]
1,14-di(3-methylimidazolium)-3,6,9,12-tetraoxopentadecane; 350	0.18	1.77	1.79	0.14	0.41	Analysis of flavor and fragrance mixture	[129]
1,14-di(3-[2-hydroxyethyl]limidazolium)3,6,9,12-tetraoxopentadecane [TfO];	0.27	1.76	2.42	0.14	0.36	Analysis of flavor and fragrance mixture	[129]
1,10-di(1-decylimidazolium)decane [NTf_2];	–0.06	1.67	1.84	0.17	0.59		[132]
1,10-di(1-decylimidazolium)decane [$FeCl_4$];	0	1.78	2.32	0.50	0.72	Hydrocarbon analysis using two-dimensional GC	[131]
Trihexyl(tetradecyl)phosphonium [NTf_2]; 300	–0.28	1.55	1.55	-0.15	0.75		
Trihexyl(tetradecyl)phosphonium [FAP]; 290	–0.29	1.39	0.49	0.25	0.63		
Trihexyl(tetradecyl)phosphonium [$FeCl_4$]; 320	–0.26	1.43	1.23	0.02	0.69		
Tetraoctylphosphonium [NTf_2^-]; 385	—	—				Flavor and fragrance analysis	[119]
Trihexytetradecyllphosphonium bis(2,4,4-trimethylpentyl)phosphinate; 300	—	—	—	—	—	Analysis of n-alkanes, esters, and alcohols	[127]

(*Continued*)

Table 11.3 (*Continued*)

Ionic liquid and maximum operating temperature (°C)	Solvation parameter model — System constants *e*	*s*	*a*	*b*	*l*	Application	Ref.
Trigonal tripropylphosphonium [TfO]	—	—	—	—	—	Trace analysis of water	[138]
1,12-di(tripropylphopshonium)dodecane [NTf_2] (SLB-IL59/60); 300	—	—	—	—	—	Analysis of haloanisoles, benzothizaoles, nitrosamines	[135–137]
1,12-di(tripropylphopshonium)dodecane[NTf_2] [TfO] (SLB-IL61); 250	—	—	—	—	—	Analysis of octa-chlorinated dibenzo-p-dioxins and dibenzofurans	[122]
Tri(tripropylphosphoniumhexamido)triethylamine (NTf_2) (SLB-IL76); 270	—	—	—	—	—	Analysis of FAMEs	[121]
1,12-di(2,3-dimethylimidazolium) dodecane [NTf_2] (SLB-IL82) 270	—	—	—	—	—	Analysis of plasticizers and synthetic musks	[134]
1,9-di(3-vinylimidazolium) nonane [NTf_2] (SLB-IL100); 230	—	—	—	—	—	Analysis of FAMEs	[121, 133]

It was reported that IL-based columns provided the best separation performance for FAMES when compared to bis(cyanopropyl)siloxane columns. Importantly, their ability to separate cis-and trans-isomers is important for the identification of trans fats in food products [21]. Recently, Do and co-workers investigated the applicability of SLB-IL111, SLB-IL61 and SLB-IL76 for the separation of 136 polychlorinated dibenzo-p-dioxins and dibenzofurans (PCDD/Fs). Owing to their high polarity and multiple solvation interactions, SLB-IL111 and SLB-IL61 separated about twice the number of PCDD/Fs congeners than conventional nonionic stationary phases [122].

11.5.3 Polymeric Ionic Liquid-Based Stationary Phases

In addition to multi-cationic ILs, polymeric ionic liquid (PIL)-based stationary phases have been investigated to improve the film stability, viscosity, and thermal stability. Commonly, imidazolium cations functionalized with polymerizable groups (e.g., vinyl and vinylbenzene) are used to prepare the PIL phases. These materials can be developed using either an in-column approach [123] or polymerized outside of the column in a suitable solvent system that can be directly coated onto the surface of the capillary [124]. Using the static coating method, a mixture of monomer, cross-linker, and a free radical initiator (e.g., AIBN) are deposited onto the wall of capillary column. Subsequently, the coated column is subjected to temperatures that can initiate the polymerization reaction. The physicochemical properties (e.g., melting point, thermal stability) of PIL phases are regulated by varying the percentage of cross-linker composition in the coating solution [125]. For example, a partially cross-linked PIL phase provided efficient chromatographic performance up to 280°C, whereas a highly cross-linked analog was stable up to temperatures as high as 360°C [123]. Recently, Anderson and co-workers developed a series of functionalized PIL columns that contain different amounts of cross-linkers. The developed PIL columns were investigated in the analysis of kerosene and diesel samples using comprehensive two-dimensional GC. It was observed that at the optimized PIL composition, the thermal stability and resolution of selected analytes are significantly higher for PIL-based stationary phases compared to PEG stationary phase [125]. The solution-based polymerization approaches are used to generate PIL phases (homopolymers) that do not contain cross-linkers. Compared to the PIL columns that are prepared using the flexible cross-linkers, homopolymers exhibited higher melting points and lower column efficiency [21, 124].

Recently, two new polymeric phosphonium ILs were prepared using an in-column polymerization approach. The thermal stability of these PIL phases was evaluated based on a series of isothermal measurements of retention time of naphthalene at different temperatures (100 to 380°C). It was demonstrated that the polymeric diphenylphosphonium cation that containing two allyl groups with [NTf_2^-] anion, demonstrated high thermal stability with no peak tailing at temperatures as high as 330°C [126]. Phosphonium-based PILs are relatively a new class of stationary phases materials that have potential to yield high thermal stability stationary phases compared to imidazolium counterparts.

11.6 Conclusions and Future Directions

Developments in column technology as well as progress in the design/production of new stationary phase materials have tremendously increased the applications of GC. The introduction of new bonding and surface modification techniques has greatly improved the thermal stability and enhanced the long-term use of poly(siloxane) and PEG stationary phases. In addition, the ability to incorporate a variety of functional groups within poly(siloxane) motifs has yielded stationary phases with a wide range of chemical selectivities. The chemical linking of chiral selectors to poly(siloxanes) and subsequent immobilization onto the capillary surface has significantly improved the separation efficiency of chiral GC columns. Although poly(siloxane) and PEG-based GC columns are useful in many cases, the stationary phases that can demonstrate high thermal stability as well as unique chemical selectivities are highly desirable.

Owing to their high thermal stability and multiple solvation properties, ILs have been exploited as potential alternatives to conventional GC stationary phases (poly(siloxanes) and PEGs). Important physicochemical properties of ILs, including negligible vapor pressure at room temperatures, tunable viscosity, and high thermal stability, are highly useful in extending the column operating temperature range and for producing low-bleed stationary phases. Functionalized IL-based stationary phases have demonstrated superior capabilities for the analysis of complex polar samples compared to nonionic stationary phases. ILs/PILs have the potential to serve as orthogonal stationary phases in multidimensional/comprehensive two-dimensional GC techniques. The unique chemical selectivities of ILs (e.g., hydrogen bond acidity and dispersive type interactions) can complement nonionic stationary phases and can provide high separation space for the analysis of complex samples. Most of the commercial GC stationary phases are relatively thermally stable and are nonbonded stationary phases. Therefore, an investigation on IL bonding chemistry and methods that can immobilize them on capillary surface can potentially extend the thermal stability as well as column lifetimes.

References

1. Wang, Z., Fingas, M. (1997). Developments in the analysis of petroleum hydrocarbons in oils, petroleum products and oil-spill-related environmental samples by gas chromatography, *J. Chromatogr. A*, **774**, pp. 51–78.
2. Lehotay, S. J., Hajšlová, J. (2002). Application of gas chromatography in food analysis, *Trends Anal. Chem.; TrAC*, **21**, pp. 686–697.
3. Begerow, J., Jermann, E., Keles, T., Koch, T., Dunemann, L. (1996). Screening method for the determination of 28 volatile organic compounds in indoor and outdoor air at environmental concentrations using dual-column capillary gas chromatography with tandem electron-capture-flame ionization detection, *J. Chromatogr. A*, **749**, pp. 181–191.
4. Verenitch, S. S., Lowe, C. J., Mazumder, A. (2006). Determination of acidic drugs and caffeine in municipal wastewaters and receiving waters by gas chromatography–ion trap tandem mass spectrometry, *J. Chromatogr. A*, **1116**, pp. 193–203.
5. Grob, R. L., and Barry, E. F. (2004) *Modern Practice of Gas Chromatography*, 4th ed. (John Wiley & Sons).
6. Rotzsche, H. (1991). *Stationary Phases in Gas Chromatography*, 1st ed. (Elsevier).

7. Herraiz, T., Reglero, G., Herraiz, M., Alonso, R., Cabezudo, M. D. (1987). Micropacked columns: A suitable alternative to very thick capillary columns, *J. Chromatogr. A*, **388**, pp. 325–333.
8. Ettre, L. (1987). MJE Golay and the invention of open-tubular (capillary) columns, *J. Sep. Sci.*, **10**, pp. 221–230.
9. Poole, C. F. (2003). *The Essence of Chromatography*, 1st ed. (Elsevier).
10. Desty, D. (1987). The history and potentiality of capillary columns in gas chromatography, *J. Chromatogr. Sci.*, **25**, pp. 552–563.
11. Dandeneau, R. D., and Zerenner, E. H. (1979). An investigation of glasses for capillary chromatography, *J. High Resolut. Chromatogr.*, **2**, pp. 351–356.
12. Peters, T. L., Nestrick, T. J., Lamparski, L. L. (1982). Etching borosilicate glass capillary columns, *Anal. Chem.*, **54**, pp. 2397–2398.
13. Grob, K., and Grob, G. (1977). The barium carbonate procedure for the preparation of glass capillary columns; further informations and developments, *Chromatographia*, **10**, pp. 181–187.
14. Grob, K. (1980). Persilylation of glass capillary columns. Part 4: Discussion of parameters, *J. High Resolut. Chromatogr.*, **3**, pp. 493–496.
15. Grob, K., Grob, G., Blum, W., Walther, W. (1982). Preparation on inert glass capillary columns for gas chromatography: A revised, comprehensive description, *J. Chromatogr. A*, **244**, pp. 197–208.
16. Hetem, M., Rutten, G., Vermeer, B., Rijks, J., van de Ven, L., de Haan, J., Cramers, C. (1989). Deactivation with polymethylhydrosiloxane: A comparative study with capillary gas chromatography and solid-state 29Si nuclear magnetic resonance spectroscopy, *J. Chromatogr. A*, **477**, pp. 3–24.
17. Novotný, M., and Bartle, K. (1974). Preparation of thick-film glass capillary columns by the dynamic coating procedure, *J. Chromatogr. A*, **93**, pp. 405–411.
18. Bouche, J., and Verzele, M. (1968). A static coating procedure for glass capillary columns, *J. Chromatogr. Sci.*, **6**, pp. 501–505.
19. Wang, D., Chong, S. L., Malik, A. (1997). Sol–gel column technology for single-step deactivation, coating, and stationary-phase immobilization in high-resolution capillary gas chromatography, *Anal. Chem.*, **69**, pp. 4566–4576.
20. Poole, C. F., and Poole, S. K. (2012). *Chromatography Today*, 5th ed. (Elsevier).

21. Poole, C. F., and Lenca, N. (2014). Gas chromatography on wall-coated open-tubular columns with ionic liquid stationary phases, *J. Chromatogr. A*, **1357**, pp. 87–109.
22. Grob, K., and Grob, G. (1978). Comprehensive, standardized quality test for glass capillary columns, *J. Chromatogr. A*, **156**, pp. 1–20.
23. Luong, J., Gras, R., Jennings, W. (2007). An advanced solventless column test for capillary GC columns, *J. Sep. Sci.*, **30**, pp. 2480–2492.
24. Anderson, J. L., and Armstrong, D. W. (2003). High-stability ionic liquids. a new class of stationary phases for gas chromatography, *Anal. Chem.*, **75**, pp. 4851–4858.
25. Qiao, L., Lu, K., Qi, M., Fu, R. (2013). Separation performance of guanidinium-based ionic liquids as stationary phases for gas chromatography, *J. Chromatogr. A*, **1276**, pp. 112–119.
26. Sun, X., Zhu, Y., Wang, P., Li, J., Wu, C., Xing, J. (2011). High temperature and highly selective stationary phases of ionic liquid bonded polysiloxanes for gas chromatography, *J. Chromatogr. A*, **1218**, pp. 833–841.
27. Abraham, M. H., Poole, C. F., Poole, S. K. (1999). Classification of stationary phases and other materials by gas chromatography, *J. Chromatogr. A*, **842**, pp. 79–114.
28. Poole, C. F., and Poole, S. K. (1989). Characterization of solvent properties of gas chromatographic liquid phases, *Chem. Rev.*, **89**, pp. 377–395.
29. Rohrschneider, L. (1998). Characterisation of GC stationary phases in multilinear retention model, *Chromatographia*, **48**, pp. 728–738.
30. Poole, C. F., Kollie, T. O., Poole, S. K. (1992). Recent advances in solvation models for stationary phase characterization and the prediction of retention in gas chromatography, *Chromatographia*, **34**, pp. 281–302.
31. Abraham, M. H. (1993). Scales of solute hydrogen-bonding: Their construction and application to physicochemical and biochemical processes, *Chem. Soc. Rev.*, **22**, pp. 73–83.
32. Anderson, J. L., Ding, J., Welton, T., Armstrong, D. W. (2002). Characterizing ionic liquids on the basis of multiple solvation interactions, *J. Am. Chem. Soc.*, **124**, pp. 14247–14254.
33. Baltazar, Q. Q., Leininger, S. K., Anderson, J. L. (2008). Binary ionic liquid mixtures as gas chromatography stationary phases for improving the separation selectivity of alcohols and aromatic compounds, *J. Chromatogr. A*, **1182**, pp. 119–127.

34. Yao, C., and Anderson, J. L. (2009). Retention characteristics of organic compounds on molten salt and ionic liquid-based gas chromatography stationary phases, *J. Chromatogr. A*, **1216**, pp. 1658–1712.

35. Atapattu, S. N., Eggers, K., Poole, C. F., Kiridena, W., Koziol, W. W. (2009). Extension of the system constants database for open-tubular columns: System maps at low and intermediate temperatures for four new columns, *J. Chromatogr. A*, **1216**, pp. 1640–1649.

36. Atapattu, S. N., and Poole, C. F. (2008). Selectivity equivalence of two poly(methylphenylsiloxane) open-tubular columns prepared with different deactivation techniques for gas chromatography, *J. Chromatogr. A*, **1185**, pp. 305–309.

37. Atapattu, S. N., and Poole, C. F. (2008). Solute descriptors for characterizing retention properties of open-tubular columns of different selectivity in gas chromatography at intermediate temperatures, *J. Chromatogr. A*, **1195**, pp. 136–145.

38. Poole, C. F., and Poole, S. K. (2008). Separation characteristics of wall-coated open-tubular columns for gas chromatography, *J. Chromatogr. A*, **1184**, pp. 254–280.

39. Vernon, F., and Ogundipe, C. O. E. (1977). Hydrocarbon stationary phases for gas-liquid chromatography, *J. Chromatogr. A*, **132**, pp. 181–185.

40. Soják, L., Krupčík, J., Janák, J. (1980). Comparison of capillary columns coated with C87 hydrocarbon and squalane in the analysis of n-pentadecene isomers, *J. Chromatogr. A*, **191**, pp. 199–206.

41. Pomaville, R. M., and Poole, C. F. (1987). Thermally stable, highly fluorinated stationary phases for gas chromatography, *Anal. Chim. Acta.*, **200**, pp. 151–169.

42. O'Mahony, T. K. P., Cox, A. P., Roberts, D. J. (1993). Gas chromatographic separation of perfluorocarbons, *J. Chromatogr. A*, **637**, pp. 1–11.

43. Varughese, P., Gangoda, M., Gilpin, R. (1988). Applications of fluorinated compounds as phases and additives in chromatography and their uses in pharmaceutical analysis, *J. Chromatogr. Sci.*, **26**, pp. 401–405.

44. Ballschmiter, K., Mennel, A., Buyten, J. (1993). Long chain alkyl-polysiloxanes as non-polar stationary phases in capillary gas chromatography, *Fresenius. J. Anal. Chem.*, **346**, pp. 396–402.

45. Haken, J. K. (1984). Developments in polysiloxane stationary phases in gas chromatography, *J. Chromatogr. A*, **300**, pp. 1–77.

46. Mayer, B. X., Zöllner, P., Lorbeer, E., Rauter, W. (2002). A new 75% diphenyl, 25% dimethyl polysiloxane coated on fused silica capillary columns for high temperature gas chromatography, *J. Sep. Sci.*, **25**, pp. 60–66.

47. Bradshaw, J. S., Schirmer, M. M., Juvancz, Z., Markides, K. E., Lee, M. L. (1991). New polar polysiloxane stationary phases containing cyano, nitrophenyl and 8-quinolinyl units attached to diethylene oxide side groups, *J. Chromatogr. A*, **540**, pp. 279–291.

48. Martin, S. D., Poole, C. F., Abraham, M. H. (1998). Synthesis and gas chromatographic evaluation of a high-temperature hydrogen-bond acid stationary phase, *J. Chromatogr. A*, **805**, pp. 217–235.

49. Blomberg, L. G. (1990). Current aspects of stationary phase immobilization in open tubular column chromatography, *J. Microcolumn Sep.*, **2**, pp. 62–68.

50. Bottoli, C. B., Chaudhry, Z. F., Fonseca, D. A., Collins, K. E., Collins, C. H. (2002). Poly (alkylmethylsiloxanes) thermally immobilized on silica as stationary phases for high-performance liquid chromatography, *J. Chromatogr. A*, **948**, pp. 121–128.

51. Markides, K., Blomberg, L., Buijten, J., Wännman, T. (1983). Cyanosilicones as stationary phases in gas chromatography: II. Gum and rubber phases on fused silica, *J. Chromatogr. A*, **267**, pp. 29–38.

52. Wright, B. W., Peaden, A., Lee, M. L., Stark, T. J. (1982). Free radical cross-linking in the preparation of non-extractable stationay phases for capillary gas chromatography, *J. Chromatogr. A*, **248**, pp. 17–34.

53. Schomburg, G., Husmann, H., Ruthe, S., Herraiz, M. (1982). Crosslinking of alkylpolysiloxane filsm on various types of glass surfaces including fused silica using γ-radiation of a60cobalt-source. Comparison to crosslinking by thermal peroxid treatment, *Chromatographia*, **15**, pp. 599–610.

54. Hubball, J., DiMauro, P., Barry, E., Lyons, E., George, W. (1984). Developments in crosslinking of stationary phases for capillary gas chromatography by cobalt-60 gamma radiation, *J. Chromatogr. Sci.*, **22**, pp. 185–191.

55. Zhou, X.-C., Yan, H., Chen, Y.-Y., Wu, C.-Y., Lu, X.-R. (1996). Chiral crown ether-anchored polysiloxanes as capillary gas chromatography stationary phases, *J. Chromatogr. A*, **753**, pp. 269–277.

56. Schurig, V. (1984). Gas chromatographic separation of enantiomers on optically active metal-complex-free stationary phases. New analytical methods (24), *Angew. Chem. Int. Ed.*, **23**, pp. 747–765.

57. Ruderisch, A., Pfeiffer, J., Schurig, V. (2003). Mixed chiral stationary phase containing modified resorcinarene and β-cyclodextrin selectors bonded to a polysiloxane for enantioselective gas chromatography, *J. Chromatogr. A*, **994**, pp. 127–135.
58. Zhang, H., Dai, R., Ling, Y., Wen, Y., Zhang, S., Fu, R., Gu, J. (1997). Resorcarene derivative used as a new stationary phase for capillary gas chromatography, *J. Chromatogr. A*, **787**, pp. 161–169.
59. Berthod, A., Li, W., Armstrong, D. W. (1992). Multiple enantioselective retention mechanisms on derivatized cyclodextrin gas chromatographic chiral stationary phases, *Anal. Chem.*, **64**, pp. 873–879.
60. Schurig, V., Schmalzing, D., Schleimer, M. (1991). Enantiomer separation on immobilized chirasil-metal and chirasil-dex by gas chromatography and supercritical fluid chromatography, *Angew. Chem. Int. Ed.*, **30**, pp. 987–989.
61. Schurig, V. (2002). Chiral separations using gas chromatography, *Trends Anal. Chem.; TrAC*, **21**, pp. 647–661.
62. Levkin, P. A., Levkina, A., Schurig, V. (2006). Combining the enantioselectivities of l-valine diamide and permethylated β-cyclodextrin in one gas chromatographic chiral stationary phase, *Anal. Chem.*, **78**, pp. 5143–5148.
63. Ding, J., Welton, T., Armstrong, D. W. (2004). Chiral ionic liquids as stationary phases in gas chromatography, *Anal. Chem.*, **76**, pp. 6819–6822.
64. König, W. A., Lutz, S., Mischnick-Lübbecke, P., Brassat, B., Wenz, G. (1988). Cyclodextrins as chiral stationary phases in capillary gas chromatography I. Pentylated α-cyclodextrin, *J. Chromatogr. A*, **47**, pp. 193–197.
65. Armstrong, D. W., Li, W., Chang, C. D., Pitha, J. (1990). Polar-liquid, derivatized cyclodextrin stationary phases for the capillary gas chromatography separation of enantiomers, *Anal. Chem.*, **62**, pp. 914–923.
66. Nowotny, H. P., Schmalzing, D., Wistuba, D., Schurig, V. (1989). Extending the scope of enantiomer separation on diluted methylated β-cyclodextrin derivatives by high-resolution gas chromatography, *J. High Resolut. Chromatogr.*, **12**, pp. 383–393.
67. Blum, W., and Aichholz, R. (1990). Gas chromatographic enantiomer separation on tert-butyldimethylsilylated β-cyclodextrin diluted in PS-086. A simple method to prepare enantioselective glass capillary columns, *J. High Resolut. Chromatogr.*, **13**, pp. 515–518.

68. Fischer, P., Aichholz, R., Bölz, U., Juza, M., Krimmer, S. (1990). Permethyl-β-cyclodextrin, chemically bonded to polysiloxane: A chiral stationary phase with wider application range for enantiomer separation by capillary gas chromatography, *Angew. Chem. Int. Ed.*, **29**, pp. 427–429.

69. Sicoli, G., Pertici, F., Jiang, Z., Jicsinszky, L., Schurig, V. (2007). Gas-chromatographic approach to probe the absence of molecular inclusion in enantioseparations by carbohydrates. Investigation of linear dextrins ("acyclodextrins") as novel chiral stationary phases, *Chirality*, **19**, pp. 391–400.

70. Huang, K., Zhang, X., Armstrong, D. W. (2010). Ionic cyclodextrins in ionic liquid matrices as chiral stationary phases for gas chromatography, *J. Chromatogr. A*, **1217**, pp. 5261–5273.

71. Conder, J. R., Fruitwala, N. A., Shingari, M. K. (1983). Thermal decomposition of polyethylene glycol 20 m and essential oils in gas—liquid chromatography and the effect of traces of oxygen, *J. Chromatogr. A*, **269**, pp. 171–178.

72. Sandra, P., and Van Roelenbosch, M. (1981). Mixed phases of Superox 20M and OV-1 in fused silica and glass capillary gas chromatography, *Chromatographia*, **14**, pp. 345–350.

73. Lindsay Smith, J. R., and Waddington, D. J. (1969). Gas chromatographic analysis of aliphatic amines: The use of ethylene glycols as stationary phases, *J. Chromatogr. A*, **42**, pp. 183–194.

74. Azzouz, A., Rascón, A. J., Ballesteros, E. (2016). Simultaneous determination of parabens, alkylphenols, phenylphenols, bisphenol A and triclosan in human urine, blood and breast milk by continuous solid-phase extraction and gas chromatography–mass spectrometry, *J. Pharm. Biomed. Anal.*, **119**, pp. 16–26.

75. He, D., Simoneit, B. R. T., Xu, Y., Jaffé, R. (2016). Occurrence of unsaturated C25 highly branched isoprenoids (HBIs) in a freshwater wetland, *Org. Geochem.*, **93**, pp. 59–67.

76. Ren, Z., Xiao, X., Chen, D., Bi, X., Huang, B., Liu, M., Hu, J., Peng, P. A., Sheng, G., Fu, J. (2014). Halogenated organic pollutants in particulate matters emitted during recycling of waste printed circuit boards in a typical e-waste workshop of Southern China, *Chemosphere*, **94**, pp. 143–150.

77. Tripathi, J., Chatterjee, S., Gamre, S., Chattopadhyay, S., Variyar, P. S., Sharma, A. (2014). Analysis of free and bound aroma compounds of pomegranate (Punica granatum L.), *LWT-Food. Sci. Technol.*, **59**, pp. 461–466.

78. Zhu, Z.-C., Chen, S.-J., Zheng, J., Tian, M., Feng, A.-H., Luo, X.-J., Mai, B.-X. (2014). Occurrence of brominated flame retardants (BFRs), organochlorine pesticides (OCPs), and polychlorinated biphenyls (PCBs) in agricultural soils in a BFR-manufacturing region of North China, *Sci. Total Environ.*, **481**, pp. 47–54.

79. An, J., Zhu, B., Wang, H., Li, Y., Lin, X., Yang, H. (2014). Characteristics and source apportionment of VOCs measured in an industrial area of Nanjing, Yangtze River Delta, China, *Atmos. Environ.*, **97**, pp. 206–214.

80. Hou, X., Zheng, X., Zhang, C., Ma, X., Ling, Q., Zhao, L. (2014). Ultrasound-assisted dispersive liquid–liquid microextraction based on the solidification of a floating organic droplet followed by gas chromatography for the determination of eight pyrethroid pesticides in tea samples, *J. Chromatogr. B*, **969**, pp. 123–127.

81. López, L. (2014). Study of the biodegradation levels of oils from the Orinoco Oil Belt (Junin area) using different biodegradation scales, *Org. Geochem.*, **66**, pp. 60–69.

82. López, L., Lo Mónaco, S., Volkman, J. K. (2015). Evidence for mixed and biodegraded crude oils in the Socororo field, Eastern Venezuela Basin, *Org. Geochem.*, **82**, pp. 12–21.

83. Maguire-Boyle, S. J., and Barron, A. R. (2014). Organic compounds in produced waters from shale gas wells, *Environ. Sci. Process. Impact.*, **16**, pp. 2237–2248.

84. Nagashima, H., Kondo, T., Nagoya, T., Ikeda, T., Kurimata, N., Unoke, S., Seto, Y. (2015). Identification of chemical warfare agents from vapor samples using a field-portable capillary gas chromatography/membrane-interfaced electron ionization quadrupole mass spectrometry instrument with Tri-Bed concentrator, *J. Chromatogr. A*, **1406**, pp. 279–290.

85. Yeh, H.-Y., Chuang, C.-H., Chen, H.-C., Wan, C.-J., Chen, T.-L., Lin, L.-Y. (2014). Bioactive components analysis of two various gingers (Zingiber officinale Roscoe) and antioxidant effect of ginger extracts, *LWT-Food. Sci. Technol.*, **55**, pp. 329–334.

86. Aghamohammadi, M., Shahdousti, P., Harooni, B. (2016). Ultrasound-assisted emulsification microextraction followed by gas chromatography–flame ionization detection for urinary methylmalonic acid determination, *Microchem. J.*, **124**, pp. 188–194.

87. Harb, J., Alseekh, S., Tohge, T., Fernie, A. R. (2015). Profiling of primary metabolites and flavonols in leaves of two table grape varieties

collected from semiarid and temperate regions, *Phytochemistry*, **117**, pp. 444–455.

88. Luo, Y.-B., Chen, X.-J., Zhang, H.-F., Jiang, X.-Y., Li, X., Li, X.-Y., Zhu, F.-P., Pang, Y.-Q., Hou, H.-W. (2016). Simultaneous determination of polycyclic aromatic hydrocarbons and tobacco-specific N-nitrosamines in mainstream cigarette smoke using in-pipette-tip solid-phase extraction and on-line gel permeation chromatography-gas chromatography–tandem mass spectrometry, *J. Chromatogr. A*, **1460**, pp. 16–23.

89. Zhang, F., Wang, H., Zhang, L., Zhang, J., Fan, R., Yu, C., Wang, W., Guo, Y. (2014). Suspected-target pesticide screening using gas chromatography–quadrupole time-of-flight mass spectrometry with high resolution deconvolution and retention index/mass spectrum library, *Talanta*, **128**, pp. 156–163.

90. Corral, S., Leitner, E., Siegmund, B., Flores, M. (2016). Determination of sulfur and nitrogen compounds during the processing of dry fermented sausages and their relation to amino acid generation, *Food Chem.*, **190**, pp. 657–664.

91. Kiene, F. E., Pein, M., Thommes, M. (2015). Orientation to determine quality attributes of flavoring excipients containing volatile molecules, *J. Pharm. Biomed. Anal.*, **110**, pp. 20–26.

92. Li, J., Peng, Y., Liu, Y., Li, W., Jin, Y., Tang, Z., Duan, Y. (2014). Investigation of potential breath biomarkers for the early diagnosis of breast cancer using gas chromatography–mass spectrometry, *Clin. Chim. Acta*, **436**, pp. 59–67.

93. Moreno, A. I., Arnáiz, N., Font, R., Carratalá, A. (2014). Chemical characterization of emissions from a municipal solid waste treatment plant, *Waste Manag.*, **34**, pp. 2393–2399.

94. Nowak, T., Graffius, G. C., Liu, Y., Wu, N., Bu, X., Gong, X., Welch, C. J., Regalado, E. L. (2016). GC-FID method for high-throughput analysis of residual solvents in pharmaceutical drugs and intermediates, *Green Chem.*, **18**, pp. 3732–3739.

95. Raich-Montiu, J., Ribas-Font, C., de Arespacochaga, N., Roig-Torres, E., Broto-Puig, F., Crest, M., Bouchy, L., Cortina, J. L. (2014). Analytical methodology for sampling and analysing eight siloxanes and trimethylsilanol in biogas from different wastewater treatment plants in Europe, *Anal. Chim. Acta.*, **812**, pp. 83–91.

96. Cantu-Jungles, T. M., Maria-Ferreira, D., da Silva, L. M., Baggio, C. H., Werner, M. F. D. P., Iacomini, M., Cipriani, T. R., Cordeiro, L. M. C. (2014). Polysaccharides from prunes: Gastroprotective activity and structural elucidation of bioactive pectins, *Food Chem.*, **146**, pp. 492–499.

97. Müller, D. C., Degen, C., Scherer, G., Jahreis, G., Niessner, R., Scherer, M. (2014). Metabolomics using GC–TOF–MS followed by subsequent GC–FID and HILIC–MS/MS analysis revealed significantly altered fatty acid and phospholipid species profiles in plasma of smokers, *J. Chromatogr. A*, **966**, pp. 117–126.

98. Watkins, B. A., Kim, J., Tamez, H., Wenger, J., Thadhani, R., Friedman, A. N. Serum phospholipid fraction of polyunsaturated fatty acids is the preferred indicator for nutrition and health status in hemodialysis patients, *J. Nutr. Biochem.*, **38**, pp. 18–24.

99. Qi, D., Fei, T., Sha, Y., Wang, L., Li, G., Wu, D., Liu, B. (2014). A novel fully automated on-line coupled liquid chromatography–gas chromatography technique used for the determination of organochlorine pesticide residues in tobacco and tobacco products, *J. Chromatogr. A*, **1374**, pp. 273–277.

100. Siddique, A., Saied, S., Mumtaz, M., Hussain, M. M., Khwaja, H. A. (2015). Multipathways human health risk assessment of trihalomethane exposure through drinking water, *Ecotoxicol. Environ. Saf.*, **116**, pp. 129–136.

101. Yang, Y., Kong, W., Zhao, L., Xiao, Q., Liu, H., Zhao, X., Yang, M. (2015). A multiresidue method for simultaneous determination of 44 organophosphorous pesticides in Pogostemon cablin and related products using modified QuEChERS sample preparation procedure and GC–FPD, *J. Chromatogr. B*, **974**, pp. 118–125.

102. Hallett, J. P., and Welton, T. (2011). Room-temperature ionic liquids: Solvents for synthesis and catalysis. 2, *Chem. Rev.*, **111**, pp. 3508–3576.

103. Ho, T. D., Zhang, C., Hantao, L. W., Anderson, J. L. (2014). Ionic liquids in analytical chemistry: Fundamentals, advances, and perspectives, *Anal. Chem.*, **86**, pp. 262–285.

104. Coker, T. G., Ambrose, J., Janz, G. J. (1970). Fusion properties of some ionic quaternary ammonium compounds, *J. Am. Chem. Soc.*, **92**, pp. 5293–5297.

105. Gordon, J. E. (1965). Fused organic salts. IV.1a characterization of low-melting quaternary ammonium salts. Phase equilibria for salt-salt and salt-nonelectrolyte systems. Properties of the liquid salt medium, *J. Am. Chem. Soc.*, **87**, pp. 4347–4358.

106. Poole, C. F. (2004). Chromatographic and spectroscopic methods for the determination of solvent properties of room temperature ionic liquids, *J. Chromatogr. A*, **1037**, pp. 49–82.

107. Bonhôte, P., Dias, A.-P., Papageorgiou, N., Kalyanasundaram, K., Grätzel, M. (1996). Hydrophobic, highly conductive ambient-temperature molten salts, *Inorg. Chem.*, **35**, pp. 1168–1178.

108. Shashkov, M. V., and Sidelnikov, V. N. (2012). Mass spectral evaluation of column bleeding for imidazolium-based ionic liquids as GC liquid phases, *Anal. Bioanal. Chem.* **403**, pp. 2673–2682.

109. Huddleston, J. G., Visser, A. E., Reichert, W. M., Willauer, H. D., Broker, G. A., Rogers, R. D. (2001). Characterization and comparison of hydrophilic and hydrophobic room temperature ionic liquids incorporating the imidazolium cation, *Green Chem.*, **3**, pp. 156–164.

110. Kroon, M. C., Buijs, W., Peters, C. J., Witkamp, G.-J. (2007). Quantum chemical aided prediction of the thermal decomposition mechanisms and temperatures of ionic liquids, *Thermochim. Acta*, **465**, pp. 40–47.

111. Lebga-Nebane, J. L., Rock, S. E., Franclemont, J., Roy, D., Krishnan, S. (2012). Thermophysical properties and proton transport mechanisms of trialkylammonium and 1-alkyl-1H-imidazol-3-ium protic ionic liquids, *Ind. Eng. Chem. Res.*, **51**, pp. 14084–14098.

112. Poole, C. F., and Poole, S. K. (2011). Ionic liquid stationary phases for gas chromatography, *J. Sep. Sci.*, **34**, pp. 888–900.

113. Payagala, T., Zhang, Y., Wanigasekara, E., Huang, K., Breitbach, Z. S., Sharma, P. S., Sidisky, L. M., Armstrong, D. W. (2009). Trigonal tricationic ionic liquids: A generation of gas chromatographic stationary phases, *Anal. Chem.*, **81**, pp. 160–173.

114. Law, G., and Watson, P. R. (2001). Surface tension measurements of N-alkylimidazolium ionic liquids, *Langmuir*, **17**, pp. 6138–6141.

115. Kilaru, P., Baker, G. A., Scovazzo, P. (2007). Density and surface tension measurements of imidazolium-, quaternary phosphonium-, and ammonium-based room-temperature ionic liquids: Data and correlations, *J. Chem. Eng. Data*, **52**, pp. 2306–2314.

116. Lantz, A. W., Pino, V., Anderson, J. L., Armstrong, D. W. (2006). Determination of solute partition behavior with room-temperature ionic liquid based micellar gas–liquid chromatography stationary phases using the pseudophase model, *J. Chromatogr. A*, **1115**, pp. 217–224.

117. Armstrong, D. W., He, L., Liu, Y.-S. (1999). Examination of ionic liquids and their interaction with molecules, when used as stationary phases in gas chromatography, *Anal. Chem.*, **71**, pp. 3873–3876.

118. Zhao, Q., Eichhorn, J., Pitner, W. R., Anderson, J. L. (2009). Using the solvation parameter model to characterize functionalized ionic liquids containing the tris (pentafluoroethyl) trifluorophosphate (FAP) anion, *Anal. Bioanal. Chem.*, **395**, pp. 225–234.

119. Breitbach, Z. S., and Armstrong, D. W. (2008). Characterization of phosphonium ionic liquids through a linear solvation energy relationship and their use as GLC stationary phases, *Anal. Bioanal. Chem.*, **390**, pp. 1605–1617.

120. Delmonte, P., Fardin-Kia, A. R., Kramer, J. K. G., Mossoba, M. M., Sidisky, L., Tyburczy, C., Rader, J. I. (2012). Evaluation of highly polar ionic liquid gas chromatographic column for the determination of the fatty acids in milk fat, *J. Chromatogr. A*, **1233**, pp. 137–146.

121. Zeng, A. X., Chin, S.-T., Nolvachai, Y., Kulsing, C., Sidisky, L. M., Marriott, P. J. (2013). Characterisation of capillary ionic liquid columns for gas chromatography–mass spectrometry analysis of fatty acid methyl esters, *Anal. Chim. Acta*, **803**, pp. 166–173.

122. Do, L., Liljelind, P., Zhang, J., Haglund, P. (2013). Comprehensive profiling of 136 tetra- to octa-polychlorinated dibenzo-p-dioxins and dibenzofurans using ionic liquid columns and column combinations, *J. Chromatogr. A*, **1311**, pp. 157–169.

123. Anderson, J. L., and Armstrong, D. W. (2005). Immobilized ionic liquids as high-selectivity/high-temperature/high-stability gas chromatography stationary phases, *Anal. Chem.*, **77**, pp. 6453–6462.

124. Hsieh, Y.-N., Ho, W.-Y., Horng, R. S., Huang, P.-C., Hsu, C.-Y., Huang, H.-H., Kuei, C.-H. (2007). Study of anion effects on separation phenomenon for the vinyloctylimidazolium based ionic liquid polymer stationary phases in GC, *Chromatographia*, **66**, pp. 607–611.

125. Zhang, C., Park, R. A., Anderson, J. L. (2016). Crosslinked structurally-tuned polymeric ionic liquids as stationary phases for the analysis of hydrocarbons in kerosene and diesel fuels by comprehensive two-dimensional gas chromatography, *J. Chromatogr. A*, **1440**, pp. 160–171.

126. Gonzalez-Alvarez, J., Arias-Abrodo, P., Puerto, M., Viguri, M. E., Perez, J., Gutierrez-Alvarez, M. D. (2013). Polymerized phosphonium-based ionic liquids as gas chromatography stationary phases, *RSC Adv.*, **3**, pp. 21377–21380.

127. Shashkov, M. V., and Sidel'nikov, V. N. (2012). Single cation ionic liquids as high polarity thermostable stationary liquid phases for capillary chromatography, *Russ. J. Phys. Chem. A*, **86**, pp. 138–141.

128. Shashkov, M. V., and Sidelnikov, V. N. (2013). Properties of columns with several pyridinium and imidazolium ionic liquid stationary phases, *J. Chromatogr. A*, **1309**, pp. 56–63.

129. Huang, K., Han, X., Zhang, X., Armstrong, D. W. (2007). PEG-linked geminal dicationic ionic liquids as selective, high-stability gas chromatographic stationary phases, *Anal. Bioanal. Chem.*, **389**, pp. 2265–2275.

130. Álvarez, J. G., Gomis, D. B., Abrodo, P. A., Llorente, D. D., Busto, E., Lombardía, N. R., Fernández, V. G., Álvarez, M. D. G. (2011). Evaluation of new ionic liquids as high stability selective stationary phases in gas chromatography, *Anal. Bioanal. Chem.*, **400**, pp. 1209–1216.

131. Hantao, L. W., Najafi, A., Zhang, C., Augusto, F., Anderson, J. L. (2014). Tuning the selectivity of ionic liquid stationary phases for enhanced separation of nonpolar analytes in kerosene using multidimensional gas chromatography, *Anal. Chem.*, **86**, pp. 3717–3721.

132. Zhang, C., Ingram, I. C., Hantao, L. W., Anderson, J. L. (2015). Identifying important structural features of ionic liquid stationary phases for the selective separation of nonpolar analytes by comprehensive two-dimensional gas chromatography, *J. Chromatogr. A*, **1386**, pp. 89–97.

133. Ragonese, C., Tranchida, P. Q., Dugo, P., Dugo, G., Sidisky, L. M., Robillard, M. V., Mondello, L. (2009). Evaluation of use of a dicationic liquid stationary phase in the fast and conventional gas chromatographic analysis of health-hazardous C18 cis/trans fatty acids, *Anal. Chem.*, **81**, pp. 5561–5568.

134. Sanchez-Prado, L., Lamas, J. P., Garcia-Jares, C., Llompart, M. (2012). Expanding the applications of the ionic liquids as GC stationary phases: plasticizers and synthetic musks fragrances, *Chromatographia*, **75**, pp. 1039–1047.

135. Domínguez, C., Reyes-Contreras, C., Bayona, J. M. (2012). Determination of benzothiazoles and benzotriazoles by using ionic liquid stationary phases in gas chromatography mass spectrometry. Application to their characterization in wastewaters, *J. Chromatogr. A*, **1230**, pp. 117–122.

136. García Pinto, C., Pérez Antón, A., Pérez Pavón, J. L., Moreno Cordero, B. (2012). Coupling of microextraction by packed sorbents with gas chromatography with ionic liquid stationary phases for the determination of haloanisoles in wines, *J. Chromatogr. A*, **1260**, pp. 200–205.

137. Reyes-Contreras, C., Domínguez, C., Bayona, J. M. (2012). Determination of nitrosamines and caffeine metabolites in wastewaters using gas chromatography mass spectrometry and ionic liquid stationary phases, *J. Chromatogr. A*, **1261**, pp. 164–170.

138. Frink, L. A., and Armstrong, D. W. (2016). Determination of trace water content in petroleum and petroleum products, *Anal. Chem.*, **88**, pp. 8194–8201.

Chapter 12

Gas Chromatography Method Development

Tien Ho and John C. Vinci

Abbvie, 1 N. Waukegan Rd,
North Chicago, Illinois 60064, USA

tiendho316@gmail.com

12.1 Gas Chromatography in the Pharmaceutical Setting

Gas chromatography (GC) is a separation technique and analytical tool that serves to provide quantitative and qualitative analysis. The separation occurs by partitioning analytes through two phases, a stationary phase and an inert gas moving phase. The stationary phase may consist of solid or a liquid coating on a solid substrate, and the column may be open tubular or packed with particles. Separation by GC is achieved by volatility differences of the analytes and by differing degrees of analyte-stationary phase interaction. This analyte–stationary phase interaction is important in distinguishing GC as a high-resolution separation technique.

Chromatographic Method Development
Edited by Gregory K. Webster and Laila Kott

ISBN 978-981-4800-53-2 (Hardcover), 978-0-429-20172-1 (eBook)
www.jennystanford.com

While classical distillation separates analytes based on vapor pressure differences, GC offers the additional advantage of separation through chemical interactions. By combing a high efficiency separation with the potential for a high degree of selectivity, the result can be a high-resolution separation that has the power to operate as a high throughput technique.

Although the separation occurs in the gas phase, the analytes amenable to GC may be solid, liquid or gas at room temperature and pressure. The analytes of interest must be stable when subjected to the GC inlet and column temperatures and must be able to volatilize such that analysis in the gas phase can occur. While these requirements create some limitations to the number of analytes that may be analyzed by GC, this limitation is overcome by the high separation efficiencies and ease of operability for the end-user, along with innovative approaches to deal with difficult to volatilize compounds (*vide infra*). When GC is applicable to a given set of analytes, it is a very powerful analytical tool worth having in the analyst's toolbox.

The modern GC system in the pharmaceutical analysis lab is fully automated with user-friendly software. A typical pharmaceutical analysis lab will have GCs equipped with an autosampler and a temperature programmable inlet, oven, and detector as the most basic components of the system. The autosampler will allow for either direct injection or headspace (HS) injection of the sample. For direct injection, the sample is typically introduced onto the GC column using a hot inlet, although cool on-column injection is also possible. With the typical hot inlet sample introduction, the sample is quickly vaporized allowing for fast transfer of sample and for the separation to occur immediately.

12.2 Instrument and Column Optimization

The optimization of gas chromatographic separations can be a cumbersome and complex task for the pharmaceutical analyst. This is primarily due to the multitude of instrumental as well as physical parameters that requires full attention to detail. The interactions observed between each parameter also drastically complicate this process. Nevertheless, analysts can gain abundant benefits when optimizing their chromatographic methods with a particular goal(s) in mind. Perhaps, the method may be optimized to achieve faster separation, resulting in higher analysis throughput, and/or to gain sufficient resolution between critical pairs. Ultimately, the optimization of chromatographic conditions enables the analyst to gain better analytical performance in their analyses (i.e., improving sensitivity, accuracy, and precision) and utilize their laboratory equipment more effectively (i.e., reducing instrument maintenance/down time and usage of consumables). This section will provide the practicing pharmaceutical analyst with a break-down of various physical and instrumental parameters that can be optimized. Additionally, we will explore the cause-and-effect(s) of varying each parameter and how they can ultimately influence the outcome of an analysis.

12.2.1 Split/Splitless Inlet

The process of introducing samples into a GC can be highly complicated, considering the variety of instrumental components involved. Varying between sample injection modes, hardware, and instrumental parameters can result in dramatic effects on both the chromatography and analytical performance of a method. In most pharmaceutical settings, the common sample introduction method for GC is by liquid injection of the sample into a split/splitless hot inlet. Other methods of liquid injection, such as on-column and programmed-temperature vaporization (PTV) are not as common and optimization for these modes will not be discussed herein.

A schematic of a split/splitless inlet is shown in Fig. 12.1. In split/splitless injections, liquid samples are withdrawn from their source and injected (manually or automatically) into the GC inlet via microsyringe. Once the liquid solution enters the hot inlet, it readily vaporizes, mixes with the carrier gas, and is transferred to a column. It should be noted that during an injection, the split/splitless inlet can only be operated in either split or splitless mode, not both.

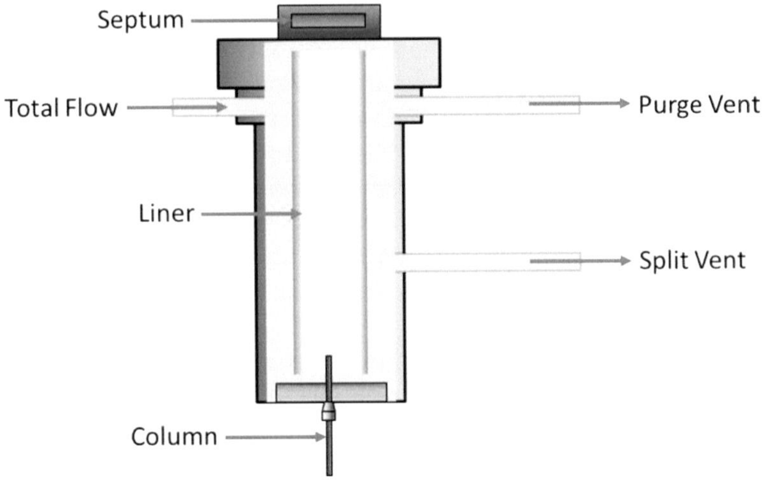

Figure 12.1 Schematic of a GC split/splitless inlet.

12.2.1.1 Split injection

Using split flow enables the analyst to address a major issue associated with conventional wall-coated open tubular (WCOT) capillary columns, namely, sample overloading. Due to their small inner diameters, capillary columns may not possess the capacity to withstand microliter(s) injections of a sample solution, leading to overload. With sample overload, the chromatographer may experience poor separation and/or asymmetrical peak shapes (i.e., peak fronting). Injections of a concentrated sample without splitting the gas flow may also result in analyte carry-over and/or contaminate various components in the GC inlet.

Applying a split injection is synonymous to performing a rapid dilution in the GC inlet. In a split injection, a fraction of the injected sample is carried to the column, while the remaining fraction is diverted to waste through a split vent. The ratio of the volumetric flow rate of the gas fraction carried into the column relative to that of the eluent exiting the split vent is deemed the split ratio. The split ratio is controlled by varying the flow rate of the carrier gas (calculated by the head pressure of the inlet) and the split vent is operated via a needle valve. Increasing the carrier gas flow rate, while opening the split vent, will result in a higher split ratio and enable a narrower analyte band to enter the column. As a result, optimizing the split ratio can also provide better peak shapes and improve signal sensitivity. It is necessary to note that the split ratio is not entirely proportional to the split flow (e.g., actual flow to column vs. split vent) due to a number of factors, such as inlet temperature, solute physico-chemical properties, injection amounts, etc. As such, the split ratio should be understood as an approximation, rather than a quantitative value [1].

Typical split ratios can range from 1:10 to 1:500. For optimization, it is recommended that the analyst start with a split ratio of 1:5 to 1:20 for columns with larger inner diameters (e.g., 0.53 mm, megabore) and higher stationary phase film thicknesses (e.g., >0.5 μm), as these columns possess better sample loading capacities. For columns with smaller inner diameters (i.e., 0.25 mm, 0.18 mm, etc.) a higher split ratio, such as 1:50 and above, is recommended as the starting point to mitigate overload.

12.2.1.2 Splitless injection

A common strategy to obtain good signal sensitivity for trace-level analysis is to inject larger quantities of the sample into the capillary column. One way that this can be achieved on a split/splitless injector is by taking advantage of the splitless injection mode. Contrary to a conventional split injection, a splitless injection transfers most (~95%) of the sample to the capillary column. In this injection mode, the split vent will remain closed for the duration of an injection, as well as an additional period immediately after the injection. When the split vent valve is closed, the sample vapor will be directly transferred to the column with minimal loss. Subsequently, the split vent is opened to flush residual sample vapor from the inlet.

Unlike split injections, the transfer of vapor in a splitless injection is significantly slower, resulting in band broadening of the analyte peak(s). This is especially true for highly volatile analytes with large thermal expansion coefficients. To address this potential issue, the analyst can optimize the oven and inlet temperatures, as well as the splitless injection time. In regards to temperature, it is recommended that the analyst set the inlet temperature to an appropriate temperature that enables vaporization of the sample. The initial oven temperature should be set at least 20°C below the solvent boiling point. By setting the temperature as recommended, the solvent/analyte band can be condensed and concentrated at the head of the column, resulting in a narrow band [2, 3]. The splitless injection time refers to the total time during an injection where the purge valve is closed. Setting the splitless injection time too short can reduce the signal sensitivity of analytes (especially those with high-boiling points). On the other hand, over-extending the splitless injection time can result in peak degradation and a broad solvent peak tail [2]. Common splitless injection times are set from 20s to 90s [3].

In addition to optimizing inlet temperature and splitless injection time, it is important to understand the thermal stability of samples capable of undergoing thermal transformations/degradations. Setting the inlet temperatures at or above the stable temperatures of the sample may result in poor accuracy and precision, in addition to inaccurate representations of its purity profile. Over-extending the splitless injection time (at or above stable sample temperatures) can result in similar issues.

12.2.2 Column Dimensions

When selecting a GC column, it is clear that there are multiple types of stationary phases commercially available for the analyst. It is even clearer that there are varieties of column dimensions available for each stationary phase. Choosing the appropriate column dimensions can be crucial for method development and transfer. Herein, we provide a brief insight on the effects of varying column dimensions and considerations for hand-picking the appropriate column for the job. Since WCOT columns are currently used for most GC analyses in the pharmaceutical industry, this section will focus on WCOT columns.

It is worthy to note that selecting the appropriate stationary phase is also vital for developing a fit-for-purpose separation. It is necessary to fully understand the selectivity of the wide variety of commercial columns currently available and how they can be properly exploited during method development. As such, the reader is referred to the in-depth discussion on column selection written by Anderson and coworkers (Chapter 11).

12.2.2.1 Column inner diameter

A factor for selecting a column I.D. is the sample capacity needs. As the sample capacity is proportional to the square of column radius, a larger column diameter exhibits higher peak capacity. For samples with analytes possessing vastly different relative concentrations, it may be beneficial to use larger diameter columns to mitigate column overload. In regards to pressure, a higher pressure drop is needed when a smaller column I.D., leading to an increase in the inlet head pressure. Another factor that may influence the selection of a suitable column I.D. is the separation efficiency. Based on Eq. 12.1, as the column I.D. (d_c) decreases, the minimum plate height (h_{min}) will also be reduced. In other words, decreasing the column I.D. can produce better separation efficiency [4]. Therefore, if the relative concentration of all analytes in the sample is similar, the analyst may choose to use a smaller I.D. column to obtain better overall separation. Equation 12.1 shows the relationship between column I.D. and plate height:

$$h_{\text{min}} = d_{\text{c}}\sqrt{\frac{1+6k+11k^2}{12(1+k)^2}}, \tag{12.1}$$

where k is the retention factor, d_c is the column diameter, and h_{min} is the minimum plate height.

In most cases, an analyst can choose a 0.25 or 0.32 mm I.D. column as a starting point for conventional split/splitless injections. However, in some occasions, it may be necessary to revert to 0.53 mm I.D. columns for volatile organic compounds (VOCs) analyses (i.e., residual solvents), as these columns can provide better robustness for method transfer and higher capacity to withstand large variations in analyte concentrations in a sample. For systems with vacuum outlets, such as those found in GC-mass spectrometry, columns with I.D. at or below 0.25 mm is recommended to maintain carrier gas flow rates in the vacuum.

12.2.2.2 Column length

Varying the column length can affect the run time, pressure drop, peak capacity, and resolution. In terms of pressure, increasing the length is similar to reducing the I.D. of a column, wherein a higher pressure drop is needed when a longer column is used, resulting in higher inlet head pressures. Increasing the length also increases the peak capacity in a separation. More importantly, increasing the column length also increases peak-to-peak resolution as a result of a higher number of theoretical plates present. However, increasing the column length proportionally affects the separation time (Eq. 12.2) [4]. Thus, it is important that the analyst choose column lengths that are "fit-for-purpose." In other words, if peaks are sufficiently resolved and their sensitivities (based on peak heights) are acceptable, shorter columns are recommended to reduce the overall analysis time. Equation 12.2 shows the relationship between analyte retention time and column length:

$$t_R = \frac{L}{\bar{u}}(k+1), \quad (12.2)$$

where t_R is the retention time, L is the column length, and $\bar{u}$ is the average linear velocity.

On the other hand, if a complex sample contains a myriad of components that require good separation resolution, it may be more beneficial to resort to longer columns. Another factor to consider is the cost of columns. Typically, the longer the column, the more expensive it is for purchase. An analyst can choose to start with column lengths at around 30 m as a starting point for general separations. For screening samples and high-throughput analyses, a shorter column may be used. For highly complex samples, the column length may be increased to 60 m.

12.2.2.3 Film thickness

The thickness of the stationary phase affects the efficiency, capacity, separation time, and in some cases, the background of a separation.

The film thickness is inversely proportional to the number of theoretical plates in a separation. As a result, when reducing the film thickness, the efficiency is enhanced. Additionally, separation time will be reduced, due to analytes retaining in the thicker film for shorter periods of time. Column bleed (background contribution) is also less pronounced in columns with smaller film thicknesses. Also, thinner films are more applicable for separating high-boiling analytes (i.e., petroleum samples, large-chain fatty acids, polycyclic aromatic hydrocarbons, etc.) as their elution temperature will be lower (faster elution) compared to larger films. Thinner films (<1.0 um) should be used for GC/MS applications to minimize column bleed and contamination of the ionization source.

On the other hand, since additional partitioning sites are available during the separation when using a thicker film, increasing the film thickness will result in an increase in the peak capacity. This inherently means that column overloading (observed as peak fronting) is less likely compared to thinner films. Moreover, this also means that thicker films enable the separation of samples containing large variations in analyte concentrations. Active sites are less frequent in columns with thicker films, which reduce the likelihood of polar compounds adsorbing to the column wall. Analytes with relatively high volatility are recommended to be separated using columns with thicker films (>1.0 um), as they can more readily interact with the stationary phase.

Taken into consideration the contributions from varying the column dimensions, the analyst can refer to Table 12.1 as guidance on the effects of altering each parameter.

As with all chromatography techniques, choosing an appropriate mobile phase can be crucial for a successful separation. Fortunately, the carrier gas choice for GC boils down to three common inert gasses, namely, H_2, He, and N_2. All three gasses possess various advantages and limitations, some of which can be exemplified by plotting their respective van Deemter curves (Fig. 12.2) [5].

As shown, the best separation efficiency (lowest height equivalent to a theoretical plate, *H*) can be obtained by using N_2 as a carrier gas. However, the speed of the separation (i.e., linear velocity) must be relatively slower than He or H_2 to achieve this efficiency. Further, the efficiency deteriorates more rapidly as the linear velocity increases for the N_2 carrier gas compared to the other carrier gases. The steeper slope of the curve (following the curve minima) means that a small change in flow rate can significantly impact the separation efficiency using N_2 [6]. On the other hand, the optimal separation efficiency using He and H_2 is slightly poorer than N_2. However, the linear velocity (at optimal efficiency) is faster using these carrier gases, meaning quicker separation times. Overall, separations using lighter carrier gases can be significantly faster, especially for H_2, and can increase the throughput of the analysis.

Table 12.1 Benefits of modifying column I.D., length, and film thickness

Parameter	Benefits of modification	
	Higher	**Lower**
Column I.D.	• Suitable for samples with large variations in concentration ranges • Higher peak capacity • Lower head-pressure • Reduced likelihood of column overloading	• Efficiency is enhanced • Suitable for GC/MS • Reduced column bleed
Column length	• Higher peak capacity • Resolution is enhanced	• Lower head-pressure • Shorter run time
Film thickness	• Higher peak capacity • Reduced likelihood of column overloading • Suitable for samples with large variations in concentration ranges • Less active sites – reduced adsorption of polar compounds to wall • Suitable for separating low-boiling compounds	• Reduced column bleed • Shorter run time • Suitable for separating high-boiling compounds • Suitable for GC/MS

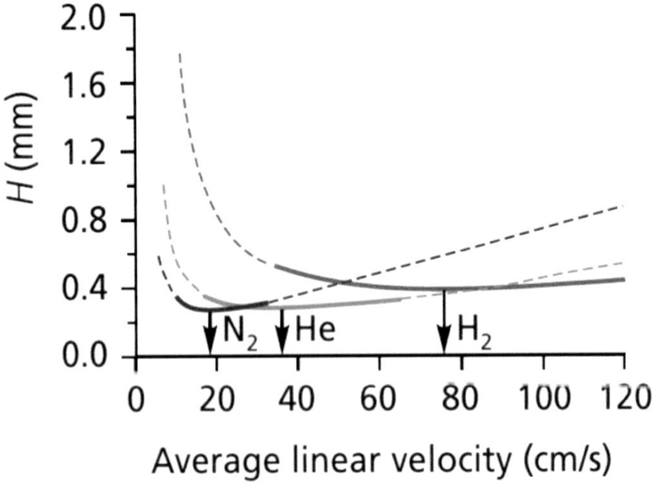

Figure 12.2 Van Deemter curves for N_2, He, and H_2. Column I.D. = 0.25 mm, analyte retention factor (k) = 10. Reproduced from [5].

The viscosity (i.e., resistance to flow) of the gases also influences their respective linear velocities at a given temperature. In GC, the viscosity of gases increases when the temperature is increased, as shown in Fig. 12.3 [5]. As depicted, H_2 possess the lowest viscosity, compared to N_2 and He, at every temperature studied. As a result, this carrier gas can produce higher linear velocities. The viscosity of the carrier gas also possesses a linear relationship with the pressure drop (difference in carrier-gas pressure from the column inlet to the outlet) of the system. A gas with a higher viscosity will require higher pressure drops to sustain a specific average linear velocity. Due to its low viscosity, H_2 would be more ideal for narrow-bore columns (<200 μm I.D.) where higher pressure drops are required to maintain a suitable linear velocity.

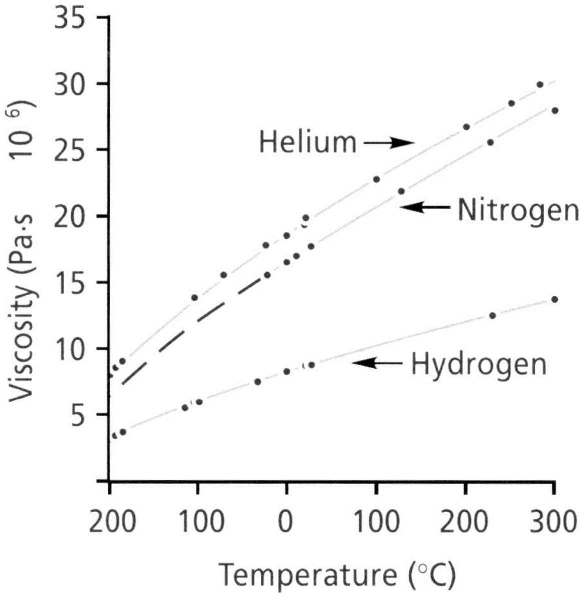

Figure 12.3 Relationship between the carrier gas viscosity and temperature. Reproduced from [5].

Generally, it is recommended that the analyst use either He or H_2 as a starting point for method development. N_2 is least preferred since its linear velocity operating range is highly limited. However, if the cost of analysis is a major concern, then using N_2 may be beneficial as this gas can be produced by generators. Also, the efficiency loss observed when using N_2 can be compensated by modifying column dimensions [7] (as discussed in Section 2.2). He is an ideal carrier gas choice for conventional GC capillary columns and can provide exceptional separation efficiency and peak resolution. H_2 is more suited for high-speed separations employing narrow-bore columns. It is important to note that the analyst recognize all safety considerations in cases where H_2 is used as a carrier gas.

12.2.3 Linear Velocity

The linear velocity of a carrier gas can dramatically affect two primary parameters in a separation, namely, efficiency and retention time. Selecting a higher linear velocity will reduce the interaction between the stationary phase and the analyte, reducing the retention time. This can be advantageous when analysis throughput is a critical factor during optimization. Conversely, lowering the linear velocity will promote the stationary phase-to-analyte interaction, resulting better separation efficiency to a certain extent (optimum level). The trick is to select a linear velocity that can strike a balance between throughput and efficiency.

At a given column length, ID, and film thickness, the separation efficiency can be simply understood as an inverse to the plate height. As shown in Fig. 12.4, the lowest *H* values can be obtained when the linear velocity is between 20–45 cm/s using He as the carrier gas. As such, it is recommended that the analyst begin their method development within this linear velocity window. Further modifications to the linear velocity can be sought after for improving the analysis throughput.

Table 12.2 displays the recommendations for selecting a flow rate using He or H_2 as a carrier gas, based on their corresponding linear velocities [8]. At a constant length of 30 m, columns with smaller IDs (i.e., narrow-bore) require lower He flow rates to achieve a linear velocity of 20–45 cm/s. This dramatically differs when increasing the ID to mega-bore type columns. It is worthy to note that increasing the linear velocity correspondingly require increasing the inlet pressure. It is recommended that the analyst review the inlet pressure capabilities of their instrumentation before modifying the linear velocity to extreme levels. In cases where high velocities are desired (inlet pressures exceedingly greater than 80 psi), the analyst can alternatively select hydrogen as carrier gas.

Table 12.2 Recommendations for selecting initial flow rates, based on the optimal linear velocity of hydrogen and helium

	Helium		Hydrogen	
Column I.D. (mm)	**Linear velocity (cm/s)**	**Flow rate (mL/min)**	**Linear velocity (cm/s)**	**Flow rate (mL/min)**
0.18	20–45	0.3–0.7	40–60	0.6–0.9
0.25	20–45	0.7–1.3	40–60	1.2–2.0
0.32	20–45	1.2–2.2	40–60	2.2–3.0
0.53	20–45	4.0–8.0	40–60	6.0–9.0

Source: Adapted from [8].

12.3 Optimization of Separation Temperature

One of the hallmarks of separation by GC is the ability to manipulate analyte retention using temperature. Fundamentally speaking, increasing the temperature during a separation will reduce analyte partitioning to the stationary phase (i.e., decrease retention), leading to shorter retention times. On the other hand, reducing the temperature will result in the opposite effect (i.e., increase retention, longer retention times). Importantly, altering the temperature will also alter analyte selectivity during a separation. This means that peak elution order can potentially be affected once temperature is manipulated during separation [4]. This section provides the analyst with practical suggestions on developing initial temperature program/conditions to achieve a "fit-for-purpose" separation. For fundamental and theoretical discussions, the reader is advised explore the outstanding library of literature available [4, 9–14], as well as the fundamental theories established by John C. Giddings [15].

12.3.1 Programmed Temperature Separations

Most GC methods today employ temperature programming over isothermal separations for a number of reasons. Temperature programming enables the separation of a mixture of analytes, with broad boiling point ranges, in the shortest run times possible. Sharper peaks can be obtained with the appropriate program as a result of thermal focusing, especially for later-eluting analytes. Also, applying a temperature gradient (and subsequent hold) near the completion of a separation ensures that high-boiling contaminants present in a sample are fully eluted and not carried-over to subsequent injections. As separation depends not only on temperature but also on a number of other parameters (i.e., stationary phase, column dimensions, carrier gas, etc.), the optimization of separation temperature program may become more or less an empirical task. However, with modern technology and third-party software available via various GC consumable suppliers, such as EZ-GC by Restek, choosing a separation temperature can be less challenging.

A general and practical temperature scouting gradient was recently described by Tony Taylor. The scouting gradient is shown in Eq. 12.3 (adapted from [16]):

40°C hold for 1 min;
Ramp at 10°C/min to 300°C, hold for 10 min (12.3)

It is worthy to mention that the final temperature (300°C) should be changed according to the maximum operating temperature of the column.

From the scouting gradient, the analyst can tweak the program according to their needs. The ramp rate should be altered within a window of 4–25°C/min, due to the limitations of some GCs. Higher ramp rates >25°C/min may sacrifice the retention time reproducibility. If analytes are closely eluting, the ramp rate should be reduced to better-resolve these peaks. On the other hand, if the selectivity of analyte pairs is high, the ramp rate can be increased to reduce separation time and peak broadening (longitudinal band broadening). Note that altering the temperature ramp rate can cause the elution order of analytes to change, as shown in the separation of a mixture of chlorinated pesticides (Fig. 12.4) [17]. As shown, drastic changes were observed when increasing the temperature ramp from 1.5 to 20°C/min. The retention times of all analytes were reduced from ~60 min to ~12 min. The elution order of peak pairs J&K, M&N, P&Q were reversed. Additionally, some peak pairs were better resolved (i.e., J&K), others loss resolution (i.e., D&E).

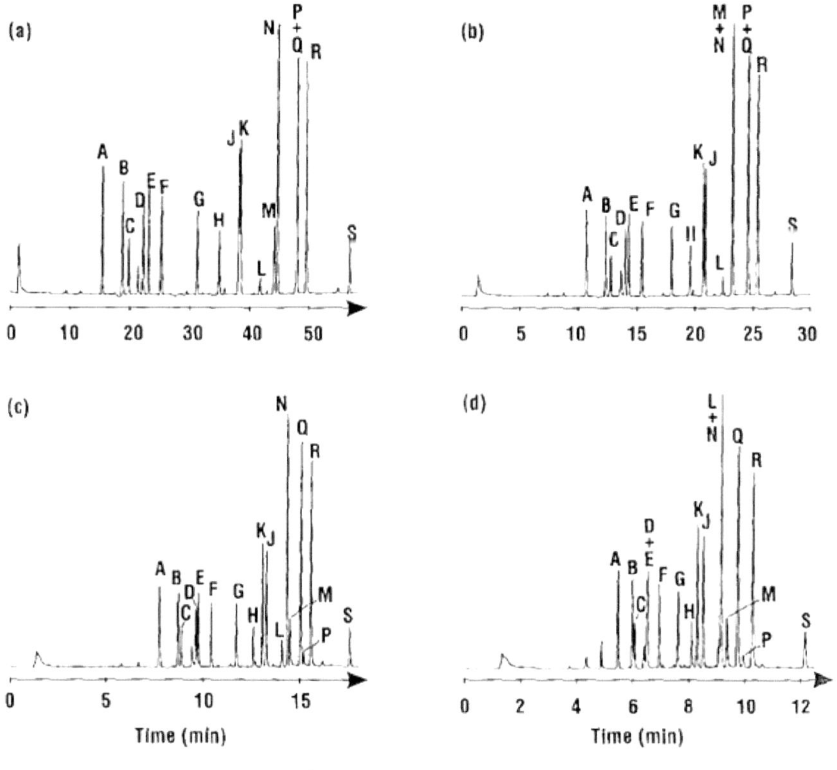

Figure 12.4 Separation of a number of chlorinated pesticides using a 25 m × 0.53 mm methyl-35% phenylsilicone capillary column. The oven temperature ramp rates were selected as (a) 1.5, (b) 4.0, (c) 8.0, (d) 16.0°C/min. (A) α-BHC, (B) γ-BHC, (C) β-BHC, (D) heptachlor, (E) δ-BHC, (F) aldrin, (G) heptachlor epoxide, (H) endosulfan I, (J) dieldrin, (K) *p,p*′-DDE, (L) eldrin, (M) endosulfan II, (N) *p,p*′-DDD, (P) endrin aldehyde, (Q) *p,p*′-DDT, (R) endosulfan sulfate, (S) decomposition product. Reproduced from [17].

In addition to changing the temperature ramp, it is recommended to add a mid-ramp isothermal hold for compounds closely eluting with each other. The isothermal hold should be around a few min, depending on empirical observations. Once all peaks have eluted, the analyst can add an additional fast ramp to below the column maximum operating temperature and hold for a few minutes to ensure other matrix components are fully eluted from the column.

12.3.2 Isothermal Separations

Isothermal separations can be viable for samples with fewer components and narrower volatility ranges. In this separation mode, system cool-down and heat-up cycles are not needed, while early peaks-of-interest can be eluted sooner. This means that the flexibility for increasing analysis throughput is greater for isothermal separation. Separation resolution can also be meticulously optimized, and therefore enhanced, for critical pairs in isothermal mode. This is especially true for chiral separations. As with temperature-programmed separations, the optimal temperature program may not be obtained without some empirical observation and development.

A recommendation for choosing a starting point would be using the scouting temperature program noted in Eq. 12.3 and selecting an isothermal temperature that is around 45°C below the elution temperature of the last analyte [16]. From the initial starting point, modify the isothermal temperature based on the method objectives. A lower temperature can be used to increase analyte retention and may lead to better resolution, although peak broadening will be more severe with longer retention times. Vice versa, a higher temperature can be applied to reduce analyte retention, run time, and peak broadening, but may sacrifice separation resolution. When increasing the isothermal temperature, it is worthy to note that the retention time of an analyte will be reduced by approximately half if the temperature is increased by 30°C [15]. Similar to temperature program, it is recommended that the analyst add a fast ramp to below the maximum operating temperature of the column with a subsequent hold, once the analyte(s) of interest has eluted, to remove other matrix components.

12.4 Fast GC: A Culmination of Instrument, Column and Temperature Optimization

As stated previously, method optimization can revolve around improving the separation efficiency and/or analysis speed. In cases where separation speed is the primary focus, the analyst can alter a variety of physical parameters to achieve "fast GC." Fast GC specifically denotes the separation of components in a sample in a few minutes with peak widths ranging from 1–3 s [18]. In addition to significant gain in analysis speed, applying fast GC for everyday analyses can provide additional benefits [19]. For example, since the capillary columns used for fast GC are typically shorter compared to conventional GC columns, they are relatively cheaper to purchase. Moreover, since the film thicknesses of typical fast GC columns are thinner, these columns contribute to lower column bleeding. Lastly, as peak elution is significantly quicker compared to conventional GC, longitudinal band broadening effects are less pronounced, resulting in better signal-to-noise ratios (i.e., sensitivity). The primary limitations for employing fast GC can be a loss in sample capacity and separation efficiency. However, some of these issues can be addressed by careful considerations in method development and/or by employing selective detection techniques, such as MS, to deliver peak resolution by mass or spectral differences. Figure 12.5 shows a chromatogram comparison of an analysis of nutmeg oil using a conventional GC method and column (60 m × 0.25 mm; 1 um HP-1) versus a fast GC method and column (20 m × 0.1 mm; 0.4 um HP-1) [20]. The chromatographic profile is highly similar; however, the retention time of the final peak was drastically reduced from 74.9 to 13.3 min. As shown, the immense gain in analysis throughput is the pinnacle of the potential benefits gained from applying a fast GC method.

Fast GC can be accomplished by: *reducing the column length, I.D., and film thickness; increasing the linear velocity of the carrier gas; using a reduced-pressure/vacuum outlet* (i.e., *GC/MS); and employing a rapid temperature program* [18, 21, 22]. As mentioned previously, reducing the column length will decrease the retention times of analytes, leading to a faster overall separation. Although the resolution may also be reduced when using shorter columns,

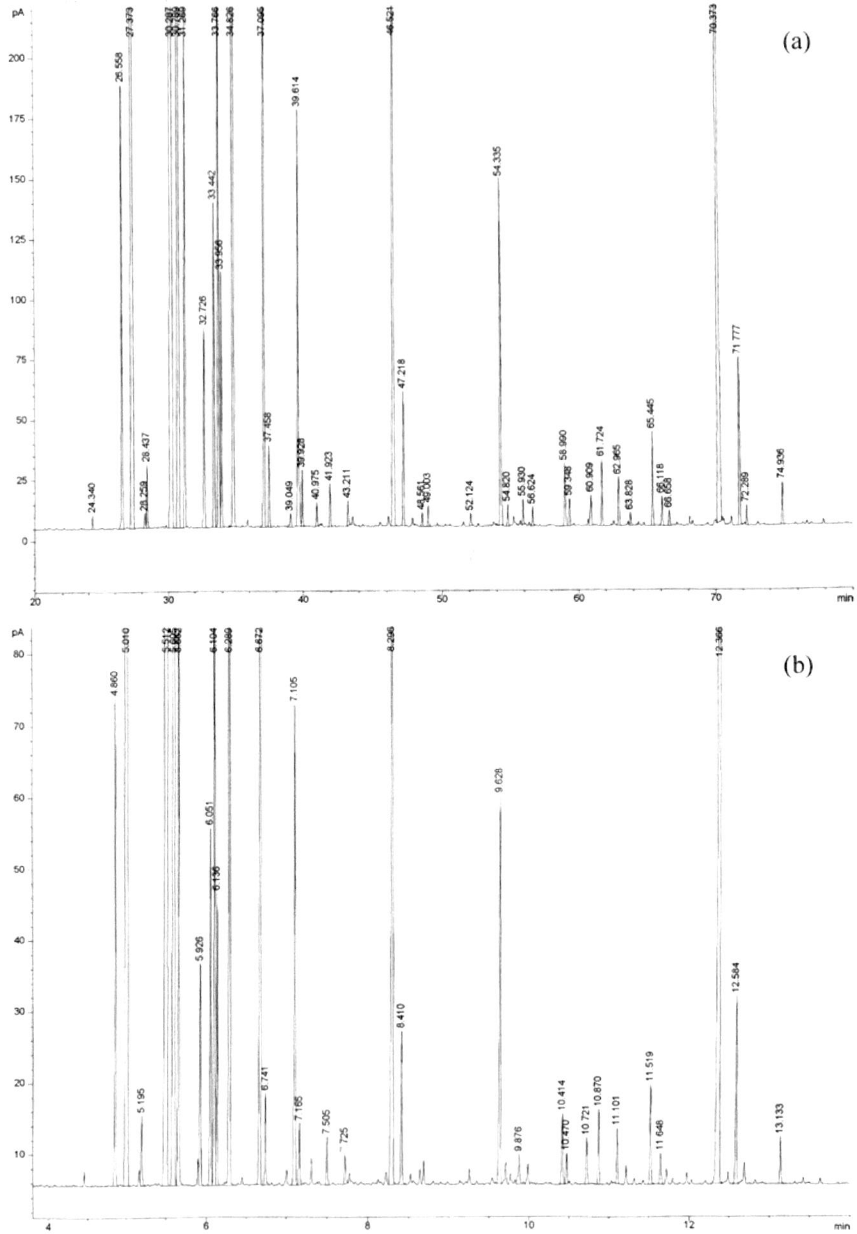

Figure 12.5 Comparison of an analysis of nutmeg oil using (A) a conventional GC method and column (60 m × 0.25 mm; 1 um HP-1) versus (B) a fast GC method and column (20 m × 0.1 mm; 0.4 um HP-1). Reproduced from [20].

careful method optimization can alleviate this potential issue [23]. On the other hand, decreasing the ID and film thickness of the column will also reduce the retention time, while providing an enhancement in the separation efficiency.

Typical GC column dimensions used in fast GC applications are as follows: column length < 20 m; I.D. < 0.25 mm; and film thickness < 0.2. It should be noted that the dimensions stated are not mutually exclusive to fast GC, as "fast GC" can theoretically be accomplished using any column dimensions. However, the recommendation provides a feasible starting point to lessen the burden of choosing a suitable column for fast GC method development.

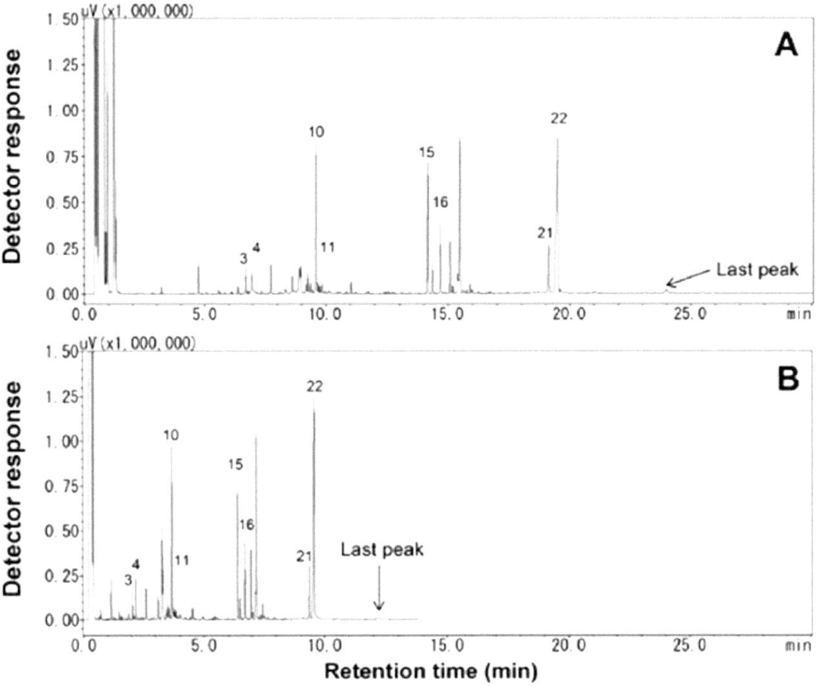

Figure 12.6 Comparison of the chromatograms obtained in the analysis of Japanese green tea extracts. (A) before and (B) after optimization of the linear velocity and temperature program. Both methods employed an RTX-5 column (10 m × 0.18 mm; 0.2 um). Reproduced from [24].

Achieving fast GC separation speeds can also be accomplished by adjusting the linear velocity of the carrier gas and developing a temperature program which enables a "fit-for-purpose" separation. This means using a rapid temperature ramp rate or a higher starting oven temperature, and/or operating the carrier gas linear velocity above the optimal/recommended speed (shown in Table 12.2). Figure 12.6 shows a comparison of the chromatograms obtained in the analysis of Japanese green tea extracts before and after optimization of the linear velocity and temperature program [24]. Using the same RTX-5 column (10 m × 0.18 mm; 0.2 um), the authors were able to reduce the retention time of the last observed peak from 24 to 12 min. This was accomplished just by modifying the linear velocity from 45 to 80 cm/s and applying a relatively faster temperature program.

As mentioned previously, employing H_2 as a carrier gas permits the analyst to operate at higher linear velocities compared to He or N_2 due to its relatively low viscosity. It is therefore recommended, but certainly not required, that the analyst exploit H_2 as a carrier gas for fast GC separations.

12.5 Conventional GC Detectors Applied in Pharmaceutical Analyses

The GC technique has many detectors available at its disposal. In fact, the wide array of detectors is a major advantage of GC over other chromatographic techniques. In the pharmaceutical industry, the flame-ionization detector (FID) and electron-capture detector (ECD) are the most commonly used. Historically, the thermal conductivity detector (TCD) is a valuable universal detector that covers most permanent gases. While there are many more GC detectors which have been reviewed [25, 26] the FID, ECD, and TCD will be covered in this section as the most common and useful that the practicing analytical chemist may routinely encounter.

12.5.1 Thermal Conductivity Detector

The TCD is a universal detector in that it responds to any analyte with thermal conductivity that differs from the carrier gas. The detector is most commonly applied to analysis of light, permanent gasses. TCD has a wide linear range capable of extending over 5 orders of magnitude and is a concentration sensitive detector.

The TCD detector is non-destructive; analyte structure is maintained after reaching the detector. In contrast to other detectors, the TCD does not require a flame or flammable gasses, thereby making it a simpler to use, safer detector. Also, the TCD power consumption is low and the detector can be miniaturized to be compatible with portable or chip-type analyses.

The TCD is based on the principle that resistance changes can be measured on a heated filament when analytes pass by. A second filament also measures this resistance for a reference gas, which minimizes drift over the course of a run. In order to maximize sensitivity of the TCD, the user should select a carrier gas with a large difference in thermal conductivity between carrier gas and analyte. Further, operating the TCD at lower temperatures can help maximize sensitivity. As temperature goes up, most analytes experience an increase in thermal conductivity. Because the common carrier gasses Helium and Hydrogen have some of the highest thermal conductivities, sensitivity decreases as analytes' thermal conductivity increases. However, operating the TCD at lower temperatures should only be employed to the point where peak shape and performance is maintained. If the TCD is operated below the maximum GC oven temperature, then the user runs the risk of peak tailing as the sample may interact with the tungsten filament.

12.5.2 Flame-Ionization Detector

The FID is also a universal detector that responds to carbon in a sample. The detector is mass sensitive. That is, the signal increases proportionally to the carbon content in an analyte but would also increase as the concentration of analyte increases. This is the workhorse detector for many industries, and the most commonly used GC detector for pharmaceutical analysis. In addition to being a universal detector for almost all volatilizable pharmaceutical compounds, the FID is relatively easy to use and maintain and cost effective. The FID also boasts a wide linear range spanning up to 7 orders of magnitude.

Further, the FID is inherently destructive; carbon in the analyte is combusted in the FID to give a response. As seen in Fig. 12.7, the schematic of the FID, analytes are flowed into the detector where they are combusted in the presence of a hydrogen-oxygen flame. Charged particles are collected by applying a voltage between the tip of the jet and a collector electrode. Signal on the order of picoamps is amplified and collected as a voltage.

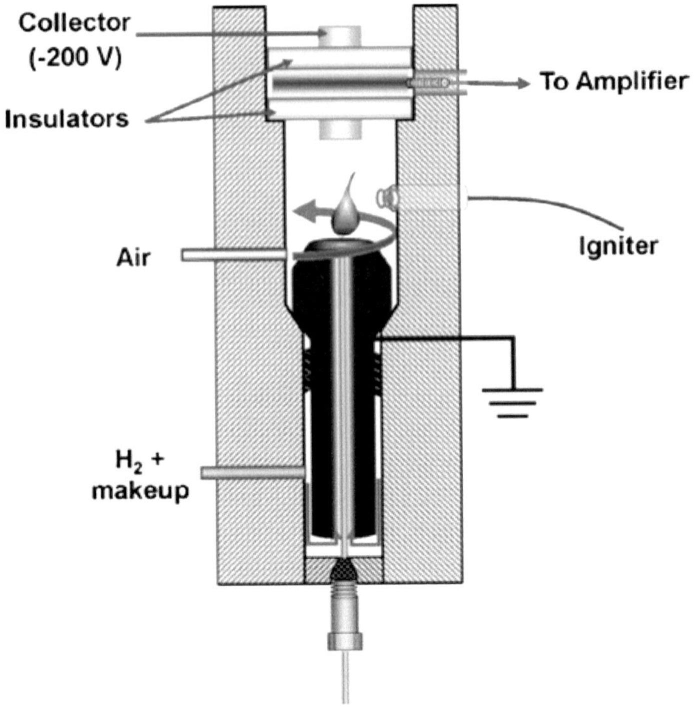

Figure 12.7 Schematic representation of the FID [27].

As carbon is combusted in the FID, the amount of carbon being combusted per unit time in the detector is proportional to signal response. As such, FID exhibits unit carbon response in the sense that for each additional carbon present in an analyte molecule, the signal response increases proportionally to the number of carbon atoms in the molecule. One advantage of this property of the FID is that response factors can be estimated, and standard calibration curves may not be needed for every analyte. In practice, however, analysts in the pharmaceutical industry would typically still fully validate all known impurities or solvents for a GC method, which would include calibration of all analytes. While the unit carbon response property of FID is a useful property to understand and apply, regulatory expectations and risk management in the pharmaceutical industry dictate that the amount of extra work involved in preparing additional calibration curves during validation is minimal against the risk of an analyte behaving in an unanticipated manner. However, for unspecified or unknown impurities that cannot be validated, the more uniform response is a major advantage to the FID. If using GC for an impurities method in the pharmaceutical industry, the response between the main component and related substance impurities is similar for a GC-FID method. Because unknown impurities in a product and method by nature cannot be validated, the similarity in response factors between related substances when using an FID helps minimize risk that an unknown impurity would be quantitated vastly inaccurately.

12.5.3 Electron Capture Detector

The electron capture detector (ECD) is a non-destructive, concentration-based detector that is suitable for analytes with electron-capturing capability. The linear range of this detector spans approximately 4–7 orders of magnitude. In principle, the detector operates based on a reagent gas with low ionizability as the background signal. A radioactive source ionizes this reagent gas, and the resulting electrons that are emitted by the reagent gas are collected as a signal. As analytes pass through the detector, they are able to capture some of these electrons that exist as a plasma from the reagent gas. The higher a given analytes' ability to capture electrons, the larger the background signal will be reduced and thus, a higher overall absolute signal will be observed. The ECD is traditionally highly sensitive toward halogenated analytes due to high electronegativity properties that are conducive to electron capture, and the signal response is greatest for iodine, followed by bromine, chlorine, and fluorine. The more that a carbon atom is substituted directly or adjacently with electronegative heteroatoms, the higher the observed signal response.

As seen in the ECD configuration in Fig. 12.8, a nickel-63 radioisotope source is frequently seen as part of ECDs due to high thermal stability as well as availability and practicality of implementing this radioactive material. While designs may have slight variations or be miniaturized, the principle of the ECD is that GC column eluent is mixed with make-up gas and brought to the radioisotope source. In this chamber, secondary electrons are generated when high-energy beta electrons collide with carrier gas. These thermal electrons form a plasma, which also includes radicals and positive ions, and a baseline current is established for the detector. When electron-capturing analytes pass through this plasma and capture the thermal electrons, negative ions are generated. After neutralization by positive ions present in the detector, the background plasma signal is altered and signal response is generated proportional to the analytes' ability to capture electrons.

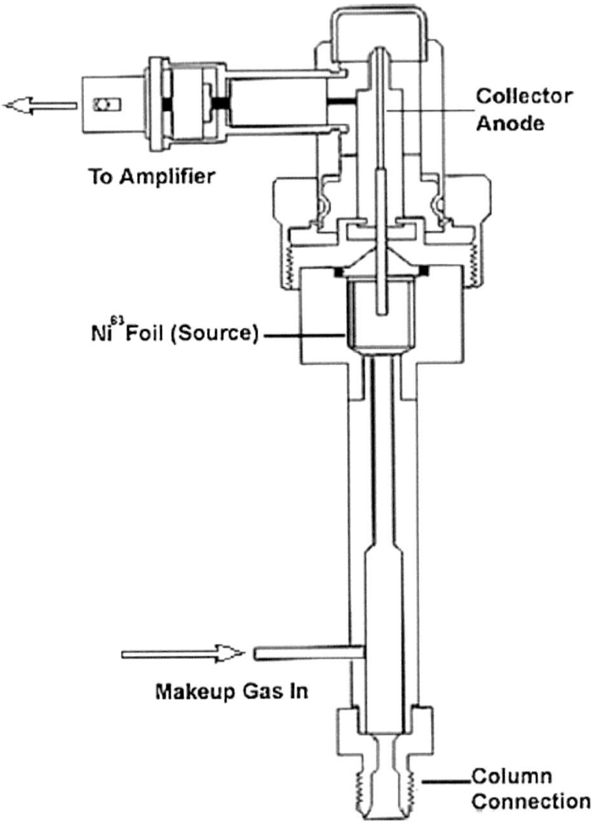

Figure 12.8 Schematic representation of the radioisotope-ECD [27].

GC instruments with ECD setup have a wide range of applications in the pharmaceutical industry due to the frequency of good electron capturing groups on drug molecules. For example, methods have been developed that can simultaneously quantitate over 50 drugs in plasma or whole blood by utilizing GC-ECD. The method was validated, and LOQs as low as 2 ng/mL were achieved [28]. Mutagenic impurity analysis was also recently accomplished by probing different ionic liquid-based diluent and using GC-ECD in order to develop highly sensitive methods down with LODs as low as 5 ppb, which is an improvement over 1,000 fold compared with GC-FID and comparable with LC/MS and GC/MS [29].

12.6 GC/Mass Spectrometry

Hyphenated techniques, such as GC/MS and LC/MS, have become essential in the pharmaceutical industry within the past decades. Compared to other detection techniques, coupling GC to MS provides the analyst with the ability to fully characterize a sample by quantification and structural elucidation. Applications of GC/MS in pharmaceutical analyses can vary from identification of impurities, to peak purity analysis, to quantification of potent/toxic compounds at trace levels in drug substances and products [30–33]. Compared to other detectors, MS detectors can provide better signal sensitivity and universal detection for most analytes.

Employing mass spectrometry also adds another dimension of separation to sample analysis. Instead of a chromatographic separation, the analyst can take advantage of separation of ions based on their mass-to-charge ratio (m/z). By selecting the proper ion(s) to analyze, it is possible to selectively detect and quantify peaks which would otherwise co-elute by chromatography [34]. For example, Xue et al. selected two distinct ion fragments observed during the analysis of penehyclidine (PH; 175 $\boldsymbol{m/z}$) and its internal standard penehyclidine-d_5 (PH-d_5; 180 $\boldsymbol{m/z}$) by GC-MS, as shown in Fig. 12.9 [35].

Although the chromatographic retention times of the drug and its internal standards were the same (~8.2 min), specificity for each analyte was obtained when only the ion fragment 175 $\boldsymbol{m/z}$ for PH and 180 $\boldsymbol{m/z}$ for PH-d_5 were extracted from the MS data (Fig. 12.10), thereby enabling quantification [35].

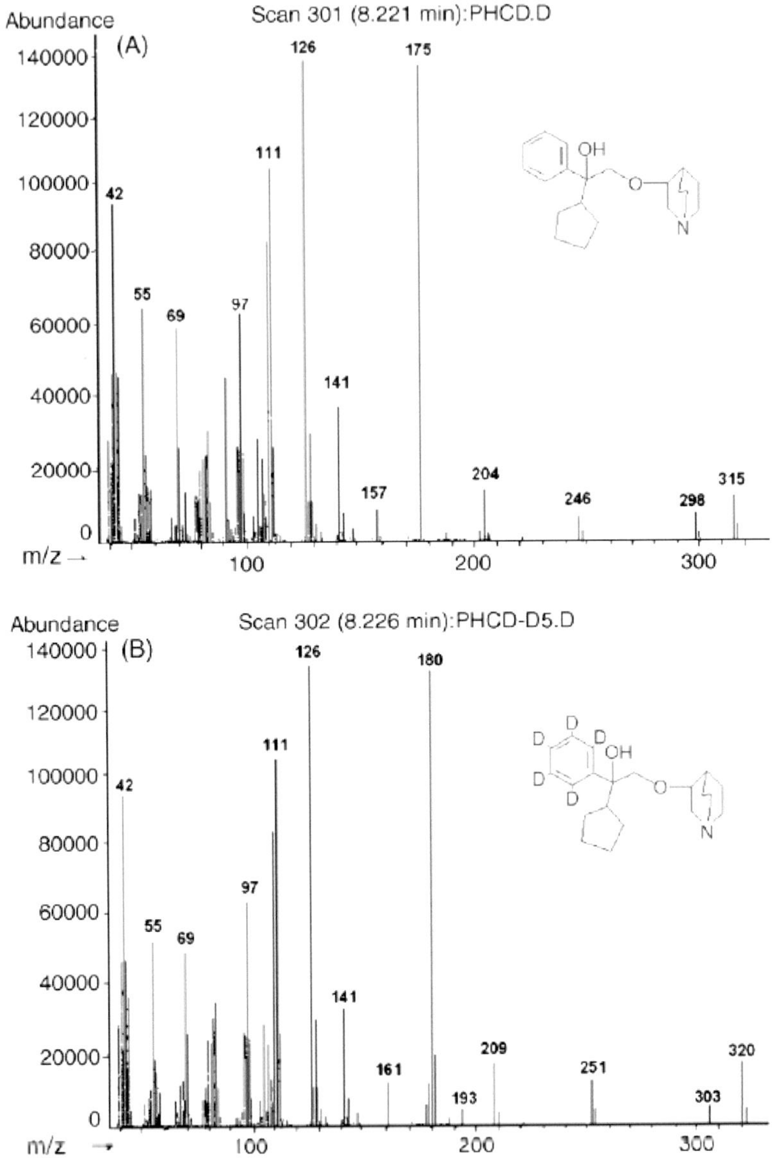

Figure 12.9 Mass spectra (electron ionization; EI) for: (A) penehyclidine and (b) the internal standard, penehyclidine-d_5. (A) EI-MS of m/z 315 → 175 for penehyclidine. (B) EI-MS of m/z 320 → 180 for penehyclidine-d_5. Reproduced from [35].

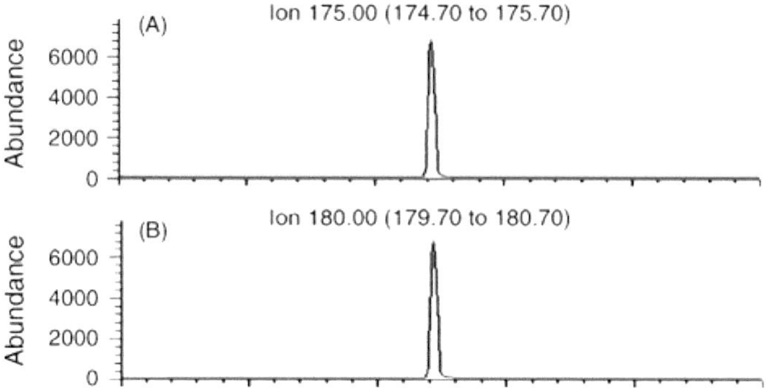

Figure 12.10 Extracted ion chromatograms of penehyclidine and penehyclidine-d_5. (A) Penehyclidine extracted at 175.00 m/z. (B) Penehyclidine-d_5 extracted at 180.00 m/z. Reproduced from [35].

12.6.1 Ionization by GC/MS

The mechanism of a GC/MS detector can be simplified to four primary functions:

1. *Ionization* of volatile matrix components as they exit the vacuum outlet of a GC column.
2. *Acceleration* and *focusing* of the ionized species and their fragments to a mass analyzer.
3. *Separation* of the ionized species based on their relative mass to charge ratios ($\boldsymbol{m/z}$) in the mass analyzer.
4. *Detection* of the ionized species as they exit the mass analyzer.

The ionization process is critical to generate detectable components in a sample. As such, it is beneficial to briefly examine the underlying principle of various ionization modes in GC/MS.

12.6.1.1 Chemical ionization

Commercially available GC/MS can be equipped with chemical ionization (CI) sources capable of producing positively charged or negatively charged ions. Ions generated via CI typically possess the same composition as the parent compound, with one more (or less) electron. These ion species are often referred to as molecular ions. As CI does not tend to generate additional ion fragments from the parent molecular ion, it is considered a soft ionization technique. Employing CI in GC/MS allows the analyst to ascertain the molecular weight of an analyte. Mass spectral data generated from CI experiments are relatively easier to interpret; however, they do not provide structural information of the compound.

12.6.1.1.1 *Positive ion*

Positive CI was introduced by Munson and Field in the mid-1960s [36, 37]. Briefly, this technique involves the continuous introduction of a reagent gas, such as methane, to the GC/MS ionization source. During its residence in the source, the reagent gas is ionized to a primary cation by an electron beam generated from applying a potential to a tungsten filament. The primary cations can also react with other the reagent gas molecules to form secondary ions (Scheme 12.1). When a cloud of sample molecules is introduced to the source, the reagent gas ions (primary and secondary) undergo various chemical reactions with these molecules to generate molecular ions (M^+), see Scheme 12.2. It is worthy to note that the molar ratio of the reagent gas is around three orders of magnitude higher than the sample molecules. This mitigates the direct ionization of the molecules by the electron source.

$$CH_4 + e^- \longrightarrow CH_4^{\bullet +} + 2e^-$$

$$\left[CH_4^{\bullet +} + CH_4 \longrightarrow CH_5^{+} + {}^{\bullet}CH_3 \right]^*$$

Scheme 12.1 Ionization of methane reagent gas by positive CI. *The primary cation ($CH_4^{\bullet +}$) can react with bulk reagent gas to form secondary cations, which can also further react.

12.6.1.1.2 *Electron capture negative ionization*

A complementary and less-understood CI technique is electron capture negative ionization (ECNI). This technique was developed by Stafford and Hunt in the mid-1960s [38, 39]. As the name implies, this ionization technique produces negative molecular ions of the sample gas. This ionization technique shares a number of similarities to positive CI by requiring the use of a reagent gas (i.e., methane, ammonia, isobutane, etc.) in a closed ionization source to aid in the production of sample ions. The primary difference is that ionization in ECNI is accomplished by the capture/reaction of thermal electrons (low-energy; 0–5 eV) expelled during the ionization of the reagent gas, rather than with the reagent gas ions themselves (see Scheme 12.3 for ions generated by ECNI). ECNI can be set up by changing the mode of detection to anion mode and following the manufacturer's recommended settings for reagent gas ionization.

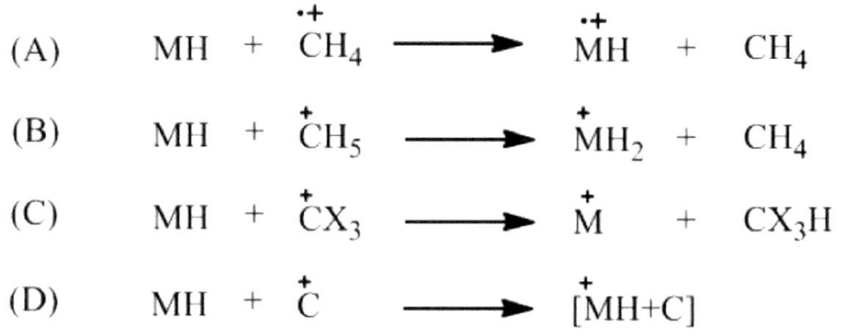

Scheme 12.2 Ionization pathways in positive CI using methane as a reagent gas. (A) Charge transfer, (B) proton transfer, (C) hydride abstraction, (D) adduct formation.

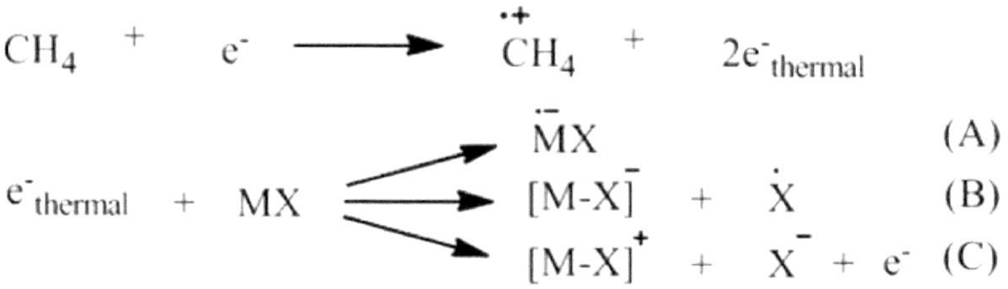

Scheme 12.3 Ionization pathways in ECNI. (A) Resonance electron capture, (B) dissociative electron capture, (C) anion formation by ion-pairing.

12.6.1.2 Electron ionization

Perhaps the most common ionization technique for GC/MS is electron ionization (EI) [40]. This technique exhibits remarkable utility for the universal detection and characterization of volatile compounds. EI is a *hard ionization* technique, which generates the sample molecular ion and its corresponding fragments. Based on the ion fragments formed during the ionization step, the analyst can elucidate the chemical structure of the sample molecule. An example mass spectra for a common synthetic precursor, epichlorohydrin (MW: 92.52 g/mol), is shown in Fig. 12.11 [41].

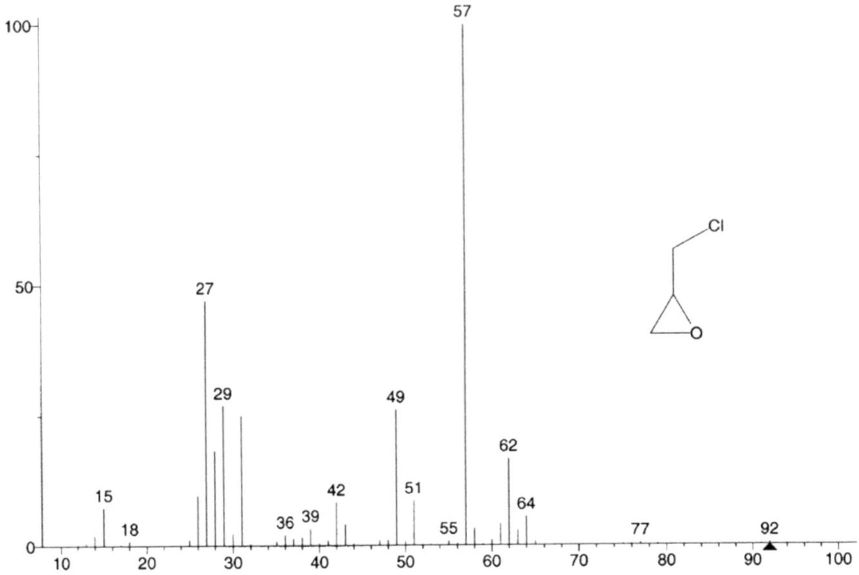

Figure 12.11 Mass spectra of epichlorohydrin generated by electron ionization. Obtained from [41]. Reprinted (adapted) with permission from Loda, C., Bernabe, E., Nicoletti, A., Bachi, S., and Dams, R. (2011) Determination of epichlorohydrin in active pharmaceutical ingredients by gas chromatography-mass spectrometry, *Organic Process Research & Development*, 15, pp. 1388–1391. Copyright 2018 American Chemical Society.

As shown, the mass spectral data obtain from EI can be relatively complex and the molecular weight of the sample may not be apparent using EI, due to fragmentation and/or rearrangement [41, 42]. Therefore, EI, CI and ECNI should be exploited as synergistic techniques to obtain the molecular weight and chemical structure/composition of a sample molecule.

A schematic illustrating the EI source is shown in Fig. 12.12. Similar to CI and ECNI, the EI source houses a heated metallic filament which emits and accelerate electrons towards an anode (via potential difference). The energy of the electrons generated is typically 70 eV, a convention that ensures the ionization of organic molecules. When sample molecules enter the source from the GC column outlet, it is immediately exposed to the high-energy electrons generated therein. The energy exerted by the electrons consequently increases the energy level of a sample molecule. To revert to a lower energy level, the molecule can expel an electron and form a radical cation $[M]^{+\cdot}$, as shown in Scheme 12.4. It is also possible for the molecule to abstract an electron (Scheme 12.4); however, ions generated in this fashion are much less abundant (~10^4 times) compared to radical cations.

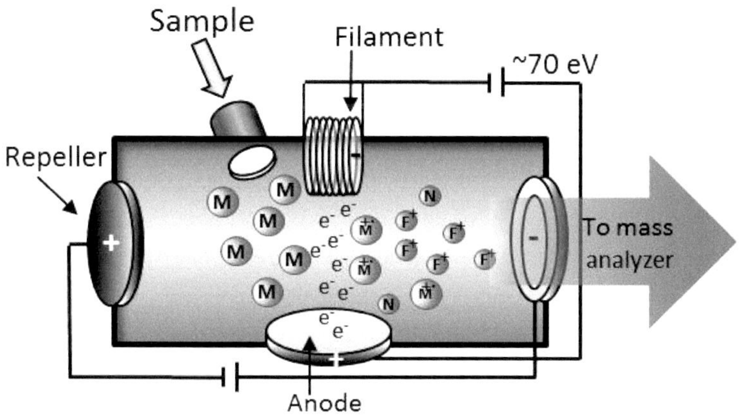

Figure 12.12 Schematic of the EI source in GC/MS.

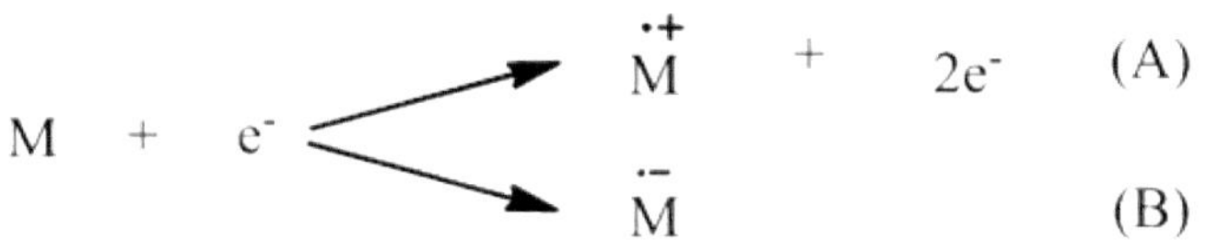

Scheme 12.4 Ionization pathways in EI. (A) Electron expulsion, and (B) electron abstraction.

Even when sample molecule expels an electron to form a radical cation and revert to a lower energy level, it may still possess significant excess energy. Therefore, the molecular ion will ultimately undergo unimolecular changes to reduce its energy level.

These chemical changes include fragmentation or rearrangements of the parent ion(s), and/or undergoing a unimolecular decomposition to form a cation and a neutral-radical species (Scheme 12.5).

$$\overset{\bullet+}{M} \longrightarrow \overset{+}{F} + \dot{N}$$

Scheme 12.5 Formation of cation and a neutral-radical species by EI.

It is important to note that different molecules will fragment in distinctive patterns, and some fragments are produced in higher abundances than others. This essentially results in a unique and highly reproducible mass spectral fingerprint for a sample molecule and enables the use of a database/library to aid in their identification [43].

12.6.2 Systematic Approaches for Chemical Structure Elucidation by EI-GC/MS

In the modern era of pharmaceutical analyses, there has been overwhelming interest on utilizing GC/MS for the rapid and accurate characterization of unknown impurities generated in a synthetic process. This is especially true for potentially toxic impurities, such as carcinogens and mutagens (i.e., mutagenic impurities) [31]. EI-GC/MS has thus become an essential tool for this continuing endeavor. Mass spectral data interpretation and fragmentation rules for EI techniques have previously been well established in literature. It is advised that the analyst re-visit these resources and consolidate information therein into their own interpretation practices. Nevertheless, even with these resources at one's disposal, it may be time-consuming and difficult to interpret unknown compounds without background knowledge of the chemical process. In this section, we discuss simple and effective strategies to tackle the challenging task of elucidating and characterizing "unknown peaks" by EI-GC/MS.

12.6.2.1 Interpretation of mass spectral via database

The simplest approach to predicting the chemical makeup of an unknown compound is to utilize the vast mass spectral libraries available commercially. Since ionization by 70 eV energy produces a highly reproducible fragmentation pattern, there has been appreciable effort to harmonize the mass spectral data of a numerous compounds into searchable databases [44, 45]. For example, the National Institute of Standards and Technology (NIST) and Wiley database store mass spectral data for hundreds of thousands of chemical structures, and these libraries are rapidly expanding. Additionally, many mass spectral data interpretation software can also provide statistical analysis on the similarities of an unknown compared to a known structure from the database in mere seconds. As a result, commercially available mass spectral databases are an outstanding starting point and a key arsenal in the interpretation of unknown structures [46].

When using mass spectral libraries to identify an unknown structure, care should be practiced in judging whether library matches actually pertain to the unknown compound. The best way to mitigate inaccurate structure identification is to obtain a chemical standard of the predicted compound and perform additional confirmation studies (i.e., comparison of retention times/factors, spiked-recovery studies, etc.) for verification. However, in situations where this is neither practical nor possible, the analyst can consider the following recommendations to utilize the mass spectral library more effectively.

(1) Perform background subtraction of the mass chromatographic peak of interest at the apex position to account for artifacts and interferences. Subtract mass spectral data of components contributing to peak asymmetry (i.e., shouldering peaks, split peaks, peak fronting/tailing) from the apex. This reduces the abundance of unrelated ion fragments, resulting in a more accurate representation of the sample mass spectrum.

(2) Consider the statistical results of the library matches and verify whether multiple matches of the same compound are reported. Multiple matches of the same compound *may* indicate that the proposed structure is indeed the correct structure. Additionally, utilize side-by-side comparisons of the sample mass spectrum with respect to the library match, such as difference and/or head-to-tail plots, to observe variations in fragmentation patterns.

12.6.2.2 Manual interpretation of mass spectral data

In cases where there are little-to-no logical library matches, a systematic approach can be followed to interpret the mass spectrum of the unknown peak [47, 48]:

(1) *Determine the molecular weight from the spectrum by deducing the molecular ion peak.*

The molecular ion peak ($M^{+\cdot}$) is an essential puzzle piece needed for accurate characterization of an unknown structure, as it pertains to the molecular weight of the structure. In general, potential $M^{+\cdot}$ ions formed in EI-MS should possess the highest ***m/z*** value, be an odd electron (OE) fragment since neutral molecules lose a radical during ionization, and rationally yield fragment masses by loss of neutral molecules/species. Although EI can fragment the molecular ion peak, residual abundances of the molecular ion may be observed, depending on the structure of the sample molecule. For example, the mass spectrum of analytes possessing unsaturated and/or aromatic cores may show a relatively high abundance of the $M^{+\cdot}$ peaks, since its molecular ion peaks can be stabilized through resonance. Conversely, the molecular ion peak may be observed at relatively low abundances (or not at all) in the mass spectrum of analytes comprising aliphatic/saturated cores. If the molecular ion peak is not observed in a mass spectrum, it may be worthwhile to pursue alternative ionization techniques (i.e., CI and ECNI) or reduce the EI electron energy to obtain this information.

(2) *Determine the elemental composition using the isotopic abundances present in the molecule. Obtain further insight by using the rings + double bonds rule and the nitrogen rule.*

Isotopic Abundances: Once an $M^{+\cdot}$ candidate has been identified, the elemental composition of the unknown can be deduced using the isotopic abundances of the molecule. Peaks in a mass spectrum are represented in groups. Within a group, the peak with the lowest ***m/z*** value is the monoisotopic peak, while successive peaks with +1 or more *m/z* are isotope peaks. Determining the relative abundance of the isotope peak(s) with respect to the monoisotopic peak give insight to the atoms comprised in the unknown compound.

To determine the elemental composition, measure the relative intensities of the isotope peaks and the monoisotopic peaks and compare the results with the relative abundances provided in Table 12.3 [47].

Table 12.3 Elements and their natural isotopic abundances

Element (A)	Abundance (%)	A+1	Abundance (%)	A+2	Abundance (%)
^{1}H	99.99	^{2}H	0.01	—	—
^{12}C	98.91	^{13}C	1.1	—	—
^{14}N	99.6	^{15}N	0.4	—	—
^{16}O	99.76	^{17}O	0.04	^{18}O	0.02
^{19}F	100	—	—	—	—
^{28}Si	92.2	^{29}Si	4.7	^{30}Si	3.1
^{31}P	100	—	—	—	—
^{32}S	95.02	^{33}S	0.76	^{34}S	4.22
^{35}Cl	75.77	—	—	^{37}Cl	24.23
^{79}Br	50.5	—	—	^{81}Br	49.5
^{127}I	100	—	—		

Source: Adapted from [47].

Typical elements (A) may possess one or two isotopes (A+1 or A+2, respectively) in significant abundances. Elements with no appreciable levels of A+1 or A+2 isotopes are H, F, P, and I. Elements with A+1 isotopes are C and N. Elements with both A+1 and A+2 isotopes are O, Si, and S. Elements with only A+2 isotopes are Cl and Br. Interestingly, carbon (^{12}C) has a distinct isotopic distribution; its A+1 isotope (^{13}C) is 1.1% of compared to the abundance of the ^{12}C monoisotope. This means that it can be fairly easy to deduce the number of carbon atoms in a fragment group. For example, a fragment group with 1 carbon atom will display an A+1 abundance of 1.1%, while a fragment group comprising 8 carbon atoms will display an A+1 abundance of 8.8%, and so forth.

The Nitrogen Rule: The parity (even or odd numbers) of nitrogen atoms comprised in the sample can be determined by using the *Nitrogen Rule*. Specifically, if a compound is made up of an even number of nitrogen atoms (or none at all), its molecular weight will be an even number. Similarly, odd molecular weights indicate an odd number of nitrogen atoms. Utilizing the isotope abundance described in the mass spectrum alongside the nitrogen rule provides sufficient discernment (in many cases) to hypothesize the elemental composition of the molecular ion.

The Rings + Double Bonds Rule: After the elemental composition is deduced using the isotope abundance and nitrogen rule, the degree of hydrogen deficiency (unsaturation) of the sample molecule can be determined using the *Rings + Double Bonds (R + dB) Rule*.

For a given formula with known valance for C, H, and N, the total number rings and double bonds is calculated as follows:

$$R + dB = C_{\#} - \tfrac{1}{2}H_{\#} + \tfrac{1}{2}N_{\#} + 1$$

Take pyridine for example: The molecular formula of pyridine is $C_5H_5N_1$. Thus, the R+dB for pyridine = 5 – 1/2(5) + 1/2(1) + 1 = 4, meaning that there can be up to three double bonds and one ring.

It is important to note additional considerations for this rule [47].

(3) *Predict and interpret the connectivity of atoms in the molecule.*

Once the chemical composition of the molecular ion peak has been established, attention should be focused on understanding the connectivity of the molecule. A strategy to elucidate the connectivity is to determine the losses of "common" fragments starting from the molecular ion peak. Table 12.4 shows common ion fragments formed by EI [49]. While identifying common ion fragment losses, postulate the possible connectivity between them and whether the connections form logical chemical substituents.

A number of references are available to provide in-depth fragmentation and rearrangement mechanisms to aid in interpretation [48, 50].

Table 12.4 Common fragment losses from the electron ionization of the molecular ion

Parent ion-loss	Fragment	Parent ion-loss	Fragment
M-1	H	M-32	CH_3OH, S
M-2	H_2	M-34	H_2S
M-15	CH_3	M-35	Cl
M-16	O, NH_2	M-43	CH_3CO, C_3H_7
M-17	OH, NH_3	M-44	CO_2, CH_3CHO
M-18	H_2O	M-45	CO_2H, OC_2H_5
M-19	F	M-46	C_2H_5OH, HCO_2H
M-27	HCN	M-60	CH_3CO_2H
M-28	CO, C_2H_4	M-73	$(CH_3)_3Si$
M-29	CHO, C_2H_5	M-79	Br
M-30	CH_2O, C_2H_6	M-89	$(CH)_3SiOH$
M-31	CH_3O	M-127	I

Source: Adapted from [49].

(4) *Finalize and hypothesize possible structure(s) and compare against known databases.*

With data in hand, tally up all results gathered from Steps 1–3 and hypothesize the structure(s) of the unknown compound. Revisit literature to investigate whether analytical/characterization data (i.e., NMR, IR, UV, XRD, etc.) exists for the structure(s) proposed. Chromatographic data, such as retention indices, may be available in the literature and can also be valuable information [51]. Finally, obtain and inject the standard compound(s) of interest to verify the findings. It is worthy to note that modern mass spectral libraries allow the user the option to input their own structures and mass spectra to the existing database. The analyst should take advantage of this option and populate their results into these libraries for future reference.

12.7 Considerations for Using GC in the Pharmaceutical Setting

GC analysis in the pharmaceutical environment is a powerful tool to tackle complex analytical challenges. Within the realm of industrial and pharmaceutical analysis, GC methods are used across all aspects of chemical synthesis, processes, and manufacturing. From starting materials, to intermediates, raw materials, and final active pharmaceutical ingredients (APIs) and drug product (DP), GC has found many applications in the pharmaceutical industry. Perhaps the most routine application of GC is for residual solvent analysis. However, GC can be useful for impurities (e.g., related substances) analysis, main component assay, and chromatographic identification methods of chemicals that may typically be analyzed by HPLC. Further, "in-process" testing, such as reaction monitoring or solvent switching, can easily be handled by GC in many cases. Lastly, as regulatory expectations tighten around the understanding and low-level control of mutagenic impurities (MIs) in products for human consumption, GC-based methods have proven to be very useful for the analysis of these highly scrutinized impurities that often require detection down to ppm levels or lower.

12.7.1 Analysis of Residual Solvents

The development and implementation of residual solvent methods for chemical analysis should be kept straightforward and simple whenever possible. When faced with development of a residual solvent method for GC, the analytical chemist should not need to place much work around method development because these types of methods are routine and typically already exist, either at the company/laboratory or in the literature. For methods that are going to be validated and implemented in a quality control (QC) setting, the recommendation is to use a GC-HS system instead of a direct injection setup. The reason for this recommendation is that the main component or matrix (e.g., the drug substance or drug product) may degrade on the GC inlet or already contain additional high-boiling components that can show up on the GC chromatogram. For a QC lab, this can be problematic depending on the quality assurance system that exists because extraneous peaks can trigger investigations, even for targeted impurities methods such as residual solvents analysis. Direct injection is still routinely used for residual solvent analysis, but care should be taken during method development to ensure that extraneous peaks are either not present or documented and explained.

A good rule of thumb is that any peak near or above the practical quantitation limit of the method (default to the lowest one in cases of multiple solvents) are addressed, documented, and noted in the method before implementation in a QC setting. However, the QC laboratory's quality and compliance systems (i.e., operating procedures) should be followed first and foremost. Again, a good practice is to adopt GC-HS as the standard setup for GC methods that will end up being validated and run in QC labs to minimize the potential for extraneous peaks. Further, GC-HS has an advantage of being more sensitive in practice. Because of the nature of GC-HS, the main component is often not volatilized and is therefore not brought to the inlet of the GC, which in the case of direct inject GC, can limit sample concentration that can be tolerated by a method. This benefit of GC-HS means that much higher sample concentrations can be achieved relative to direct injection, thus increasing the observed LOQ for the analyst developing the method.

An example of a GC-HS method used for testing of residual solvents in pharmaceutical products is shown below in Fig. 12.13. In this example, the original proposed method (Panel A) is for the quantitation of seven residual solvents using NMP as the diluent. This GC-HS method achieves good resolution of the solvents of interest. However, residual solvent analyses are often desired in short time frames to determine if further drying is needed to remove more a given solvent during in-process production, or to quickly complete product release testing. As such, a 30 min method is workable but far from ideal considering that one injection sequence may include a few blanks, LOQ standard, six working standards and multiple sample injections. By developing a fast GC-HS method in 8 min (Panel B), the analysts were able to achieve a method with excellent resolution and similar accuracy and LOQs. The chromatographic conditions of the fast GC-HS method are shown in Table 12.5. As seen, instead of a method where the chromatography occurs over 30 min plus re-equilibration time, the fast GC-HS allowed for the chromatography to occur in under 8 min. In changing from the longer to the faster method, one big change was the column length: 20 m instead of 60 m. Additionally, a 0.18 mm ID column was used for the fast method versus a 0.25 mm ID column for the longer method. Compared to larger inner diameter columns, smaller inner diameters will afford higher plate counts. Therefore, one strategy for developing faster GC methods is using smaller inner diameter columns, while also using shorter length columns. While shorter length columns will have less total theoretical plates, this is partially offset by the plate count boost by using smaller inner diameter columns.

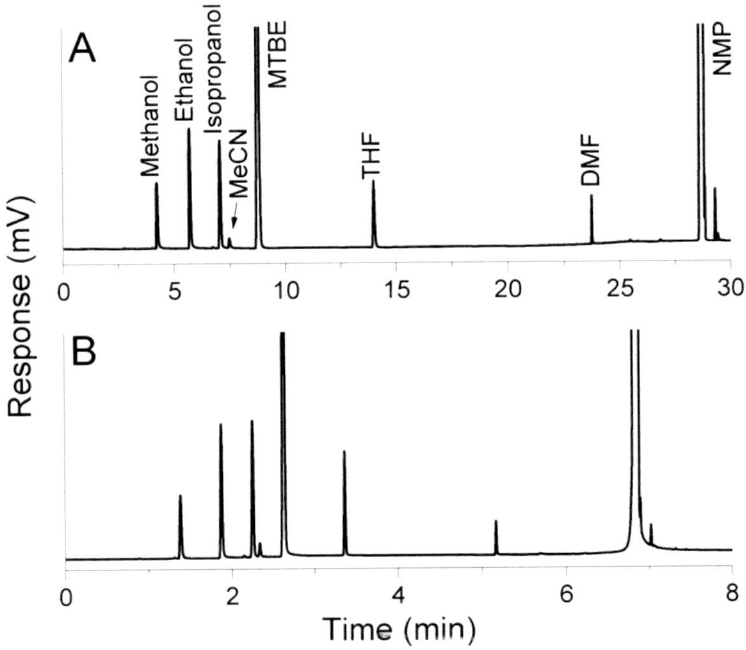

Figure 12.13 (A) Original GC-HS method for in-process testing of residual solvents, and (B) Fast HS-GC method for in-process testing of residual solvents. Elution orders are the same of solvents peaks in (B) as (A). Credit: Harry Patel, Umesh Patel, Tien Ho (Process Research & Development, AbbVie, Inc.).

Table 12.5 GC-HS conditions for the optimized (fast) GC-HS method

Carrier gas (mL/min)	1.3 (constant)	**Temperature program**
Split Ratio	30-1	35°C hold for 1.5 min, ramp at 25°C/min to 90°C, hold for 0 min, ramp 35°C/min to 235°C, hold for 3.16 min
Injector (°C)	210	
Detector (°C)	250	
FID hydrogen flow (mL/min)	40	Vial incubation time: 10 min
FID air flow (mL/min)	40	
FID nitrogen makeup (mL/min)	25	GC run time: 11 min/GC cycle time: 17 min
		Column
GC-HS oven (°C)	100	624-type phase; 20 m × 0.18 mm × 1.0 μm
GC-HS loop (°C)	110	
GC-HS transfer line (°C)	130	

Additionally, fast direct inject GC methods are also useful depending on equipment availability and ease of use and preparations. As seen in Fig. 12.14, 37 solvents, including the diluents, were separated in less than 7 min. Most solvents were well resolved, which would allow for a single method to be used to solve a variety of analytical challenges and with short runtime. The chromatographic conditions for the fast GC direct inject method are shown in Table 12.6.

For work outside of the QC environment, fast and user friendly methods have been developed to aid in R&D activities in the pharmaceutical development space. For example, Hamilton and co-workers have implemented and automated GC-HS system for the determination of residual solvents [52]. By applying this method to both isolated solids and reaction slurries of pharmaceutically relevant compounds, solvent analysis can be accomplished in a fast and user-friendly manner. Recently, innovative approaches to residual solvent analysis have allowed for much lower detection limits to be obtained using a standard GC-FID configuration.

Nacham and co-workers [53] showed that by using room-temperature ionic liquids (RTILs), residual solvent analysis in drug substances could be achieved with up to 25 times improvements in LODs. By exploiting the high thermal stability and low vapor pressure of RTILs as diluents, GC-HS methods could be pushed to higher incubation temperatures with less interfering peaks from the diluent over conventional organic solvent diluents. LODs in the range of 6–20 ppm were achieved, an impressive accomplishment that approaches LODs needed for low-level detection of mutagenic impurities.

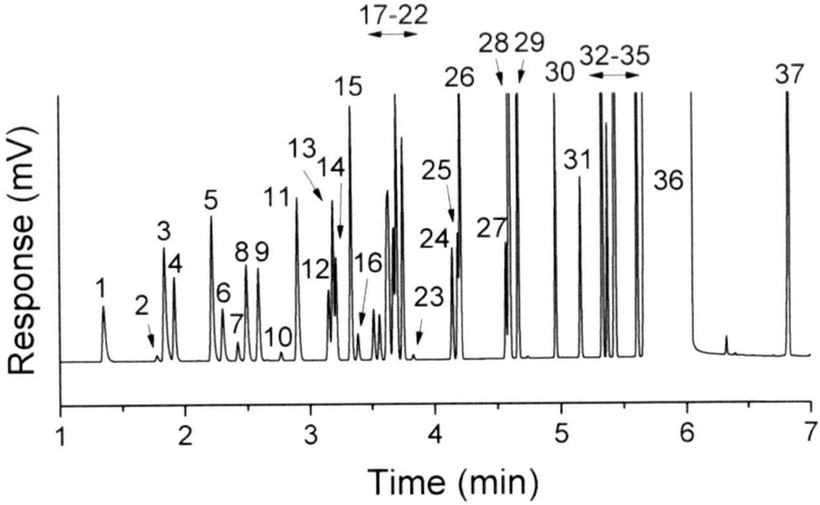

Figure 12.14 Fast GC direct inject method for residual solvent analysis. Solvents: (1) Methanol, (2) Pentane, (3) Ethanol, (4) Diethylether, (5) Ethyl Formate, (6) Acetonitrile, (7) Dichloromethane, (8) t-Butanol, (9) MTBE, (10) n-Hexane, (11) 1-Propanol, (12) Nitromethane, (13) Methyl Ethyl Ketone, (14) Ethyl Acetate, (15) Tetrahydrofuran, (16) Chloroform, (17) Cyclohexane, (18) Carbon Tetrachloride, (19) 1,2-Dimethoxy ethane, (20) Benzene, (21) Isopropyl Acetate, (22) 2-Methyl Tetrahydrofuran, (23) n-Heptane, (24) Methyl Cyclohexane, (25) 1,4,-Dioxane, (26) n-Propyl Acetate, (27) MIBK, (28) Pyridine, (29) Toluene, (30) 2-Hexanone, (31) Dimethylformamide, (32) Chlorobenzene, (33) Ethyl benzene, (34) Propylene Glycol/m-Xylene/p-Xylene, (35) o-Xylene, (36) DMSO, (37) NMP. Credit: Harry Patel, Umesh Patel, Tien Ho (Process Research & Development, AbbVie, Inc.).

Table 12.6 GC conditions for fast GC direct injection method for residual solvent analysis

Carrier gas (mL/min)	1.3	Temperature program
Split ratio	25–1	35°C hold for 1.5 min, ramp at 25°C/min to 90°C, hold for 0 min, ramp 35°C/min to 200°C, hold for 3.16 min
Injector (°C)	210	
Detector (°C)	240	
FID hydrogen flow (mL/min)	40	GC run time: 10 min
FID air flow (mL/min)	400	
FID nitrogen makeup (mL/min)	25	
Injection volume (μL)	1	**Column**
		624-type phase; 20 m × 0.18 mm × 1.0 μm

12.7.2 Analysis of Ordinary Impurities

Impurity analysis by GC encompasses a wide range of possible methods. In this section, ordinary impurities refer to non-MI, non-solvent/reagent impurities whereas main component refers to product (e.g., a starting material, intermediate or drug substance). This includes related substances analysis, targeted chromatographic impurities, and assay methods. In pharmaceutical analysis, one of the most critical quality controls is measuring related substances (impurities) of the main component, as well as assay of the main component. Testing is performed against specifications in a QC environment. These related substances measurements are critical to patient safety and meeting regulatory requirements because their allowable limit is directly related to process rejection capability as well as toxicological data for safety.

Related substance measurements typically occur at many points along the synthesis of a pharmaceutical product from starting materials and raw materials, to intermediates, drug substance, and drug product. Traditionally, HPLC-UV instrumentation and detection is the standard scheme for a related substance method. Understandably so, as the many column choices, good chromatographic resolution, robustness of the UV detector, and good instrument precision allow for most related substance analysis needs to be met. However, as pharmaceutical products become more advanced and complex, there is a trend for regulatory agencies to strengthen their oversight of the Good Manufacturing Process (GMP) and pushback drug makers further upstream to simpler, smaller molecule starting materials, often times commodity chemicals. For example, a simple amino acid could be designated a regulatory starting material (RSM) by pharmaceutical regulators, even for pharmaceutical molecules that are much more complex in structure. As such, validated, regulatory test methods are needed for these small, simple starting materials and their respective downstream intermediates. Traditional HPLC-UV methods are not always the best choice, particularly when a good chromophore may be lacking or there is low retention and resolution with typical HPLC methods. In these cases, GC methods are more sensitive, can provide more uniform response factors, and can provide for better chromatographic resolution for related substances analysis.

Consider a main component on an HPLC-UV method contains an aromatic ring, and one unspecified impurity that was not validated appears after a few years of commercial manufacture of a pharmaceutical product. Because of response factor differences, if this new unspecified impurity had a disruption or loss of aromaticity, a 1% impurity could possibly be inaccurately quantitated at 0.1% or 0.01%, for example. However, if the method were GC-FID, then a 1% impurity would likely be quantitated much closer to 1%, because the number of carbons between a main component and a related substance impurity of the main component are very similar, and FID offers more uniform response. While electronegative functional groups, for example, can alter the unit-carbon response, GC-based detection is still more uniform than UV many cases where UV chromophore of a main component or impurity are drastically different. While GC-FID for impurity analysis of pharmaceuticals is rarely used for full profiling and testing of related substance impurities, this is an area for expanded research and implementation that should be further explored as a means to mitigate risk of severely under- or over-quantitating impurities by HPLC-UV methods.

A risk-based approach by analyzing the main component and potential impurities should first be performed, and in cases where GC-FID makes sense, this should be considered over HPLC-UV. One example of where GC-FID would excel is for a starting material or intermediate impurities method where the main component has no significant chromophore. While an HPLC-UV detector at 200 nm might be a typical path forward, GC-FID offers significant advantages in higher sensitivity and more uniform response. If an aromatic impurity is present in the HPLC-UV method and present at 0.01%, it is not unreasonable that due to response factor differences this impurity would be quantitated at 1.0% and cause unnecessary failures of product. With add-ons available to GC-FID detectors to create more uniform response, such as the Polyarc reactor that converts all species to into methane before reaching the FID, there is much room for additional implementation of GC-FID for standard impurities methods.

GC coupled with detectors other than the FID can also be useful tools for impurity and related substance analysis. For example, Belal and co-workers developed a GC-MS with direct injection that was used for the analysis of Trimetazidine Dihydrochloride in order to quantitate two of its related substances with a stability-indicating method. In this method, an RTX-1 column was employed, and an electron impact (EI) ionization source was used to generate ions for MS analysis [54]. In other work from Balal and co-workers [55], they showed the GC-MS chromatogram of the antihistamine drug dimenhydrinate (DMH) and separation from six of its related substances. Further, Balal and co-workers [55] found that their GC-MS methodology was also potentially suitable for main component assay analysis. By using GC-based methodology, the analytical chemist has a powerful tool at his disposal to tackle analytical challenges where HPLC-UV may have significant deficiencies.

Reaction products and impurities are also well suited for GC analysis. By using GC-MS technology, unknown products can be fragmented with an electron impact (EI) ionization source and fragments can be interpreted or compared against a library to search for likely structures. Chemical ionization (CI) can also be used to observe the molecular ion and pull a mass on intact analytes to help identify a structure. Van den Berg and coworkers showed that acid-mediated transformation of cholesterol and phospholipids can be studied by GC-MS. The authors notably found that no chlorohydrin species were detected as would be expected with unsaturated fatty acids. Instead, hydroxy- and keto-derivatives were confirmed by GC-MS [56].

12.7.3 Main Component (Assay) Methods

In pharmaceutical analysis, an assay specification is critical to ensure the quality of starting materials, intermediates, and drug substance/product. Whereas related substance analysis will test for the presence of low-level impurities, assay testing is a catch-all approach to determine any impurity. For example, if a drug was 100% pure from all related substances, solvents, metals, and so forth, but there was 15% of sodium chloride present, then the assay testing would return a result of roughly 85%. If this drug were being tested against an assay specification of not less than 98.0%, for example, then this 85% result would clearly fail. Again, typically HPLC-UV would be used for main component assay analysis, but the in certain cases, GC is more appropriate.

The most important consideration when developing a GC method for assay is if the method relative standard deviation (RSD) is suitable for the intended testing. As a good rule of thumb, the statistical principle that three standard deviations on each side will encompass 99.7% of normal variance should be applied because this will ensure that specification failures do not routinely occur due to normal method variance. For example, if a method being developed is intended to be used for an assay specification of not less than 95%, then the assay gap is 100% – 95% = 5%. Then, calculate the standard deviation such that the assay gap equals 3 standard deviations. In this example, 5%/3 = 1.7%. This exercise tells the analyst that a method is needed that can reliably achieve area %RSD of the standard of not more than 1.7%. This is challenging to achieve for low concentration standards with GC analysis, and can also be challenging to achieve due to inherent variability in some detectors, such as MS where ionization and signal capture of ions through a quadrupole is less repeatable than GC-FID.

With GC-FID, and an appropriate concentration of standard, area %RSD of around 1% can routinely be achieved to accomplish a main component assay method. However, for tighter assay specifications, the analyst will run into repeatability issues when using GC. For example, when developing and assay by GC method to test against a not less than 98% specification, the analyst would want a method than can reliably hit area RSD of not more than 0.67%. In this case, even with an optimized sample concentration, inlet liner, column, peak shape, and so forth, this criterion would still be challenging to regularly achieve by GC analysis. In these cases where low area %RSD is needed, GC analysis with an internal standard is recommended. Ideally, a structurally similar compound that is resolved away from the main component for assay would be used as an internal standard. By invoking an internal standard, an optimized GC method that typically would return a result of 1% RSD might now be expected to reliably achieve results of <0.5% RSD, even as low as 0.1–0.2% as would be possible on an HPLC-UV assay method. Because the internal standard experiences similar variability as the main component (e.g., injection precision, inlet liner interactions), the area ratio of main component to internal standard provides a much better degree of repeatability and thus low %RSD. Internal standards can similarly be applied to impurities methods when traditional GC impurities methods struggle to achieve repeatability, linearity, accuracy, and so forth.

12.7.4 In-Process Testing

GC has been used in various industries as a means to monitor various synthetic, chemical, or physical operations. These methods are typically run "in-process" industrially, whereby the analytical result is needed quickly to gain insight into the progress of an operation in order to move on to the next step. For such methods, time is of the essence, and GC is often a preferred choice. For example, for a solvent switching operation, one solvent may be distilled off while another solvent is added in to replace it. This operation would be a natural fit for GC analysis because these solvents might be transparent to a UV detector, and GC can provide fast and accurate analysis.

In industrial processes, reaction products can also be monitored by GC methods. For example, in the pharmaceutical industry, most commercial drug manufacturing processes require methodology that monitors for reaction conversion at various points along the chemical synthesis. GC has also been applied in thermodynamic studies of complexation reactions. For example, Skorka and co-workers found that chiral GC columns were useful for studying complexes of monoterpenes by alpha- and beta-cyclodextrins and experimentally determined distribution constants, stability constants and thermodynamic parameters [57]. In another study, Fiamegos and co-workers monitored the reaction of amino acids with a derivatization-extraction approach. The amino acids were converted to N(O,S)-ethoxycarbonyl amino acid ethyl esters followed by a single-drop micro extraction technique. By using GC-FID and GC-MS, detection limits as low as 0.01 μg/mL and 0.26 ng/mL were achieved, respectively. The kinetics of the reaction were studied by using GC data, which allowed the authors to gain insight into the progress of the reaction as shown in Fig. 12.15 [58].

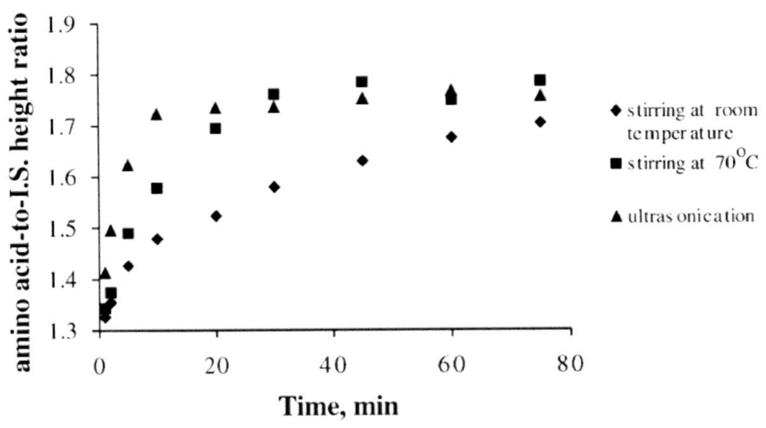

Figure 12.15 Plot of the progress of an amino derivatization reaction using GC data. The ratio of the amino acid derivatives peak height to internal standard (n-Pentadecane) peak height was plotted over time. Adapted from Fiamegos and co-workers [58].

12.7.5 Mutagenic Impurities (MIs)

MIs, which are synonymous with genotoxic impurities (GTIs), are known to cause mutations of genes or rearrangements of chromosomes. MIs are impurities that are also potentially carcinogenic to humans [59], and adequate control of MIs by pharmaceutical companies in their products has received heightened scrutiny from regulators over recent years. This trend has led to increased interest in analytical methods capable of low detection limits for MIs. Further, both MIs and impurities that are *potentially* MIs need to be addressed, which opens up potentially hundreds of compounds to scrutiny during the development of a drug candidate.

A typical workflow for addressing MIs and potential MIs during the development of a drug would involve an in silico, expert rule-based, and orthogonal statistical model in order to identify impurities that are known MIs or potential MIs. After screening up to hundreds of known and potential impurities, intermediates, reagents, by-products, and starting materials, a list of known and potential MIs would be compiled. Then, pharmaceutical companies would address these alerting structures by a variety of mechanisms such as Ames testing to prove or disprove mutagenicity, lab-based process spike experiments to demonstrate adequate purge of an impurity, or scientific justification based on process rejection capability such as removal of impurities during crystallizations, extractions, or reactivity arguments, for example.

For MIs and potential MIs that cannot be addressed adequately by the previous mechanisms, then analytical testing is often a last-line but highly important means to address these impurities. Analytical testing is not preferred because of the high cost and time commitment associated with specialty, targeted impurity methods. However, when MIs and potential MIs need to be tested by analytical methods, GC is considered the standard method of choice for analysis of small molecule mutagenic impurities (MIs) [60], and recent advances in GC for pharmaceutical applications have seen breakthroughs in these types of analysis. LC-MS should also be considered for MI analysis, such as when analytes are not susceptible to GC analysis or due to instrument limitation in a laboratory. A decision tree for when to use LC-MS and GC-MS based methods for MI analysis can be helpful in this regard [61]. Lastly, the importance of having appropriate analytical methods with low enough detection limits for the testing of MIs and potential MIs cannot be understated due to patient safety, and the case of the withdrawal of Viracept (Nefinavir mesilate) in 2007 is an example where elevated levels of a known MI, ethyl methanesulfonate, were the cause of the drug withdrawal from the market [62].

Based on current International Conference on Harmonisation (ICH) guidelines for MIs, the maximum allowed MI in a drug substance for a lifetime of dosing is 1.5 μg per day. For a hypothetical 2 g per day drug dose, this is an allowable amount of 0.75 ppm in the drug substance. As such, it is easy to see why GC is a preferred technique due to its wide array of column stationary phases available, headspace format, and sensitive detectors that are at the analyst's disposal. A range of stationary phases allows the user to select a column with good selectivity whereby the MI is well resolved from main component and any potentially interfering matrix components. Further, headspace format allows for sensitive analysis with reduced potential for matrix interference. In order to achieve good method sensitivity toward a MI, high sample concentrations are often employed (e.g., 100 mg/mL). As a result, direct injection can become problematic as sample builds up on the inlet of the GC, which can lead to issues with accuracy, tailing, extraneous peaks, and deteriorated detection limits.

Headspace-GC solves many of these challenges. And lastly, a myriad of detectors in the GC realm allows the user flexibility to select an appropriate detection scheme for a required challenge. For GC analysis of MIs, a GC-FID setup is a good starting point due to simplicity and ease of use in many labs. However, for low-level detection needs, such as analysis in the range of lower ppm and below, typically more sensitive detectors such as TCD or GC-MS are more appropriate due to lower levels of quantitation (LOQ) achievable compared to GC-FID. Each MI analysis need is different and depends on the daily dose of the drug, the duration of dose, as well as the analyte at hand and the matrix in which it is present.

For compounds that are not volatile enough to be amenable to GC analysis, or in order to improve analysis, derivatization approaches have been successful to afford GC analysis with useful LOQs for MIs [63]. For example, pentafluorothiophenol derivatives have been used by Alzaga and co-workers [64] in order to afford some common alkylating agents analysis by headspace GC with improved analysis time, LOQ, and analyte stability. Figure 12.16 shows various PFTP derivatives for the headspace GC-MS analysis of various carbon chain length alkyl tosylates. Good LOQ is reached at 1 ppm with impressive signal-to-noise. Another major advantage of this approach is the headspace setup, whereby potentially interfering sample matrix components are eliminated from GC analysis through volatility discrimination.

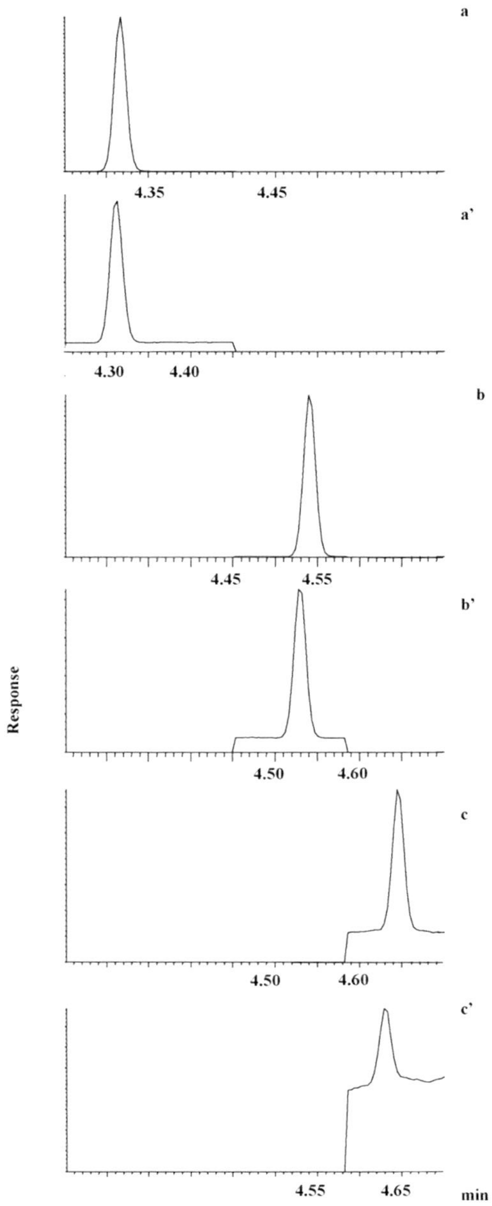

Figure 12.16 HS-GC-MS selected ion chromatograms showing the PFTP derivatives: (a) methyl ($m/z+ = 214$); (b) ethyl ($m/z+ = 228$); (c) isopropyl ($m/z+ = 242$), and their corresponding deuterated IS's (a′) ($m/z+ = 217$), (b′) ($m/z+ = 233$) and (c′) ($m/z+ = 249$). Selected analyte was alkyl tosylate at 1.0 μg g^{-1} (1 ppm) relative to 50 mg of sample [64].

Other strategies for MI analysis have employed GC-HS with an ECD for detection of halogenated MIs. By using ionic liquids as the diluent, Ho and coworkers were able to show LODs from 5–500 ppm for two MI classes: alkyl/aryl halides and nitro-aromatics. These MIs are relatively high boilers, and by employing ionic liquid-based diluent, the HS oven temperature could be increased and tuned to allow for low-level detection compared with a conventional DMSO diluent, for example [65]. In another interesting study around GC for the quantitation of halogenated MIs, Byun and coworkers used electron impact MS to study halogenated cytosines and deoxycytidines. GC-EI-MS was compared to GC with fast atom bombardment and electrospray (ESI)-MS along with liquid chromatography-ESI. In this study, GC-EI-MS was determined to be superior for low-level detection of these halogenated MIs due to low sensitivity and wide dynamic range [66]. Outside of the pharmaceutical industry, Kataoka and coworkers have used GC along with a nitrogen-phosphorous detector (NPD) to detect mutagenic heterocyclic amines (HCAs). These MIs are created during heating of protein-based foods through boiling, frying, or grilling, and they can be formed in meats or fishes. The authors used a derivatization approach to afford GC analysis, and found that HCAs were present in higher amounts in certain fishes compared with meats [67]. In another example of innovative approaches to MI analysis by GC, David and coworkers employed a two-dimensional GC system with MS detection. The heart-cut MI fraction coming off of a 5% diphenyl-based column was transferred to a more polar 50% diphenyl column, whereby the MI fraction could further separate without interference from drug substance, solvents or derivatizing agents that afforded the GC separation. Michael acceptors and haloalcohols were quantitated with LODs near the 0.01–0.1 ppm range [68].

12.8 Final Remarks

GC is a powerful analytical tool with a variety of uses across the pharmaceutical industry and beyond. The range of columns, detectors, modes of operation, and sample preparation techniques all contribute to the overall versatility of this indispensable equipment. Optimization of these parameters enables GC to be an essential means for rapid determination of a wide variety of small molecules, with superb sensitivity and selectivity.

Within the pharmaceutical realm, GC has been well utilized for residual solvent analysis, and exploited for determining potent/toxic molecules, such as MIs, at trace and ultra-trace levels. Recently, applications in the pharmaceutical industry have also expanded towards assay determinations, impurities and related substance characterizations, and in-process testing/monitoring. As innovations in GC continue to cultivate and develop within the pharmaceutical world, it is critical that we take full-advantage of these progressions to guarantee quality in our products, minimize supply chain risk, and ultimately ensure patient safety.

References

1. Hinshaw, J. V. (2015) Split-vent traps: GC's dirty secret, *LCGC N. Am.*, **33**, pp. 758–763.
2. Snow, N. H. (2004) Inlet systems for gas chromatography, in *Modern Practice of Gas Chromatography* (Grob, R. L., and Barry, E. F., eds.), John Wiley & Sons, Inc., Hoboken, New Jersey, pp. 461–489.
3. Watson, D. W. (2016) How to optimize key variables in GC analysis: Sample introduction, *LCGC N. Am.*, **34**, p. 446.
4. Hinshaw, J. V. (2004) Optimization of separations and computer assistance, in *Modern Practice of Gas Chromatography* (Grob, R. L., and Barry, E. F., eds.), John Wiley & Sons, Inc., Hoboken, New Jersey, pp. 193–228.
5. Hinshaw, J. V. (2001) Selecting carrier gases and conditions, *LCGC N. Am.*, **19**, pp. 1056–1064.
6. Bartram, R. J. (2004) Gas management systems for gas chromatography, in *Modern Practice of Gas Chromatography* (Grob, R. L., and Barry, E. F., eds.), John Wiley & Sons, Inc., Hoboken, New Jersey, pp. 491–543.
7. Watson, D. W. (2017) Advantages of using nitrogen in capillary GC, *LCGC N. Am.*, **35**, p. 142.
8. Barry, E. F. (2004) Columns: Packed and capillary; column selection in gas chromatography, in *Modern Practice of Gas Chromatography* (Grob, R. L., and Barry, E. F., eds.), John Wiley & Sons, Inc., Hoboken, New Jersey, pp. 65–191.

9. Sun, Y., Zhang, R., and Wang, Q. (1993) Programmed-temperature gas chromatographic retention index, *J. Chromatogr. A*, **657**, pp. 1–15.
10. Cavalli, E. J., and Guinchard, C. (1995) Emperical approach to the dependence of plate height on temperature and gas velocity in gas chromatography, *J. Chromatogr. Sci.*, **33**, pp. 573–577.
11. Blumberg, L. M. (2012) Theory of gas chromatography, in *Gas chromatography* (Poole, C. F., ed.), Elsevier, Oxford, UK.
12. Blumberg, L. M. (2010) *Temperature-Programmed Gas Chromatography*, Wiley-VCH, Weinheim, Germany.
13. Castello, G., Moretti, P., and Vezzani, S. (2009) Retention models for programmed gas chromatography, *J. Chromatogr. A*, **1216**, pp. 1607–1623.
14. Blumberg, L. M., and Klee, M. S. (2001) Quantitative comparison of performance of isothermal and temperature-programmed gas chromatography, *J. Chromatogr. A*, **933**, pp. 13–26.
15. Giddings, J. C. (1962) Elementary theory of programmed temperature gas chromatography, *J. Chem. Ed.*, **39**, pp. 569–573.
16. Taylor, T. (2015) GC temperature programming-10 things you absolutely need to know, *LCGC N. Am.*, **33**, p. 438.
17. Hinshaw, J. V. (1991) Programmed-temperature optimization, *LCGC*, **9**, pp. 470–473.
18. Korytár, P., Janssen, H.-G., Matisová, E., and Brinkman, U. A. T. (2002) Practical fast gas chromatography: Methods, instrumentation and applications, *TrAC, Trends Anal. Chem.*, **21**, pp. 558–572.
19. Matisová, E., and Dömötörová, M. (2003) Fast gas chromatography and its use in trace analysis, *J. Chromatogr. A*, **1000**, pp. 199–221.
20. David, F., Gere, D. R., Scanlan, F., and Sandra, P. (1999) Instrumentation and applications of fast high-resolution capillary gas chromatography, *J. Chromatogr. A*, **842**, pp. 309–319.
21. Klee, M. S., and Blumberg, L. M. (2002) Theoretical and practical aspects of fast gas chromatography and method translation, *J. Chromatogr. Sci.*, **40**, pp. 234–247.
22. Cramers, C. A., and Leclercq, P. A. (1999) Strategies for speed optimisation in gas chromatography: An overview, *J. Chromatogr. A*, **842**, pp. 3–13.
23. Snow, N. H. (2004) Fast gas chromatography with short columns: Are speed and resolution mutually exclusive?, *J. Liq. Chromatogr. Related. Technol.*, **27**, pp. 1317–1330.

24. Jumtee, K., Bamba, T., and Fukusaki, E. (2009) Fast GC-FID based metabolic fingerprinting of Japanese green tea leaf for its quality ranking prediction, *J. Sep. Sci.*, **32**, pp. 2296–2304.
25. Yan, X. (2006) Unique selective detectors for gas chromatography: Nitrogen and sulfur chemiluminescence detectors, *J. Sep. Sci.*, **29**, pp. 1931–1945.
26. Ragunathan, N., Krock, K. A., Klawun, C., Sasaki, T. A., and Wilkins, C. L. (1999) Gas chromatography with spectroscopic detectors, *J. Chromatogr. A*, **856**, pp. 349–397.
27. Poole, C. F. (2015) Ionization-based detectors for gas chromatography, *J. Chromatogr. A*, **1421**, pp. 137–153.
28. Gunnar, T., Mykkänen, S., Ariniemi, K., and Lillsunde, P. (2004) Validated semiquantitative/quantitative screening of 51 drugs in whole blood as silylated derivatives by gas chromatography-selected ion monitoring mass spectrometry and gas chromatography electron capture detection, *J. Chromatogr. B*, **806**, pp. 205–219.
29. Ho, T. D., Yehl, P. M., Chetwyn, N. P., Wang, J., Anderson, J. L., and Zhong, Q. (2014) Determination of trace level genotoxic impurities in small molecule drug substances using conventional headspace gas chromatography with contemporary ionic liquid diluents and electron capture detection, *J. Chromatogr. A*, **1361**, pp. 217–228.
30. Kosjek, T., Heath, E., Petrović, M., and Barceló, D. (2007) Mass spectrometry for identifying pharmaceutical biotransformation products in the environment, *TrAC, Trends Anal. Chem.*, **26**, pp. 1076–1085.
31. David, F., Alegre, M. R., Vanhoenacker, G., and Sandra, P. (2013) Determination of genotoxic impurities in pharmaceuticals, *LCGC Eur.*, **26**, pp. 31–34.
32. Girijan, R. (2012) Gas chromatography-mass spectrometry technique and its applications in pharmaceutical industry, *Chem. Ind. Dig.*, **25**, pp. 59–63.
33. Gillespie, T. A., and Winger, B. E. (2011) Mass spectrometry for small molecule pharmaceutical product development: A review, *Mass Spectrom. Rev.*, **30**, pp. 479–490.
34. Ji, C., Dean, J. L., Flores, R. A., and Boisvert, S. M. (2009) Quantitative determination of co-eluting compounds by using extract ion chromatogram in GC-MS, *Chem. Educator.*, **14**, pp. 212–214.
35. Xue, M., Yuan, S., Ruan, J., Qiao, J., Xu, Y., Zhang, Z., and Liu, K. (2006) Determination of penehyclidine by gas chromatographic-mass

spectrometry and its application to pharmacokinetics in humans, rabbits and mice, *J. Chromatogr. B*, **843**, pp. 234–239.

36. Munson, B. (2000) Development of chemical ionization mass spectrometry, *Int. J. Mass Spectrom.*, **200**, pp. 243–251.
37. Munson, M. S. B., and Field, F. H. (1966) Chemical ionization mass spectrometry, *J. Am. Chem. Soc.*, **88**, pp. 2621–2630.
38. Hunt, D. F., and Crow, F. W. (1978) Electron capture negative ion chemical ionization mass spectrometry, *Anal. Chem.*, **50**, pp. 1781–1784.
39. Stafford, G. C. (1980) Instrumental aspects of positive and negative ion chemical ionization mass spectrometry, *Environ. Health Perspect.*, **36**, pp. 85–96.
40. Vaclavik, L., Krynitsky, A. J., and Rader, J. I. (2014) Mass spectrometric analysis of pharmaceutical adulterants in products labeled as botanical dietary supplements or herbal remedies: A review, *Anal. Bioanal. Chem.*, **406**, pp. 6767–6790.
41. Loda, C., Bernabe, E., Nicoletti, A., Bachi, S., and Dams, R. (2011) Determination of epichlorohydrin in active pharmaceutical ingredients by gas chromatography-mass spectrometry, *Org. Proc. Res. Dev.*, **15**, pp. 1388–1391.
42. Taylor, T. (2015) Understanding electron ionization processes for GC-MS, *LCGC N. Am.*, **33**, p. 290.
43. Belas, F. J., and Blair, I. A. (2002) Mass spectrometry in pharmaceutical analysis, in *Handbook of Pharmaceutical Analysis* (Ohannesian, L., and Streeter, A. J., eds.), Marcel Dekker, Inc., New York, New York, pp. 151–185.
44. Mclafferty, F. W. (2009) *Registry of Mass Spectral Data and Registry Combined with NIST*, 9th ed., Wiley-Blackwell, Hoboken, NJ.
45. Mclafferty, F. W. (2011) A century of progress in molecular mass spectrometry, *Ann. Rev. Anal. Chem.*, **4**, pp. 1–22.
46. Little, J. L., Cleven, C. D., and Howard, A. S. (2013) Identifying "known unknowns" in commercial products by mass spectrometry, *LCGC N. Am.*, **31**, pp. 114–125.
47. Sparkman, D. O., Penton, Z. E., and Kitson, F. G. (2011) Mass spectral data interpretation, in *Gas Chromatography and Mass Spectrometry: A Practical Guide*, Elsevier, Oxford, UK, pp. 149–205.
48. Mclafferty, F. W., and Tureček, F. (1993) *Interpretation of Mass Spectra*, University Science Books, Sausalito, CA.
49. Masucci, J. A., and Caldwell, G. W. (2004) Techniques for gas chromatography/mass spectrometry, in *Modern Practices of Gas*

Chromatography (Grob, R. L., and Barry, E. F., eds.), John Wiley & Sons, Inc., Hoboken, New Jersey, pp. 339–401.

50. Smith, R. M. (2004) *Understanding Mass Spectra: A Basic Approach*, 2nd ed., John Wiley & Sons, Inc., Hoboken, New Jersey.
51. Zhang, L., Tang, C., Cao, D., Zeng, Y., Tan, B., Zeng, M., Fan, W., Xiao, H., and Liang, Y. (2013) Strategies for structure elucidation of small molecules using gas chromatography-mass spectrometric data, *TrAC, Trends Anal. Chem.*, **47**, pp. 37–45.
52. Hamilton, S. E., Rossington, M. D., and Bertrand, A. (2016) Development of an automated headspace gas chromatography instrument for the determination of residual solvents in pharmaceutical compounds and reaction mixtures, *Org. Proc. Res. Dev.*, **20**, pp. 189–194.
53. Nacham, O., Ho, T. D., Anderson, J. L., and Webster, G. K. (2017) Use of ionic liquids as headspace gas chromatography diluents for the analysis of residual solvents in pharmaceuticals, *J. Pharm. Biomed. Anal.*, **145**, pp. 879–886.
54. Belal, T. S., Awad, T., and Clark, C. R. (2014) Stability-indicating determination of trimetazidine dihydrochloride in the presence of two of its related substances using a direct GC/MS method, *J. AOAC Int.*, **97**, pp. 1514–1518.
55. Belal, T. S., Abdel-Hay, K. M., and Clark, C. R. (2016) Selective determination of dimenhydrinate in presence of six of its related substances and potential impurities using a direct GC/MS method, *J. Adv. Res.*, **7**, pp. 53–58.
56. van den Berg, J. J. M., Winterbourn, C. C., and Kuypers, F. A. (1993) Hypochlorous acid-mediated modification of cholesterol and phospholipid: Analysis of reaction products by gas chromatography-mass spectrometry, *J. Lipid Res.*, **34**, pp. 2005–2012.
57. Skórka, M., Asztemborska, M., and Żukowski, J. (2005) Thermodynamic studies of complexation and enantiorecognition processes of monoterpenoids by α- and β-cyclodextrin in gas chromatography, *J. Chromatogr. A*, **1078**, pp. 136–143.
58. Fiamegos, Y. C., Nanos, C. G., and Stalikas, C. D. (2004) Ultrasonic-assisted derivatization reaction of amino acids prior to their determination in urine by using single-drop microextraction in conjunction with gas chromatography, *J. Chromatogr. B*, **813**, pp. 89–94.
59. McGovern, T., and Jacobson-Kram, D. (2006) Regulation of genotoxic and carcinogenic impurities in drug substances and products, *TrAC, Trends Anal. Chem.*, **25**, pp. 790–795.

60. Reddy, A. V. B., Jaafar, J., Umar, K., Majid, Z. A., Aris, A. B., Talib, J., and Madhavi, G. (2015) Identification, control strategies, and analytical approaches for the determination of potential genotoxic impurities in pharmaceuticals: A comprehensive review, *J. Sep. Sci.*, **38**, pp. 764–779.

61. David, F., Alegre, M. R., Vanhoenacker, G., and Sandra, P. (2013) Determination of genotoxic impurities in pharmaceuticals, *LCGC Eur.*, pp. 31–34.

62. Elder, D. P., Delaney, E., Teasdale, A., Eyley, S., Reif, V. D., Jacq, K., Facchine, K. L., Oestrich, R. S., Sandra, P., and David, F. (2010) The utility of sulfonate salts in drug development, *J. Pharm. Sci.*, **99**, pp. 2948–2961.

63. Liu, D. Q., Sun, M., and Kord, A. S. (2010) Recent advances in trace analysis of pharmaceutical genotoxic impurities, *J. Pharm. Biomed. Anal.*, **51**, pp. 999–1014.

64. Alzaga, R., Ryan, R. W., Taylor-Worth, K., Lipczynski, A. M., Szucs, R., and Sandra, P. (2007) A generic approach for the determination of residues of alkylating agents in active pharmaceutical ingredients by in situ derivatization–headspace–gas chromatography–mass spectrometry, *J. Pharm. Biomed. Anal.*, **45**, pp. 472–479.

65. Ho, T. D., Joshi, M. D., Silver, M. A., and Anderson, J. L. (2012) Selective extraction of genotoxic impurities and structurally alerting compounds using polymeric ionic liquid sorbent coatings in solid-phase microextraction: Alkyl halides and aromatics, *J. Chromatogr. A*, **1240**, pp. 29–44.

66. Byun, J., Henderson, J. P., and Heinecke, J. W. (2003) Identification and quantification of mutagenic halogenated cytosines by gas chromatography, fast atom bombardment, and electrospray ionization tandem mass spectrometry, *Anal. Biochem.*, **317**, pp. 201–209.

67. Kataoka, H., Nishioka, S., Kobayashi, M., Hanaoka, T., and Tsugane, S. (2002) Analysis of mutagenic heterocyclic amines in cooked food samples by gas chromatography with nitrogen-phosphorus detector, *Bull. Environ. Contam. Toxicol.*, **69**, pp. 682–689.

68. David, F., Jacq, K., Sandra, P., Baker, A., and Klee, M. S. (2010) Analysis of potential genotoxic impurities in pharmaceuticals by two-dimensional gas chromatography with Deans switching and independent column temperature control using a low-thermal-mass oven module, *Anal. Bioanal. Chem.*, **396**, pp. 1291–1300.

Chapter 13

Method Transfer between HPLC and UHPLC

Susanne Fabel

Thermo Fisher Scientific, Dornierstr. 4, D-82110 Germering, Germany

Susanne.fabel@thermofisher.com

13.1 Introduction

The method transfer between conventional HPLC and UHPLC is done either to accelerate the method or to gain more resolution compared to the HPLC method. The opposite transfer from UHPLC to HPLC could also be a requirement, for example, if in the R&D department only UHPLC equipment is available and the methods have to be transferred to HPLC equipment in the QC laboratory. Conveniently, method transfer for UHPLC to HPLC can be achieved by applying the same rules and principles for HPLC to UHPLC methods. In this chapter, the focus is on the method transfer from conventional HPLC to UHPLC. The theory behind

Chromatographic Method Development
Edited by Gregory K. Webster and Laila Kott

ISBN 978-981-4800-53-2 (Hardcover), 978-0-429-20172-1 (eBook)
www.jennystanford.com

the method transfer from HPLC to UHPLC is explained along with instrument tips and tricks.

13.2 Method Transfer Strategy

The goals for HPLC method transfer to an UHPLC method can focus on either maintaining or improving the HPLC separation efficiency, or accelerating the HPLC method to increase sample throughput (Fig. 13.1).

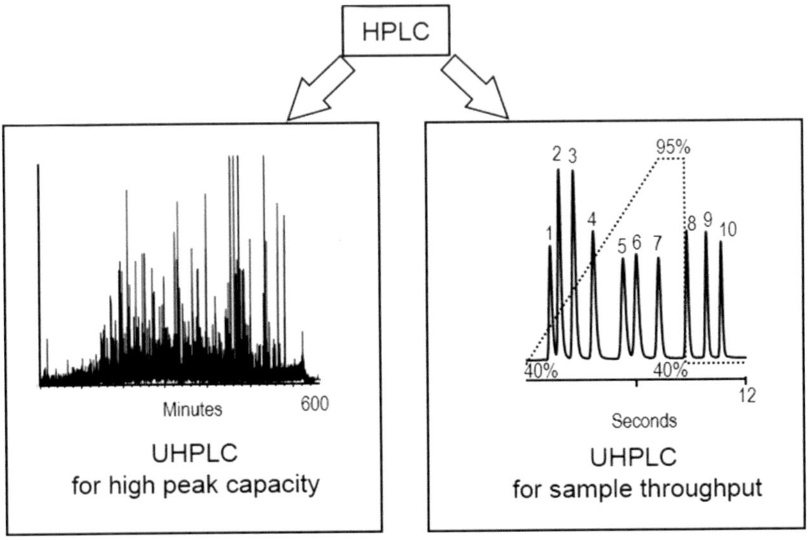

Figure 13.1 The focus of method transfer from HPLC to an UHPLC method can be on high resolution for high peak capacity (left, up to 1000 peaks separated in a single run) or on high speed for high sample throughput (right, 10 peaks separated in 10 s). Reproduced with permission of Thermo Fisher Scientific.

Theoretical HPLC to UHPLC calculations allow you to either improve efficiency of the separation within the same analysis time or shorten the analysis time while maintaining the same efficiency. Compared to HPLC separations, UHPLC methods are more lenient with method acceleration due to the ability to maintain resolution. However, the trade-off for improved efficiency or throughput is higher back pressure.

13.2.1 Geometric Rules for Transfer from HPLC to UHPLC

Method transfer calculations differ between isocratic and gradient separations. As a start, the basic geometric rules applying to isocratic separations are discussed in this section. The arithmetical continuation of the rules to gradient methods is discussed later in Section 13.2.4.

One motivation for the transfer from an HPLC to UHPLC method is to obtain better separation of the analytes. The quality of a separation between two compounds is described with the Resolution R of two neighboring eluting compounds, described by the Purnell equations Eq. 13.1 [1], where N is the number of theoretical plates, k the retention factor of the second peak, and α the separation factor between two peaks.

$$R = \frac{1}{4} \cdot \sqrt{N} \cdot \frac{k_2}{1 + k_2} \cdot \frac{\alpha - 1}{\alpha} \tag{13.1}$$

The separation factor of two adjacent peaks is defined by their k-values:

$$\alpha = \frac{k_2}{k_1} \tag{13.2}$$

The separation factor α is influenced by column chemistry, mobile phase chemistry and temperature. The retention factor k is primarily influenced by mobile phase strength and temperature. The number of theoretical plates N, which is also defined as the efficiency factor, refers to the column conditions like length, particle size, and flow rate. An increase of the efficiency by a factor of 2 would generate a 1.4-fold increase in resolution due to the square root dependency.

The number of theoretical plates N is correlated with the length of the column L and the height equivalent to a theoretical plate height H as described in Eq. 13.3.

$$N = \frac{L}{H} \tag{13.3}$$

Therefore, by keeping the column length constant, lower height equivalents will generate a higher efficiency of the separation and a better resolution. The factors influencing the theoretical plate height are explained by considering the diffusion processes during the separation in the column, which are expressed by the Van Deemter equation (Eq. 13.4).

$$H = A + \frac{B}{u} + C \cdot u \tag{13.4}$$

The *A*-term, eddy dispersion, describes the mobile phase movement along random paths through the stationary phase resulting in band broadening of the analyte. The longitudinal diffusion of the analyte in and against the flow direction is expressed by the *B*-term. The resistance of the analyte to the mass transfer into the pores of the stationary phase is described by the *C*-term. u is the linear velocity of the mobile phase in the column. The connection between plate height H and particle size d_p can be simplified according to Halász [2]. This approximation is valid for fully porous material.

$$H = 2 \cdot d_\text{p} + \frac{6}{u} + \frac{d_\text{p}^2 \cdot u}{20} \tag{13.5}$$

The Van Deemter plots of plate height H against velocity u, depending on the particle size d_p, help visualize the role of small particle sizes in UHPLC method transfer strategy (see Fig. 13.2).

The optimum flow rate and minimum achievable plate height can be estimated by setting the first derivative of the Eq. 13.5 to zero. The optimal linear velocity u_opt (in mm/s) can then be estimated by Eq. 13.6.

$$u_\text{opt} \approx \frac{10}{d_\text{p}} \tag{13.6}$$

The minimum achievable plate height H_min is subsequently calculated by insertion of Eq. 13.6 in Eq. 13.5.

$$H_\text{min} \approx 3 \cdot d_\text{p} \tag{13.7}$$

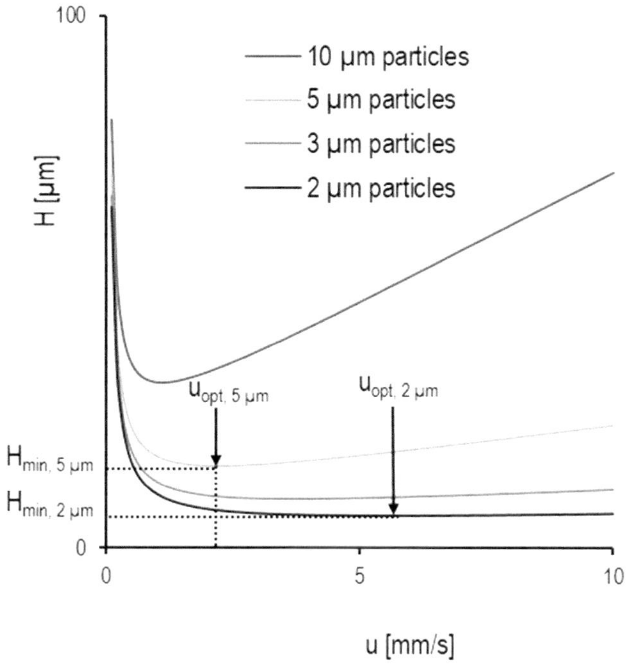

Figure 13.2 Smaller particles provide more theoretical plates as demonstrated by the smaller minimum plate height *H* in the Van Deemter plot and considering the relation of *N* and *H* in Eq. 13.3. Reproduced with permission of Thermo Fisher Scientific.

The factor 3 in Eq. 13.7 is a conservative calculation. Excellent packed modern UHPLC columns operated in a modern UHPLC system could generate plate heights proportional to a factor 2 of the particle size. With Eq. 13.7 it can be concluded that smaller particles generate lower plate heights and higher efficiencies according to Eq. 13.3 and better resolutions according to Eq. 13.1. This is visualized in a simulated chromatogram in Fig. 13.3 for 5 and 2 µm particle size.

Following Eq. 13.3, shorter columns with smaller particles can be used while maintaining efficiency. The use of shorter column allows for analysis times to be shortened—achieving the second motivating factor for UHPLC method, faster separations.

The analysis time is proportional to the column dead time, which is proportional to the column length. Moving from an HPLC method with column length L_1, particle size d_{p1} and analysis time

t_1, the calculated analysis time of the UHPLC method t_2 with new column parameters is described by Eq. 13.8.

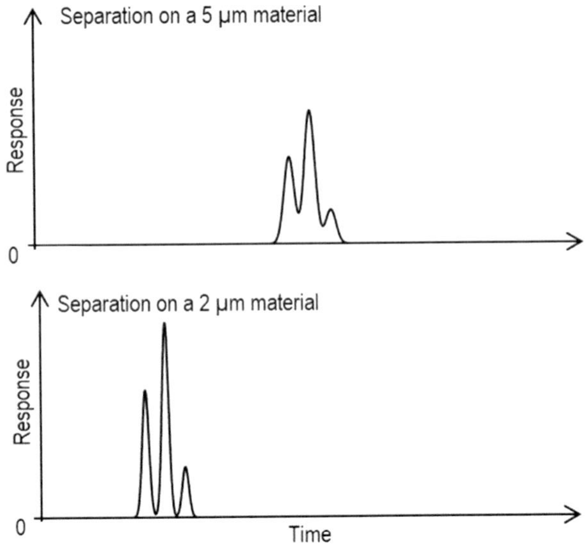

Figure 13.3 The simulated chromatograms show that smaller particles, under the same separation conditions, provide improved resolution. Reproduced with permission of Thermo Fisher Scientific.

$$t_2 = t_1 \cdot \frac{d_{p,2}}{d_{p,1}} \cdot \frac{L_2}{L_1} \tag{13.8}$$

In addition to saving time, reducing solvent consumption is another strong motivator for transferring conventional methods to UHPLC. Solvent consumption is reduced by using shorter columns with smaller void volumes and shorter analysis time. Additional solvent can be conserved by scaling down the column inner diameter. However, the linear velocity must be kept constant to operate with the optimum Van Deemter conditions. Subsequently, the flow rate has to be decreased with the new geometric cross section of the column. By using smaller inner diameter d_c and shorter length L columns, the reduction in solvent consumption can be calculated by Eq. 13.9.

$$V_2 = V_1 \cdot \frac{d_{c,2}^2}{d_{c,1}^2} \cdot \frac{L_2}{L_1} \tag{13.9}$$

Another benefit of transferring a conventional method to UHPLC is the ability to gain more mass sensitivity or consume less sample by using more efficient column material and scaling the injection volume accordingly, which will be discussed later in Section 13.2.2.4.

While there are benefits to transferring an HPLC method to UHPLC (i.e., improvements in efficiency or a decrease in analysis time), there are additional factors that must be considered such as the increase in back pressure in UHPLC. The pressure drop across a column filled with porous medium is described by Darcy's law (Eq. 13.10) as a function of the linear velocity u, viscosity η, length L, porosity Φ, and particles size d_p.

$$\Delta P = \frac{u \cdot \eta \cdot L \cdot \ddot{\mathrm{O}}}{d_\mathrm{p}^2} \quad (13.10)$$

The pressure drop generated by different particle sizes is shown in Fig. 13.4. Operating a column filled with 5 μm particles at 160 bar pressure drop would mean that the same column filled with 2 μm particles, would in theory generate 1000 bar pressure drop assuming that the other parameters like linear velocity, column length, viscosity, and porosity are kept constant.

The optimum flow rate for HPLC to UHPLC method transfer is influenced by the particle size (see Eq. 13.6). This dependency, combined with Darcy's law, allows the conclusion that the added back pressure is inversely proportional by the third power of particle size and directly related to the column length. The added pressure drop for the method transfer from HPLC column formats to UHPLC column formats can be calculated by Eq. 13.11.

$$\Delta P_2 = \Delta P_1 \cdot \frac{d_{\mathrm{p},1}^3}{d_{\mathrm{p},2}^3} \cdot \frac{L_2}{L_1} \quad (13.11)$$

UHPLC instrumentation operating with smaller particles has to be capable of operating at higher pressures.

Comparison of theoretical plates, savings in analysis time and solvent consumption, and the expected back pressure increase by changing the method from a conventional 4.6 × 150 mm column filled with 5 μm particle to various UHPLC column formats is shown in Table 13.1.

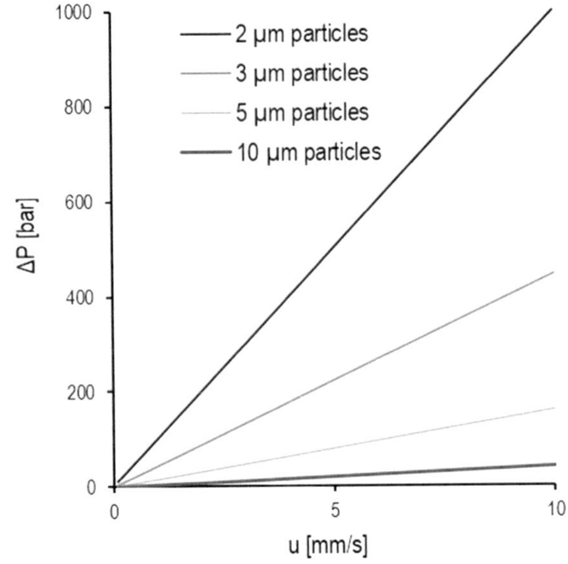

Figure 13.4 The simulated pressure drop on the column rises with inverse square of the particle size according to Eq. 13.10 demonstrates the trade-off for UHPLC separation targeting higher peak capacity or higher analysis speed. Reproduced with permission of Thermo Fisher Scientific.

Table 13.1 Estimated efficiency, time saving, solvent reduction, and back pressure increase for method transfer from HPLC method to UHPLC method for different column geometries generating comparable efficiencies and resolution

Column format						
Length [mm]	ID [mm]	Particle size [μm]	Theoretical plates N	Relative time saving	Reduction in solvent consumption	Back pressure increase factor
150	4.6	5	10000	—	—	—
100	4.6	3	11111	60%	33%	3.1
75	4.6	2.2	11363	78%	50%	5.9
50	4.6	2.2	7575	85%	67%	3.9
50	4.6	1.5	11111	90%	67%	12.3
150	2.1	5	10000	0%	79%	1.0
100	2.1	3	11111	60%	86%	3.1
75	2.1	2.2	11363	78%	90%	5.9
50	2.1	1.5	11111	90%	93%	12.3

13.2.2 System Requirements

The instrument and column design as well as method details like mobile phase, flow rate, and injection volume have to fulfill certain requirements to accomplish an HPLC to UHPLC method transfer by geometric rules.

13.2.2.1 Column choice

For successful HPLC to UHPLC method transfer, an adequate column packed with smaller particles has to be carefully selected. The column chemistry and selectivity of the stationary phase should be maintained to get the closest prediction of the UHPLC method.

Numerous tools can be found from column manufactures and on the internet to get the most appropriate UHPLC column to match the original HPLC column's characteristics in hydrophobicity, steric hindrance, hydrogen-bonding acidity, hydrogen-bonding basicity, and cation-exchange capacity. An overview of 300 reversed-phase columns by dedicated selectivity parameters has been described by Snyder et al. [3]. If no selectivity equivalent UHPLC column can be found, simple geometric method transfer rules may not be sufficient to predict the method transfer.

The preferable dimension for the UHPLC column should be 3 or 2.1 mm. Inner diameters of 4.6 mm filled with small particles are not recommended for UHPLC columns due to the high flow rate needed to operate at the optimal linear velocity and the resulting heating effects (see Section 1.4.1). Smaller inner diameters than 2.1 mm need special care for minimizing the extra column volume (see Section 1.2.2.4) and might disrupt the HPLC to UHPLC method transfer by simply applying geometric transfer rules.

The careful consideration of column length and suitable particle size is required to meet the column efficiency needed for the separation [4, 5].

If the analysis time is kept constant and the resolution of a critical peak pair should be doubled, then the efficiency of the column must be increased fourfold, due to the square root dependency of efficiency and resolution (Eq. 13.1). If the objective is to improve efficiency by reducing the particle size, then the particle size must be reduced fourfold (Eq. 13.3). Keeping the flow rate and column length constant results in a pressure drop that is elevated 16-fold (Eq. 13.10), possibly succeeding the pressure of the UHPLC instrumentation.

Another way to double the resolution is to increase the column length fourfold; however, this would also increase the analysis time four-fold, which is not wanted in most cases. Depending on the analytes, separation efficiency could also be increased by using high temperature chromatography or changing the stationary phase morphology [6]. However, application of the geometric method transfer rules to predict separations using high temperature chromatography or adjustment of the stationary phase morphology is difficult and beyond the scope of this chapter. The combined information on separation quality and analysis time, including information on permeability for different stationary phase morphologies and geometries, can be combined to kinetic plots to help find the most suitable column [7]. In this chapter, the main focus is on choosing columns with comparable or slightly better efficiency by using smaller particles to accelerate the chromatographic method (see Table 13.1).

13.2.2.2 Mobile phase

For UHPLC methods, care has to be taken in the preparation of the mobile phase to avoid clogging of the flow path. Smaller particles in the UHPLC column require smaller frits than those for the conventional HPLC columns, making UHPLC setups more susceptible to clogging. HPLC columns generally have inlet frits with a pore size of 2 μm, whereas the inlet frits of UHPLC columns have a pore size of 0.5 to 0.2 μm for 3 μm and smaller particles. Next to the column hardware, UHPLC instrument hardware is optimized for low dispersion connections and therefore more receptive to particle accumulation than conventional HPLC instrumentation. Therefore, high-grade organic solvents and high quality water should be used. Solvents should be filtered through a 0.2 μm membrane, if it has not been done already by the supplier or the lab water system. Buffer preparation and mobile phase additives should also be added in high quality grade to achieve the best UHPLC performance and robustness. Special care should be taken to prevent organic growth by the usage of buffers, as it is already the case for HPLC methods. This becomes even more important for UHPLC methods to avoid problems. Inlet frits in the solvent supply line of the UHPLC have to be used and changed from time to time.

13.2.2.3 Flow rate

The mobile phase flow rate should be near the optimal linear velocity value of the Van Deemter curve. The flow rate is directly proportional to the square of the column diameter and depends on the particle size of the stationary phase. A theoretical calculation of the optimum flow rate is possible, but other parameters like temperature and viscosity also influence the Van Deemter curve. A rough estimation of the optimal linear velocity according to Eq. 13.6 is possible. If that is not sufficient, recording a curve with experimental data should be completed.

When transferring an HPLC method to an UHPLC method, prediction of the new UHPLC flow rate F_2 can be calculated considering the original HPLC flow rate F_1, the change in column cross section d_c and particle size d_p according to Eq. 13.12.

$$F_2 = F_1 \cdot \left(\frac{d_{c,2}}{d_{c,1}}\right)^2 \cdot \frac{d_{p,1}}{d_{p,2}} \tag{13.12}$$

In addition, the flat C-term of the Van Deemter curve allows for an acceleration of the flow rate without compromising resolution with smaller particles. The new analysis time is then calculated by Eq. 13.13.

$$t_2 = t_1 \cdot \frac{F_1}{F_2} \cdot \frac{L_2}{L_1} \cdot \left(\frac{d_{c,2}}{d_{c,1}}\right)^2 \tag{13.13}$$

The increasing back pressure generated by the higher flow rates limit method acceleration. The back pressure is directly proportional to the flow rate and the resulting analysis time is indirectly proportional to the pressure drop. With a system capable of 1500 bar, it is in theory possible to accelerate a 400 bar method 3.75-fold.

13.2.2.4 Injection volume

Compared to longer columns, shorter columns with smaller particles cause a reduced zone dilution, and consequently sharper peaks. However, an equivalent mass loading onto columns with smaller inner diameters is recommended for method transfer. Scaling down the injection plug adjusting to the change in the column cross section avoids mass overloading and peak dispersion. The new injection volume $V_{\text{inj},2}$ can be calculated by taking into account the changed cross section and reduced band broadening by changing the particle diameter and column length according to Eq. 13.14.

$$V_{\text{inj},2} = V_{\text{inj},1} \cdot \left(\frac{d_{\text{c},2}}{d_{\text{c},1}}\right)^2 \cdot \sqrt{\frac{L_2 \cdot d_{\text{p},2}}{L_1 \cdot d_{\text{p},1}}} \qquad (13.14)$$

By adjusting the injection volume to load the smaller column with a similar mass, sensitivity should be consistent in theory. In praxis, the sensitivity could differ by changes in the HPLC and UHPLC instrumentation like improved detector technology. For example for UV detection, using the fiber-optics technology compared to a normal flow-cell design, leads to sensitivity increases.

13.2.2.5 Extra-column volume

The extra-column volume plays an important role for the HPLC to UHPLC method transfer. The extra-column volume contributes to band broadening of the analyte peak and is considered the volume in the system through which the sample passes except the column volume (Fig. 4.5). With the reduction of the column internal volume by shorter lengths and smaller inner diameters, the extra-column volume should be reduced proportionally to limit band broadening effects and to maintain the efficiency of the UHPLC stationary phase [8].

The total band broadening of the analyte peak is the sum of the band broadening caused by the column and the extra-column volume. Factors contributing to the extra-column volume are the injection, the capillary connection and the detector flow cell volume. The instrumentation has to be optimized to generate the separation efficiency capable with small-particle stationary phases [9].

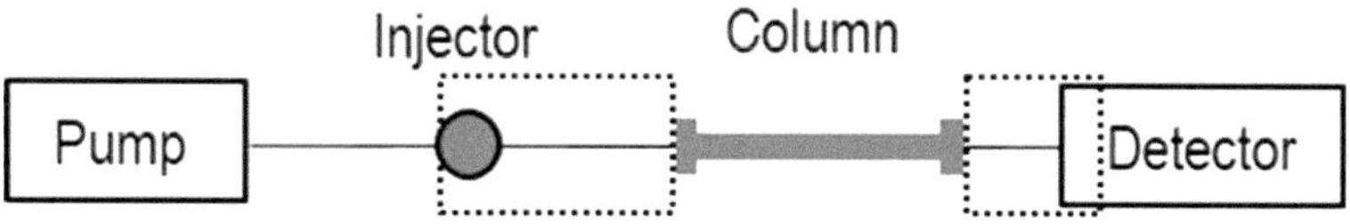

Figure 13.5 The extra-column volume contributes to the band broadening of the analyte peak. Extra-column volume should be minimized to exploit the maximum possible efficiency of UHPLC columns. Reproduced with permission of Thermo Fisher Scientific.

The injection volume should be scaled according to the geometric rules as discussed above (see Eq. 13.14). As a rule of thumb, the injection volume should not exceed 1–5% of the column volume.

The capillary connections should be dead-volume free, the length and the radius of the tubing should be kept as small as possible to minimize the extra-column band broadening. A compromise between generated back pressure and low volume of the connecting capillaries has to be found (see Table 13.2). For particles smaller than 2 µm, 0.1 mm inner diameter capillaries should be used, and for particles smaller than 3 µm capillaries with 0.13 mm inner diameter are appropriate for UHPLC methods. The back pressure generated by the column capillary inner diameter and other fluidic components, like the flow cell, cannot be ignored during UHPLC method transfer. When accelerating an UHPLC method, the backpressure generated from all the fluidic components becomes even more relevant when increasing the flow rate and operating in the flat C-term of the Van Deemter curve.

In addition to adjusting the injection volume and the tubing volume for UHPLC methods, the detector flow cell volume has to be adapted to the peak volumes eluting from the UHPLC column. As a rule of thumb, the detector flow cell volume should be no larger than approximately 10% of the smallest eluting peak volume. UHPLC columns with sub-2 µm particles, an inner diameter of 2.1 mm and operated around 600 µL/min produce very narrow peaks with peak widths around 1 to 3 s at the base. The peak volumes are therefore around 10 to 30 µL; accordingly, a typical flow cell volume would need to be 2.5 µL or less in UHPLC.

Table 13.2 The pressure drop generated by flow through the capillary is proportional to the fourth power of the inner diameter of the capillary according to the Hagen–Poiseuille equation

ID Capillary [µm]	Pressure drop [bar] (1 mL/min, 0.894 cp, 1 m)	Internal volume [µL]
75	192.0	4.4
100	60.7	7.9
130	21	13.3
180	6	25.4
250	2	49.1

Note: Small extra column volume generates a trade-off in pressure (prerequisites viscosity of water at 25°C, flow rate 1 mL/min, capillary length 1 m).

The extra-column volume can be evaluated to verify the compatibility of the UHPLC column with the instrumentation [10, 11]. As a rule of thumb, the extra-column volume should not be more than 10% of the column volume.

The extra-column volume should not be mixed up with the gradient delay volume. The gradient delay volume is important for the transfer of gradient methods and will be discussed in Section 1.2.4.1.

13.2.2.6 Data collection rate, injection cycle

Small eluting peaks from the UHPLC column have an impact on the detector settings. The data collection rate and time constant have to be adapted to the narrow eluting bands. It is good practice that every peak should be at least defined by 30 raw data points. The detector settings have to be set to optimize the data points per peak while reducing the short-term noise, but still maintain the peak height, symmetry, and resolution. Figure 13.6 shows an example how the detector settings influence the quality of the chromatogram.

Assuming the optimal data rate setting for the HPLC method is known, the data collection rate D of the UHPLC method can be calculated taking into account the changes in column length, particle diameter and peak dispersion [12].

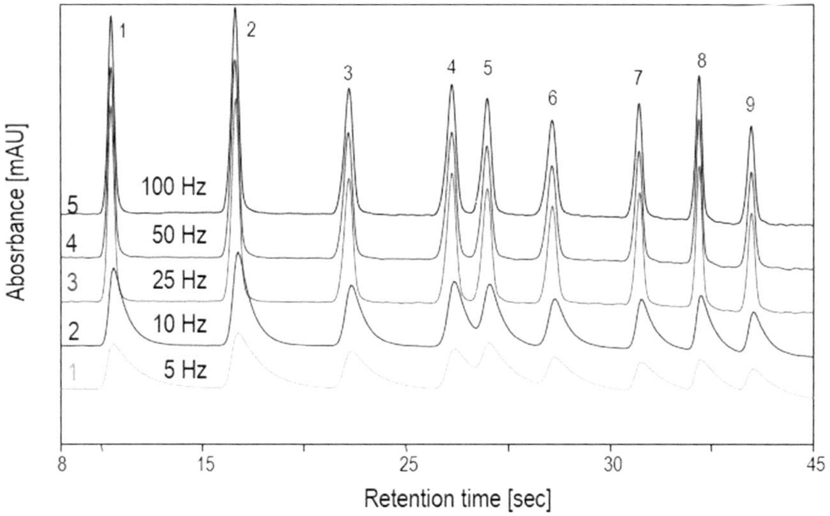

Figure 13.6 Ideally 30 data points should be recorded for each peak between the limits of integration. The time constant should be fit to the set data collection rate. The best settings should be determined by the narrowest peak. Poor peak symmetry can be improved by reducing the time constant. High baseline noise can be improved by increasing the time constant. Reproduced with permission of Thermo Fisher Scientific.

$$D_2 = \frac{D_1}{\sqrt{\frac{L_2 \cdot d_{p,2}^3}{L_1 \cdot d_{p,1}^3}}} \quad (13.15)$$

Chromatography data systems software wizards can help find the optimal detector settings.

13.2.3 Method Transfer Applied to Isocratic Methods

The rules and equations discussed in Section 1.2.1 are all valid for isocratic method transfer. Following the strategy to focus on obtaining more efficiency or reducing analysis time, new column dimensions for the UHPLC method must be chosen. Examples describing the process of the isocratic method transfer in detail can be found in literature [13].

An example of a method transfer from conventional HPLC to UHPLC is described below and shown in Fig. 13.7. A separation of three analytes according to United States Pharmacopeia (USP) was accelerated using the HPLC to UHPLC guidelines described above. Compared to the original HPLC method, the solvent consumption was reduced 95% and the analysis time was decreased 91% of the original method. The efficiency of the UHPLC column was less than the HPLC column but still sufficient to get the desired resolution between analytes.

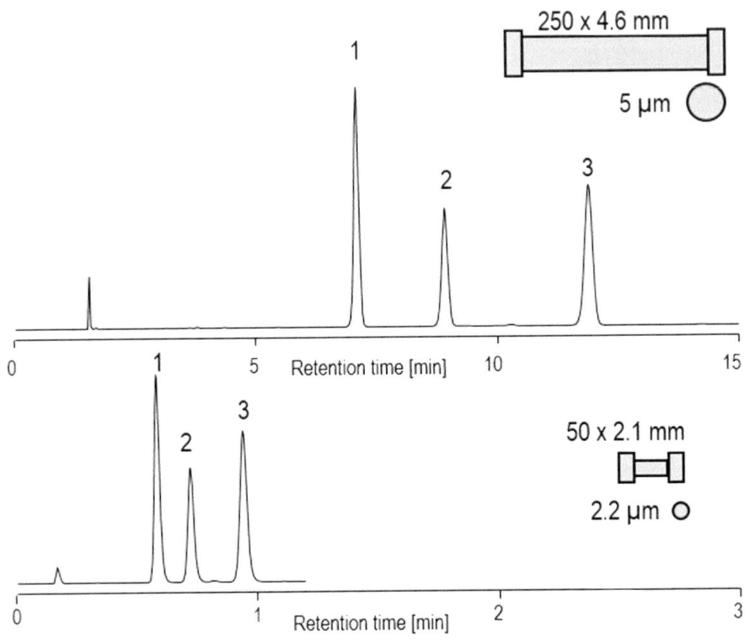

Figure 13.7 Isocratic HPLC method (top) transferred to the UHPLC method (bottom) applying geometric transfer rules. Separation of mevastatin (1), lovastatin (2), and simvastatin (3), mobile phase 38 mM phosphate buffer, pH4.5/acetonitrile (35/65 v/v), T = 45°C; top: Acclaim C18, 250 × 4.6 mm, 5 µm, 1.5 mL/min, V_{inj} = 10 µL; bottom: Acclaim C18, 50 × 2.1 mm, 2.2 µm, 0.8 mL/min, V_{inj} = 0.6 µL. Reproduced with permission of Thermo Fisher Scientific.

13.2.4 Method Transfer Applied to Gradient Methods

Beyond the simple geometric transfer rules for column dimension, flow rate, and injection volume that apply to isocratic methods, the scaling of the gradient profiles for new column formats and flow rates are unavoidable to maintain separation performance. The theoretical background is known as the gradient volume principle, introduced by *L.* Snyder [14]. The gradient volume V_G is defined as the mobile phase volume flowing through a column at a defined gradient time t_G.

$$V_G = F \cdot t_G \tag{13.16}$$

To maintain the selectivity and relative position of the analyte in the gradient, the ratio between gradient volume and column volume is kept constant [15] when transferring the method to a column of different dimensions. With this rule and assuming the total porosity is constant for both columns, the gradient can be transferred to different column formats. The chromatographic separation only changes in terms of analysis time for different gradient time intervals.

$$t_{g,2} = t_{g,1} \cdot \frac{F_1}{F_2} \cdot \frac{L_2}{L_1} \cdot \left(\frac{d_{c,2}}{d_{c,1}}\right)^2 \tag{13.17}$$

The transfer of a conventional gradient separation of seven soft drink additives to an UHPLC method is shown as example in Fig. 13.8. As a first consideration, a column providing sufficient efficiency has to be selected to resolve the most critical pair, peak 5 and peak 6. Following the geometric transfer rules, the 11,000 theoretical plates of the conventional column should be considered. With a current resolution of $R_{5/6}$ being nearly 3.5 choosing a column with slightly fewer theoretical plates while maintaining full resolution should be sufficient. Therefore, a 2.1 × 50 mm, 2.2 μm column generating calculated 7600 plates was selected. Using the relationship between R and the square root of N, a resolution increase to 2.87 can be calculated. Accelerating the flow rate to a higher linear velocity takes full advantage of the optimal separation efficiency of the new UHPLC column. By scaling down the inner diameter of the column, the flow rate is decreased accordingly to keep the linear velocity constant. The gradient steps are calculated by the gradient volume principle (see Eq. 13.17). The resulting chromatogram shows a successful separation of the critical peak pair with a resolution of 2.91 (see Fig. 13.8B). A further acceleration of the flow rate is possible, as the influence of the linear velocity on the plate height is smaller than the influence of the linear velocity with larger particles. A further acceleration of the flow rate by a factor of 2.5 generates a satisfactory resolution of $R_{5/6}$ = 2.56 (see the zoom in Fig. 13.8C).

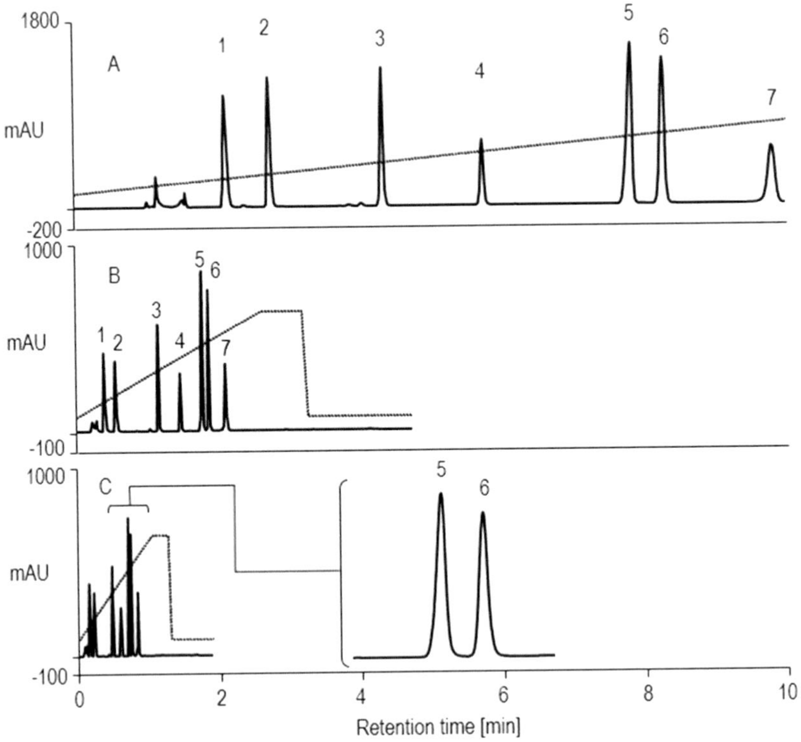

Figure 13.8 Method acceleration of seven soft drink analytes on to an Acclaim C18 column applying gradient transfer rules from HPLC to UHPLC and subsequent acceleration of flow rate [12]: Aceculfame K (1), saccharin (2), caffeine (3), aspartame (4), benzoate (5), sorbate (6), benzaldehyde (7). A: conventional column, 4.6 × 150 mm, 4.5 µm, 1.5 mL/min; B: UHPLC column, 2.1 × 50 mm, 2.2 µm, 0.639 mL/min; C: UHPLC column and accelerated flow rate, 2.1 × 50 mm, 2.2 µm, 1.599 mL/min. Reproduced with permission of Thermo Fisher Scientific.

13.2.4.1 Gradient-delay volume

The gradient delay volume plays an important role for HPLC to UHPLC method transfer. The gradient delay volume could result in significant changes in selectivity if it is not correctly accounted for. The gradient delay volume is also referred to as dwell volume and corresponds to the volume between the mixing point of solvents to the inlet of the column. As a result, the programmed eluent compositions reach the column head after the dwell volume has been flushed (see Fig. 13.9). Therefore, the sample is confronted with an undesired isocratic step that could cause retention-time changes, which may result in the loss of resolution for early eluting peaks.

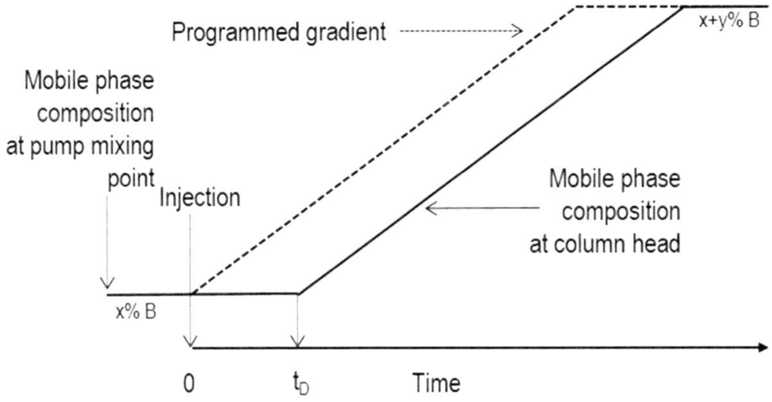

Figure 13.9 The dwell volume creates a delay before the programmed solvent compositions reaches the column inlet. This adds an isocratic hold at the beginning of the gradient. Reproduced with permission of Thermo Fisher Scientific.

To overcome this, the gradient delay volume must also follow the gradient volume principle of Snyder. Therefore, in a correct gradient method transfer, the ratio between the gradient delay volume and the column volume is kept constant, while changing the column format and flow rate. The new gradient delay volume can be calculated by

$$V_{\mathrm{GDV},2} = \frac{V_{\mathrm{GDV},1} \cdot V_{\mathrm{c},2}}{V_{\mathrm{c},1}}. \tag{13.18}$$

Examples describing the process of the gradient method transfer in detail, considering differences in column formats and gradient delay volume, can be found in literature [16].

The volume going from a typical HPLC column, 4.6 mm × 150 mm, to a 2.1 × 50 mm UHPLC column reduces the column volume by a factor of 14 when considering the same porosity. Therefore, the gradient delay volume should also be reduced by a factor of 14.

The main contributors to the gradient delay volume are the pump-mixing volume, the autosampler fluidics and all capillary connections before the column inlet. The determination of the gradient delay volume is described in the literature [17] and can help optimize the UHPLC method.

For UHPLC gradient methods it is imperative to work with UHPLC systems with low gradient delay volume. By its technical design, high pressure mixing pumps are the preferred as they have smaller dwell volume (see Fig. 13.10).

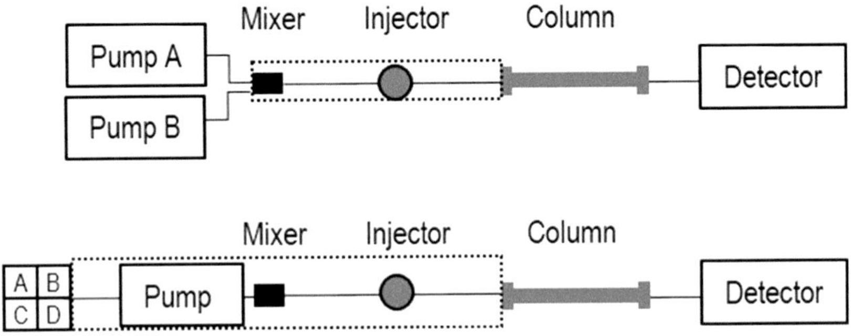

Figure 13.10 The gradient delay volume (dotted line) is defined as the volume from the mixing of solvents to the column head. By its technical operating principle the gradient delay volume is lower for high pressure gradient pumps (top) than for low pressure gradient pumps (bottom). Gradient delay volume must be considered for optimal gradient method transfer from HPLC to UHPLC. Reproduced with permission of Thermo Fisher Scientific.

The biggest contributor to the gradient delay volume is the volume of the mixer. Therefore, the mixer must be considered when optimizing the dwell volume. Instrument manufacturers offer systems with adjustable gradient delay volumes to ease method transfer.

In practice, it is difficult to scale down the gradient delay volume to exactly match the geometric calculation according to Eq. 13.18. For high differences in the gradient delay volumes of HPLC and UHPLC system the solution is either to add an isocratic step or to delay the injection after the start of the gradient.

$$t_D = \frac{V_{GDV,2} - V_{GDV,opt}}{F_2} \quad (13.19)$$

The influence of different gradient delay volumes to a gradient separation can be seen in Fig. 13.11. Transferring a method from an instrument with a high gradient delay volume (bottom) to a system with less gradient delay volume (top) shifts the peaks toward earlier retention times. This shift in the early eluting polar analytes is especially critical as they may be wrongly integrated with peak picking automation.

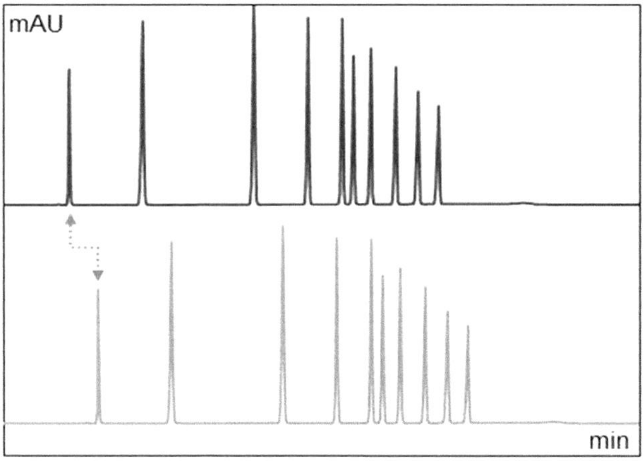

Figure 13.11 The pseudo isocratic hold while transferring from an HPLC system with high gradient delay volume (bottom, grey) to a system with lower gradient volume (top, black) has to be considered for the smooth transition from HPLC to UHPLC to avoid problems. Reproduced with permission of Thermo Fisher Scientific.

13.3 Method Transfer Tools

The method transfer process from HPLC to UHPLC considering all the geometric rules can be easily integrated into calculators.

These calculators are available from several sources. An overview can be found in reference [18]. The original column dimensions, flow rate, injection volume, and gradient profile can be programmed in by the user to automatically determine the new UHPLC parameters to be used. Additionally, parameters like resolution prediction, adjusted data collection rate, back pressure prediction, and consideration of gradient delay volumes are also integrated to aid in successful method transfer.

These calculators are available as online tools [19], downloadable offline tools [12, 20] or even embedded into the chromatography data software, allowing for easy method transfer by generating the transfer method parameters automatically [21].

13.4 Limitations of Method Transfer

Method transfer from HPLC to UHPLC has different limitations. Working at very high pressures, the frictional heating phenomena may affect method transfer. Furthermore, transferring the stationary phase to a different morphology and the transfer of instrumental parameters settings can be challenging. Other theoretical method transfer options might be limited by regulatory affairs and related restrictions.

13.4.1 Frictional Heating

Thermostatting the separation column and mobile phase is very important for achieving a robust and reproducible separation method. Small changes in temperature can have a huge effect on the retention factor and in turn affect the selectivity of the method as expressed by the logarithmic relationship of retention factor k and temperature T in the Van't Hoff equation (see Eq. 13.20).

$$\ln k = \frac{\Delta H^0}{R} \cdot \frac{1}{T} + \frac{\Delta S^0}{R} + \ln \beta \tag{13.20}$$

In addition to external thermostatting of the UHPLC column, internal heat generation must be considered. UHPLC separations using columns packed with fine particles experience frictional heating throughout the column. When mobile phase enters the column, frictional heat is generated inside the column by the rubbing of viscous eluents over the packed particles at high pressures and flow rates. The generated frictional heat within the column can influence the retention time prediction of analytes being transferred from an HPLC to UHPLC method [22, 23].

The dissipation of heat occurs along and across the column, generating axial and radial temperature gradients, which can influence not only the efficiency of the separation but also the retention and selectivity [24]. The magnitude of the axial and radial temperature distribution in the column depends on the thermostatting conditions in the column compartment.

A radially heterogeneous distribution of temperature is mainly observed with isothermal or forced air column thermostatting, because the column wall temperature is controlled but the temperature in the center of the column is increased. As a result, the peak shape is distorted and a loss of efficiency may be observed.

An axially or longitudinally heterogeneous distribution of temperature is mainly observed when the temperature at the column wall is not kept constant because of adiabatic or still-air thermostatting. As a result, the temperature at the column outlet is higher than at the column inlet and changes in retention time and selectivity may occur [25, 26].

An example of loss in efficiency, by using forced air thermostatting compared to still air thermostatting while operating under frictional heating conditions, is shown in Fig. 13.12.

Frictional heating also occurs in HPLC methods operating at relatively low pressures. However, the influence of frictional heat is more significant for methods using sub-2 μm particles and column lengths of 50 to 100 mm operating at pressures above 400 bar. Frictional heat should not be neglected when aiming for a smooth HPLC to UHPLC method transfer. This is especially true if the thermostatting operating principles have to be adjusted during method transfer due changing instrumentation. For UHPLC methods working under frictional heat conditions it could be suitable to set the column thermostatting at a lower temperature to overcome the heat addition if allowed by the method [26]. The eluent pre-heating also influences the formation of the temperature gradient. An example how the thermostatting of column and pre-heating of mobile phase can influence the method transfer result is shown in Fig. 13.13.

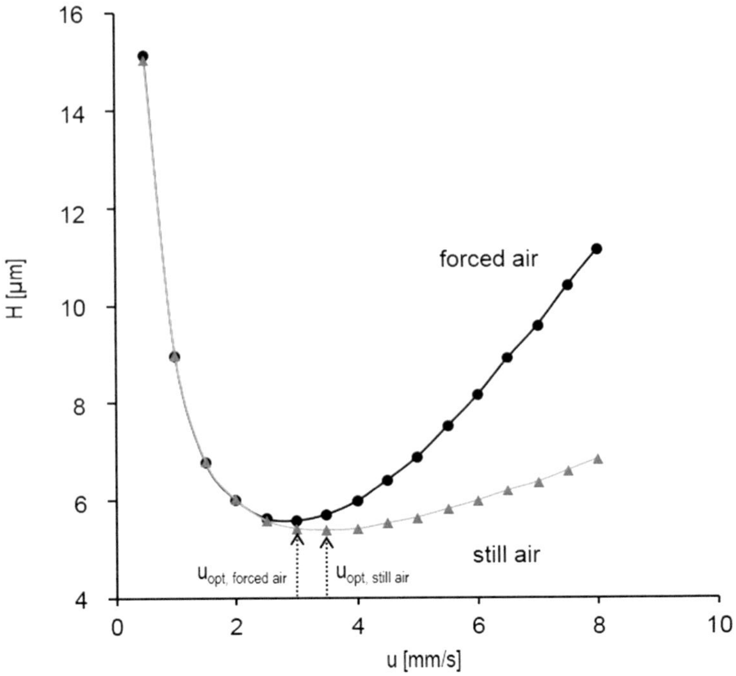

Figure 13.12 Frictional heating effects and generation of a longitudinal temperature gradient in the column shows that still air operation mode results in increased optimal flow rate and increased efficiency with a slightly lower pressure drop of 860 bar versus 880 bar [25]. Column: Thermo Scientific Hypersil Gold, 1.9 µm, 2.1 × 100 mm, acetonitrile/waters (50:50 v/v). Thermostatting: 30°C in still air or forced air mode, active eluent pre-heating: 30°C, sample: hexanophenone. At a linear velocity of 8 mm/s the column efficiency by forced air mode is 40% lower than with still air mode. The optimum linear velocity in still air mode 20% higher and has 10% better efficiency than in forced air mode. Reproduced with permission of Thermo Fisher Scientific.

13.4.2 Stationary Phase Differences

Transferring methods onto a column with similar stationary phase chemistry, but different particle size can cause selectivity issues. This is mostly related to a difference in silanol activity. Particle size distribution of the stationary phase is another important aspect to consider when applying geometric rules for transferring methods. The effective and nominal particle size might differ. For example, a difference in 5 or 4.5 µm particle size can generate different separation results.

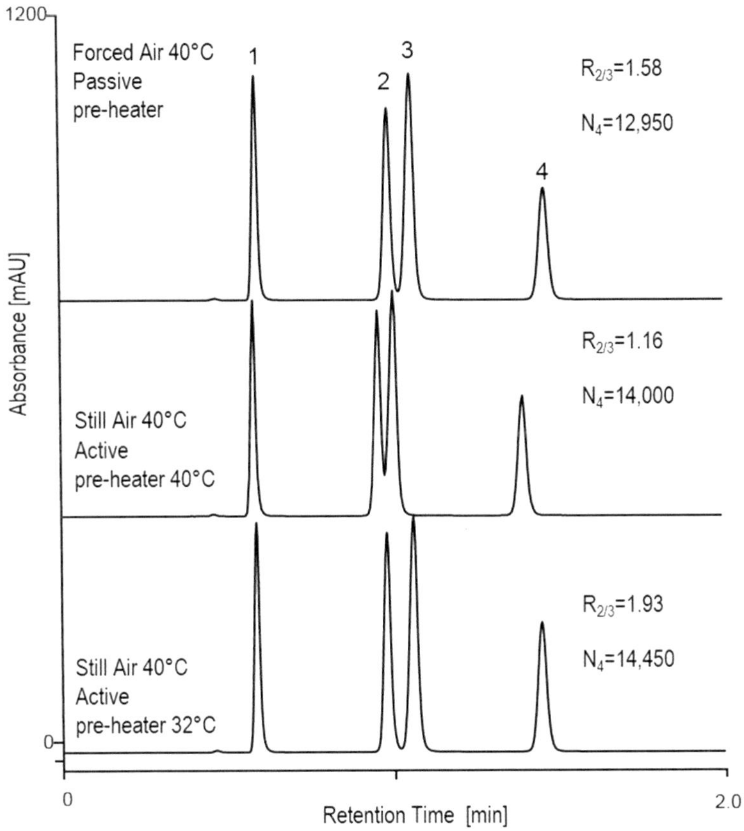

Figure 13.13 The effect of thermostatting conditions on an UHPLC separation experiencing frictional heating (>750 bar) Column: Acclaim RSLC PA2, 2.1 × 150 mm, 2.2 µm, 0.55 mL/min, resulting in 760 bar back pressure (Sample: Uracil (1), Dimethyl phthalate (2), Methyl Paraben (3), methyl benzoate (4). Forced air thermostatting with passive pre-heating resulted in a resolution of the critical peak pair above 1.5. However, the plate number was below 13,000 plates. With still-air thermostatting and active-preheater, the resolution of the critical peak pair was insufficient, but the efficiency was improved. By setting the active pre-heater at a lower temperature independent of the column thermostatting resulted in an increase in both the resolution and the plate number [26]. Reproduced with permission of Thermo Fisher Scientific.

Another potential method transfer challenge may occur when attempting to transfer a method from fully porous to core–shell stationary phase. The change in the particle morphology should be considered as well, especially when transferring gradient methods. In the gradient volume equation (see Eq. 13.18), the internal column volume filled with mobile phase should be taken into consideration and adjusted by Eq. 13.21.

$$V_{\mathrm{M}} = \pi \cdot r^2 \cdot L \cdot \varepsilon \tag{13.21}$$

The parameter ε is the column interstitial porosity percentage. As a rule of thumb ε is 0.68 for fully porous material and 0.55 for core–shell particles. When transferring an UHPLC method from a column filled with fully porous material to a core–shell material change of the particle morphology effects, the volume of mobile phase in the column and should be accounted for in method transfer calculations [27].

13.4.3 Regulatory Limitations

The method transfer from HPLC to UHPLC or vice versa from UHPLC to HPLC can be made by applying the rules discussed in this chapter. However, validated methods and methods described in monographs with the need to meet specific requirements are somewhat restricted to HPLC to UHPLC method transfer. The European Pharmacopeia (EP) and the United States Pharmacopeia (USP) limit some modifications to the chromatography methods. Chapter <621> of the USP defines an acceptable range where method parameters can be adjusted without needing to revalidate the method. The USP prerequisites include that the transferred method still meets the system suitability requirements and passes verification tests after the changes.

It is relatively straightforward to transfer isocratic methods within certain range of variations. However, the transfer of gradient methods allows very limited adjustments. Changes in length, column inner diameter and particles size as well as changes in flow rate are not allowed for gradient separations. Changes in column packing while maintaining the chemistry of stationary phase, changes in the duration of a prescribed isocratic hold and adjustments of the dwell volumes are in theory allowed, but practically those changes alone do not support an HPLC to UHPLC method transfer without the allowance of changing column properties. Furthermore, changes in the gradient table are perceived as uncertain and are more strongly limited for modifications. The limitations described in USP chapter <621> [28] are summarized in Table 13.3.

Table 13.3 Allowances and limitations for method transfer according to USP [28]

Chromatographic parameter	Allowed adjustments by USP 38-NF 32 (December 1, 2015) Chapter <621>	
	Isocratic	**Gradient**
Particle size	L/dp ratio constant or N: –25% to +50%	No changes allowed
Column length	L/dp ratio constant or N: –25% to +50%	No changes allowed
Flow rate	$F_2 = F_1 \times [(dc_2^2 \times dp_1)/(dc_1^2 \times dp_2)]$ and ± 50%	No changes allowed
Column ID	Any allowed if linear velocity is constant	No changes allowed
Injection volume	Can be adjusted as consistent with precision, linearity, and detection limits	
Column temperature	±10°C	
Mobile phase pH	±0.2 units	
Concentration of buffer salts	±10%	
Composition of mobile phase	±30% relative of the minor solvent (no other component is altered by more than 10% absolute)	

Any changes of the method parameters beyond the ranges described above require the full re-validation. A gradient method transfer from HPLC to UHPLC with the motivation to speed-up the method requires therefore a re-validation. However, this can also be taken as an opportunity to fully optimize the method with modern sub-2 μm particle columns by doing a fast UHPLC method development on up-to-date UHPLC instrumentation and validation supported by latest chromatography data systems.

13.5 Conclusion

The motivation for transferring a conventional HPLC method to UHPLC can be better resolution or faster analysis time. Following simple geometric rules and some practical guidelines allows for easy transfer the conventional method. A systematic method speed-up can be made by reducing the particle size, shortening the column length and increasing the linear velocity of the mobile phase. For high-resolution chromatography, all of the geometric rules can greatly impact the separation and should be considered accordingly. Method transfer tools embedded in chromatography data software and available online or offline help to easily apply the geometric rules.

Not only does the pressure capability of the UHPLC instrument performance for method transfer matter, but the extra-column volume, gradient delay volume, detection options, and thermostatting characteristics have to be considered as well. While there are many facets to consider for successful method transfer, UHPLC separations give a greater freedom to exploit the principles of chromatography and get the optimized method to solve the analytical problem.

Acknowledgements

My thanks go to my colleagues for contribution to the chapter, especially Holger Franz, Frank Steiner, Sabrina Patzelt, Michael Heidorn, and Jenny Wong.

References

1. Purnell, J. H. (1959) Comparison of efficiency and separating power of packed and capillary gas chromatographic columns, *Nature,* **184**, p 2009.
2. Halász, I., Endele, R., and Asshauer, J. (1975) Ultimate limits in high-pressure liquid chromatography, *J. Chromatogr. A,* **112**, pp. 37–60.

3. Snyder, L. R., Dolan, J. W., and Carr (2004) The hydrophobic-subtraction model of reversed-phase column selectivity, *J. Chromatogr. A*, **1060**, pp. 77–116.
4. Guillarme, D., Grata, E., Glauser, G., Wolender, J. L., Veuthey, J. L., and Rudaz, S. (2009) Some solutions to obtain very efficient separations in isocratic and gradient modes using small particles size and ultra-high pressure, *J. Chromatogr. A*, **1216**, pp. 3232–3243.
5. Dolan, J. W. (2017) Increasing resolution by increasing column efficiency, *LCGC Eur.*, **30**(3), pp. 138–142.
6. Guillarme, D., Ruta, J., Rudaz, S., and Veuthey, J. L. (2010) New trends in fast and high-resolution liquid chromatography: A critical comparison of existing approaches, *Anal. Bioanal. Chem.*, **397**, pp. 1069–1082.
7. Desmet, G., Cabooter, D., and Broeckhoven, K. (2015) Graphical data representation methods to assess the quality of LC column, *Anal. Chem.,* **87**, pp. 8593–8602.
8. Fountain, K. J., Neue, U. D., Grumbach, E. S., Diehl, D. M. (2009) Effects of extra-column band spreading, liquid chromatography system operating pressure, and column temperature on the performance of sub-2-μm porous particles, *J. Chromatogr. A*, **1216**, pp. 5979–5988.
9. Gritti, F., Guichon, G. (2010) On the extra-column band-broadening contributions of modern, very high pressure liquid chromatographs using 2.1 mm I. D. columns packed with sub-2 μm particles, *J. Chromatogr. A,* **1217**, pp. 7677–7689.
10. Gritti, F., Guichon, G. (2014) Accurate measurements of the true column efficiency and of the instrument band broadening contributions in the presence of a chromatographic column, *J. Chromatogr. A*, **1327**, pp. 49–56.
11. Vanderlinden, K., Broeckhoven, K., Vanderheyden, Y., Desmet, G. (2016) Effect of pre- and post-column band broadening on the performance of high-speed chromatography columns under isocratic and gradient conditions, *J. Chromatogr. A*, **1442**, pp. 73–82.
12. Franz, H., Fabel, S. (2013) A universal tool for method transfer from HPLC to UHPLC, Technical Note 75, Thermo Fisher Scientific, USA.
13. Guillarme, D., Nguyen, D. T. T., Rudaz, S., Veuthey, J. L. (2007) Method transfer for fast liquid chromatography in pharmaceutical analysis: Application to short columns packed with small particle. Part I: Isocratic separation, *Eur. J. Pharm. Biopharm.*, **66**, pp. 475–482.
14. Snyder, L. R., Dolan, J. W., Gant, J. R. (1979) Gradient elution in high-performance liquid chromatography: I. Theoretical basis for reversed-phase systems, *J. Chromatogr. A*, **165**, pp. 3–30.

15. Schellinger, A. P., Carr, P. W. (2005) A practical approach to transferring linear gradient elution methods, *J. Chromatogr. A*, **1077**, pp. 101–119.
16. Guillarme, D., Nguyen, D. T. T., Rudaz, S., Veuthey, J. L. (2007) Method transfer for fast liquid chromatography in pharmaceutical analysis: Application to short columns packed with small particle. Part II: Gradient experiments, *Eur. J. Pharm. Biopharm.*, **68**, pp. 430–440.
17. Dolan, J. W. (2006) Dwell volume revisited, *LCGC Eur.*, **24(5)**, pp. 458–466.
18. Majors, R. E. (2011) Method translation in liquid chromatography, **29**(6), pp. 476–484.
19. HPLC Method Transfer Calculator, Thermo Fisher Scientific, (https://www.thermofisher.com/de/de/home/global/forms/industrial/lc-method-transfer-calculator.html).
20. Easy HPLC to UHPLC Transfer (xlm–file), Thermo Fisher Scientific (http://www.thermoscientific.com/content/dam/tfs/ATG/CMD/cmd-documents/oper/tech/chrom/lc/sys/RSLC-Method-Transfer-Tool-2-3.xlsm).
21. Hillbeck, D. (2016) Rapid screening method for statins using an advanced UHPLC column and system, Application Note 21496, Thermo Fisher Scientific, USA.
22. McCalley, D. V. (2014) The impact of pressure and frictional heating on retention, selectivity and efficiency in ultra-high-pressure liquid chromatography, *Trends Anal. Chem.*, **63**, pp. 31–43.
23. Grinias, J. P., Keil, D. S., Jorgenson, J. W. (2014) Observation of enhanced heat dissipation in columns packed with superficially porous particles, *J. Chromatogr. A*, **1371**, pp. 261–264.
24. Heidorn, M. (2016) The role of temperature and column thermostatting in liquid chromatography, White Paper 71499, Thermo Fisher Scientific, USA.
25. Heidorn, M. (2016) A more flexible column thermostatting technique in LC method transfer, White Paper 71500, Thermo Fisher Scientific, USA.
26. Steiner, F., Patzelt, S., De Pra, M., Martin, M. M. (2016) Optimizing selectivity and efficiency in critical UHPLC methods by separating eluent pre-heating from column thermostatting, Poster Note 72103, Thermo Fisher Scientific, USA.
27. Taylor, T. (2015) HPLC column transfer from fully porous to core–shell in three simple equations, *Column*, **11** (9), pp. 2–4.
28. United States Pharmacopeia, USP 38-NF 33, chromatography <621>.

Chapter 14

Small-Molecule Pharmaceutical Impurities Test Method Validation: Precision Acceptance Criterion

Russell L. Hertzler,[a] Mark D. Johnson,[a] Teodora Moldovan,[a] Michael D. Oberlander,[a] James Reynolds,[a] Holger van Lishaut,[b] and Yanbing Zheng[a]

[a]*AbbVie, 1 N. Waukegan Rd, North Chicago, Illinois 60064, USA*
[b]*AbbVie Deutschland GmbH & Co. KG*
Knollstrasse, 67061 Ludwigshafen, Germany

michael.d.oberlander@abbvie.com

14.1 Introduction

The International Conference on Harmonization (ICH) guideline [1] for pharmaceutical small-molecule impurities test method validation does not provide any guidance in setting numerical validation performance criteria for linearity, accuracy, and precision. It specifies a numerical minimum acceptance criterion

Chromatographic Method Development
Edited by Gregory K. Webster and Laila Kott

ISBN 978-981-4800-53-2 (Hardcover), 978-0-429-20172-1 (eBook)
www.jennystanford.com

for signal-to-noise performance respective of the test method's limit of detection and limit of quantitation. The lack of ICH defined acceptance criteria for the other performance attributes is most likely intentional, so the analytical procedures developed can be validated for their intended purpose. However, for regulatory submissions, the test method acceptance criteria used to demonstrate that analytical procedures are suitable are required to be reported. Regulatory Agency CMC reviewers may request the sponsor to justify their test method validation acceptance criteria.

For the main component (active ingredient assay) quantitation test method, sponsors could select to set absolutely the precision performance acceptance criterion as the tolerable variation predefined as a statistically based fraction of the active ingredient specification range [2]. A specification-based approach to setting precision performance acceptance criterion for the active ingredient assay test method is sensible and straight forward from the scientific perspective and is less likely to be met with objection during regulatory review.

Setting an objective, predefined, numerical acceptance criterion for impurities test method precision performance is more challenging in advance of the initial marketing application in conjunction with Chemistry, Manufacturing, and Controls (CMC) development of proprietary drugs. The challenge in using a specifications-based approach to set precision acceptance criterion for impurities and degradation products test methods is due in part to uncertainties in negotiating the final impurities and degradation products specifications with the regulatory bodies. Additional limitations to using the specifications-based approach could include a statistically limited knowledge of the independent variation in the manufacturing process batch to batch performance, as well as the product stability profile variation, within their respective design space, at the time of test method validation and method technical transfer which typically occur in advance of the initial marketing application date.

A critical factor that should not be overlooked is that one of the test method validation acceptance criteria should account for the combined effect of precision and accuracy performance [3], regardless of the nature of the quantitated analyte, active ingredient, impurity, or inactive ingredient critical to the product quality. For example, a test method systematic accuracy bias would require corresponding improvement in the test method precision performance in order to avoid excessive incidence of reporting false passing or false failing results versus the material specification limits.

The focus of this chapter does not include the accounting of any systematic quantitation or manufacturing process bias, or combinations of these, on test method precision requirements.

Instead this chapter, irrespective of quantitation and manufacturing process bias, focuses on impurities and degradation products test method validation precision in terms of the comparative performance of a set of AbbVie test method validations to an empirical observation that the literature credits as being first investigated by Horwitz [4–6]. Without regard to an approach of setting test method precision performance requirements as a function of material specification limits, Horwitz observed test method(s) inter-laboratory precision to generally exhibit a systematic trend where the relative variability of the analytical determination increases exponentially with decreasing analyte concentration expressed as absolute mass fraction.

In terms of current trends applicable to small-molecule pharmaceutical quality, Horwitz's work has not been forgotten. Without any direct explicit reference to Horwitz's work, the 2015 Chinese Pharmacopoeia (Chapter 9101 Guidelines for Validation of Analytical Method Adopted in Pharmaceutical Quality Specification) can be reasonably inferred to have adopted the Horwitz function as the tolerable level of test method inter-laboratory variability and thus as the Chinese Pharmacopoeia Acceptance Criterion for Reproducibility. A previously published review of small-molecule pharmaceutical industry test method precision performance has also recognized the work of Horwitz [7].

Hence, given our recent internal experiences in the context of some external trends potentially signaling more comprehensive expectations for objective predefined acceptance criteria, AbbVie elected to make an assessment of our historical "minors" ("minors" include manufacturing impurities, degradation products, and functionally critical formulation excipients requiring testing for control) quantitation test method performance against the published Horwitz Acceptance Criteria as the performance parameter indicating the acceptability of methods of analysis with respect to inter-laboratory precision (Reproducibility) [6].

This chapter does not present a comprehensive review of the topic of defining suitable precision performance of impurities test methods, which lies at the interface of statistical sciences and the pragmatic implementation of robust quality control analytical test method technology in the industrial pharmaceutical CMC development and subsequent commercial manufacturing setting.

This chapter presents a meaningful sample of AbbVie impurities and degradation products test method validation precision performance against the Horwitz Reproducibility acceptance criterion.

14.2 Horwitz's Empirical Observation of Test Method Variability as a Function of Analyte Concentration

William Horwitz published in 1980 an empirically observed correlation between test method coefficient of variation and analyte concentration from a collection of 7502 observations from inter-laboratory collaborative studies (Reproducibility) spanning multi-decades of analytical progress. The examined test method sample matrices included pharmaceuticals, food, cosmetics, and environmental samples of many kinds [4] (see Fig. 14.1).

Horwitz and coworkers found an empirical relation between the absolute mass fraction concentration of the analyte quantitated, *c*, and the observed value of the corresponding inter-laboratory *percentage* relative standard deviation of the analyte determination.

$$\mathrm{RSD}_{\mathrm{R,\,Horwitz}} = 2^{1-0.5\log c} = 2c^{-0.5\log 2} = 2c^{-0.1505}$$

where log is base 10.

Equivalently in its linear transformation:

$$\log(\mathrm{RSD}_{\mathrm{R,\,Horwitz}}) = \log(2) - 0.1505\log(c) = 0.3010 - 0.1505\log(c).$$

Note: The concentration, *c*, was expressed in terms of *absolute mass fraction* to establish the relationship. For example, a concentration of 1 g analyte per 100 g matrix corresponds to $c = 0.01$, and using base 10, $\log c = -2$. The above equations give an $RSD_{R,\ Horwitz}$ of 2%, for c equal to 1; and an $RSD_{R,\ Horwitz}$ of 16%, for a matrix concentration c of 0.000001, which corresponds to an analyte concentration of one part per million.

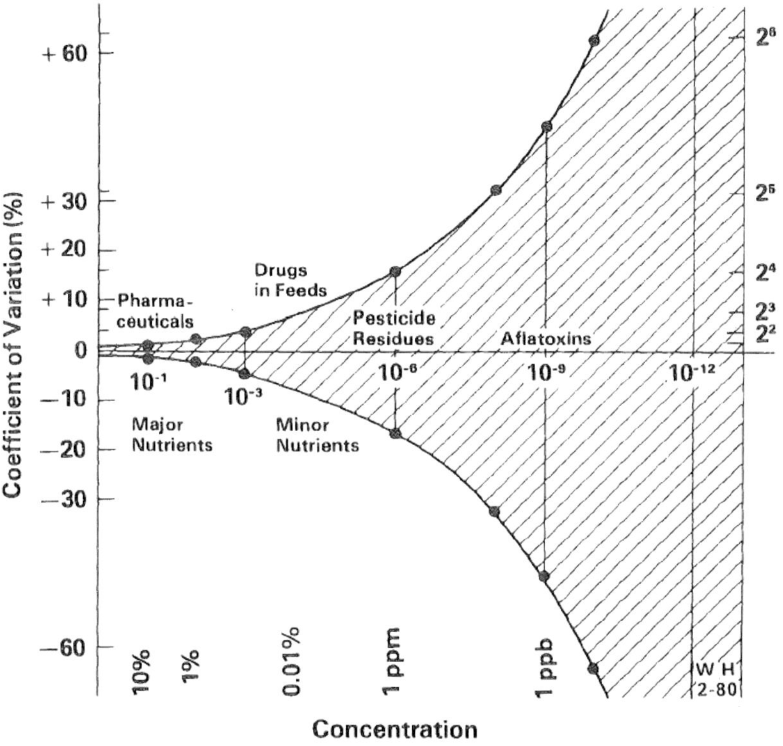

Figure 14.1 The conceptual "Horwitz horn". Inter-laboratory coefficient of variation as a function of concentration. [4] Reprinted with permission from AOAC INTERNATIONAL.

The formula became known as the "Horwitz horn" or "Horwitz formula". Qualitatively, the property reflected by the Horwitz function [4–6] is familiar and intrinsic to analytical test method precision performance as a function of a multi-fold reduction of analyte concentration. The relative variation (Relative Standard Deviation) of the analyte determination is generally found to systematically increase with decreasing analyte concentration, c.

See Fig. 14.2, noting that it plots the log of *absolute standard deviation* versus log *c*.

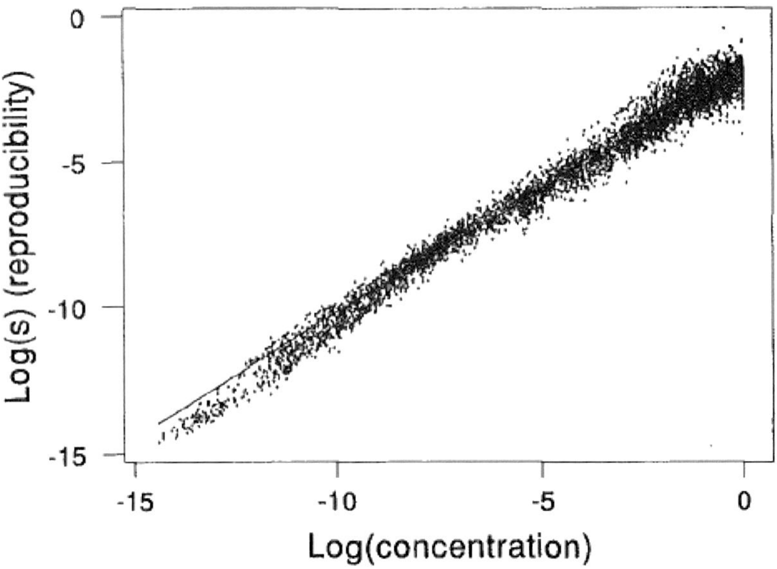

Figure 14.2 Reproducibility standard deviation and concentration of analyte in collaborative trials (7,502 observations). The line is the Horwitz function Reprinted with permission from AOAC INTERNATIONAL [5].

In our view, we note that the y-intercept (the %RSD criterion corresponding to $c = 1$) of the log linear Horwitz function (or equivalently the coefficient of *c* for the Horwitz function expressed in its power form) could be justifiably re-scaled to value(s) less than 2% RSD. Horwitz and Albert [6] do not cite any specific experimental observations as forming the justification for their assignment of this intercept. Hofer et al. [8] cite a pooled RSD of 0.61% and an RSD range of 0.29% to 1.0% as representative of intermediate precision performance for an HPLC-based drug substance potency assay.

As well, in terms of the Horwitz function, a re-scaled intercept value could be justifiably assigned based on the tolerable variability predefined from a specification-based Capability Analysis, C_{pk}, to setting the inter-laboratory precision acceptance criterion.

$$C_{\text{pk,assay test}} = \text{Minimize}\,\sigma_{\text{assay test}} \left[\frac{\text{USL} - c_{(i)\text{active; process}}}{3\sigma_{\text{assay test}}}, \frac{c_{(i)\text{active; process}} - \text{LSL}}{3\sigma_{\text{assay test}}} \right] \geq 1.33$$

$$C_{\text{pk,impurity test}} = \text{Minimize}\,\sigma_{\text{impurity test}} \left[\frac{\text{USL} - c_{(i)\text{impurity; process}}}{3\sigma_{\text{impurity test}}}, \frac{c_{(i)\text{impurity; process}} - \text{LoQ}}{3\sigma_{\text{impurity test}}} \right] \geq 1.33$$

The validated test method precision standard deviation, s_i, according to the precision validation design and degrees of freedom, is an estimate of the respective test method true standard deviation σ_i, $\sigma_{\text{impurity test method}}$ or $\sigma_{\text{active ingredient test method}}$. USL, LSL and LoQ, respectively, are the Upper Specification Limit, Lower Specification Limit and the impurities or degradation products test method Limit of Quantitation applicable to the determined analyte concentration level, c_i, whose true value is, in general, manufacturing process (i.e., variation batch to batch) and stability storage time and temperature dependent in the case of degradation products. Of course, the setting of the specification limits, USL(s), for impurities and degradation products is strictly according to the results of toxicological qualification studies and, or permitted daily exposures specified by the respective ICH Regulations [9–11], and in conjunction with the assessment of manufacturing process capability. From the test method performance perspective, C_{pk} is optimally ≥1.33.

For an unbiased drug substance manufacturing process whose batch content is subject to an assay specification of 98.0 to 102.0%w/w, a suitable predefined acceptance criterion for the validation inter-laboratory precision of an unbiased test method would be $s \leq 0.50\%$ RSD based on a C_{pk} of 1.33 and a statistically robust precision validation design and degrees of freedom.

Returning to the topic of impurities test method precision performance, although analytical technology has very obviously improved over time in terms of absolute concentration sensitivity, Horwitz's empirical observation indicates that the relative variability of analytical determinations is meaningfully independent of these sensitivity improvements. Given that a first principles theoretical model to support Horwitz's empirical observation does not appear to exist, we can only postulate, similarly to Ermer et. al. [7] who stated, *"the standard deviation decreases less rapidly than the concentration, resulting in an increase of the relative standard deviation for lower concentrations."*

At present, a broad multi-fold concentration range collaborative inter-laboratory precision re-assessment might be indicated, represented by the most contemporary practiced analytical technology and classified according to the test method intended purpose. For example, it can be reasoned that the set of reproducibility performance observations evaluated by Horwitz through 1980 would not have likely included a meaningful sample of quantitative instrument-based chromatographic pharmaceutical impurities and degradation product test methods. The expectation for routine impurities quality control for small-molecule pharmaceutical products was not firmly in place until initiated by ICH in 1996 [12].

The essential elements of the Horwitz observed empirical correlation between inter-laboratory relative variability of the determination and the analyte weight fraction, *c*, are the following:

1. The relative variability (%CV or %RSD) is a power function of *c*.
2. For a trained analyst, the relative variability is largely independent of further experience.
3. There is no overriding influence of routinely practical analytical technology progress on variability measured according to the relative standard deviation of the analyte determination.
4. There is no overriding reduction in the relative variability performance over a broad multi-fold range of *c* as a function of the selected analytical technique beyond the obvious course of action that more sensitive detectors and sophisticated techniques, for example sample pre-concentration extraction, are selected for the determination of lower analyte absolute mass concentrations.

The Horwitz function can be considered as an objective basis for the acceptance criterion for variability in inter-laboratory trials, also referred to as test method Reproducibility. A complete summary can be found in Horwitz and Albert [6].

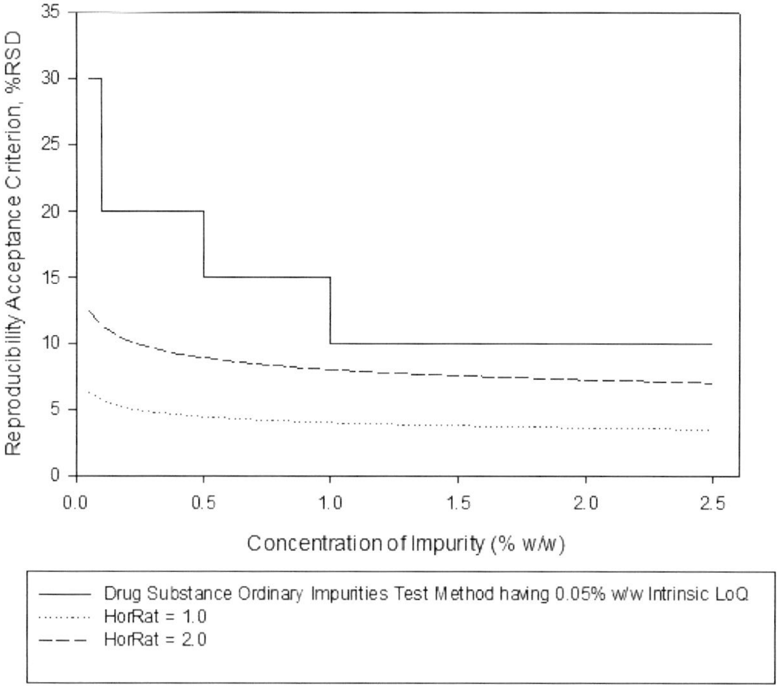

Figure 14.3 Example of a published inter-laboratory precision (Reproducibility) acceptance criteria [13] for Drug Substance and that corresponding to HorRat = 1.0 and HorRat = 2.0.

In addition and alternate to comparing the AbbVie test method performance to the Horwitz acceptance criterion, an arbitrarily selected example from a published [13] impurities test method Reproducibility acceptance criterion is presented in Figs. 14.3 and Figs. 14.4. Although we present only two specific cases, the Reproducibility acceptance criterion of Crowther et al. [13] is general, based on the assignment of specified %RSD(s) criterion respectively to analyte concentration ranges which are distinguished as multiples of the test method's *intrinsic* Limit of Quantitation. This approach sets the Reproducibility acceptance criterion as a function only of the test method intrinsic LoQ and therefore, in our view, can be regarded as a Signal to

Noise-based approach to setting impurities test method precision performance. If our interpretation is appropriate, then presumably the acceptable Signal to Noise at the intrinsic LoQ is that defined by ICH [1], namely Signal to Noise equal to and *not meaningfully exceeding* 10 to 1. Accordingly then, for any impurity test method, Reproducibility performance for quantitation very near the method intrinsic LoQ would be subject to an acceptance criterion of ≤30%RSD [13], regardless of the impurity absolute concentration corresponding to the intrinsic LoQ. As would be expected, Crowther et al., [13] indicate a series of more restrictive Reproducibility RSD(s) step wise as the concentration of analyte increases above the test method's intrinsic LoQ.

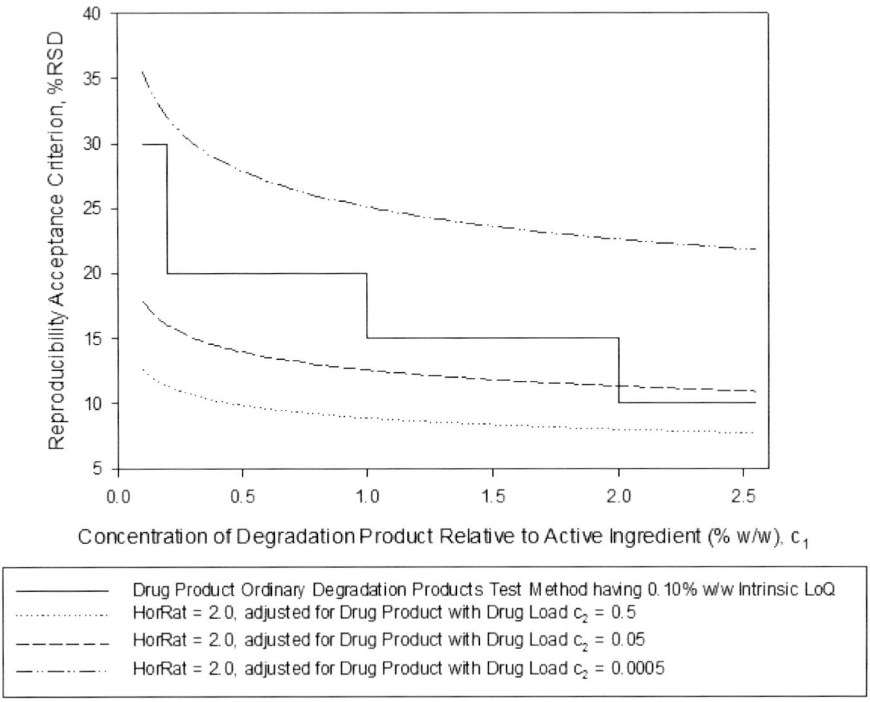

Figure 14.4 Example of a published inter-laboratory precision (Reproducibility) acceptance criteria [13] for Drug Product and those corresponding to Horwitz showing the dependence on formulation drug load fraction, c_2. Cases plotted for HorRat = 2.0 and c_2 fractions of 0.5, 0.05 and 0.0005.

In Fig. 14.4, our depiction of Crowther's et al. [13] drug product Reproducibility acceptance criterion in direct comparison with those of Horwitz must be viewed with caution. This is due to the fact that for the Horwitz criterion, the concentration axis, c, is to reflect the mass fraction of analyte in the whole matrix mass, whereas we expect certainly that Crowther's intended concentration scale for drug product degradation products is the mass fraction, or mass percent, of degradation product relative to the mass of active ingredient formulated.

As reflected in the examples plotted in Fig. 14.4, in our view a generalization of the Horwitz function can be considered when analyte concentration is expressed relative to the active ingredient mass, which is the expected basis of pharmaceutical impurity and degradation products reporting for Regulatory purposes.

The Horwitz mass fraction of analyte relative to the total matrix mass, c, can be expressed as the product of the fraction of impurity or degradation product mass to drug mass, c_1, times the fraction of drug mass to the total matrix mass, c_2. The Horowitz equation then becomes

$$\mathrm{RSD}_{\mathrm{R,\,Horwitz}} = 2c^{-0.1505} = 2(c_1 \cdot c_2)^{-0.1505}.$$

For drug substance, c_2 equals one, and by definition for drug product c_2 is less than one.

In Section 14.2.1 we introduce Horwitz's Ratio (HorRat) [6].

14.2.1 Horwitz's Acceptable Observed Test Method Reproducibility: $\mathrm{RSD}_{\mathrm{R,\,Observed}}$

The "Horwitz-Ratio" or "HorRat" describes the quotient [6]:

$$\text{Horwitz Ratio} = \text{HorRat} = (\mathrm{RSD}_{\mathrm{R,\,Observed}})/(\mathrm{RSD}_{\mathrm{R,\,Horwitz}}).$$

Horwitz allowed up to $2 \times \mathrm{RSD}_{\mathrm{R,\,Horwitz}}$ or HorRat ≤ 2.0 as the acceptable upper limit of variability, $\mathrm{RSD}_{\mathrm{R,\,Observed}}$

$$\text{HorRat} = \frac{\mathrm{RSD}_{\mathrm{R,\,Observed}}}{\mathrm{RSD}_{\mathrm{R,\,Horwitz}}} \leq 2.0$$

Therefore the upper limit of acceptable inter-laboratory precision (Reproducibility) is:

$$\mathrm{RSD}_{\mathrm{R,\ Observed}} \leq 2\mathrm{xRSD}_{\mathrm{R,\ Horwitz}} = 4c^{-0.1505} = 4(c_1 \cdot c_2)^{-0.1505}.$$

Even though a rigorous justification of his cited 0.5 ≤ HorRat ≤ 2.0 acceptance range is not explicitly provided by Horwitz, one can rationalize acceptable HorRat values falling within the specified acceptance range 0.5 ≤ HorRat ≤ 2.0 as in likeness to the statistical principle for calculation of confidence intervals around the estimate, s, of the true standard deviation, σ, of a population. Generally the standard deviation estimate of the population true value is associated with a confidence interval calculated from the Chi-squared distribution according to a chosen specified confidence and the finite degrees of freedom of the precision experiment. See Table 4 in Horwitz and Albert [6].

14.2.2 Evaluation of AbbVie Test Method Validation Repeatability, Intermediate Precision Performance

The selected AbbVie test methods evaluated included chromatographic ordinary and targeted impurities, and minor formulation component (i.e., antioxidant, functional excipient) test method validations. Here, a targeted impurity can be distinguished as that, for example, requiring control to a very stringent specification limit, such as a carcinogenic or mutagenic impurity [11]. Ordinary impurities, for example, are those which have undergone specific toxicological qualification, and, or impurities which are to be quantified if present at or above the respective ICH Reporting Threshold [14, 15], and therefore do not require as highly sensitive analytical test methodology. New Chemical Entity Drug Substance (DS), Drug Substance Manufacturing Intermediates (DS-I), Drug Product Intermediates (DP I) and Drug Product (DP) test methods were included. The evaluated test methods composition break down is summarized as follows:

Impurity-degradation product analytical method type	Number of instances included
HPLC-UV	53
HPLC-MS	6
GC-FID	11
GC-MS	2
Subtotal of (DS + DS-I) test methods	45
Subtotal of (DP + DP-I) test methods	27
Total of Number of test methods	72
Total number of analyte Concentration levels in evaluation	813

RSDs corresponding to impurity accuracy study percentage recoveries were also included, which are typically performed intra-laboratory. Therefore the data is composed of inter-laboratory precision, combined with data from accuracy studies, the latter of which are properly described as "repeatability type" designs. The test method variability assessment derived from the RSDs from multi-level analyte recovery studies (Accuracy validation) would be expected to have reduced variability as compared to the variability that would be expected from an inter-laboratory determination conducted at the same concentration level of analyte. The data fractional composition of RSDs derived from test method Accuracy validation is 644 of 813, or about 79%, and the balance of 21% is data generated from inter-laboratory precision.

For the formulated products, the range of drug load fractions (w/w) included in the data is 3.38×10^{-6} to 0.938, and the median drug load is 0.048.

The data fractional composition of impurity or minor component concentration(s) that were adjusted for respective drug product drug load in calculating the concentration, *c*, is 520 of 813, or about 64%.

In Fig. 14.5 there are no active ingredient potency test methods, for drug substance or for drug product, plotted in our data evaluation versus Horwitz. If the active ingredient assay test methods "class" were plotted, they would exhibit very distinct superiority versus HorRat ≤ 1.0 acceptance criterion, even without drug load adjustment for drug product active ingredient assay test methods.

The AbbVie test method data is plotted in Fig. 14.5 along with lines indicating a HorRat of 2.0 and the 95% Quantile Regression line in conjunction with its evaluation according to a statistical Quantile Regression (QR) analysis [16, 17]. The QR objective function line and the HorRat = 2.0 line are nearly superimposed for a high proportion of the data (i.e., for the quantile τ = 0.95). The vast majority of the AbbVie data is below both of the lines, with only 2% of the data above the HorRat = 2.0 line. Some of the points with HorRat > 2.0 are described in Section 14.2.3. Adequate precision performance against the criterion of HorRat ≤ 2.0 is supported by the quantile regression of the AbbVie data set.

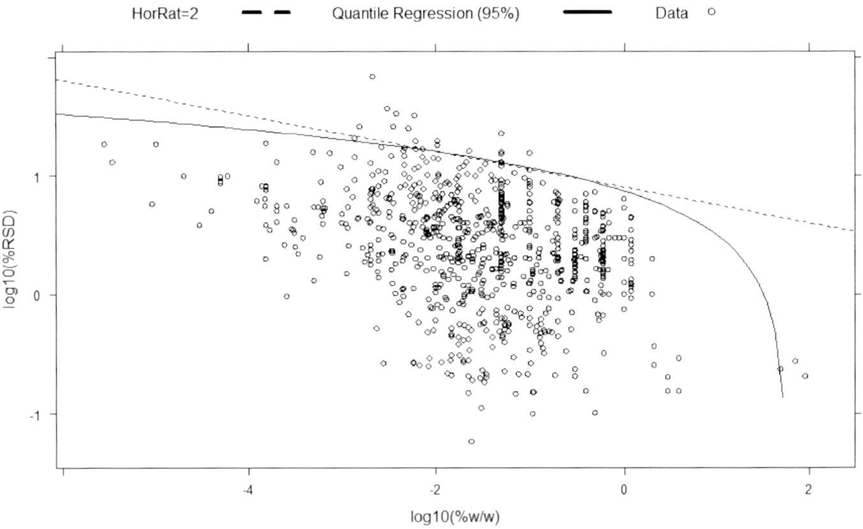

Figure 14.5 HorRat = 2.0 line and the quantile regression of the AbbVie data of seventy two impurity, degradation product or minor component test method precision validations. The analyte concentration, c (%w/w), is relative to the whole matrix mass.

14.2.3 Discussion and Conclusion

While a small fraction of the data RSDs (15 of 813) exceed the HorRat ≤ 2.0 (the Horwitz Acceptance Criterion), these instances generally exhibit an "attributable" cause. For example in one instance, the subject impurity UV Response Factor was not found repeatable inter-laboratory between different HPLC instrument manufacturer(s) and in another, the cause is attributed as intrinsic to the test method which employs a complex liquid-liquid extraction sample preparation. In all cases, the respective test method performance was suitable for its intended Regulatory use.

The following conclusions can be drawn from the data evaluation versus Horwitz (HorRat ≤ 2.0 Acceptance Criterion).

The test method validation precision performance for both drug substance and drug product targeted and ordinary impurities and degradation products, and formulation critical excipient component(s), is predominately in-line with, or superior to, HorRat ≤ 2.0 acceptance criterion.

An advantage of HorRat ≤ 2.0 is principally that the slope of the Horwitz function (or equivalently the exponent of c expressed in power form of the Horwitz function) is an objective externally recognized basis for acceptable test method Reproducibility, i.e., as inter-laboratory co-validated Intermediate Precision or alternatively as test method transfer performance. The Horwitz Acceptance Criterion also offers an advantage that the precision acceptance criteria are defined as a continuous function of c. Additionally the Horwitz Acceptance Criterion offer a means to objectively define inter-laboratory Intermediate Precision (Reproducibility) Acceptance Criterion in particular for cases of "exceedingly low" drug load drug products, or for any drug product drug load, c_2, for that matter.

In general, however, the quality control and Regulatory reported level, and the respective specification limits of impurity, degradation product or formulated functional minor component would remain calculated and specified relative to the weight of active ingredient (or relative to the intended weight of formulated functional minor component), c_1, in the formulated drug product.

Example of Formulation Drug Load Adjusted Degradation Products Test Method Inter-laboratory Precision Acceptance Criteria

%RSD criteria for an example of $c_2 = 3.38 \times 10^{-6}$ weight fraction drug load formulation

(3.38 micrograms drug per gram formulation)

Impurity or Degradation Product Level (%w/w, Relative to Active Ingredient Weight, c_1)	HorRat ≤ 2.0 Acceptance Criterion for Co-Validated Intermediate Precision (or Reproducibility), %RSD Limit
0.50%	≤59%
0.10% (the drug product applicable ICH Reporting Threshold [15])	≤75%

Acknowledgments

The authors thank the many Abbott and AbbVie analytical chemists who endeavored to develop and validate the robust pharmaceutical test methods reflected by the test method validation data set characterized herein. The authors also thank AbbVie New Chemical Entity Analytical Chemistry R&D and Non-Clinical Statistics department senior management for their support of resources for this standing functional interface within the AbbVie Global Pharmaceutical R&D Development Sciences CMC function.

References

1. International Conference on Harmonisation of Technical Requirements for Registration of Pharmaceuticals for Human Use. Validation of Analytical Procedures: Text and Methodology Q2(R1), *Step 4*, November 2005.
2. Debesis, E., Boehlert, J. P., Givand, T. E., Sheridan, J. C. (1982) Submitting HPLC methods to the compendia and regulatory agencies. *Pharm. Tech.*, September 1982, pp. 120–137.
3. The United States Pharmacopeia, Chapter <1210>. http://app.uspnf.com accessed 22 September 2018.
4. Horwitz W., Kamps L. R., Boyer K. W. (1980) Quality assurance in the analysis of foods for trace constituents. *J. Assoc. Off. Anal. Chem.* 63(6), pp. 1344–1354.

5. Thompson, M., Lowthian, P. J. (1997) The Horwitz function revisited. *J. AOAC Int.,* 80(3), pp. 676–680.
6. Horwitz, W., Albert, R. (2006) The Horwitz ratio (HorRat): A useful index of method performance with respect to precision. *J. AOAC Int.* 89(4), pp. 1095–1109.
7. Ermer, J., et al. (2005) Precision from drug stability studies. Investigation of reliable repeatability and intermediate precision of HPLC assay procedures. *J. Pharm. Biomed. Anal.,* 38, pp. 653–663.
8. Hofer J. D., Olsen B. A., Rickard E. C. (2007) Is HPLC assay for drug substance a useful quality control attribute? *J. Pharm. Biomed. Anal.,* 44, pp. 906–913.
9. International Conference on Harmonisation of Technical Requirements for Registration of Pharmaceuticals for Human Use. ICH Harmonized Guideline. Impurities: Guideline for Residual Solvents Q3C(R6), Current *Step 4* version, October 20, 2016.
10. International Conference on Harmonisation of Technical Requirements for Registration of Pharmaceuticals for Human Use. ICH Harmonized Guideline. Guideline for Elemental Impurities Q3D, Current *Step 4* version, December 16, 2014.
11. International Conference on Harmonisation of Technical Requirements for Registration of Pharmaceuticals for Human Use. ICH Harmonized Guideline. Assessment and Control of DNA reactive (Mutagenic) Impurities in Pharmaceuticals to Limit Potential Carcinogenic Risk. M7(R1), Current *Step 4* version, March 31, 2017.
12. ICH Topic Q3B, Impurities in New Medicinal Products, Step 4, Consensus Guideline, November 6, 1996.
13. Crowther J. B., Jimidar M. I., Niemeijer N., Salomons P. (2000) Qualification of laboratory instrumentation, validation, and transfer of analytical methods. In *Analytical Chemistry in a GMP Environment: A Practical Guide.* eds., Miller J. M., Crowther J. B., Wiley, New York, pp. 423–458.
14. International Conference on Harmonisation of technical requirements for registration of pharmaceuticals for human use. ICH harmonized tripartite guideline. Impurities in New Drug Substances Q3A(R2), Current *Step 4* version, October 25, 2006.
15. International Conference on Harmonisation of technical requirements for registration of pharmaceuticals for human use. ICH harmonized tripartite guideline. Impurities in New Drug Products Q3B(R2), Current *Step 4* version, June 2, 2006.

16. Gilchrist, W. G. (2000) *Statistical Modelling with Quantile Functions*, Chapman & Hall/CRC Press.
17. Koenker, R., Hallock, K. F. (2001) *J. Econ. Perspect.* 15(4), pp. 143–156.

Chapter 15

Method Development Strategies for Ion Exchange Chromatography Using pH Gradient Separation for mAbs, Antibody Drug Conjugates (ADCs) and Other Complex Bio-Pharmaceuticals

Norman L. Fischer and Scott Allen

Biotherapeutics Pharm Sci, Pfizer Inc.,
700 Chesterfield Parkway West, Chesterfield, Missouri 63017, USA

Norman.L.Fischer@Pfizer.com

15.1 Introduction

Recombinant proteins and monoclonal antibodies (mAbs), although relatively complex, are becoming even more complex with the advent of mAb conjugates (mAb + conjugated peptide or protein) and antibody drug conjugates (ADCs) consisting of a mAb + conjugated small molecules. These more complex molecules are typically characterized by an array of orthogonal analytical

Chromatographic Method Development
Edited by Gregory K. Webster and Laila Kott

ISBN 978-981-4800-53-2 (Hardcover), 978-0-429-20172-1 (eBook)
www.jennystanford.com

and biochemical methods which are repeatedly used as part of the ongoing control system that ensures product quality and manufacturing consistency throughout the commercial life of the product [1, 2]. Ion exchange chromatography (IEC) has long been one of the primary separation modes used for characterization and stability monitoring of the charge heterogeneity within these molecules. This approach has been used to help identify and characterize product heterogeneity, including C-terminal processing of lysine residues [3–5], N-terminal pyroglutamate formation [6, 7], deamidation [8, 9] glycation (non-enzymatic glucose addition) [10], amino acids sequence variation [4], and non-covalent complexes [11]. With the increasing complexity of these types of biological therapeutics, subsequent improvements in separation technology are becoming necessary to meet the additional challenges associated with the characterization and stability monitoring of these molecules.

IEC chromatography represents a commonly used method for profiling charge heterogeneity of mAbs and ADCs, and is the most suitable approach for purification of charged variants to support these characterization and validation activities. A 2008 review of charge heterogeneity within mAbs, by Vlasak and Ionescu [12], called cation exchange chromatography the gold standard for charge-sensitive antibody analysis. Multiple approaches to this type of separation exist including salt gradient, chromatofocusing, and pH gradient. While salt gradient separations are most prevalent, and may demonstrate sufficient resolving power and robustness, they tend to be product specific and can be time consuming to develop, requiring optimization of multiple factors including mobile phase buffer and pH, column, mobile phase additives, and salt concentration gradient. Additionally, the sensitivity, resolution, and reproducibility of these salt gradient separations may be unknowingly compromised due to several factors such as buffer choice, buffer concentration, and subsequent pH adjustment.

The approaches discussed in this chapter will focus on strategies around mobile phase selection for the development of pH gradients. This chapter will also cover the evaluation and preparation of mobile phases, and incorporate molecule information, such as molecule type and pI into method development of pH gradients. The theoretical pI of a molecule is determined from the number of acidic and basic residues in the molecule and is representative of the pH at which the molecule has a net charge of zero. In theory, during pH gradient separations, the pH at which a molecule is eluted from the stationery phase is similar to the pI of the molecule. Additionally, in this chapter, mobile phase gradient strategies will also be discussed which will provide a basis for developing both sensitive and robust separation of the more complex molecules such as mAbs, ADC's, and fusion proteins.

15.1.1 Salt and pH Gradient IEC Separations

The basic principle of IEC separation, as it relates to this chapter, uses electrostatic interactions between charged variants (of a bio-molecule) and an oppositely charged stationery phase (column). The charged variants of the bio-molecule are bound, to the column, via electrostatic interactions which are facilitated by the mobile phase pH and ionic strength. The charged variants are eluted using mobile phases with increasing levels of counter ions to either displace the molecule (salt gradient) or alter the charge of the molecule (pH gradient) to enable separation of the charged variants from the stationery phase.

Ion exchange chromatography can be accomplished with mobile phases using salt gradients at a fixed pH, a pH gradient, a single-buffer ionic strength gradient, or various combinations of each. In general, the distinction between salt gradient at a fixed pH and a pH gradient is as follows:

Salt Gradient: The salt gradient uses mobile phases at a fixed pH with increasing levels of a counter ion, typically sodium or chloride from NaCl, to displace the molecule from the stationery phase.

pH Gradient: The pH gradient uses a change in mobile phase pH to alter the charge state of the molecule in order to facilitate its release from the stationery phase. However, depending on how the mobile phases are prepared, the elution of the charged variants, in pH gradient separations may occur through a combination of both a change in molecule charge state and displacement mechanisms similar to salt gradients. In this chapter, we will focus on method development of pH gradient and single-buffer ionic strength separations due to their flexibility and ease of development.

15.2 Approaches to PH Gradient Separations

This chapter will focus on IEC method development using pH and ionic strength gradients, since these strategies offer great flexibility and represent the simplest approach with regard to method development. The use of pH and ionic strength gradients also provides a good foundation for developing more appropriate mobile phases relative to the needs of the molecule. This strategy starts with a dilute buffer solution suitable for retaining the molecule, then gradually shifting the charge of the mobile phase, in the form of pH and or ionic strength increases, as necessary, to facilitate the separation. In this way, the addition of unnecessary ions, which can compromise the resolution of the separation, is limited. This simple approach to buffer selection and mobile phase development will provide the basis for sensitive yet robust IEC separations of these more complex molecules.

The development of pH gradients for use in IEC separations can be achieved in several different ways.

15.2.1 Individual Acidic and Basic Buffers/No pH Adjustment

Definition: The use of multiple buffers covering a wide pH range (e.g., 4.5–10.5) consisting of a set of acidic buffers (e.g., acetic acid, 2-(N-morpholino)ethanesulfonic acid (MES), (4-(2-hydroxyethyl)-1-piperazineethanesulfonic acid) (HEPES), ((1,1-dimethyl-2-hydroxyethyl)amino)-2-hydroxypropanesulfonic acid (AMPSO), and N-cyclohexyl-3-aminopropanesulfonic acid CAPS) and a corresponding set of basic buffers (e.g., 1-methylpiperazine, pyridine, Bis-Tris propane, AMP, piperidine) which cover a similar pH range.

The advantage of this approach is that a relatively wide pH range can be achieved which is useful as a platform method, or for separation of mAb mixtures with a wide variance of pIs.

Additionally, for this approach using both acidic and basic buffers, no pH adjustment is required. A disadvantage of this approach is that care must be taken to ensure that ionic strength of the individual buffers within the mobile phases are minimized to avoid compromising the resolution of the separation due higher ionic buffer strength from the combined buffers. Another disadvantage of this approach is that it requires multiple acidic buffers and multiple basic buffers, making mobile phase preparation a more complex process requiring multiple acidic and basic buffers. Additionally, more experiments are required during development to insure that the buffering strength of the acidic and basic buffers is appropriately balanced so that a reasonably stable pH gradient is achieved. In general, the use of a wider pH range may compromise the resolution of the separation, and will generally require the use of shallower gradients to achieve adequate separation. This is due to the increased rate of pH change across the stationery phase associated with the wider pH gradient. However, this drawback is easily overcome by reducing the pH range of the gradient. The advantages of narrower pH gradients will be discussed later in this chapter.

15.2.2 Combined Buffers with pH Adjustment

Definition: The use of multiple buffers covering a wide pH range consisting of a set of either acidic or basic buffers combined in a single bulk mobile phase which is then split and pH adjusted to cover the buffering range of the buffers in the mobile phase.

This approach is presented as an alternative to the approach listed in option one. In this case, only the acidic or basic buffers would be used for both mobile phases (e.g., Acetic Acid, MES, HEPES, AMPSO, CAPS at pH 4.5 and pH 10.5). This approach would significantly reduce the number of buffers required for the mobile phase preparation. The advantage of this approach over the first is that the mobile phase development and preparation are simpler, since the same buffers are used for both mobile phases. As with the first option, the use of a wider pH range is useful as a platform method, or for separation of mAb mixtures with varying pIs. The disadvantages of this approach are similar to those listed in option 15.2.1. Also, care must be taken to ensure that the ionic strength of the individual buffers within the mobile phases is minimized to avoid compromising the resolution of the separation due higher ionic strength from the combined buffers and subsequent pH adjustment. As with option 15.2.1 the drawbacks associated with a wider pH gradient range are easily overcome by reducing the pH range of the gradient in the mobile phases. This approach is outlined in the following section.

15.2.3 Buffers Covering a Narrower pH Range

Definition: The use of fewer buffers covering a narrower pH range consisting of a single acidic and or basic buffer in each mobile phase so that the pH and buffering range of the mobile phases covers the pI of the molecule.

This option represents a scaled down version of the first two options. The advantage of this approach over the first two is that the mobile phase development and preparation are the simplest, since fewer buffers are used for the mobile phases. The mobile phases can use either a single buffer which is pH adjusted to cover the desired pH range, (e.g., HEPES adjusted to pH 6.8 and pH 8.2), or an acidic buffer in one mobile phase and a basic buffer in the other mobile phase to generate the pH gradient. Because of the narrower pH range, this approach will generally yield better resolution. The primary disadvantage of a narrower pH range is this separation may not be suitable for separating multi-mAb mixtures if all of the mAb pIs do not fall within the pH range of the buffers.

In this chapter, we will provide illustrations from these options with the primary focus on the development of narrower pH gradients (option 15.2.3), as well as ionic strength gradients since these approaches are the simplest, from a development standpoint, and may yield greater resolution and sensitivity than the other approaches. We will also provide further discussion on when the use of ionic strength gradients is most appropriate.

15.2.4 General Concept of pH Gradient Separations

Figure 15.1 provides a simple illustration of the concept of pH gradient separations. Using a pH gradient approach for cation exchange, the balance of positively charged H^+ ions from the acidic buffer in mobile phase A is shifted toward increasing levels of electronegative ions from the basic buffers in mobile phase B. This charge shift occurs through the delivery of the pH gradient via the HPLC pump and serves to alter the charge of the bound molecule on the stationery phase, from positive to negative for cation exchange and negative to positive for anion exchange, thus facilitating elution and separation of the charged variants. For cation exchange separations, the starting pH of mobile phase A will be below the pI of the molecule, and the pH of mobile phase B will be above the pI of the molecule. For anion exchange, the opposite is true, with the pH of mobile phase A being above the pI of the molecule, and the pH of mobile phase B below the pI of the molecule.

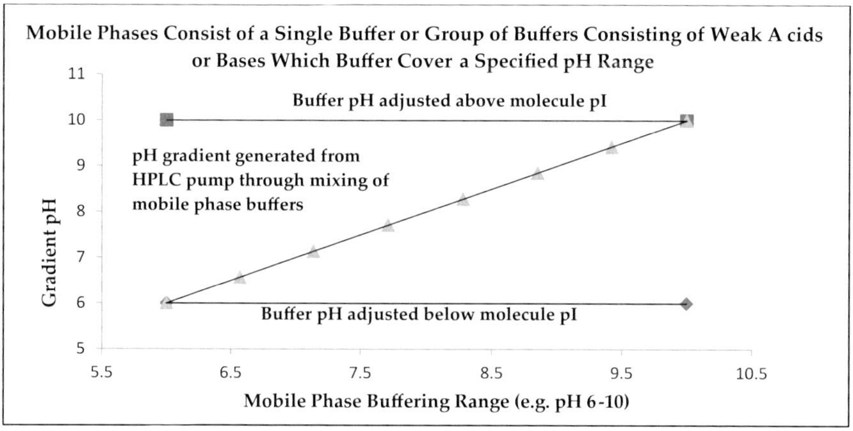

Figure 15.1 General concept of pH gradient separations.

As with salt gradients, the addition of ions through pH adjustment can adversely impact resolution of the separation with pH gradients as well.

15.3 Impact of Ions on IEC Separations

For IEC separations in general, an inverse correlation exists between mobile phase ionic strength and molecule retention time. For example, an increase in ionic strength of the mobile phases will decrease molecule retention time while a decrease in mobile phase ionic strength will increase molecule retention time. The ion concentration in the mobile phase must be sufficient to elute the molecule, but not so high as to cause poor retention or resolution of the charged variants. Excessive ions in the mobile phase will necessitate the use of shallower gradients to compensate for the increased ion concentration in order to achieve adequate separation. The need for use of shallow gradients (< 0.5%/min), to achieve adequate separation, may be indicative of excessive ions in the mobile phases that could result in method robustness problems including significant retention time shifts and changes in or loss of resolution. These problems with method robustness are, in most cases related to a combination of excessive ions in the mobile phase along with variance in mobile phase preparation, and not the inability of the pump to accurately deliver a mobile phase gradient of < 0.5 %/min.

15.3.1 Impact of Ions from pH Adjustment on Salt Gradient Separations

The retention time shift associated with variations in mobile phase preparation is illustrated in the experiment shown in Fig. 15.2, for a cation exchange separation using a salt gradient. The molecule in this experiment is a mAb with a theoretical pI of ~8.1. In this experiment the mobile phases consisted of 10 mM HEPES, pH 8.0 ± 0.1 using 6N NaOH (to adjust pH) for mobile phase A, and mobile phase B the same as A with 200 mM NaCl added. The gradient slope was 1%/min.

In this method, the retention time shift associated with the allowable pH adjustment (pH 8.0 ± 0.1) in the method could result in a retention time shift of approximately 6 min, from 26–32 min. Additionally, the retention time shift between pH 7.8 and 8.2 is also associated with a loss in resolution of the acidic species. This experiment highlights the importance of maintaining control of ion concentrations in the mobile phases which are added during pH adjustment of the mobile phases. This control is essential to develop robust methods which deliver consistent results. The following experiment illustrates the impact of additional of ions from pH adjustment for pH gradient separations.

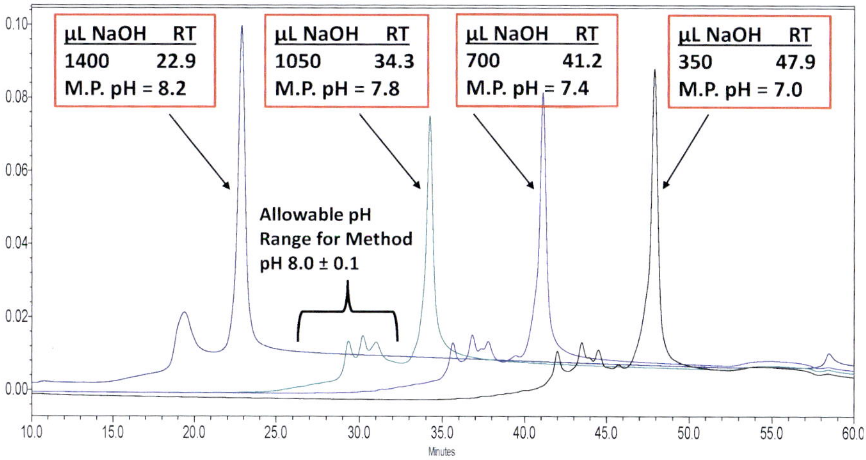

Figure 15.2 Impact of ion level in mobile phases associated with pH adjustment.

15.3.2 Impact of Ions from pH Adjustment on pH Gradient Separations

In this experiment, four molecules were separated with identical IEC pH gradient methods using pH adjusted buffers and non pH-adjusted buffers that covered a comparable pH range. The purpose of this experiment was to evaluate the impact of ions from pH adjustment on the resolution of the separation.

Three IgG2 mAbs and one IgG1 mAb-conjugate were evaluated in this experiment. The theoretical pIs of these molecules (labeled as mAb 1–4) range from 7.7–8.1.

Three buffers, listed in Table 15.1, were used for this experiment. Two of the buffers were acidic and one was basic. The buffering range for both pH adjusted and non-pH adjusted mobile phases was comparable ranging approximately from pH 6.5–9.5. The mobile phase preparations were evaluated in the cation exchange (CEX) separation mode. In this experiment, 2 combinations of mobile phases were evaluated as listed in Table 15.2. Combination 1 uses pH adjustment and combination 2 does not use pH adjustment. Mobile Phase A for both combinations is the same, consisting of 5 mM HEPES and 5 mM AMPSO, with no pH adjustment. Mobile Phase B in combination 1 is 5 mM HEPES and 5 mM AMPSO, pH adjusted to 9.3 with 2.5N NaOH.

Mobile Phase B in combination 2 is 5 mM Bis-Tris propane without any pH adjustment. The sodium concentration associated with the addition of NaOH to mobile phase B, in combination 1, amounts to 9.4 mM. The mobile phase combinations are shown in Table 15.2. For each mobile phase combination, a pH titration curve was generated as shown in Fig. 15.3. The pH titration curves for combinations 1 and 2 are nearly identical, and if evaluated solely on pH gradient, one would expect that the mobile phases with and without pH adjustment would yield similar chromatographic profiles when used with the same method.

Table 15.1 Mobile phases for pH adjustment experiment

Mobile phase #	Buffer description
Mobile phase 1	5 mM HEPES, 5 mM AMPSO, pH adjusted with 3.75 mL of 2.5N NaOH
Mobile phase 2	5 mM Bis-Tris propane, with no pH adjustment
Mobile phase 3	5 mM HEPES, 5 mM AMPSO, with no pH adjustment

Table 15.2 Mobile Phase combinations to assess influence of pH adjustment with NaOH on separation

Combination #	pH adjustment	Mobile phase A	Mobile phase B
Combination 1	Yes	Mobile phase 3	Mobile phase 1
Combination 2	No	Mobile phase 3	Mobile phase 2

The column used in this experiment is a Dionex ProPac WCX-10, 4 mm × 250 mm, 10 μm. The chromatographic profiles for separation of mAbs 1–4, with all mobile phase combinations are shown in Fig. 15.4. The separation with combination 2 (non-pH adjusted) is consistently better than that obtained with combination 1 (pH adjusted).

This result suggests that the resolution difference between pH adjusted and non-pH adjusted mobile phases may be related to the addition of ions via pH adjustment of the mobile phases and not a difference in the pH gradient. In this case, the pH adjustment with NaOH may be facilitating molecule displacement similar to a salt gradient, in addition to altering the molecule charge on the stationery phase via the pH gradient. Also, this result further suggests that minimizing the addition of ions from pH adjustment in the mobile phases is a plausible strategy for significantly improving the separation.

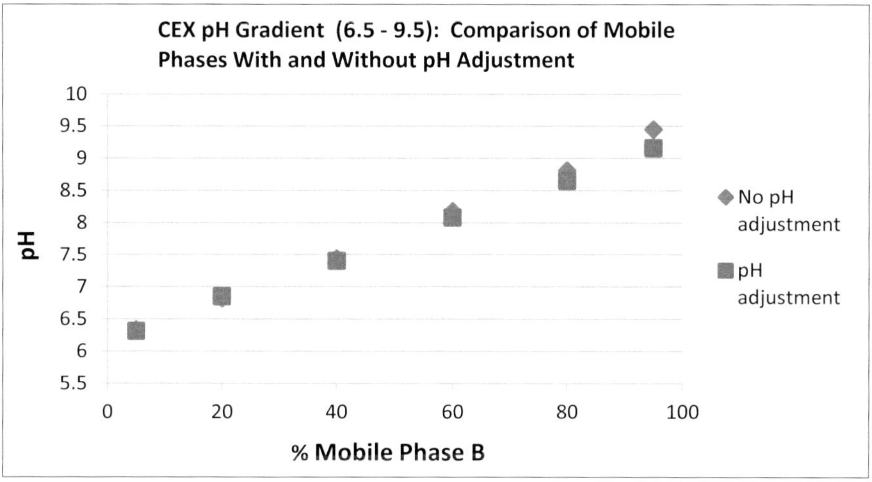

Figure 15.3 The pH titration curves for combination 1 (pH adjustment) and combination 2 (no pH adjustment) are shown. The pH gradients from the two approaches are nearly identical.

15.3.3 Impact of Ions from Buffer Concentration on pH Gradient Separations

The chromatographic profiles shown in Figures 15.2 and 15.4 A–D illustrate the potential impact of additional ions in the mobile phases associated with pH adjustment for both salt and pH gradients. The following experiment demonstrates the impact of buffer concentration on a pH gradient separation using cation exchange.

In this experiment, two sets of mobile phases were prepared without pH adjustment using acidic buffers in mobile phase A (HEPES and AMPSO) and a basic buffer in mobile phase B (Bis-Tris Propane). The first buffer set comprised mobile phases with buffer components at a concentration of 5 mM while the second buffer set comprised the same acidic and basic mobile phases with the buffer components at a concentration of 50 mM. The mobile phase compositions are listed in Table 15.3.

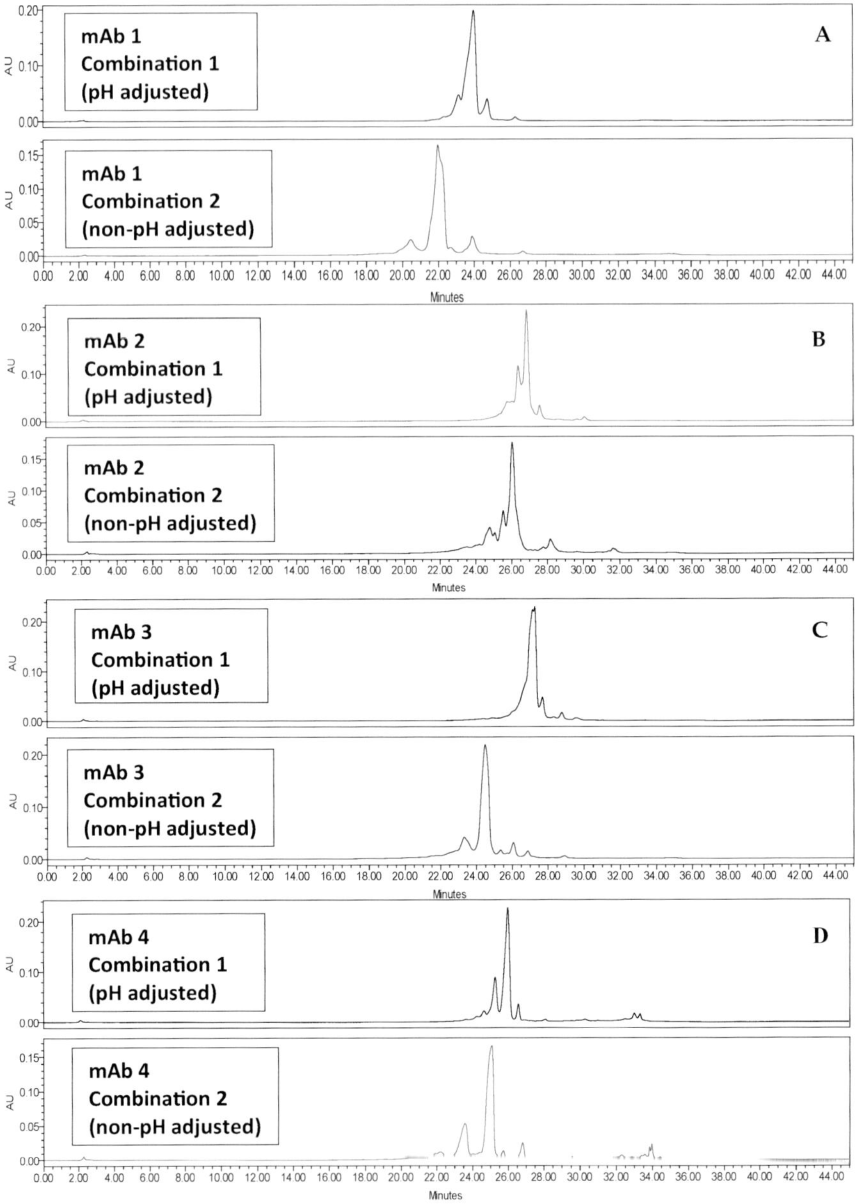

Figure 15.4 (A–D) Three mAbs (labeled 1–3) and one mAb conjugate (labeled 4) were compared using combination 1 and combination 2. The separation is improved for all molecules using combination 2, the non-pH adjusted buffers.

Table 15.3 Mobile phase composition for ionic strength experiment

Acidic components (A)	Basic components (B)
5 mM HEPES, 5 mM AMPSO Buffering range 6.8–9.7	5 mM BIS-TRIS propane Buffering range 6.3–9.5
50 mM HEPES, 50 mM AMPSO Buffering range 6.8–9.7	50 mM BIS-TRIS propane Buffering range 6.3–9.5

The column used in this experiment is a Dionex ProPac WCX-10, 4 mm × 250 mm, 10 μm. A gradient slope, of 2.4%/min, was used in this experiment. Mobile phase compositions were adjusted to yield buffer concentrations of 5, 10, 25, 40, and 50 mM for each buffer component of the mobile phase. Figure 15.5 shows the influence of mobile phase ionic strength on the resolution of charged variants from mAb 2 in Fig. 15.4B.

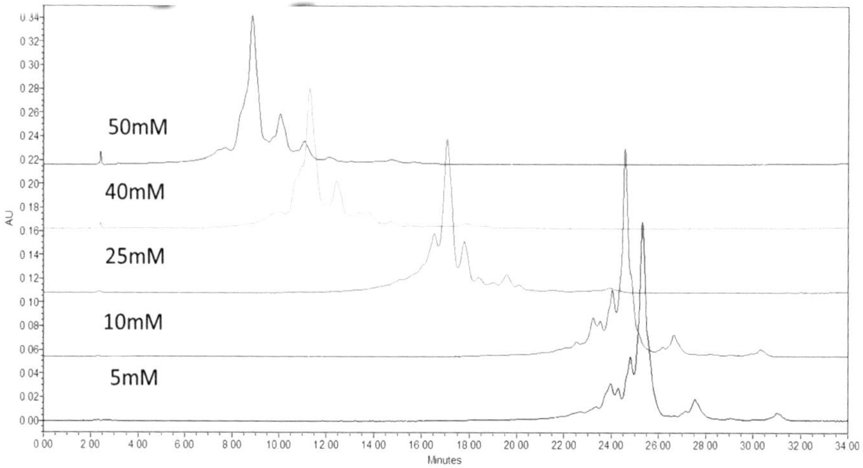

Figure 15.5 Chromatographic overlay showing influence of mobile phase ionic strength on a pH gradient separation of a monoclonal antibody.

For this experiment in general, the acidic components were best resolved at lower concentrations of (5–10 mM) while the basic species were resolved better at the higher concentrations (25–50 mM). As a result, another condition was evaluated at a buffer concentration of 16 mM, using the same gradient slope to evaluate resolution. The chromatography is shown in Fig. 15.6.

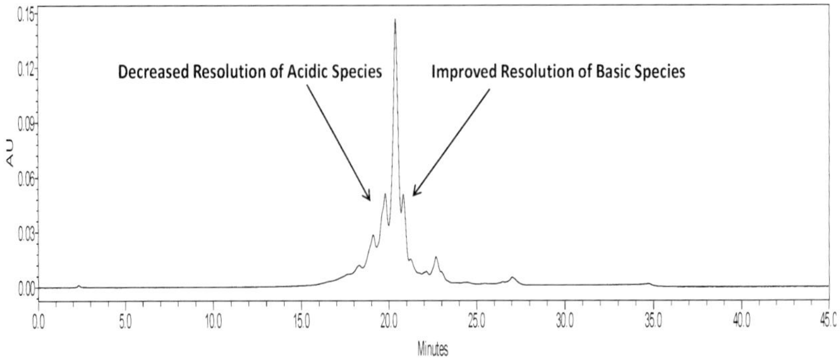

Figure 15.6 mAb separation at 16 mM with gradient slope of 2.4%/min, showing improved separation of the basic species relative to 10 mM buffer concentration.

The chromatographic profile from the 16 mM buffer concentration yielded the best overall separation and proved to be a good compromise between the lower and higher buffer concentrations, providing necessary balance between resolution of acidic and basic species with a gradient slope of 2.4%/min. Using the buffer concentration of 16 mM, the gradient slope was reduced to 1.2%/min, and lower buffer concentrations down to 5 mM were again evaluated. The profile from the optimized conditions for this separation is shown in Fig. 15.7. Optimization of buffer concentration and gradient slope resulted in acidic and basic species which are more resolved from the previous separation conditions.

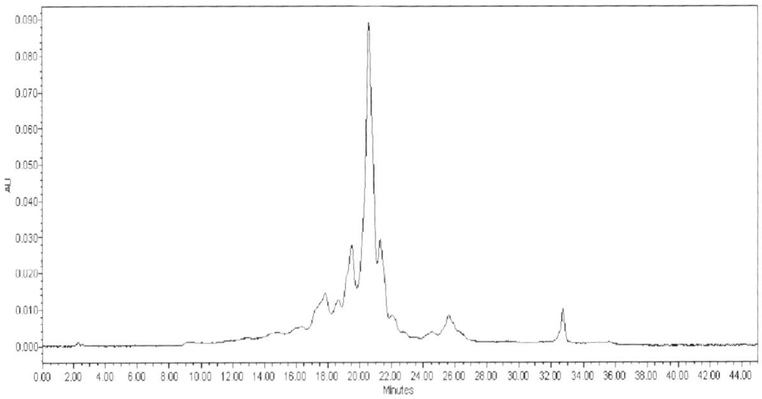

Figure 15.7 Optimized separation at 7 mM buffer concentrations and gradient slope of 1.2%/min.

This experiment demonstrates the relationship between species charge and ionic buffer strength and shows how buffer concentration and gradient slope can be used to optimize the separation of acidic and basic species. It further illustrates a strategy for identifying method conditions which yield suitable separation of both the acidic and basic species.

While further optimization activities could be performed, assessment of total area% for acidic, main, and basic peaks for the various separation conditions evaluated, are shown in Table 15.4. The results for the area% for acidic, main, and basic species show the robustness of this separation to be good at various buffer concentrations ranging from 5–16 mM, and gradient slopes of 1.2%–2.4%/min with different run times.

Table 15.4 Area% data for acidic, basic, and main peak at various separation conditions

Name	**Run time (Min)**	**Buffer conc. (mM)**	**Grad. slope (%/min)**	**%Acidic**	**% Main**	**% Basic**
mAb 2	45	16	2.4	35.64	37.63	26.73
mAb 2	75	16	1.2	34.40	37.46	28.14
mAb 2	75	10	1.2	36.80	36.68	26.52
mAb 2	75	7	1.2	36.12	37.33	26.56
mAb 2	75	5	1.2	37.06	36.89	26.05
mAb 2	45	7	1.2	36.72	37.36	25.92
			Mean	36.12	37.23	26.65
			Stdev	0.99	0.36	0.79
			%RSD	2.74	0.97	2.98

This experiment illustrates the influence of the mobile phase strength on the separation, and provides a viable strategy for evaluating and establishing the optimal conditions for separating the molecule. After identifying the appropriate mobile phase conditions, the HPLC pump can be used to assess different slope conditions to achieve optimal separation of both acidic and basic variants.

15.3.4 Further Evaluation of Buffer Concentration

A titration curve of the pH gradient, for the 5 mM buffers and the 50 mM buffers used in Fig. 15.5, is shown in Fig. 15.8. The titration curves show nearly identical pH values at various buffer concentrations throughout the curve, and if evaluated solely on pH gradient, one would expect that the mobile phases at both concentrations would yield similar chromatographic profiles when used with the same method.

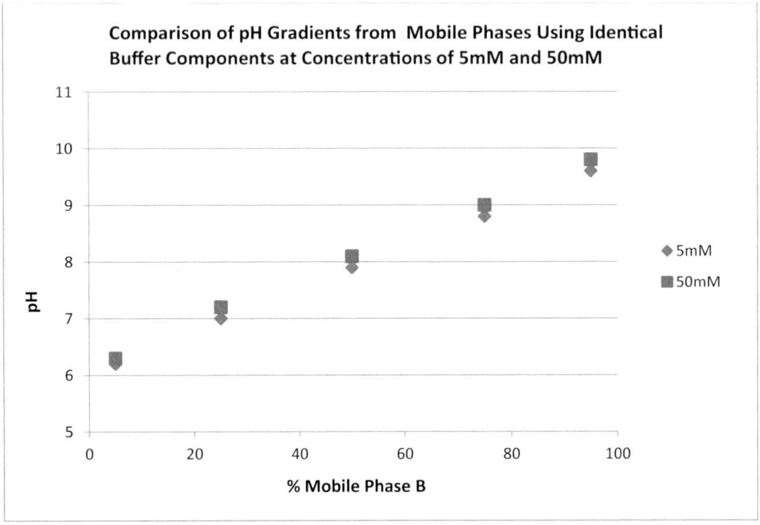

Figure 15.8 Titration curve showing pH gradients from mobile phases using the same buffers at 5 and 50 mM concentration.

The consistency of the titration curves at 5 and 50 mM buffer concentrations suggests the significant retention time shift (16.2 min in Fig. 15.5) is not related to differences in mobile phase pH but rather driven by the ionic strength of the mobile phase being delivered by the HPLC pump. The results from this experiment show that the ionic strength required to facilitate the elution of the molecule from the stationary phase, occurs in an earlier part of the pH gradient due to increased buffer concentration in the in the mobile phases.

The overlay in Fig. 15.5 shows the impact of mobile phase ionic strength on resolution of acidic and basic species. However, there is also a clear impact of ionic strength on the elution pH of the molecule. The information in Table 15.5 contains data from the separation shown in Fig. 15.5.

Table 15.5 Impact of buffer concentration on retention time and subsequent elution pH from Fig. 15.5

MAB # (TABLE 15.1)	Molecule Pi	Buffer Conc. (Mm)	Main Peak Rt, (Min) (Fig. 15.5)	%B (Based On Main Peak Rt)	Elution pH* (Fig. 15.5)
2	8.1	5	25.1	78	9.1
2	8.1	10	24.6	77	9.0
2	8.1	25	17.1	59	8.3
2	8.1	40	11.3	45	7.8
2	8.1	50	8.9	39	7.6

*Elution pH is calculated from main peak RT data in Fig. 15.5 with the gradient slope of 2.4%/min to determine the mobile phase composition and subsequent elution pH which is calculated from the average slope – intercept data from Fig. 15.8.

15.3.5 Buffer Composition

The buffers used in this experiment are zwitterionic, which are neutral molecules containing both positive and negative charges. The list and number of possible charged groups for each molecule are listed in Table 15.6.

Table 15.6 Potential charged groups for zwitterionic buffers (HEPES, AMPSO, and Bis-Tris propane

Buffer	Sulfonic acid (RSO_3H) groups/molecule	Hydroxy groups (OH)/molecule	Amine groups (N)/ molecule
HEPES	1	1	2
AMPSO	1	2	1
Bis-Tris propane	0	6	2

Based on the number of potentially charged groups per molecule in Table 15.6 along with the calculated mobile phase composition at the elution pH in Table 15.5, one can calculate the approximate concentration of charged groups at the point of elution based on main peak retention time. Figure 15.9 shows the sulfonic acid (RSO_3H) groups from Table 15.5 as a percentage of total charged groups in the mobile phases at the calculated elution pH. This result shows a correlation between increased sulfonic acid groups and lower calculated pH at the point of elution. This result is to be expected considering the mobile phase composition at the time of elution. However, this result is contrary to the general expectation with pH gradient separations where elution is expected at a pH near the pI of the molecule.

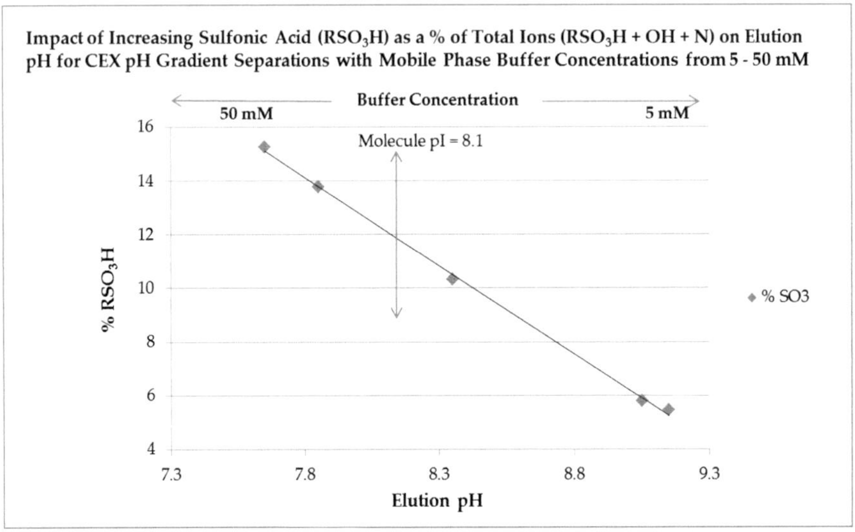

Figure 15.9 Impact of sulfonic acid concentration on elution pH at 5–50 mM buffer concentrations.

This data further confirms that pH is not the primary factor driving molecule elution in this separation. A closer look at the concentration of all possible charged groups is shown in Fig. 15.10.

This figure shows increased sulfonic acid groups at the lower retention times associated with increased levels of mobile phase A at the higher buffer concentrations. However, under these conditions, the concentration of the amine (N) and hydroxy (OH) groups are more prevalent due to the higher buffer concentrations along with the fact that the mobile phase comprises two buffers which both contain these groups. It is plausible that the increased levels of these groups are facilitating the decrease in retention time in this experiment and may further explain the result of decreasing elution pH as buffer concentration is increased. Given the fact that there are a finite number of charged groups in the sample matrix based on the primary structure and concentration of the molecule, one could speculate that only a certain number of oppositely charged groups in the mobile phase are necessary to alter the charge on the molecule to facilitate its elution earlier in the separation. This result further highlights the importance of the HPLC pump in combination with minimizing ions in the mobile phases, and thereby facilitating delivery of ions to the stationery phase at the appropriate rate to yield optimal resolution during the separation.

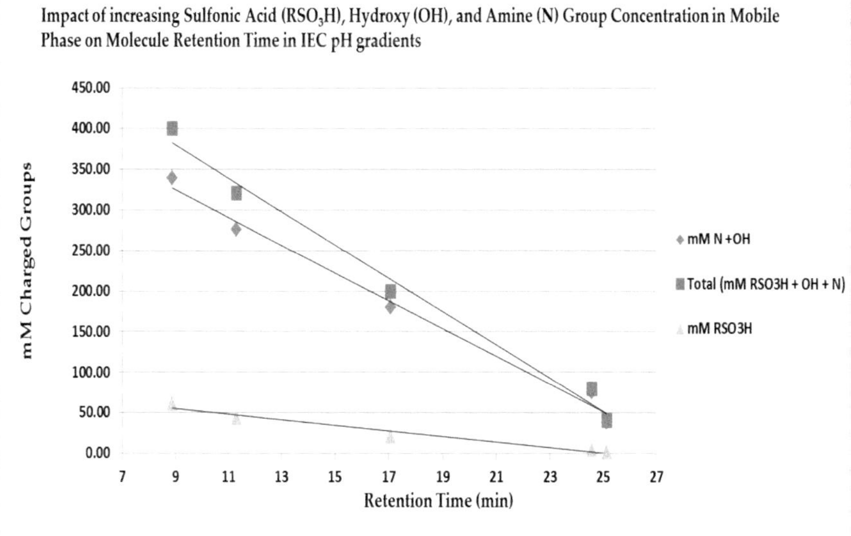

Figure 15.10 Impact of concentration of charged groups on molecule retention time in CEX pH gradient.

This section illustrates the potential impact of excess ions in the mobile phase associated with pH adjustment or excessive buffer concentration. It also provides a framework for experimental strategies to assess and optimize mobile phase concentrations along with gradient slope to achieve a robust separation. The following sections will begin to incorporate specific molecule information into the method development strategy in order to quickly develop a robust method with sufficient resolution of acidic and basic species.

15.4 Molecule Information and Method Development

Before starting the development of a pH gradient IEC separation method, it is important to gather information about the molecule in order to establish optimal separation conditions. The following information will be helpful during the method development process.

(1) pI of the molecule
(2) Molecule type: mAb (IgG1, IgG2) ADC (antibody drug conjugate), fusion protein, vaccine, etc.
(3) Molecular weight
(4) Presence and abundance of glycosylated species

These molecule characteristics can be useful in guiding several aspects of the method development including buffer selection for mobile phases, establishing buffer strength of mobile phases, selection of the appropriate separation mode (anion or cation), and whether or not to consider incorporating alternative separation strategies into the method. The information listed above will be discussed in further detail later in this chapter.

15.4.1 Mobile Phase Considerations for pH Gradient Separations

This section will look further at buffers and mobile phases with a focus on mobile phase composition/concentration as it relates to mAb subclass and pI. Mobile phase composition is most important, since this will have the greatest impact on resolution of the charged variants with pH gradient (IEC) separations. If care is not taken to select the appropriate buffer and optimize the mobile phases during method development, the separation may be compromised to the point where altering any of the other possible variables (e.g., stationary phase, sample diluent, column temperature, and gradient slope) will have minimal or no positive impact. Additionally, poorly developed mobile phases will result in methods which can be difficult to validate due the extreme requirements placed on the instrument to compensate for excessive ions in the mobile phases. A common example of this are methods which use very shallow gradients (<0.5%/min) in combination with mobile phases containing high levels of ions. In most cases, a small portion of the mobile phases eluting capacity is used for the separation, typically occurring within a small percentage of the gradient, over a longer period of time. As a result, a shallow gradient becomes necessary to compensate for the excessive ions in the mobile phase in order to achieve adequate resolution. Additionally, these methods may exhibit significant, run to run, shifts in retention time along with inconsistent peak resolution.

The first step in mobile phase development is choosing a buffer with a buffering range that will bracket the pI of the molecule being separated. The list of buffers for use in pH gradient separations shown in Tables 15.1 and 15.7, is by no means exhaustive, but provides some examples of suitable buffers for pH gradient separation which will cover the pH range of most mAbs.

Table 15.7 List of example buffers suitable for use in pH gradient separations

Buffer	pH range
MOPSO	6.2–7.6
HEPES	6.5–7.9
AMPSO	8.3–9.7
BIS-TRIS propane	6.3–9.5
Ammonium acetate	3.8–5.8
Ammonium formate	2.8–4.8

15.4.2 mAb pI, Subclass, and Buffer Concentration

The approach described in this chapter uses low molecular weight acidic and or basic buffers to generate a pH gradient. The buffers do not need to be pH adjusted if one of the buffers in the separation is acidic and the other is basic (e.g., HEPES, BIS-TRIS propane). However, if a single buffer is used for both mobile phases, then at least one of the mobile phases will need to be adjusted so that the mobile phase pH range covers the pI of the molecule. It is acceptable to adjust the pH in both mobile phases. However, the amount of acid or base used to adjust the pH of mobile phase should be minimized, keeping in mind that the mobile phase pH will need to bracket the pI of the molecule. Some examples of this will be shown later in this chapter. In either case, the pH is generated through the HPLC mixing of acidic and basic mobile phases.

The primary goal, as it relates to mobile phase development for IEC separations, is to minimize excessive ions in the mobile phases to improve the sensitivity and resolution of the separation. This is critical in developing a highly resolving, sensitive and robust method for separation of these more complex bio-pharmaceuticals. At this point, we will begin to discuss buffer concentration with regard to mAb subclass and molecular weight.

The isoelectric point (pI) of the molecule is a critical piece of information which is important to the buffer selection and should also be considered with regard to the separation mode (cation or anion) that is used. In theory, for pH gradient separations, the pI of a molecule should correlate somewhat with the elution pH of the molecule. However, the data in Table 15.5 shows the calculated elution pH at the various buffer concentrations to be well above and below the pI of the molecule. This indicates that buffer selection and concentration plays a greater role in molecule elution than buffer pH.

Additionally, there are other factors which dictate the elution pH of a molecule.

Figure 15.11 shows a graphical illustration from an actual separation of an IgG1 and an IgG2 mAb with identical theoretical pIs that were run with the same pH gradient method.

Table 15.8 shows the retention time and elution pH data from the illustration in Fig. 15.11.

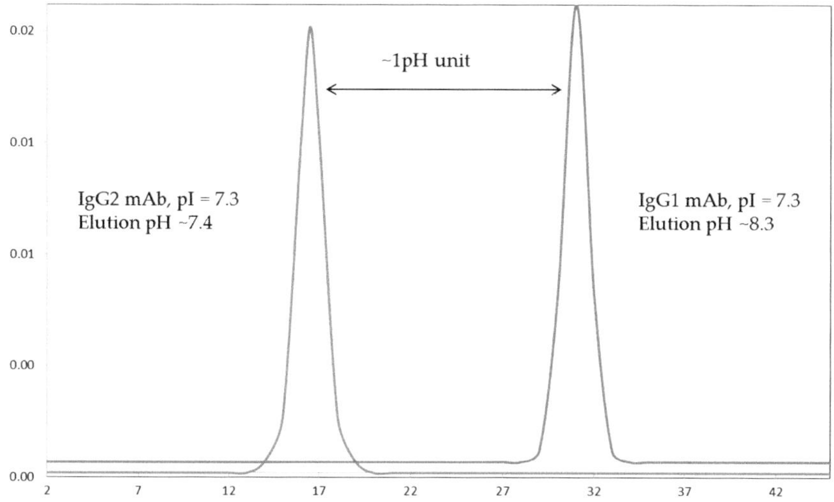

Figure 15.11 Graphic Illustration from an actual separation of two different mAbs with the same theoretical pI of 7.3 using the same pH gradient method (actual profiles not disclosed).

Table 15.8 Retention time and elution pH information from 2 different mAbs with the same theoretical pI of 7.3, which were run with Bis-Tris propane mobile phases at 5 mM

mAb subclass	pI	Buffer conc. (mM)	Acidic RT (min)	Main RT (min)	Basic RT (min)	Elution pH
IgG1	7.3	5 mM	30.0	31.0	32.5	8.3
IgG2	7.3	5 mM	12.5	16.5	25.0	7.4

In this experiment, both mAbs were run using the same pH gradient (cation exchange) separation method. The mobile phases used 5 mM Bis-Tris propane buffers, with mobile phase A pH adjusted to 6.3 and mobile phase B at pH 9.5 (non-pH adjusted). The pH gradient was run at 2%/min. The column used in this experiment is a Thermo MAbPac SCX-10, 4 mm × 250 mm, 10 μm.

The significant retention time difference between these 2 mAbs with identical pIs suggests that there are additional molecule related factors which are also playing a role in the retention of the charged variants during pH gradient separations. These factors can include differences such as primary structure, amino acid composition, and conformation of the molecule. In this case, the early eluting mAb is an IgG2 and the late eluting mAb is an IgG1. This data suggests that characteristics associated with mAb subclass may play a role on retention time of the molecule.

Figure 15.12 shows the graphical illustration, and Table 15.9 shows retention time and elution pH data from the same molecules in Fig. 15.11 and Table 15.8 separated using the same method, but at an increased buffer concentration consisting of 10 mM Bis-Tris propane.

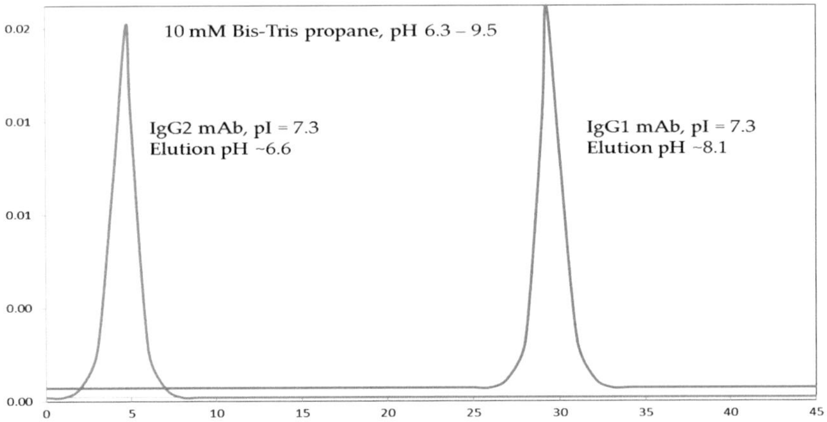

Figure 15.12 Graphic Illustration from an actual separation of two different mAbs with the same theoretical pI of 7.3, which were run using the same pH gradient (cation exchange) method with Bis-Tris propane mobile phases at 10 mM (actual profiles not disclosed).

Table 15.9 Retention time and elution pH information from two different mAbs with the same theoretical pI of 7.3, which were run with Bis-Tris propane mobile phases at 10 mM

mAb subclass	pI	Buffer Conc. (mM)	Acidic RT (min)	Main RT (min)	Basic RT (min)	Elution pH
IgG1	7.3	10 mM	27.75	29.25	30.5	8.1
IgG2	7.3	10 mM	3.75	4.75	6.75	6.6

In this experiment, the same two mAbs from Table 15.8 were run using the same pH gradient (cation exchange) method. However the mobile phase concentration was increased from 5 mM to 10 mM. Mobile phase A was adjusted to pH 6.3 and Mobile phase B was pH 9.5 (non-pH adjusted). The pH gradient was run at 2%/min using the same column as shown in Table 15.8.

Results from Tables 15.8 and 15.9 show that a 5 mM increase in buffer strength negatively impacted the retention of the IgG 2 mAb while having little impact on the retention of the IgG 1 mAb. Also, the resolution of the IgG2 peaks was negatively impacted by the 5 mM increase in buffer strength (actual profiles not shown). These results from Tables 15.8 and 15.9 show the influence of mAb subclass, on IEC separations, and how subsequent molecule retention and resolution is impacted by mobile phase concentration. Figure 15.13 shows the chromatographic profile comparison of two mAbs consisting of an IgG1 (pI 8.5) and an IgG2 (pI 8.7) that were run.

In this experiment, 2 different mAbs with theoretical pIs of 8.5 (IgG1) and 8.7 (IgG2) were run using the same mobile phase buffers, column, and separation method from Tables 15.8 and 15.9. Mobile phases used in the top overlay were at 5 mM while mobile phases used in the bottom overlay were at 10 mM. The buffer was Bis-Tris propane adjusted to pH 6.3 for mobile phase A, and pH 9.5 (non-pH adjusted) for mobile phase B. The pH gradient was run at 2%/min. Results from Fig. 15.11 show the best separation for the IgG1 occurring at 10 mM while the best separation for the igG2 occurred at a buffer concentration of 5 mM.

The comparison in Fig. 15.13 confirms the data shown in Tables 15.8 and 15.9, showing that mAb characteristics related to subclass have significant influence on molecule retention and resolution. Tables 15.8 and 15.9, along with Fig. 15.13, show that IgG 2's are more sensitive to smaller changes in buffer concentration, highlighting the need for consideration of mAb subclass and careful preparation of mobile phases. These results indicate that the mobile phase concentration for optimal separation of IgG 2's is around 5 mM and is lower than the concentration required for optimal separation of IgG 1's which is around 10 mM or higher.

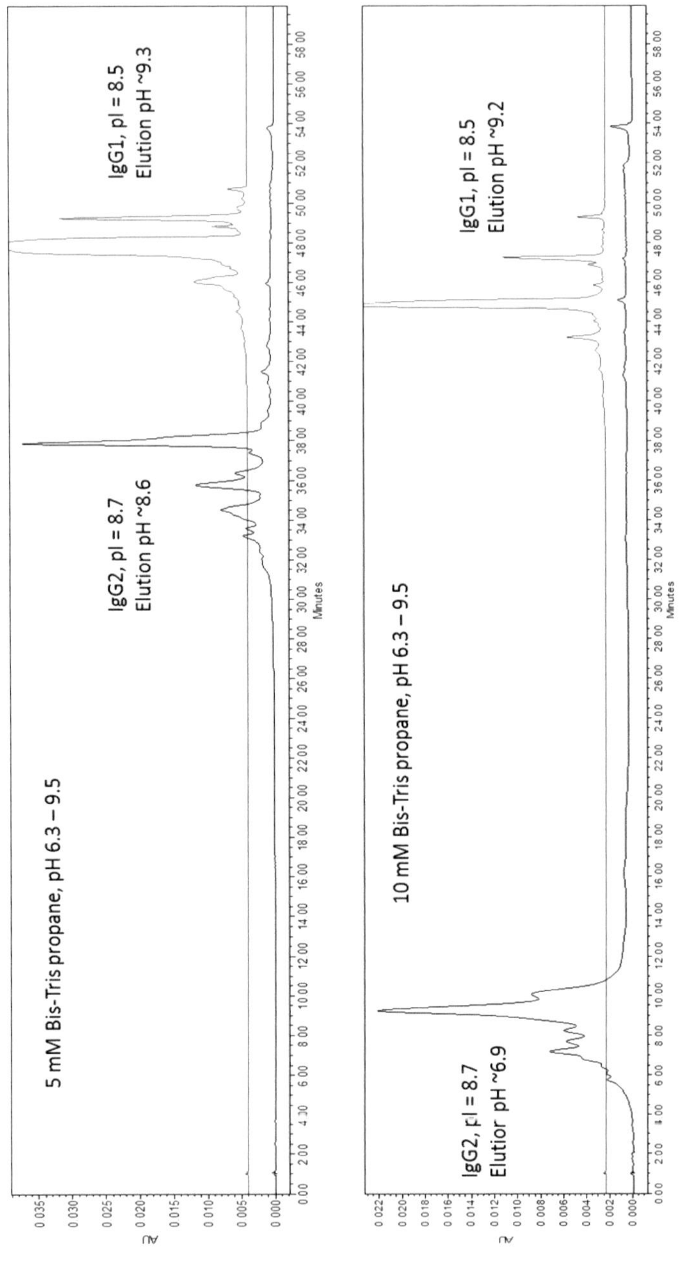

Figure 15.13 Chromatographic overlays of two different mAbs with theoretical pIs of 8.5 and 8.7. Both were run using the same mobile phases and pH gradient (cation exchange) method used in Table 15.8 and 15.9.

15.4.3 mAbs with Higher pIs and Ionic Strength Gradients

However, in addition to mAb subclass, overall mAb charge as it relates to the molecule pI can also influence the separation necessitating the need for alternative strategies related to mobile phase development and usage. Figure 15.14 shows another IgG 2 mAb with a pI of 9.1. In this experiment, a pH gradient with the same buffer concentrations in both mobile phases was evaluated, along with a combination approach using pH and ionic strength gradients (e.g., 5 mM in MP A, and 10 or 20 mM in mobile phase B). Mobile phases used 5, 10, or 20 mM Bis-Tris propane buffers: Mobile phase A = pH 6.3 and Mobile phase B = pH 9.5 (non-pH adjusted). The pH gradient was run at 2%/min. The column used in this experiment is a Dionex MAbPac SCX-10, 4 mm × 250 mm, 10 μm. In this case, we see that the mAb was not eluted from the column at buffer concentrations up to 20 mM in mobile phase B. This result is in direct contrast with the results from Tables 15.8 and 15.9 and Fig. 15.13 for an IgG 2 mAb. This result further illustrates the need for altering mobile phase composition based on molecule characteristics.

Figure 15.15 shows a comparison of the different buffer concentrations that were used in Fig. 15.14, for an IgG 1 molecule with a pI of 8.9. In this case the molecule was not eluted at buffer concentrations of 5 mM and 5–10 mM. At a buffer concentration of 5–20 mM, the molecule was eluted at the end of the gradient, indicating that conditions using higher buffer concentrations should be evaluated.

Figure 15.16 shows optimized separation for both molecules in Figs. 15.14 and 15.15. In this case, an ionic strength gradient was used which consisted of low concentration, non pH adjusted Bis-Tris propane in mobile phase A and higher concentration non-pH adjusted Bis-Tris propane in mobile phase B. The column used to generate the top profile in Fig. 15.16 is a Dionex MAbPac SCX-10, 4 mm × 250 mm, 10 μm. The column used in the bottom profile is a Waters Protein Pak, Hi Res, CM, 7 μm, 4.6 × 100 mm. The use of an ionic strength gradient resulted in greatly improved separations for both molecules. The advantage of the ionic strength gradient applied in this manner is that it facilitates a very efficient delivery of ions to the column due to the simplicity of the mobile phases.

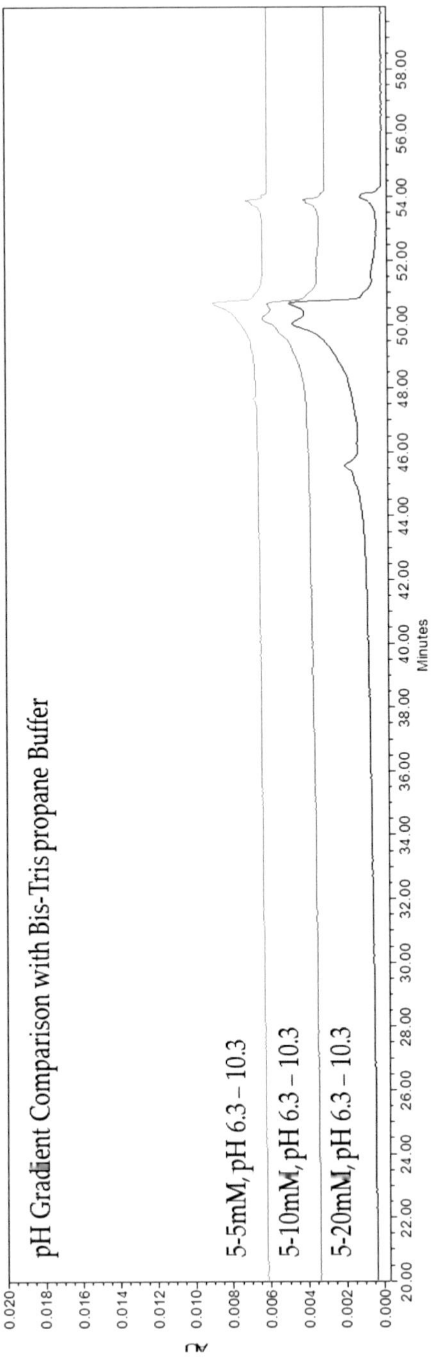

Figure 15.14 Chromatographic overlay of a single mAb (IgG 2) with a theoretical pI of 9.1 run using a 5 mM pH gradient, or a 5–10 or 5–20 mM pH/ionic strength gradient (combination) in cation exchange separation mode.

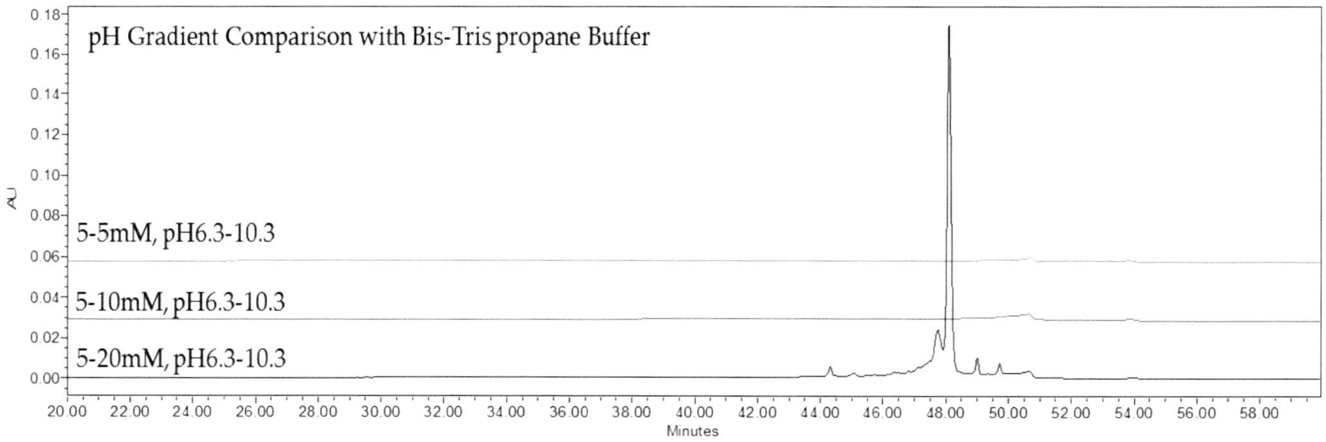

Figure 15.15 Chromatographic overlay of a single mAb (IgG 1) with a theoretical pI of 8.9 run using a 5 mM pH gradient, a 5–10 or 5–20 mM pH/ionic strength gradient (combination) with cation exchange separation mode.

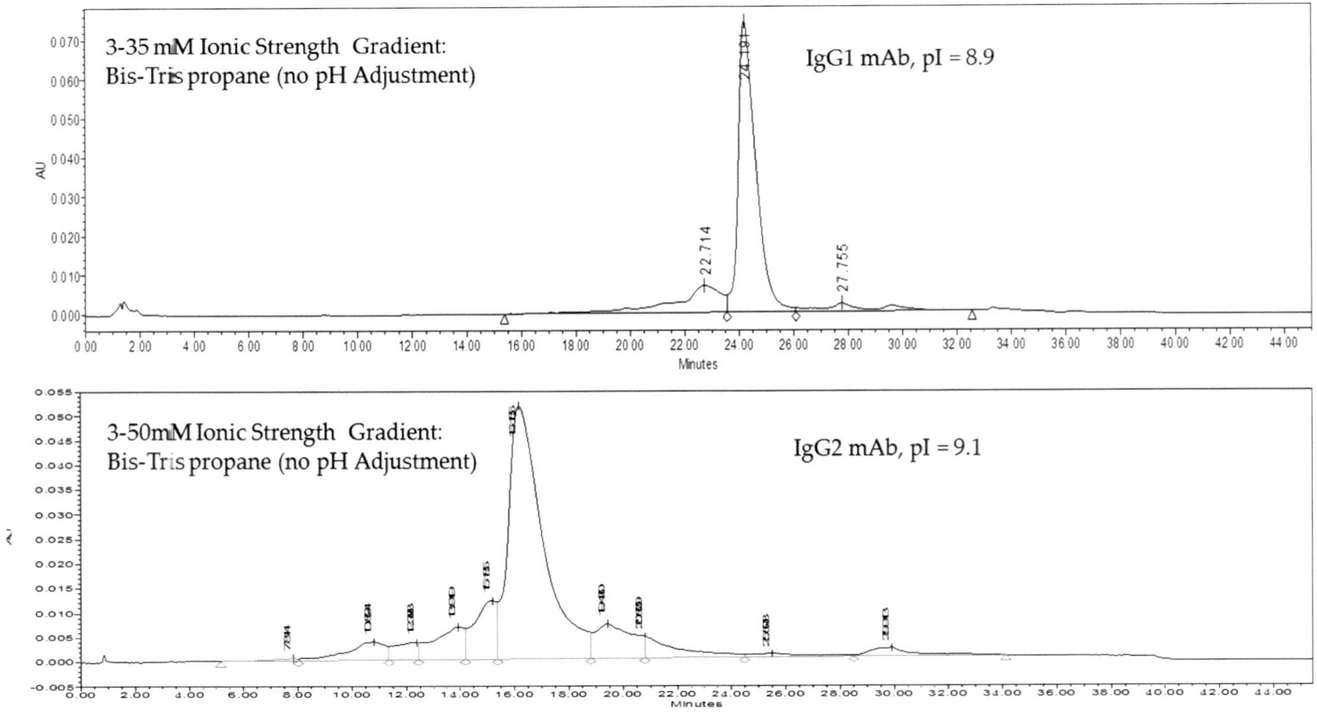

Figure 15.16 Chrcmatographic overlay of two mAbs from Figs. 15.14 and 15.15 (IgG 1, Top and IgG 2, Bottom) with theoretical pIs of 8.9 and 9.1 respectively, run using a 3–35 or 3–50 mM ionic strength gradients in cation exchange separation mode.

This experiment demonstrates the influence of molecule pI on separation strategy. In this case we see that as pI increases to an approximate pI of 9, the requirements for buffer strength also increases significantly for both IgG 1's and IgG 2's. Additionally, at the higher molecule pIs, an ionic strength gradient can be used in place of a pH gradient. In fact, at pIs of ~9, even small amounts of acidic buffers (HEPES and AMPSO) in the mobile phases can significantly increase the molecule retention as well as negatively impact the resolution. Evaluation using an ionic strength gradient is recommended for molecules with a pI ~9, if a suitable separation cannot be achieved using a pH gradient.

15.4.4 Ionic Strength Gradients and Fusion Proteins

The use of ionic strength gradients is appropriate for other mAb like proteins as well. Figure 15.17 shows an overlay of the separation of a 100 kDa fusion protein with a theoretical pI of 8.35. In this separation a combination pH gradient/ionic strength separation was developed using 3 mM Bis-Tris propane + 2 mL/L of 1N HCl in mobile phase A and 60 mM Bis-Tris propane (no pH adjustment) in mobile phase B. The gradient slope for this separation was 4%/min. The column used in this experiment is a Dionex MAbPac SCX-10, 4 mm × 250 mm, 10 μm. The top profile in Fig. 15.17 was generated using the same separation, as the bottom profile, with the addition of 5 mM NaCl in mobile phase B. In this case the addition of 5 mM NaCl resulted in co-elution of homo and hetero-dimer species. In the bottom chromatograph, which did not use any NaCl in mobile phase B, we see well-resolved separation of the hetero- and homo-dimer species.

This separation demonstrates that ionic strength and pH gradients can be incorporated in the same method. In this case, HCl was added to mobile phase A to facilitate retention of the molecule at a low buffer concentration. It also shows that ionic strength gradients can be successfully applied to molecules with lower pIs. Another point about this separation is that the pI for this molecule is lower (pI 8.35) but the concentration requirement for mobile phase B was 60 mM. This is in direct contrast to mAbs, which at a similar pI will require a much lower buffer concentration to facilitate the separation. This requirement for increased buffer strength may possibly be attributed to the lower mass of the molecule (100 kDa) as compared to a mAb, or it may be related to the overall number and availability of charged groups, as it relates to conformation, on the molecule. Finally, it clearly shows the negative impact on resolution that even a small amount of additional ions can have on the separation. It further emphasizes the need for minimizing ions in the mobile phases to facilitate molecule retention, then adding ions as necessary, during method development, to achieve an acceptable separation.

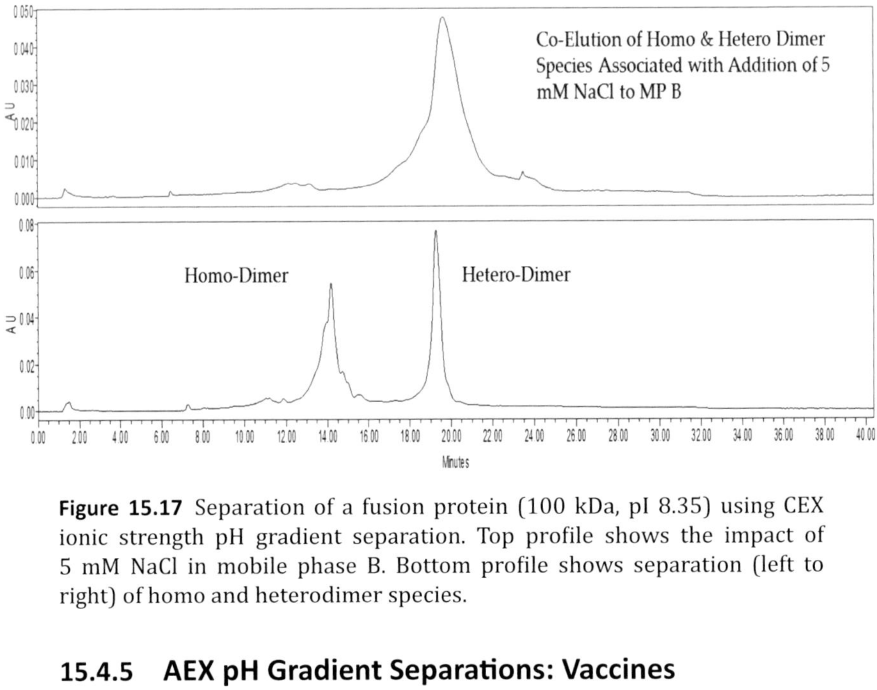

Figure 15.17 Separation of a fusion protein (100 kDa, pI 8.35) using CEX ionic strength pH gradient separation. Top profile shows the impact of 5 mM NaCl in mobile phase B. Bottom profile shows separation (left to right) of homo and heterodimer species.

15.4.5 AEX pH Gradient Separations: Vaccines

The next profile in Fig. 15.18 shows a profile from an AEX pH gradient separation of a vaccine with a theoretical pI of ~4.0 and molecular weight of ~55 kDa. The mobile phases used ammonium formate with formic acid to adjust the pH. Mobile phase A was 50 mM ammonium formate + 300 μL/L of formic acid (approximate pH ~4.4). Mobile phase B was 50 mM ammonium formate + 1.5 mL/L of formic acid (approximate pH ~3.4). These mobile phases will buffer over an approximate pH range of 2.5–5. The column used in this separation was a Dionex, ProPac, SAX-10, 4 × 250 mm.

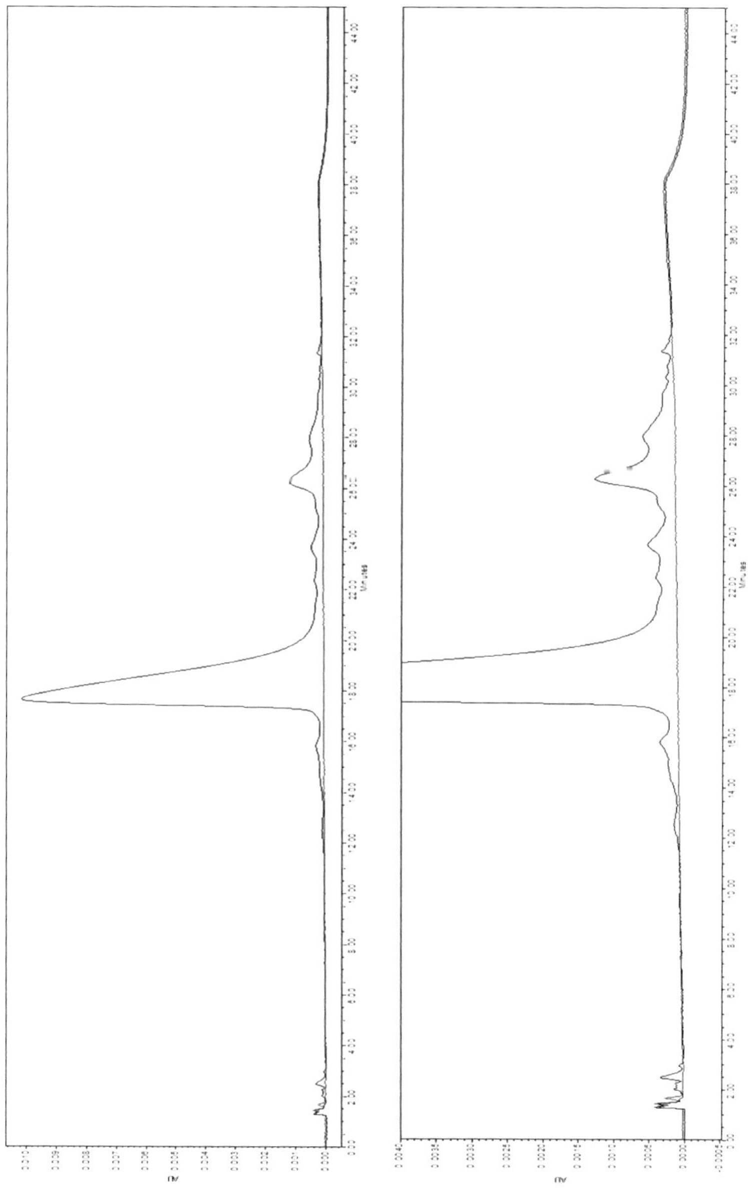

Figure 15.18 AEX profile and expanded plot of vaccine with a theoretical pI of ~4.0, and molecular weight of ~55 kDa.

In this separation, the buffer concentration required is somewhat higher (50 mM), than what is required for mAbs with pIs of ~ 9. This information again suggests that the requirement for higher buffer concentration may be attributed to the lower molecular weight in combination with the number of charged sites, along with molecule conformation. This separation was further optimized and subsequently validated, confirming that pH gradient separations if appropriately developed represent a simple, scientifically sound strategy for developing robust IEC methods.

15.4.6 Parent mAb and ADC Separations

The profiles in Figs. 15.19 and 15.20 show profile comparisons from CEX pH gradient separations of parent mAb and the subsequent ADC for mAbs consisting of an IgG 4 and an IgG 1 with theoretical pIs of 8.1 and 8.9, respectively. Additionally, the payload used for each mAb is also different.

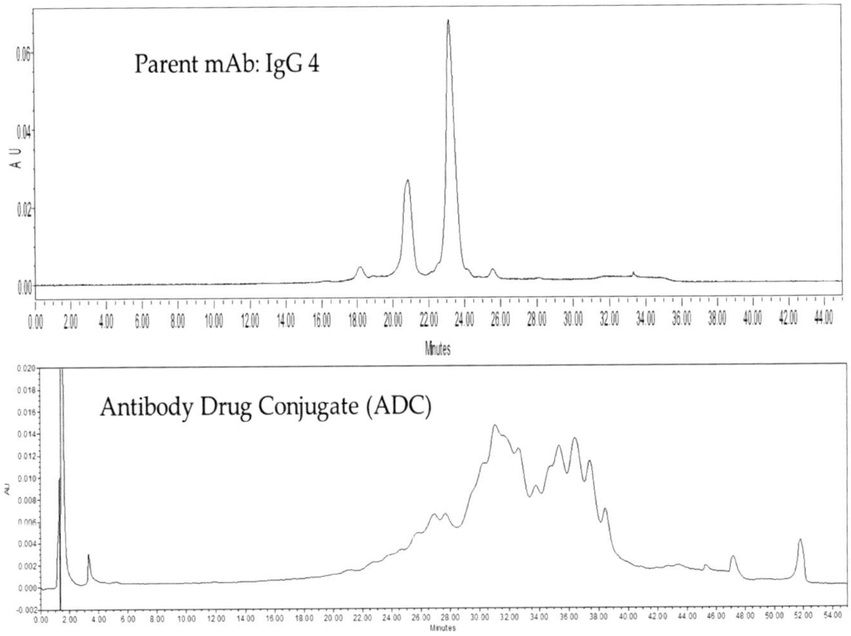

Figure 15.19 AEX IgG 4 parent mAb and ADC with pH gradient CEX separation.

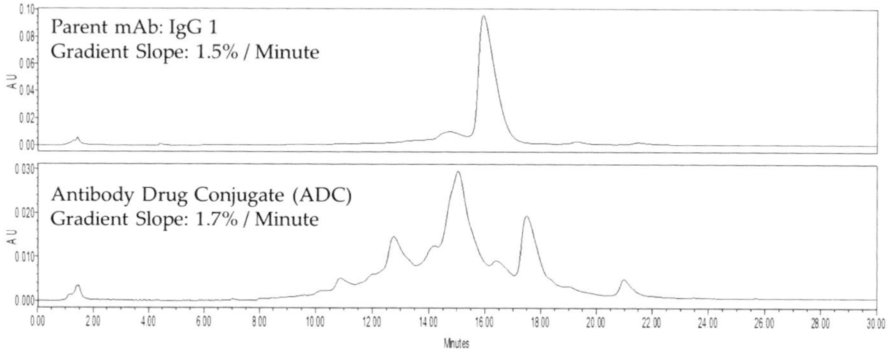

Figure 15.20 IgG 1 parent mAb and ADC with pH gradient CEX separation.

Both of the profiles in Fig. 15.19 were run using CEX separation mode. The mobile phases were the same for both separations. Mobile phase A was 5 mM HEPES, 5 mM AMPSO no pH adjustment. Mobile phase B was 5 mM Bis-Tris propane no pH adjustment. The gradient slope for both was 1.5%/min. The column used for the mAb was a Dionex ProPac WCX-10, 4 × 250 mm. The column used for the ADC was a MabPac SCX-10, 4 × 250 mm.

The mAb/ADC comparison clearly shows the influence of the conjugate on the separation. However, the separation conditions in terms of mobile phases and gradient slopes are the same for both showing in this case that the conjugation may significantly impact the chromatographic profile, but separation conditions are comparable.

Both of the profiles in Fig. 15.20 were also generated in the CEX separation mode. Based on the higher pI of the molecule (pI 8.9), an ionic strength gradient was used for these separations. The mobile phases were the same for both the mAb and the ADC. Mobile phase A was 3 mM Bis-Tris propane, no pH adjustment. Mobile phase B was 50 mM Bis-Tris propane no pH adjustment. The gradient slope for the mAb was 1.5%/min, while the gradient slope used for the ADC separation was slightly greater at 1.7%/min. The column used for the mAb and ADC was a MabPac SCX-10, 4 × 250 mm

In this comparison as with the comparison in Fig. 15.19 we see the influence of conjugation on the separation. The separation conditions in terms of mobile phases are the same for both molecules and the gradient slopes are comparable. This further demonstrates that while the conjugation may significantly impact the chromatographic profile, the necessary method conditions to facilitate a suitable separation, for the mAb and ADC, may be comparable.

15.5 Benefits of Narrow vs. Wide pH Gradients

One of the benefits of using pH gradient separations is related to the ease of optimizing the separation during development. The important parameter to consider during optimization is the rate of delivery of ions across the stationery phase, which has been briefly discussed earlier in this chapter. Subsequently, there are two variables which need to be considered during development in order to optimize rate of ion delivery to achieve the best resolution.

First and most, and importantly, is the amount of ions which are included in the mobile phases, which has been discussed previously in this chapter. Failure to consider this parameter can compromise the resolution to the point that all other attempts to optimize the separation will be minimized if not completely ineffective. Further guidance on this parameter for the various molecule types evaluated in this chapter is included in a summary table (Table 15.12) later in this chapter. Following the guidelines for buffer concentration, listed in the summary table in most cases will provide a good starting point for optimizing the ions in the mobile phases.

After the appropriate buffer concentration is established, the gradient slope can be further decreased to improve the resolution of the separation. However, another strategy should be considered at this point in addition to decreasing the gradient slope.

This strategy is related decreasing the pH gradient range, and can be easily accomplished in two ways. The first and most effective is by decreasing the number of buffers used in the mobile phase so that the pH range of your mobile phases is decreased, for example, reducing the pH range of the buffers from 7 pH units down to 3 pH units.

The following experiment demonstrates the impact of decreasing the buffer pH range through the removal of buffers in the mobile phases. This experiment compared narrow and wide pH gradients to assess their impact on chromatographic profile in the cation separation mode. Figure 15.21 shows overlays of 2, IgG 2 mAbs run with the same method and column (Dionex MAbPac, SCX-10, 4 × 250 mm). The gradient slope for the separation of both mAbs was 2%/min. The mobile phases for the wide pH gradient covered a pH range of ~7 pH units. Mobile phase A consisted of MES, HEPES, AMPSO, and CAPS all at a 6 mM concentration. Mobile phase B consisted of Piperazine, Bis-Tris propane, AMP, and Piperidine all at 3 mM concentration. The mobile phases for the narrow pH gradient covered a pH range of ~3 pH units. Mobile phase A consisted of HEPES and AMPSO both at 5 mM concentration. Mobile phase B consisted of Bis-Tris propane at 5 mM concentration.

The above results show improved resolution associated with narrower pH gradients and demonstrate that narrower pH gradients, in general, would be preferable to wider pH gradients. Additionally, the use of narrower pH gradients will facilitate use of gradient slopes that are within a more preferable operating range for the pump.

Another strategy for accomplishing this is through mobile phase blending. Blending is accomplished by combining mobile phase A and B to establish mobile phase conditions closer to the start and end of molecule elution. For example, if the separation is completed between 40% and 70% mobile phase B, then the mobile phases can be blended so that mobile phase A is actually a mix of A and B (e.g., 80% M.P. A, and 20% M.P. B).

Likewise, mobile phase B would also be a mix of mobile phase A and B (e.g., 90 M.P. B & 10% M. P. A).

Mobile phase blending should be the last step in mobile phase optimization, and may be not necessary in many cases. It should be performed after the optimization of buffer concentration and reducing the pH range of the mobile phases. The primary benefit from mobile phase blending is that it will essentially further stretch the gradient over the operating range of the pump, thereby facilitating some further, but usually minor, improvement in resolution or allowing the use of increased gradient slopes within a more suitable range of the instrument.

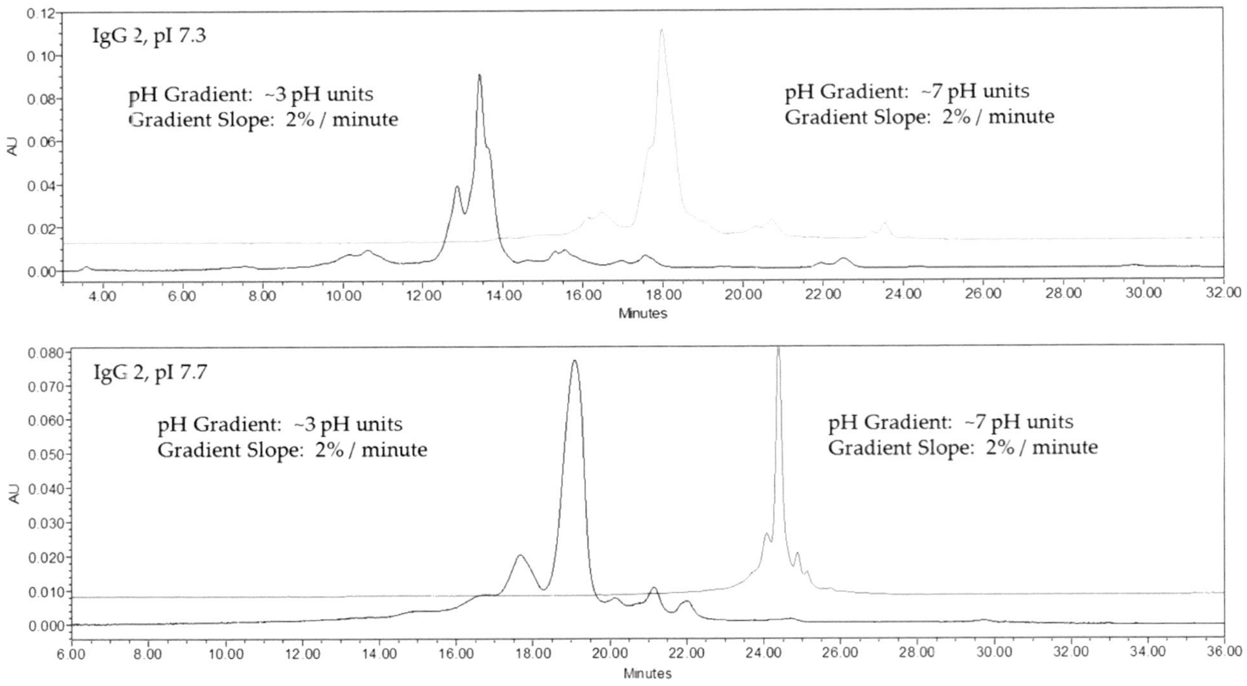

Figure 15.21 Separate overlays of two IgG 2 mAbs comparing a wider pH gradient (~7 pH units) to a narrower pH gradient (~3 pH units).

15.6 Platform Application of pH Gradients

This experiment was designed to demonstrate the platform nature of pH gradient separations. In this experiment, 12 different mAbs were separated using the same mobile phases and instrument method. The mobile phases used mobile phases 2 and 3 from Table 15.1. Mobile phase A: 5 mM HEPES, 5 mM AMPSO. Mobile phase B: 5 mM Bis-Tris propane. The gradient slope for this separation was 2.1% min. The column used in this experiment was a Waters Protein-Pak, Hi Res CM, 7 μm, 4.6 × 100 mm column. The chromatographic profiles along with a scatter plot with linear regression, showing the correlation between retention time and pI, associated with this pH gradient, are shown in Fig. 15.22. The mAb subclass, retention time, and pI values for this experiment are listed in Table 15.10.

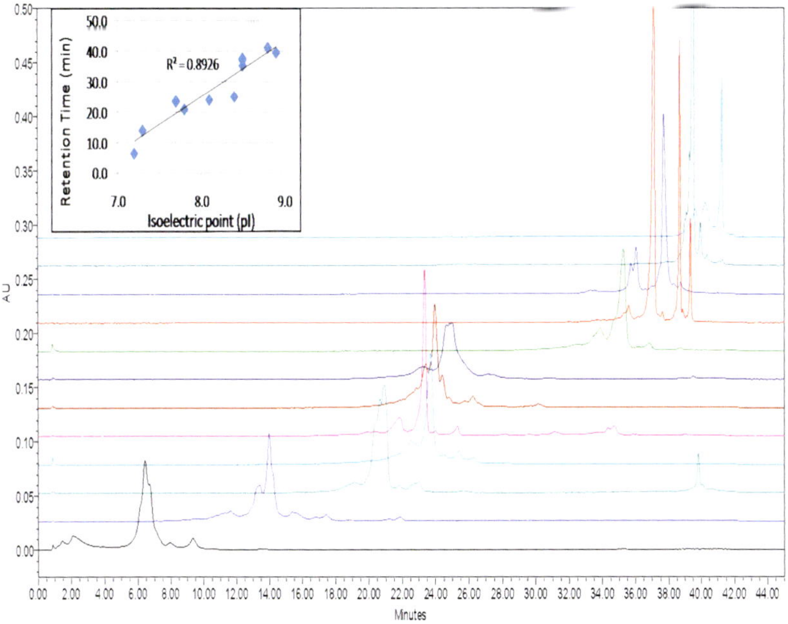

Figure 15.22 Shows chromatographic profiles, along with RT/pI plot (r^2 = 0.8926), for 12 different mAbs separated using the same pH gradient in CEX separation mode.

Results from this experiment clearly show the platform nature of pH gradient IEC separations. It also, shows the general correlation between main peak retention time and molecule pI, further demonstrating the ability of this approach to separate based on pI. This correlation between retention time and pI for different molecules demonstrates the methods viability for separation of charged variants within the molecule, based on their pI. However, it is also clear, due to the lack of complete agreement (r^2 = 0.8926) for the linear regression, that there are additional factors, which have been discussed earlier in this chapter, that influence the retention time of the molecules relative to their pI. These factors could include molecule conformation, size, and the presence of conjugates or glycosylated species on the molecule among other things. These observed differences in retention time for molecules with the same theoretical pI, point to the power of this approach to separate intact molecules and charged variants, within a molecule, based in part, on some characteristics which are not easily measured. Finally, the ability of this approach to separate based on these additional factors could prove useful for gaining further information about the molecule. This experiment further demonstrates the utility of this approach for separating complex mixtures containing molecules with similar or even identical pIs.

Table 15.10 Shows the list of mAbs, used in this experiment, with respective retention time and pI in order of elution

Type	mAb subclass	Retention time (min)	pI
mAb	IgG-2	6.4	7.2
mAb	IgG-2	14.0	7.3
mAb	IgG-2	20.9	7.8
ADC	IgG-2	23.7	7.7
ADC	IgG-1	23.4	7.7
mAb	IgG-2	24.0	8.1
ADC	IgG-2	25.1	8.4
mAb	IgG-2	35.3	8.5
mAb	IgG-1	37.1	8.5
mAb	IgG-1	37.7	8.5
mAb	IgG-1	39.5	8.9
ADC	IgG-1	41.2	8.8

15.7 Purification Using pH Gradient Separations

The use of pH gradient separations is also suitable for purification of charged variants to support validation and characterization of charged variants. Figure 15.23 shows the separation from 2 different columns, with the top overlay showing the separation from a column which could be used for purification of charged variants (Agilent, PL-SCX, 4000Å, 150 × 4.6 mm, 8 μm) with a 5 mg load. The bottom profile shows the separation from an analytical column (Dionex, MabPac, SCX-10, 4 × 250 mm) with 25 μg load run with the same method. The mobile phases used in this experiment were HEPES, and Bis-Tris propane. Mobile phase A was 5 mM HEPES, and mobile phase B was 1.5 mM HEPES and 3.5 mM Bis-Tris propane. The mAb is an IgG 2 with a theoretical pI of 7.3. The method and gradient slopes used were the same for both columns.

This method was originally developed for purification purposes to support both characterization and possible validation activities. The top profile from this method shows the separation of a 5 mg load on an Agilent PL-SCX column. The bottom profile is the same method run on a Dionex MAbPac column with a 25 μg load. The result from this experiment shows comparable profiles from the different columns with significantly different loads (5 mg vs. 25 μg). This experiment further demonstrates the robustness of pH gradients for both analytical and purification purposes.

15.8 Mobile Phase Preparation

One of the most critical parameters related to developing well-resolved and robust IEC separations is the mobile phase. With regard to mobile phase development, the choice of buffer and buffer concentration must be made with consideration for the following information which includes molecule pI, mAb subclass, or molecule type, and molecular weight.

These factors have been discussed earlier in this chapter and will not be discussed further.

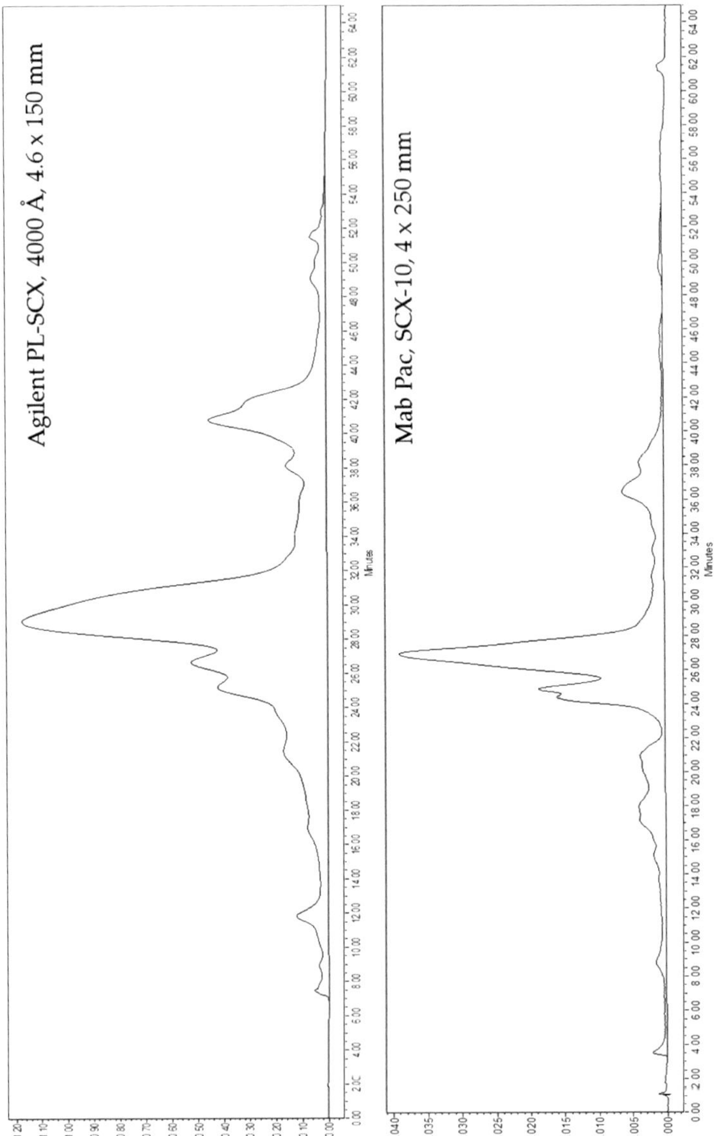

Figure 15.23 Comparison of two profiles from the same method using columns from different manufacturers with loads of 5 mg (top profile) and 25 μg (bottom profile).

Another critical factor with regard to mobile phases is the preparation. Failure to adequately control the level of ions in the mobile phases, during preparation, can lead to excessive retention time shifts and loss of resolution. As a result, all factors which could contribute to variability of the ion concentration in the mobile phases should be considered when writing mobile phase preparation steps in the method.

The sources of variability during mobile phase preparation include analyst, pH meter, balance, pipettes, and glassware. The analyst represents the largest potential source of variability in the mobile phase preparation. However, the majority of this variability is related to either analyst training, or ambiguity in the preparation instructions, leading to variability associated with different mobile phase preparation techniques. Removing most of the variability associated with differences in preparation technique can be accomplished by adding an appropriate amount of detail in mobile phase preparation to facilitate consistent analyst-to-analyst mobile phase preparation. These variations can be identified and minimized by having several analysts prepare the mobile phases and running the method during method development.

Another source of variability that can be minimized for IEC separations, is the variability associated with the pH meter. As seen in Fig. 15.2, a method with a pH criterion of ± 0.1 pH units can result in significant retention time shifts and possible loss of resolution. There are several factors which can influence the accuracy of the pH meter including ambient temperature, calibration standards, pH calibration range, variations in analyst calibration procedure, improper maintenance of pH probe, in addition to minor differences between manufacturers and models. All of these factors are sources of variability which can influence the accuracy of pH meter and subsequently the amount of pH adjuster (ions) which are added to the mobile phases.

One of the primary advantages of pH gradients which use acidic and basic buffers is that the mobile phase pH is self-adjusting and occurs with the mixing of the mobile phases during the separation. This approach to pH gradient separations will provide the simplest mobile phase preparation along with the greatest control over the amount of ions in the mobile phases. However, if pH adjustment of the mobile phases is required, the analytical method section for preparation of mobile phases should not incorporate the use of a pH meter. The pH, and more importantly, the ion concentration in the mobile phases can be precisely controlled through the use of analytical balances, pipettes, and calibrated glassware (e.g., graduated cylinders or volumetric flasks). The pH meter should be limited to the development phase of IEC methods and the pH should be linked to the weight or volume of buffer in combination with the weight or volume of pH adjuster added to the mobile phases. Once the appropriate weights and volumes of mobile phase components are established, during the development phase, there is no need to reference the mobile phase pH in the analytical method. In other words, the mobile phase composition should be always determined by the amount of buffer and the amount of pH adjuster added as measured by an analytical balance and a calibrated pipette, respectively.

For example, mobile phase preparation instruction might look like this for mobile phase 1 in Table 15.1:

Mobile Phase B: 5 mM HEPES, 5 mM AMPSO + 3.75 mL of 2.5 N NaOH

Add ~800 mL of HPLC grade water to a 1 L graduated cylinder. Add 1.19 ± 0.01 g of HEPES, and 1.14 ± 0.01 g AMPSO. Using a calibrated pipette, add 3.75 mL of 2.5 N NaOH. Q.S. to 1 L with water and mix well. Filter through a 0.2 μm filter.

In this example the concentration of ions in the mobile phase, will be more precisely controlled, from one preparation to the next, than if one were to place a pH criteria of 9.5 ± 0.1 pH units on the mobile phase preparation. In the experiment in Fig. 15.2, the criteria of ± 0.1 pH units could result in an approximate range of ± 100 μL of 6N NaOH adding unnecessary variability to the mobile phase preparation.

15.9 Sample Preparation

Sample preparation is very important with regard to consistent performance of the method. The goal for sample preparation is to make sure that the sample is adequately diluted to reduce excessive levels of ions in the sample matrix which can interfere with the separation. Additionally, one must consider sample stability in the diluent as well as considering the final diluted sample concentration as it relates to column loading. Failure to identify the appropriate sample preparation instructions relative to the samples being tested can result in, significant retention time shifts, poor profile and loss of resolution.

The sample diluents used in the experiments in this chapter consist of water or mobile phase A. Water is a good choice of diluent since this will reduce the amount of ions in the sample matrix while still maintaining some sample buffering from the formulation buffer. However, if the pH of mobile phase A is comparable to that of the sample matrix, then mobile phase A can also be a good option. In either case, sample stability on the instrument should be evaluated with the goal of maintaining sample stability for an acceptable period of time.

For analytical samples, diluting to a concentration of 1 mg/mL is adequate. However, if the starting sample concentration is close to 1 mg/mL, it may be necessary to dilute the sample to a lower concentration to eliminate matrix interference with the separation. Typically a 2–3 fold dilution is adequate, but a lower sample dilution should be more thoroughly evaluated to assess the impact of increased levels of sample formulation matrix on the separation. The dilution must be evaluated while keeping in mind the necessity of loading an adequate amount of material on the column. A typical range for loading is ~ 25–100 μg on the column for analytical testing. If the initial sample concentration is too low to achieve sufficient column loading with the necessary dilution, lower column loads can be evaluated with more efficient columns (e.g., narrow bore and or smaller particle size columns).

For purification purposes, it is important to maintain the highest sample concentration while diluting just enough to minimize the influence of ions from the sample matrix. The goal in this case is to maximize the load on the column per injection which may require one to minimize the dilution. A dilution as low as 2–3-fold will usually be adequate to eliminate matrix interference. However, a lower dilution, again, should be more thoroughly evaluated for the negative profile influences which can be associated with increased levels of sample matrix.

If the sample concentration is not high enough to achieve the desired load on the column from a single injection, a short loading method can be used load multiple injections. In this case, starting conditions should be evaluated to insure that the sample is retained at the column head throughout the loading injections. If the sample is not adequately retained under starting conditions, the profile will be negatively impacted, resulting in repeating/ overlapping profiles from the multiple injections.

For in-process samples, it is most important to adequately dilute the sample to eliminate or minimize the influence of the sample matrix. For this reason, it is especially important to know the sample concentration and the sample matrix of the samples being tested. For in-process samples, if the sample concentration is not high enough to achieve the necessary load on the column from a single injection, a short loading method can be used to load multiple injections. As mentioned above for multiple loading injections, starting conditions should be evaluated to insure that the sample is retained at the column head throughout the loading injections.

15.10 Columns and Column Temperature

Table 15.11 is by no means exhaustive but is reflective of the columns used for the experiments presented in this chapter as well as other pH gradient–related work by the author.

Table 15.11 List of columns which have been used for IEC separations in this chapter

Manufacturer	Column desc.
Dionex	ProPac WCX-10, 4 × 250 mm, 10 μm
Dionex	MabPac SCX-10, 4 × 250 mm, 10 μm
Dionex	ProPac SAX-10, 4 × 250 mm, 10 μm
Waters	Protein-Pak, Hi Res CM, 4.6 × 100 mm, 7 μm
Agilent	PL-SCX 4000 Å, 4.6 × 150 mm, 8 μm
Agilent	PL-SAX 4000Å, 4,6 × 50 mm, 5 μm

Column temperature control should be specified in the method since pH is influenced by temperature. Failing to control temperature in the method will result in retention time shifts as well as some minor changes in peak resolution which could result in variations in quantitated peak area percent results.

Overall, failing to control temperature will result in a less robust method with greater variability with regard to precision. For all of the separations listed in this chapter the column temperatures were controlled at 35–40°C.

Figure 15.24 shows an example of the impact of column temperature on retention time for an AEX pH gradient separation. The profiles show that decreasing the column temperature was associated with decreasing retention time.

Figure 15.25 shows the impact of column temperature on resolution for an AEX pH gradient separation. The profiles show that decreasing the column temperature was associated with decreased resolution of the first acidic peak after the main peak.

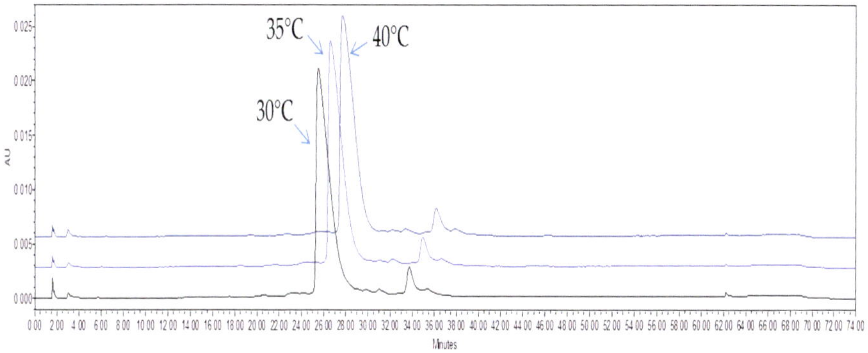

Figure 15.24 Overlay showing impact of column temperature on general retention time.

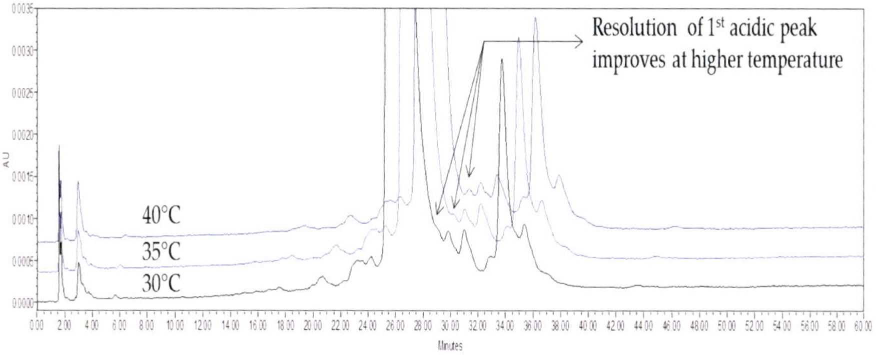

Figure 15.25 Overlay showing impact of column temperature on peak resolution.

These results illustrate the impact of column temperature on both retention time and resolution, and further demonstrate the need for evaluating and controlling column temperature during IEC separations.

15.11 Choosing the Separation Mode

The primary consideration for choosing the separation mode during initial development is based on molecule pI. For example, a molecule with a pI at < 7 would initially be evaluated using AEX while a molecule with a pI of ~7 or above would be evaluated using CEX. Molecules with pIs in the 6.5–7.5 range may separate equally well in either mode and could be evaluated in both.

An interesting advantage of using mobile phases which contain both an acidic buffer and a basic buffer with no pH adjustment (e.g., HEPES, AMPSO, and Bis-Tris propane), is that these mobile phases can be run in either the cation or anion exchange modes by simply switching which mobile phase is used for eluting the molecule.

This represents an advantage in method development capabilities allowing for an efficient and more thorough evaluation of both separation modes. In most cases, following the guidelines above for choosing the separation mode will yield the best separation. Additionally, due to the higher pI for most mAbs, the preferred separation, based on profile, will be in the cation mode. However, the presence of negatively charged structures on the molecule, in the form of sialylated glycans or conjugates, can facilitate adequate separation in either separation mode.

Figure 15.26 shows the IEC profile from a highly sialylated mAb that was separated in both cation and anion exchange modes using the same buffers. The mAb was an IgG2 with a pI of 7.3. The mobile phases used in this separation were the same as what was used in Fig. 15.22 for the wide pH gradient. The column used in the separation for CEX was a Varian PL SCX 4000 Å, 4.6 × 150 mm, 5 μm. The column used in the separation for AEX was a Varian PL-SAX 4000 Å, 4.6 × 150 mm, 5 μm.

Based on the profiles in Fig. 15.26, it appears that the profiles from either separation mode are acceptable and could be evaluated for further development.

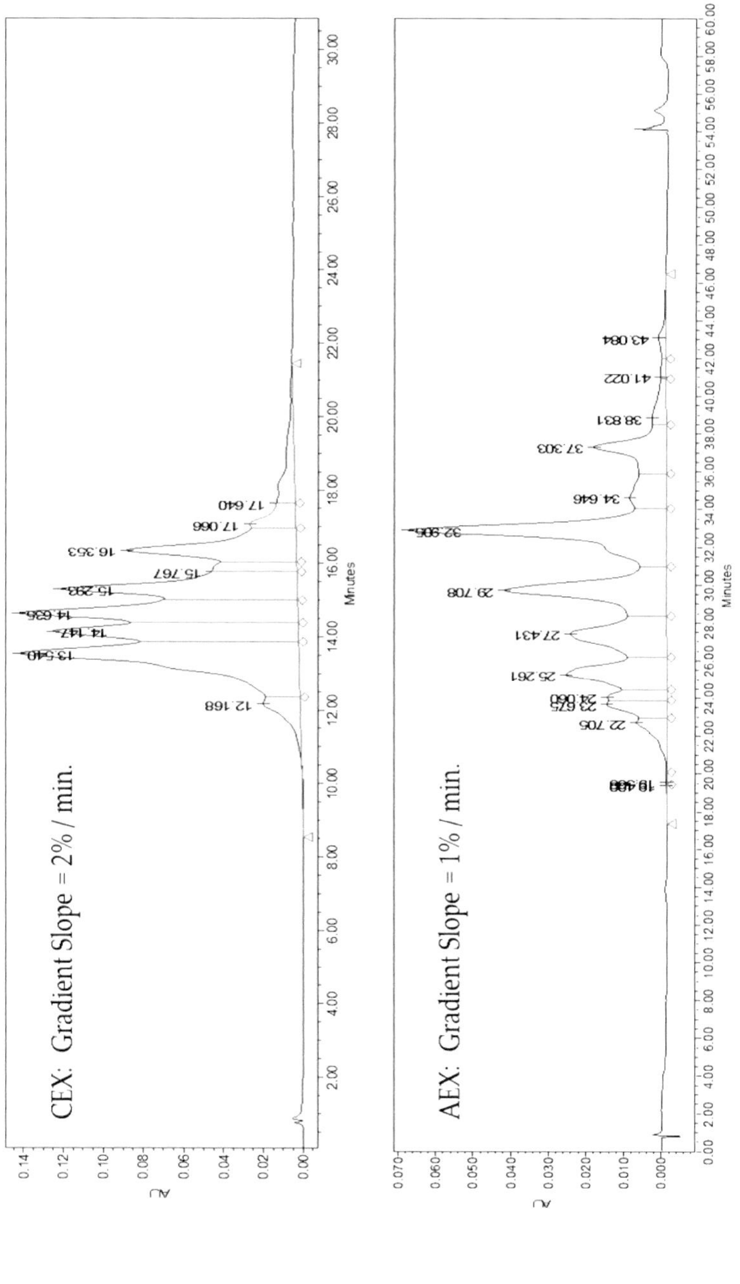

Figure 15.26 CEX (Top) and AEX (Bottom) profiles for a highly sialylated IgG2 with a theoretical pI of 7.3.

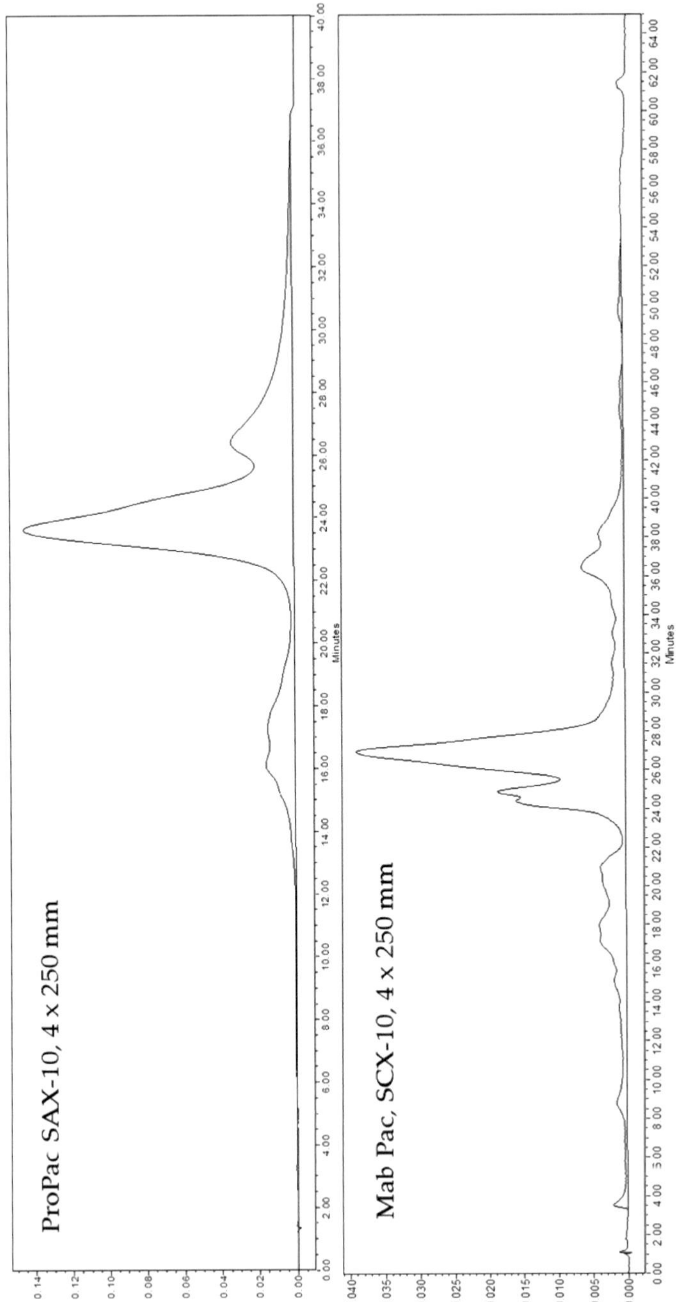

Figure 15.27 CEX (Bottom) and AEX (Top) profiles for an IgG2 mAb with a theoretical pI of 7.3.

Figure 15.27 shows the IgG2 from Fig. 15.23, with a theoretical pI of 7.3 which has been separated using the same mobile phases used in Fig. 15.22 in both CEX and AEX separation modes.

In this case, the CEX separation is shown to be preferable over the AEX separation mode. The gradient slope for the CEX mode was 0.7%/min while the gradient slope for the AEX mode was 0.5%/min.

15.12 System/Column Equilibration

For IEC separations proper column equilibration is very important in achieving consistent separation both within a sample sequence, and from one instrument to another. There are several factors which influence the time it takes to re-equilibrate a column, including column size, flow rate, and system volume. All of these parameters must be considered to establish an appropriate re-equilibration time.

Symptoms of poor re-equilibration include retention time shifts during the run or when switching from one instrument to another, change in retention time, profile, and or loss of resolution.

However, proper re-equilibration is easily achieved by adding a re-equilibration step as part of the method. The re-equilibration step should occur after the separation gradient is complete, and prior to the next injection. The re-equilibration time is established based on system dead volume or volume of liquid in the system, to include the column. In this case, the retention time of the void peak from the chromatogram can be used to determine the system void. The re-equilibration step in the method should be at least 5 times the retention time of the void peak in the chromatogram. The re-equilibration step can be easily evaluated during a sequence by checking main peak retention times to insure that the retention time has stabilized by the second or third injection of the sequence, and that it is consistent with the expected retention time in the method. If the retention time continues to shift to the left after the second or third injection, or if the profile is changing, additional time may need to be added to the re-equilibration step in the method to stabilize retention time.

Table 15.12 Summary Table of separation conditions for mAbs, ADCs, and other bio-therapeutics

Molecule	pI	MW KDa	Buffer (low pH)	Buffer (high pH)	Buffer strength	Mode	Gradient type	Figures in Chapter
mAb, IgG1	7 to < 9	150	HEPES/AMPSO	Bis-Tris propane	~10–20 mM	CEX	pH	4D, 11, 12, 13, 23,
mAb IgG1	≥ 9	150	N/A	Bis-Tris propane	3–35 mM	CEX	Ionic Strength	16
ADC, IgG1	~ 9	150	N/A	Bis-Tris propane	3–35 mM	CEX	Ionic Strength	21
mAb IgG2	7 to < 9	150	HEPES/AMPSO	Bis-Tris propane	~5–10 mM	CEX	pH	4A, 4B, 4C, 10, 11, 13, 23, 24,
mAb IgG2	≥ 9	150	N/A	Bis-Tris propane	3–35 mM	CEX	Ionic strength	16, 21
mAb, IgG4	8.1	150	HEPES/AMPSO	Bis-Tris propane	~5 mM	CEX	pH	20
ADC, IgG4	8.1	150	HEPES/AMPSO	Bis-Tris propane	~5 mM	CEX	pH	20
Vaccine	4	55	Ammonium Formate + Formic Acid	Ammonium Formate + Formic acid	50 mM	AEX	pH	19, 25
Fusion Protein	~8.4	100	Bis-tris propane + HCL	Bis-tris propane	3–60 mM	CEX	pH-Ionic strength combination	17
Fusion Protein	6.2	40	Ammonium acetate + Formic acid	Ammonium acetate + Formic acid	60 mM	AEX	pH	18

15.13 Summary

Table 15.12 provides a summary of the information from the separations in this chapter and should serve as a resource for method development of other mAbs, ADCs, as well as other bio-therapeutics. Matching the general information in the molecule and pI columns to another mAb, ADC, or other molecule should provide useful starting information for developing a method.

For example, if one wanted to develop a separation for a mAb, IgG1 subclass, with a pI of 8.1, the conditions listed in the first row of Table 15.12 will likely be suitable for generating an acceptable separation to start with, and are summarized in Table 15.13. Further information regarding the separations associated with each row can be obtained through the figure references in the far right hand column of Table 15.12.

Table 15.13 Shows initial method parameters using information from Table 15.10 for an IgG 1 mAb with a pI of 7 to <9

Parameter	Specification
Separation mode	CEX
Columns	Dionex, Mab Pac, SCX-10, 4 × 250, Waters, Protein-Pak, Hi Res CM, 4.6 × 100 Agilent, PL-SCX 4000 Å, 4.6 × 150 mm, 8 μm
Column temp.	35–40°C
Autosampler temp.	2–8°C
Flow rate	~1 mL/min
Detector wavelength	280 nm
Injection volume	50 μL
Sample dilution	1 mg/mL (water or mobile phase A)
Mobile phase A	15 mM HEPES/15 mM AMPSO
Mobile phase B	15 mM Bis-Tris propane

15.14 Conclusions

This chapter describes a simple approach to developing highly resolving and robust pH gradient IEC methods for mAbs, ADC's, and other bio-therapeutics such as fusion proteins and vaccines. The flexibility, ease of development and use, along with the versatility of pH gradients gives this type of IEC method an advantage over traditional salt gradients for the separation of these more complex molecules. The experiments in this chapter demonstrate the negative impact from excessive ions in the mobile phases providing rationale and strategies for easily controlling them to optimize the separation. Additionally, the use of pH gradients for IEC separations as described in this chapter provides additional information regarding specific development strategies related to mAb subclass, molecule pI, and molecular weight.

Because these strategies take into account additional information with regard to molecule characteristics, they represent an improved approach over current practices used for IEC pH gradient separations. This approach will prove useful as a tool for supporting analysis of bio-therapeutics, as well as supporting characterization and validation activities. This approach also has the potential for application within process development.

References

1. Flatman, S., Alan, I., Gerard, J., Mussa, N. *J. Chromatogr. B,* 2007, 848, 79–87.
2. Schenerman, M., Sunday, B. R., Kozlowski, S., Webber, K., Gazzano-Sanotora, H., Mire-Sluis, A. *BioProcess Int.,* 2004, 42–52.
3. Harris, R. J. *J. Chromatogr., A,* 1995, 705, 129–134.
4. Weitzhandler, M., Farnan, D., Horvath, J., Rohrer, J. S., Slingsby, R. W., Avdalovic, N., Pohl, C. *J. Chromatogr. A,* 1998, 828, 365–372.
5. Santora, L. C., Krull, I. S., Grant, K. *Anal. Biochem.,* 1999, 275, 98–108.
6. Moorhouse, K. G., Nashabeh, W., Deveney, J., Bjork, N. S., Mulkerrin, M. G., Ryskamp, T. J. *J. Pharm. Biomed. Anal.,* 1997, 16, 593–603.
7. Chelius, D., Jung, K., Lueras, A., Rehder, D. S., Dillon, T. M., Vizel, A., Rajan, R. S., Li, T., Trauheit, M. J., Bondarenko, P. B. *Anal. Chem.,* 2006 78, 2370–2376.
8. Donato, A., Ciardiello, M., Nigris, M., Piccoli, R., Mazzarella, L., D'Alessio, G. *J. Biol. Chem.,* 1996, 268, 4745–4751.
9. Hsu, Y.–R., Chang, W.-C., Mendiaz, E. A., Hara, S., Chow, D. T., Mann, M. B., Langley, K. E., Lu, H. S. *Biochemistry,* 1998, 37, 2251–2262.

10. del la Guntiñas, M., Wissiack, R., Bordin, G., Rodrigez, A. R. *J. Chromatogr. B,* 2003, 791, 73–83.
11. Santora, L., Kaymakcalan, Z., Sakorafas, P., Krull, I., Grant, K. *Anal. Biochem.,* 2001, 299, 119–129.
12. Vlasak, J., Ionescu, R. *Curr. Pharm. Biotechnol.,* 2008, 9, 468–481.

Chapter 16

Size Exclusion Chromatography Method Development for Therapeutic Proteins

Alexandre Goyon, Jean-Luc Veuthey, Davy Guillarme, Szabolcs Fekete

University of Geneva, University of Lausanne,
CMU—Rue Michel-Servet 1, Geneva, 1211, Switzerland

davy.guillarme@unige.ch, Szabolcs.fekete@unige.ch

16.1 Introduction to Size Exclusion Chromatography

Size exclusion chromatography (SEC) is a mode of liquid chromatography in which separation is based on the partial exclusion of the solutes from the pores of the stationary phase [1]. The larger molecules elute earlier than the smaller ones, because large solutes can penetrate into the pores only by a smaller fraction

Chromatographic Method Development
Edited by Gregory K. Webster and Laila Kott

ISBN 978-981-4800-53-2 (Hardcover), 978-0-429-20172-1 (eBook)
www.jennystanford.com

(partially) [1]. Very large solutes (larger than the pores) can be totally excluded from the pores and will elute within the interstitial column volume. Based on its mechanism, SEC is applied to separate solutes of different sizes (molecular weights). Hence, SEC is commonly used to separate protein aggregates (size variants) such as monomers, dimers, trimers, and other oligomers or high-molecular-weight species (HMWS).

Among the different techniques applied for the characterization of aggregates—such as sodium dodecylsulfate polyacrylamide gel electrophoresis (SDS-PAGE), asymmetrical field flow fractionation (AF4), fluorescence spectroscopy, circular dichroism (CD) or light-scattering-based methods, like multi-angle laser light scattering (MALLS)—SEC is considered today as the reference technique for the quantification of protein aggregates. The main advantage of this approach is the mild elution conditions (non-denaturing) allowing the characterization of proteins with minimal impact on their conformational structure and the local environment. In SEC, we strive for separating the proteins under inert conditions. Indeed, the separation should ideally depend only on the size of the solutes. However, in practice, some secondary interactions (hydrophobic and/or electrostatic) may occur between the proteins and the stationary phase [2]. To limit such interactions and work under inert conditions, the pH and ionic strength of the mobile phase should be carefully controlled. Ideally, the mobile-phase pH should be set close to the isoelectric point (pI) of the proteins but it is not applied in practice for basic proteins and a physiological pH is usually applied. Salts (sodium or potassium chloride) or other additives (arginine) are also often added to the mobile phase to avoid or at least limit electrostatic interactions, while a small proportion of organic modifier (e.g., 10% 2-propanol) can also be used to limit hydrophobic interactions as much as possible.

The aim of this chapter is to provide some guidelines to develop the SEC method for proteins and to show the possibilities of modern SEC (UHP-SEC) separations.

16.2 Theoretical Aspects

In SEC, biomolecules are separated based on their hydrodynamic radius. The stationary phase consists of spherical porous particles (silica or polymer based) with a carefully controlled pore size and pore size distribution, through which the solutes diffuse based on their molecular size differences, using an aqueous buffer as the mobile phase. Entropic processes without any physical or physicochemical adsorption ideally drive the partitioning between the internal pores and interstitial volume.

To describe the elution of a peak in SEC, the so-called thermodynamic retention factor (k'')—or zone-retention factor—has to be used, since there is no retention and therefore conventional retention factor k' comprised between −1 and 0 is not a good measure of elution. Thus, k'' expresses the ratio of the probabilities of the sample staying in the stagnant mobile phase inside the pores and in the moving mobile phase between the particles (interstitial volume) [3]. It is easy to understand that this k'' value is limited (k''_{max}), and its maximum value is given by the ratio of the pore volume (V_p) and the interstitial volume (V_e) of a column, or similarly by the pore porosity (ε_p) and the interstitial porosity (ε_e):

$$k'' = \frac{V_{elution} - V_e}{Ve} = \frac{t_{elution-t_e}}{t_e} \tag{16.1}$$

$$k''_{max} = \frac{V_p}{V_e} = \frac{\varepsilon_p}{\varepsilon_e} = \frac{t_0 - t_e}{t_e}, \tag{16.2}$$

where $V_{elution}$ and $t_{elution}$ are the elution volume and elution time, respectively, while V_e and t_e are the total exclusion volume and time (determined by the interstitial volume) and t_0 is the column dead time (determined by the sum of interstitial and pore volumes). The k''_{max} value then defines the possible elution window (t_W) for the solutes. The larger the k''_{max}, the better the probability to separate the compounds. Figure 16.1 shows a typical SEC chromatogram and corresponding t_e, $t_{elution}$, t_0 and t_W. Using currently available state-of-the-art columns, the k''_{max} is around 1.4–1.8. This parameter—and the associated selectivity—can be altered by packing a column more densely (or decreasing particle size) or by using particles with larger pore volumes.

Another important characteristic of SEC columns is the calibration curve, since it illustrates the size (weight) of analytes that can be separated on a given column. The calibration curve can also be used to estimate the molecular weight of an unknown analyte. Typical calibration curves are based on proteins or polymers of known molecular weights (M). By plotting log M vs. the retention volume, one typically obtains a third order polynomial with a linear region providing the highest resolution and molecular weight accuracy [4]. The linear part of the curve indicates the range of molecular weights that can be adequately separated on the given stationary phase.

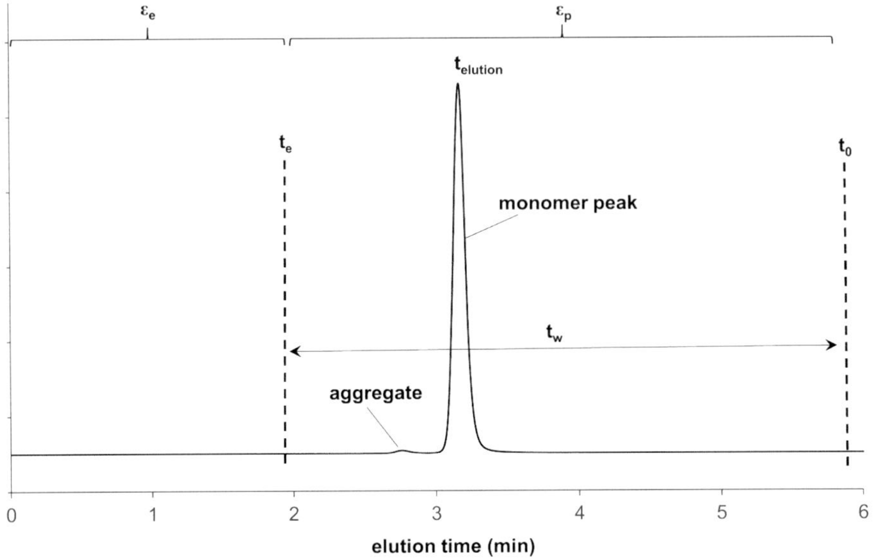

Figure 16.1 Typical SEC chromatogram of monoclonal antibody monomer and aggregate, using a 150 × 4.6 mm column.

The slope of the line in the linear portion of the calibration curve is a measure of the stationary-phase selectivity, which can be defined by the following relationship:

$$\log M = A \cdot V_{\text{elution}} + B, \tag{16.3}$$

where A and B are the slope and intercept of the line, respectively. As the pore size distribution becomes narrower, the slope becomes shallower which results in a greater selectivity to discriminate analytes of similar size. The intercept of the linear part depends on the average pore diameter. Figure 16.2 shows some model calibration curves related to different pore size distributions and pore diameters.

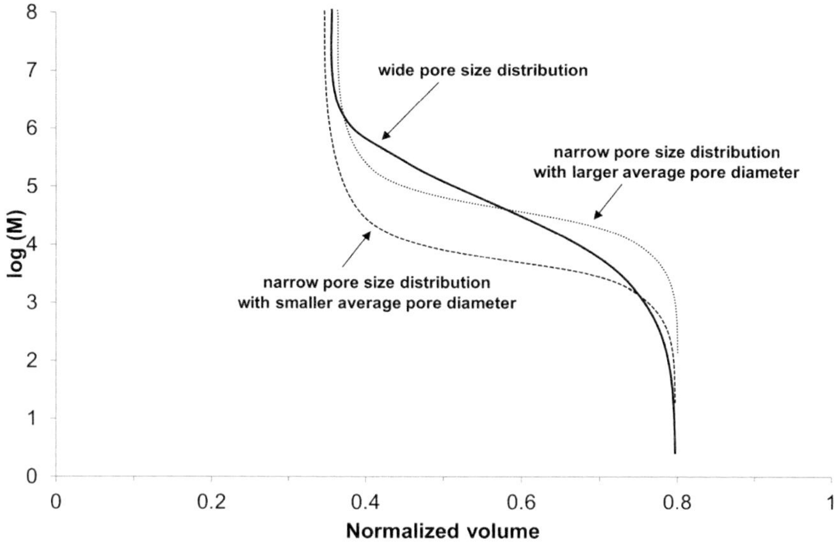

Figure 16.2 Illustration of SEC calibration curves as log M vs. normalized elution volume for columns possessing different average pore size and pore size distribution.

There are some potential sources of error when measuring proteins molecular weight based on a SEC calibration curve. First, since proteins shapes could vary (e.g., globular, rod-like or flexible chains), their Stokes radii do not correlate exactly with molecular weight. Second, non-ideal adsorption may alter the retention volume [5–7].

Ideally, SEC involves solely size-based separation. Thus, there is no true retention compared to other sorptive techniques. Nevertheless, apparent plate numbers (N) can be used as a measure of the column performance, as introduced by Engelhardt and used by others [8–10]. The efficiency (N) in SEC can also be studied and compared by constructing plate height (H) vs. linear velocity (u) plots.

$$H = A \cdot u_0^{1/2} + \frac{B}{u_0} + C \cdot u_0, \quad (16.4)$$

where A corresponds to eddy-dispersion, B to longitudinal diffusion and C to mass transfer resistance, which in this case is related to the partitioning of the solute between the internal pores and the interstitial volume. We thought that it is useful to illustrate the expected shape of an H–u plot for large proteins in SEC, since it could be quite different compared to the frequently presented van Deemter curve shapes observed with small molecules. Figure 16.3 shows the total plate height (H_{total}) and the plate height contributions of eddy-dispersion (H_A), longitudinal diffusion (H_B), and mass transfer resistance (H_C) as a function of mobile-phase velocity.

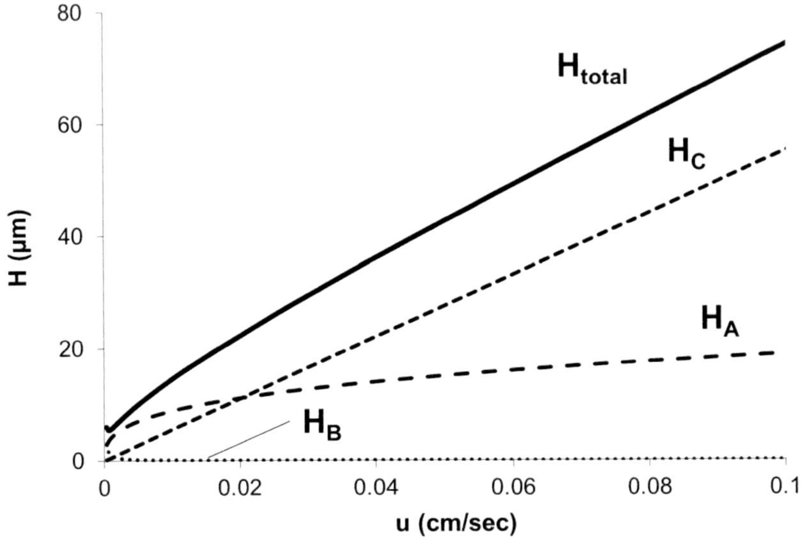

Figure 16.3 Calculated H–u curve in SEC for a 150 kDa protein using a column packed with 3.0 µm particles operating at 25°C. Reproduced with permission from Elsevier.

As expected, the longitudinal diffusion (B term) is negligible with large proteins. The plate height is determined by the eddy-dispersion (A term), especially for more excluded species, and mass transfer resistance (C term). To conclude, the lowest plate height (highest efficiency, N) is reached at low mobile-phase velocity. The higher the velocity (mobile-phase flow rate), the lower the efficiency.

By using current state-of-the art 150 × 4.6 mm SEC columns (packed with sub-3 μm particles), $N \sim 2000$–10,000 can be reached for a 150 kDa protein, while using conventional 300 mm × 6–8 mm column (packed with 5 μm particles), $N \sim 4000$–15,000 can be performed in the practically useful analysis time range. Such separations typically take 5–10 min on 150 × 4.6 mm and 10–40 min on 300 × 6–8 mm columns, respectively. For smaller compounds (MW < 1 kDa, e.g., protein formulation excipients), a higher efficiency is expected in most cases, while very large proteins often elute in very broad and tailed peaks. Small analytes eluting at t_0 (totally penetrated through the pores) possess similar plate numbers as can be observed in reversed-phase liquid chromatography (RPLC). In average, an efficiency of ~100,000 plates/meter can be attained with a column packed with 5 μm particles, while a SEC material packed with 2 μm particles may perform up to 250000 plates/meter for common small molecules such as uridine.

16.3 Method Development of Proteins

16.3.1 Stationary-Phase Selection

16.3.1.1 Pore size, particle size, and column dimensions

First of all, a stationary phase with appropriate pore size has to be selected. The pore size should fit to the size (or molecular weight) of the investigated protein and its aggregates. The log M of the solutes should be included in the linear part of the calibration curve. For most protein separations, typical pore sizes between 150 and 500 Å are often employed. For the separation of common proteins, a pore size of 150–200 Å is generally well suited in the 0.5–450 kDa range, while a 200–300 Å pore size is commonly applied for monoclonal antibodies or antibody drug conjugates (MW ~ 150 kDa), to separate HMWS up to 1000 kDa. For very large proteins (MW > 200 kDa, e.g., PEGylated proteins), the 500–1000 Å phases offer probably the best selectivity up to 7500 kDa. Obviously, not only the average pore size, but also the pore size distribution has to be considered depending on the range of molecular size/weight. If only the protein monomer and dimer need to be separated, a stationary phase with narrow pore size distribution is preferred. On the other hand, if a broad range of compounds needs to be separated, a wide pore size distribution material is more adequate.

Resolution between the aggregates and native protein peaks can be improved by decreasing the column particle size. The achievable plate number is indeed inversely proportional to the particle size, while the resolution is inversely proportional to the square root of the particle size. Historically, columns packed with 5–10 μm particles were commonly used in routine SEC applications, but several providers offer now 3.0, 2.7, 2.5 μm or even sub-2 μm particles packings. By using these state-of-the-art columns packed with such small particles, the resolution can be significantly improved and/or the analysis time can be shortened by applying shorter columns. Figure 16.4 shows experimentally measured *H–u* curves for β-lactoglobulin (18.7 kDa) and a therapeutic monoclonal antibody (150 kDa) when using SEC columns packed with 5.0, 2.7, 2.0, and 1.8 μm particles. As shown, the smaller the particle, the lower the plate height. For a monoclonal antibody, about three times lower plate heights can be obtained with sub-2 μm materials compared to conventional 5 μm packing.

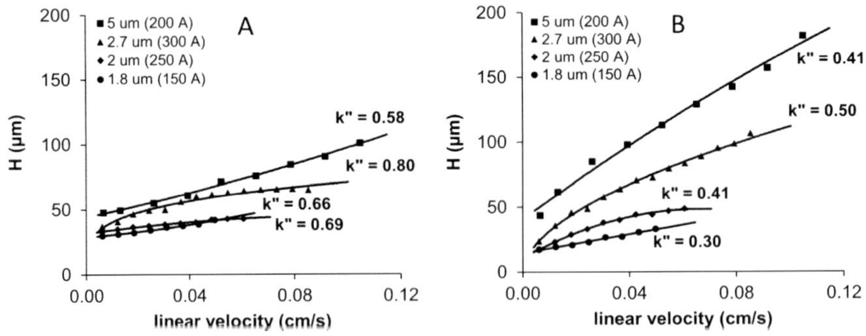

Figure 16.4 Measured *H–u* curves for β-lactoglobulin (A) and a therapeutic monoclonal antibody (B) by using columns packed with 5.0, 2.7, 2.0, and 1.8 μm particles, at ambient temperature. Reproduced with permission from Elsevier.

Table 16.1 Some popular state-of-the-art SEC columns and their properties

Name (provider)	Material	Particle size (μm)	Pore size (Å)	Max. temperature (°C)	pH range	Max. pressure (bar)
Acclaim SEC-300 (Thermo)	hydrophilic polymethacrylate resin	5, 7	300, 1000	60	2–12	80
Acquity UPLC BEH SEC (Waters)	diol modified hybrid-based	1.7, 2.5	125, 200, 450	60	2–8	400
AdvanceBio SEC (Agilent)	surface-coated silica-based	2.7	130, 300	80	2–8.5	400
Bio SEC-3 (Agilent)	surface-coated silica-based	3	100, 150, 300	80	2–8.5	240
Bio SEC-5 (Agilent)	surface-coated silica-based	5	100, 150, 300, 500, 1000, 2000	80	2–8.5	240
SRT-SEC (Sepax)	surface-coated silica-based	5	100, 150, 300, 500, 1000, 2000	80	2–8.5	240
TSKgel Ultra SW aggregate (Tosoh)	diol modified silica-based	3	300	30	2.5–7.5	120
TSKgel SW mAb (Tosoh)	diol modified silica-based	4	250	30	2.5–7.5	120
TSKgel UP-SW3000 (Tosoh)	diol modified silica-based	2	250	30	2.5–7.5	250
Yarra (Phenomenex)	silica-based	1.8, 3, 5	145, 150, 290, 300, 500	60	2.5–7.5 (except for SEC-X150, X300: 1.5–8.5)	200 (except for 1.8 μm: 480 bar)
YMC-Pack Diol-SEC (YMC)	diol modified silica-based	3, 5	60, 120, 200, 300	40	5–7.5	200
Zenix-SEC (Sepax)	surface-coated silica-based	3	80, 100, 150, 300	80	2–8.5	240

The very high pressure generated by small particles (P > 250 bar) in SEC may be critical for some pressure sensitive proteins, which may suffer from on-column aggregation, and therefore misleading results can be obtained [11]. However, this pressure induced aggregation (or degradation) is protein dependent and has to be experimentally evaluated. Columns packed with 2.0–3.0 μm particles are good compromise between efficiency (or analysis time) and pressure. Some characteristics of the most popular SEC columns on the market were summarized in Table 16.1.

In conventional SEC, 300 mm long columns with internal diameters (I.D.) of 6, 7.8, 8, or 10 mm are generally applied. These SEC columns are referred to as standard bore columns. Now, several vendors offer narrow-bore columns with 4.6-mm I.D. and 150 mm length, packed with small particles. By using these shorter columns, similar separation quality can be attained as with 5 μm particles in 300 mm standard bore columns, but the analysis time can be reduced by a factor of 3 to 4 [11].

When using the 4.6 mm I.D. columns, the extra-column (system) band broadening could have a serious effect on the observed efficiency in SEC. Those particular columns require the use of optimized UHPLC systems with very low extra-column volume (<10 μL). Otherwise, 30–60% loss of column efficiency may occur on conventional HPLC systems, which possess large extra-column band variance up to 100 μL^2 [12]. Then, conventional HPLC systems are only compatible with standard bore SEC columns, to attain their full performance.

16.3.1.2 Stationary-phase chemistry

SEC columns packed with porous cross-linked dextran particles were introduced in 1953 by Wheaton and Bauman [13]. Polyacrylamide-based gels were subsequently developed by Hjertén and Mosbach in the 1960s [14]. The porosity of those columns was controlled by adjusting the degree of cross-linking degree. These polymeric stationary phases were effective to limit non-specific interactions between the polymer (stationary phase) and sample molecules (proteins). Several of these phases were developed among the years, using various organic polymers such as polymethacrylate, polystyrene, polydivinylbenzene, polyamide, or agarose. However, their performance was extremely limited due to their low mechanical stability, preventing the use of small particles and high mobile-phase flow rate. Therefore, these phases were solely considered for low-resolution separations. In addition, the polymeric resins used in SEC can also shrink and/or swell in various solvents.

To overcome the mechanical stability of the polymeric-based stationary phases, porous silica particles were introduced later to perform SEC separations at higher pressures (few hundred bars). However, the acidic residues of silica surfaces still required modification to minimize the possibility of strong ionic interactions between the proteins and the silanolates. The regular improvement of silica materials (type B silica or hybrid silica) enabled further developments of silica particles for routine SEC analysis. Surface modification includes unbonded, diol (1,2-propanediol) or tetramethylsilane (TMS) modifications to minimize the amount of available residual silanolates. Residual acidic silanols are detrimental in SEC applications since electrostatic interactions may occur with the sample, influencing both elution time and peak tailing. Bouvier et al. have proved that TMS was more effective than diol to reduce the acidity of the silanols [15], but shorter column lifetime may be experienced. Therefore, diol bonding remains the most predominantly used silica-surface chemistry modification in SEC when analyzing proteins. In addition, diol modification increases the hydrophilic nature of the stationary phase. This ligand blocks or reacts with many of the acidic silanol groups and neutralizes the surface. Having a hydrophilic surface in SEC is quite important, since many of the proteins include some hydrophobic parts, which may interact with the stationary phase through hydrophobic interactions, and may affect peak tailing and elution time. It can be noted that some differences can always be observed between two diol-bonded phases in terms of possible electrostatic and hydrophobic interactions, due to differences in the properties of the silica itself and also different bonding density. A recent study compared the inertness of current state-of-the-art silica based SEC columns, and found some important differences between them, especially in their capability to form electrostatic interactions with monoclonal antibodies [16].

As mentioned, non-specific interactions may result in protein adsorption, elution time shifting or asymmetrical band elution (tailing). A shift in elution time may lead to false interpretation of molecular weights, while peak tailing can decrease the resolution between consecutively eluted peaks. Non-specific interactions can include hydrophobic, electrostatic and hydrogen-bond interactions. Hydrophobic interactions occur between the protein and hydrophobic sites of the stationary phase. Electrostatic interactions can be classified in two categories: (a) ion-exchange when the protein and the stationary phase carry an opposite charge (b) ion-exclusion when the protein and the stationary phase carry the same charge. As an example, most monoclonal antibodies are basic proteins possessing a pI around 7–9 and ion-exchange often occurs between the positively charged proteins and the negatively charged residuals silanolates of the stationary phase at physiological mobile-phase pH conditions.

To select a suitable stationary phase, the change in peak elution time, tailing factor and recovery (peak area) has to be investigated using different amounts of mobile-phase salt additives in order to evaluate non-specific interactions. When plotting the elution time or peak tailing as a function of salt concentration or buffer concentrations for several SEC columns, the optimal stationary phase can be selected. Figure 16.5A shows SEC chromatograms of an IgG1 monoclonal antibody, in presence of various salt concentrations, while Fig. 16.5B shows the impact of salt concentration for four different stationary phases by means of relative change in k''.

As shown in Fig. 16.5, the elution time of the IgG1 monomer peak increased with the salt concentration. However, Fig. 16.5B clearly illustrates some significant differences between SEC stationary phases. As example, no significant change was observed with increasing ionic strength in column B, while an important shift in elution time was observed with columns C or D. Obviously, this behavior is protein dependent, and has to be studied individually during the method development. The same methodology can be applied to compare possible hydrophobic interactions too, by plotting k'' (or its relative change) as a function of mobile-phase organic modifier content.

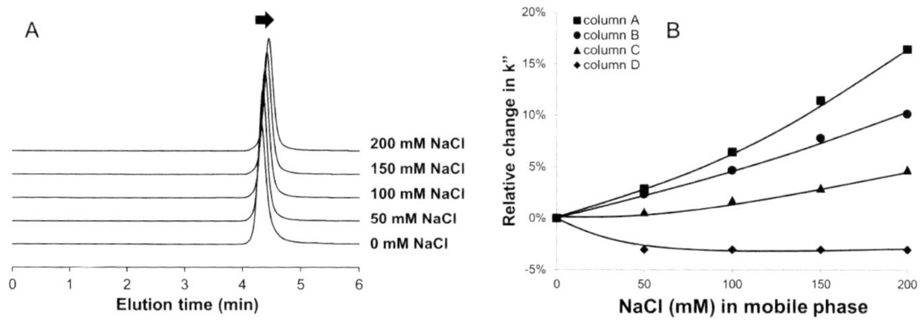

Figure 16.5 Effect of salt (NaCl) concentration on the elution time of an IgG1 monoclonal antibody. Overlaid chromatograms correspond to different salt concentrations on column C (A) while plots of relative change in k'' vs. salt concentrations have been prepared for four different columns (B). Reproduced with permission from Elsevier.

16.3.2 Mobile-Phase Selection (Salts, Buffers, Organic Modifier)

To work under non-denaturing conditions in aqueous SEC, the pH of the mobile phase is generally adjusted between 6.2 and 7.4 using phosphate buffer. Mobile-phase salt additives play a major role to limit possible electrostatic interactions between the proteins and the stationary phase, as well as to maintain proteins under their native conformation.

To decrease protein–stationary phase interactions, the nature and the concentration of the salt have to be carefully optimized. A combination of phosphate buffer and sodium chloride (NaCl) or potassium chloride (KCl) has been widely applied for protein analysis in SEC [2, 11, 17]. The evaluation of three common buffers with a wide range of proteins has proved that potassium- and sodium-salts were more effective than Tris-HCl to improve the recovery of proteins [18]. In addition, the authors evaluated various ionic species ($(NH_4)_2HPO_4$, Na_2SO_4, $(NH_4)_2SO_4$, $MgCl_2$, NaCl) and found that better resolution factors were obtained for six different protein samples with NaCl [18]. When dealing with the analysis of monoclonal antibodies, the addition of NaCl in the mobile phase can significantly increase the recovery of the HMWS, as shown in Fig. 16.6 with rituximab (pI = 9.2).

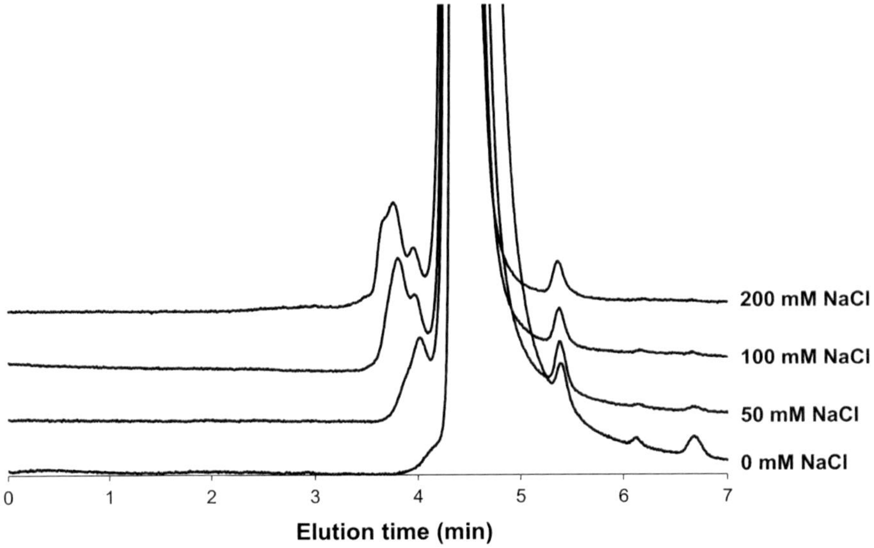

Figure 16.6 Impact of salt concentration added to the mobile phase on the recovery of rituximab HMWS. HMWS elute in the first peak on the chromatogram and their amount varies with salt concentration, while the amount of the monomer and fragment remains constant. Reproduced with permission from Elsevier.

Recently, we demonstrated that potassium-based salts were more effective than sodium-based salts to limit non-specific interactions between the stationary phase and monoclonal antibodies [19]. Indeed, larger cations such as potassium have a higher ability to lose their hydration layer than sodium cations and to interact strongly with the residual silanolates of the stationary phase. The study highlighted an improvement in selectivity between HMWS and the protein monomer when using a concentration of potassium salt higher than 0.2 M, in comparison with sodium salts. In addition, the use of potassium salt resulted in higher protein recovery especially for the HMWS. These observations are in agreement with the fact that electrostatic interactions are often enhanced for HMWS vs. protein monomers [20].

However, high salt concentration can cause aggregates formation that were not originally present in the sample [21]. In addition, high ionic strength increases the risk of salting-out effects and can promote hydrophobic interactions between the proteins and the stationary phase. As an example, an increase in elution time of very hydrophobic mAb (ipilimumab) has been observed when increasing the amount of NaCl in the mobile phase [16]. Finally, some buffers have also been reported to promote non-specific interactions even on inert materials [6].

Except salt cations, the effect of salt anions (phosphate, chloride) is also worth studying. A mobile phase containing 100 mM potassium phosphate buffer and 200 mM potassium chloride is, in our opinion, an effective combination to minimize both electrostatic and hydrophobic interactions. Increased salt concentration can also improve the resolution between dimer and monomer species, since dimers carry more charges and possess a more hydrophobic nature than the monomer. Therefore, in most cases, the elution time of dimer peaks decreases to a larger degree than the one of monomers, when increasing salt concentration and finally results in higher selectivity.

Arginine was also proposed as a suitable mobile-phase additive to improve protein recovery during SEC analysis. A concentration of 0.20–0.75 M arginine has been reported as an effective range to increase proteins recovery and separate soluble oligomers [22]. With a heat-stressed mAb sample, 22% and 67% HMWS were observed in absence and in presence of arginine, respectively. Arginine is able to limit both electrostatic and hydrophobic interactions with most stationary phases [22]. Arginine stabilizes proteins by refolding their structure, it solubilizes proteins from insoluble pellets, suppresses protein aggregation and limit non-specific adsorption [22]. While in the past the use of arginine raised concerns as protein-denaturant, lately it has been proved that arginine does not cause denaturation of proteins and was therefore suitable for SEC applications [23]. However, one limitation of arginine is that it strongly absorbs at UV wavelengths below 220 nm.

With the latest generation of SEC materials, hydrophobic interactions seem to be more critical than electrostatic interactions [16]. To limit possible hydrophobic interactions, small amounts (up to 15% (v/v)) of organic modifiers (2-propanol, methanol, ethanol, or acetonitrile) can be added to the mobile phase for the analysis of the most hydrophobic proteins. Antibody–drug conjugates (ADCs) are particularly challenging samples for SEC, due to the hydrophobicity of their cytotoxic drugs and peptide linkers. In the work of Wakankar et al., the addition of 15% 2-propanol significantly reduced hydrophobic interactions, when analyzing trastuzumab emtansine and resulted in symmetrical peak and slightly decreased elution time [24]. However, 2-propanol drastically increases the mobile-phase viscosity and generates high pressure which may increase the risk of on-column protein aggregation. In addition, with very viscous mobile phase, the peak width is often larger than with mobile phases of low viscosity (water or acetonitrile). To overcome this limitation with 2-propanol, less viscous organic modifiers may be preferred. In 1984, Malawer et al. evaluated the effect of various polar solvents on the SEC of PVP K-15, 30, 60, and 90 using diol-derivatized gel silica columns and found that methanol was more efficient then acetonitrile to limit adsorption [25]. Finally, the addition of organic solvent has to be carefully evaluated as it may alter the native conformation of proteins and interfere with SEC aggregate analysis.

Figure 16.7 shows an example on the effect of organic modifier (acetonitrile) on the recovery and peak shape of an ADC (trastuzumab emtansine).

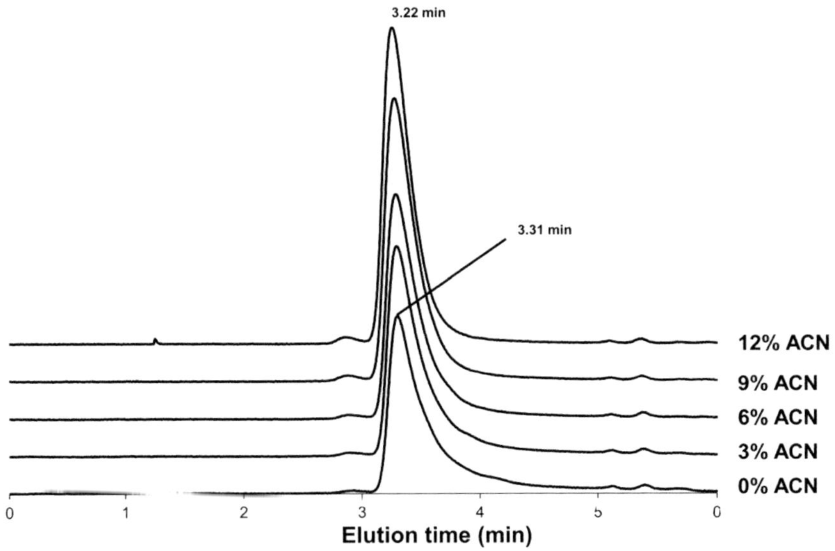

Figure 16.7 Impact of mobile-phase organic modifier (acetonitrile) content on the recovery of ADC (trastuzumab emtansine). By increasing the acetonitrile content, the recoveries of both monomer and dimer species increase and peak symmetry improves.

16.3.3 Effect of Mobile-Phase Temperature

To increase the mass transfer rate, mobile-phase temperature could be a valuable parameter. As the temperature increases, the mobile-phase viscosity decreases, and the analytes diffusivity improves. It is especially important for large molecules, which possess very low diffusivity. Please note that in Eq. 16.3, the B term is proportional to diffusion coefficient, while the C term is inversely proportional to the diffusion coefficient. Furthermore, elevated temperature allows the use of higher flow rates for a given column dimension, and therefore faster analysis can be performed.

As expected, the symmetry of protein peak is often better at higher temperature. It is not exclusively due to the improved mass transfer properties, but also to the reduced secondary interactions. Even if attempts are made to work under inert SEC conditions, secondary interactions are often observed with large therapeutic proteins. As the temperature increases, the strength of secondary interactions becomes weaker. The decrease of non-specific interactions manifests in less tailed peaks.

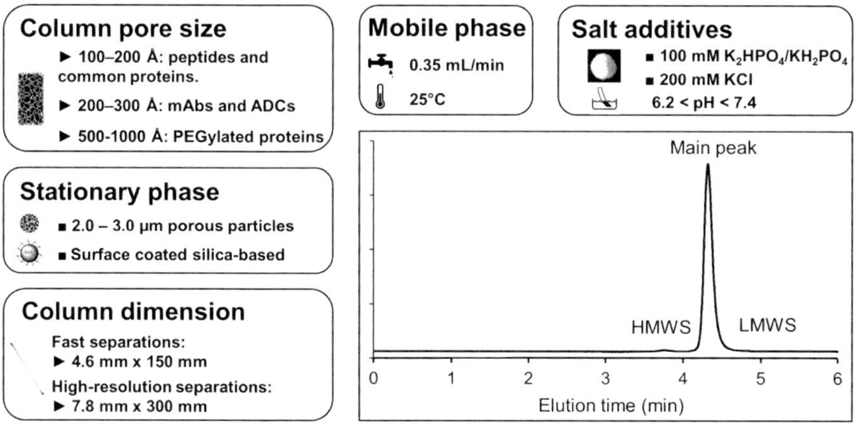

Figure 16.8 Recommended mobile-phase and stationary-phase conditions for the SEC of proteins.

Another important effect of temperature in SEC is the possible shift in elution time when changing the mobile-phase temperature. In most cases, the higher the temperature, the lower the elution time. However, in SEC, there is no adsorption (and retention); therefore, the decrease of elution time with temperature may be unexpected. The possible explanation is the conformational change of proteins with temperature. At higher temperature, proteins tend to unfold and their apparent size becomes larger; then they will be more excluded from the pores.

On the other hand, working at elevated temperature in SEC can be risky. Indeed, many proteins are thermo-labile and suffer from aggregation and/or fragmentation at high temperature. It has already been reported that the observed aggregate amount is often larger at high temperature due to temperature induced on-column aggregation [11].

Finally, Fig. 16.8 summarizes the conditions for a successful SEC method development, when characterizing protein samples.

16.4 Applications for Protein Analysis

The main application for SEC is the relative quantitation of biopharmaceuticals' aggregates. In 2016, the United States Pharmacopeia (USP) released the monograph <129> Analytical Procedures for Recombinant Therapeutic Monoclonal Antibodies [26]. This chapter contains biopharmaceuticals purity assessments by SEC and provides validated procedures. SEC has been used in numerous stability studies to understand the effect of temperature, pH, formulation, light and glycosylation on the aggregation of therapeutic proteins [27–31]. Except the formulation study from Nicoud et al., where a polymeric stationary phase was selected (dextran covalently bonded to agarose), others studies were generally performed on silica-based SEC materials. The separation of protein fragments (reduced protein, partially digested or degraded) has also been successfully performed using sub-2 μm UHP-SEC columns [17]. Columns packed with 1.7 μm BEH particles were also employed to separate the aggregates of insulin within 7 min [32], and some sub–3 min separations (UHP-SEC) of test proteins and mAb aggregates were reported with very small particles based columns (Fig. 16.9) [11]. The very hydrophobic ADC trastuzumab emtansine was analyzed using 15% of 2-propanol on a silica-based stationary phase [24]. More recently, the HMWS of trastuzumab emtansine were successfully separated using the Tosoh TSKgel UP-SW3000 diol-bonded and Agilent AdvanceBioSEC (proprietary bonding) columns using purely aqueous mobile phase [16]. Finally, a combination of preparative protein A and SEC analysis has been reported for the analysis of mAb aggregates and host cell proteins (HCPs) [33].

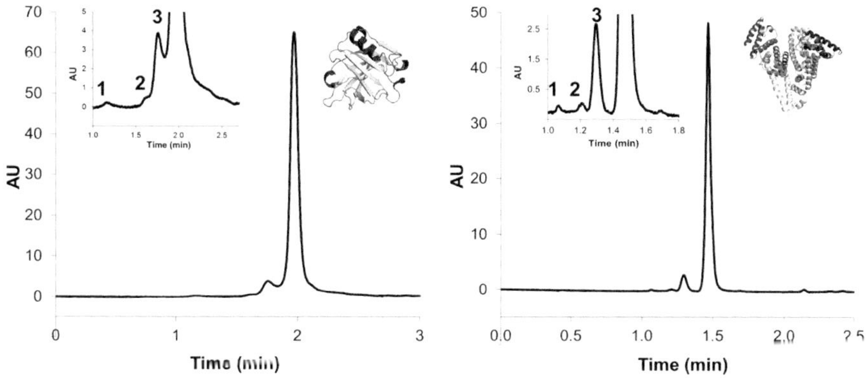

Figure 16.9 UHP-SEC separation of B-lactoglobulin (A) and chicken egg ovalbumin (B) HMWS using 150 × 4.6 mm columns packed with 1.7 μm particles operated at relatively high flow rate (0.85 mL/min). Reproduced with permission from Elsevier.

The use of volatile salts has attracted attention in the last few years, to make SEC compatible with mass spectrometry (MS). Typically, 75 mM ammonium acetate is used as mobile-phase additive for this hyphenation [34]. Combination of low amounts of volatile salts (<200 mM, e.g., ammonium formate) and organic modifier (<20%, e.g., acetonitrile) has also been investigated for SEC-MS analysis of peptides and proteins under non-denaturing conditions [35]. This allows direct coupling of SEC to electrospray ionization time-of-flight mass spectrometry (ESI-TOF-MS) and online characterization of the species. However, if quantitation is required, the effect of the volatile salts on non-specific interactions should be evaluated, as there might be a risk of underestimating the HMWS due to the nature and the low concentrations of the salt [20]. Liu et al. have successfully performed SEC-MS of a reduced mAb using a mobile phase containing 20% acetonitrile with TFA and formic acid as mobile-phase additives [36].

Lately, two-dimensional liquid chromatography (2D-LC) emerged as an effective strategy for the analysis of complex samples. 2D-LC methodologies have been applied using SEC in the first dimension and RPLC in the second dimension, to quantify the amount of free cytotoxic drug (small molecule) in antibody–drug conjugate sample [37–38]. The first dimension also provided information about the HMWS (monomer, dimer, oligomers). In these setups, 20% (v/v) of acetonitrile was systematically added to the mobile phase to improve the chromatographic performance. Finally, on-line comprehensive 2D-LC using SEC in the first dimension, mixed-mode chromatography in the second dimension and evaporative light scattering detector (ELSD) have been reported for the separation and detection of various analytes (including protein, anionic, cationic, zwitterionic, and nonionic excipients) in a biopharmaceutical drug product [39].

16.5 Perspectives

Since the introduction of UHPLC systems and columns, which enabled high resolution, sensitivity, and peak capacity, there have been a variety of new analytical advancements in LC and LC–MS based on the use of low dispersion systems [40]. In SEC, the trend is the same as in the other modes of liquid chromatography, namely the decrease of particle sizes and column dimensions. In conventional SEC, 300 mm × 6–8 mm columns are commonly used, while the recent trend in UHP-SEC is to work with 150 mm × 4.6 mm columns. Many column providers offer such UHP-SEC columns and efforts are also made to improve the inertness of the stationary phases. However, further decrease in SEC column dimension is not expected, due to the loss of intrinsic column efficiency caused by commercially available LC systems. This behavior is related to the nature of SEC separations. Indeed, the band variance depends on the solute retention factor k', which in SEC ranges between –1 and 0. It inherently results in very low peak variances for columns of low volumes, which would be too strongly affected by the system variance [12].

To improve the sensitivity of SEC separations or work with very small amounts of available samples, the use of capillary columns can be a promising approach. However, important modifications to a commercially available LC systems are required to drastically reduce the system volume. Until now, the number of applications in this field is rather limited, but 300 mm × 300 µm I.D. SEC capillary columns were successfully applied for the separation of mAb fragments [41]. Hyphenation of SEC with MS is expected to attract still more attention in the coming years for the characterization of biopharmaceutical proteins. With this perspective, 2D-LC applications should rapidly grow to overcome the current challenges, e.g., the analysis of HCPs in therapeutic protein samples.

A possible way to improve throughput in SEC is to work with two columns in parallel. A further reduction of analysis time can be obtained by combining interlaced sample injections with parallelization of two narrow bore columns packed with small particles. With this strategy, both holdup times before and after the peaks of interest can successfully be eliminated, resulting in significant decrease of analysis time [42]. This methodology is based on injecting a new sample before the ongoing analysis of a previous sample has ended. To perform parallel interlaced SEC—with two columns, two switching valves are required to direct the flow alternately between the autosampler, the two columns and the detector. The use of two columns and switching valves require two distinct programs on which pumps, autosampler and column compartment, including the switching valves, are controlled.

References

1. Uliyanchenko, E., Schoenmakers, P. J., and Van der Wal, S. (2011). Fast and efficient size-based separations of polymers using ultra-high-pressure liquid chromatography. *Journal of Chromatography A*, **1218**(11), 1509–1518.
2. Fekete, S., Beck, A., Veuthey, J. L., and Guillarme, D. (2014). Theory and practice of size exclusion chromatography for the analysis of protein aggregates. *Journal of Pharmaceutical and Biomedical Analysis*, **101**, 161–173.
3. Barth, H. G., Saunders, G. D., and Majors, R. E. (2012). The state of the art and future trends of size-exclusion chromatography packings and columns. *LCGC North America*, **30**(7), 544-563.
4. Hong, P., Koza, S., and Bouvier, E. S. (2012). A review size-exclusion chromatography for the analysis of protein biotherapeutics

and their aggregates. *Journal of Liquid Chromatography & Related Technologies*, **35**(20), 2923–2950.

5. Kopaciewicz, W., and Regnier, F. E. (1982). Nonideal size-exclusion chromatography of proteins: Effects of pH at low ionic strength. *Analytical biochemistry*, **126**(1), 8–16.
6. Arakawa, T., Ejima, D., Li, T., and Philo, J. S. (2010). The critical role of mobile phase composition in size exclusion chromatography of protein pharmaceuticals. *Journal of Pharmaceutical Sciences*, **99**(4), 1674–1692.
7. Golovchenko, N. P., Kataeva, I. A., and Akimenko, V. K. (1992). Analysis of pH-dependent protein interactions with gel filtration medium. *Journal of Chromatography A*, **591**(1), 121–128.
8. Engelhardt, H., and Ahr, G. (1983). Optimization of efficiency in size-exclusion chromatography. *Journal of Chromatography A*, **282**, 385–397.
9. Mori, S., and Barth, H. G. (2013). *Size Exclusion Chromatography*. Springer Science & Business Media.
10. Popovici, S. T., and Schoenmakers, P. J. (2005). Fast size-exclusion chromatography—Theoretical and practical considerations. *Journal of Chromatography A*, **1099**(1), 92–102.
11. Fekete, S., Ganzler, K., and Guillarme, D. (2013). Critical evaluation of fast size exclusion chromatographic separations of protein aggregates, applying sub-2 μm particles. *Journal of Pharmaceutical and Biomedical Analysis*, **78**, 141–149.
12. Goyon, A., Guillarme, D., and Fekete, S. (2017). The importance of system band broadening in modern size exclusion chromatography. *Journal of Pharmaceutical and Biomedical Analysis*, **135**, 50–60.
13. Wheaton, R. M., and Bauman, W. C. (1953). Non-ionic separations with ion exchange resins. *Annals of the New York Academy of Sciences*, **57**(3), 159–176.
14. Hjertén, S., and Mosbach, R. (1962). "Molecular-sieve" chromatography of proteins on columns of cross-linked polyacrylamide. *Analytical Biochemistry*, **3**(2), 109–118.
15. Bouvier, E. S., and Koza, S. M. (2014). Advances in size-exclusion separations of proteins and polymers by UHPLC. *TrAC Trends in Analytical Chemistry*, **63**, 85–94.
16. Goyon, A., Beck, A., Colas, O., Sandra, K., Guillarme, D., and Fekete, S. (2016). Evaluation of size exclusion chromatography columns packed with sub-3 μm particles for the analysis of biopharmaceutical

proteins. *Journal of Chromatography A,* DOI: 10.1016/j.chroma.2016.11.056.

17. Yang, R., Tang, Y., Zhang, B., Lu, X., Liu, A., and Zhang, Y. T. (2015). High resolution separation of recombinant monoclonal antibodies by size-exclusion ultra-high performance liquid chromatography (SE-UHPLC). *Journal of Pharmaceutical and Biomedical Analysis*, **109**, 52–61.
18. Wu, C. S. (1999). *Column Handbook for Size Exclusion Chromatography*. Academic Press.
19. Goyon, A., Beck, A., Veuthey, J. L., Guillarme, D., and Fekete, S. (2016). Comprehensive study on the effects of sodium and potassium additives in size exclusion chromatographic separations of protein biopharmaceuticals. *Journal of Pharmaceutical and Biomedical Analysis*. DOI: 10.1016/j.jpba.2016.09.031.
20. Carpenter, J. F., Bain, D. L., and Johnson, G. R. (2016). Use of analytical ultracentrifugation as an orthogonal method for size exclusion chromatography: Assuring quality for therapeutic protein products and meeting regulatory expectations. *Analytical Ultracentrifugation,* Springer Japan, pp. 389–395.
21. Philo, J. S. (2006). Is any measurement method optimal for all aggregate sizes and types?. *The AAPS journal*, **8**(3), E564–E571.
22. Ejima, D., Yumioka, R., Arakawa, T., and Tsumoto, K. (2005). Arginine as an effective additive in gel permeation chromatography. *Journal of Chromatography A*, **1094**(1), 49–55.
23. Ishibashi, M., Tsumoto, K., Tokunaga, M., Ejima, D., Kita, Y., and Arakawa, T. (2005). Is arginine a protein-denaturant?. *Protein Expression and Purification*, **42**(1), 1–6.
24. Wakankar, A., Chen, Y., Gokarn, Y., and Jacobson, F. S. (2011). Analytical methods for physicochemical characterization of antibody drug conjugates. *mAbs*, **3**(2), 161–172.
25. Malawer, E. G., DeVasto, J. K., Frankoski, S. P., and Montana, A. J. (1984). Size exclusion chromatography of poly (vinylpyrrolidone): I. The chromatographic method. *Journal of Liquid Chromatography*, **7**(3), 441–461.
26. Monograph <129> Analytical Procedures for Recombinant Therapeutic Monoclonal antibodies, USP-NF PF39(3).
27. Hawe, A., Kasper, J. C., Friess, W., and Jiskoot, W. (2009). Structural properties of monoclonal antibody aggregates induced by freeze–thawing and thermal stress. *European Journal of Pharmaceutical Sciences*, **38**(2), 79–87.

28. Liu, B., Guo, H., Xu, J., Qin, T., Xu, L., Zhang, J., and Dai, J. (2016). Acid-induced aggregation propensity of nivolumab is dependent on the Fc. *mAbs*, **8**(6), 1107–1117.

29. Nicoud, L., Cohrs, N., Arosio, P., Norrant, E., and Morbidelli, M. (2015). Effect of polyol sugars on the stabilization of monoclonal antibodies. *Biophysical Chemistry*, **197**, 40–46.

30. Hernández-Jiménez, J., Salmerón-García, A., Cabeza, J., Vélez, C., Capitán-Vallvey, L. F., and Navas, N. (2016). The effects of light-accelerated degradation on the aggregation of marketed therapeutic monoclonal antibodies evaluated by size-exclusion chromatography with diode array detection. *Journal of Pharmaceutical Sciences*, **105**(4), 1405–1418.

31. Zheng, K., Bantog, C., and Bayer, R. (2011). The impact of glycosylation on monoclonal antibody conformation and stability. *mAbs*, **3**(6), 568–576.

32. Koza, S., Hong, P., and Fountain, K. J. (2012). Size-exclusion ultra performance liquid chromatography analysis of insulin (Application note No. 720004271EN). Retrieved from Waters Corporation website: http://www.waters.com/webassets/cms/library/docs/720004271en.pdf.

33. Gjoka, X., Schofield, M., Cvetkovic, A., and Gantier, R. (2014). Combined protein A and size exclusion high performance liquid chromatography for the single-step measurement of mAb, aggregates and host cell proteins. *Journal of Chromatography B*, **972**, 48–52.

34. Haberger, M., Leiss, M., Heidenreich, A. K., Pester, O., Hafenmair, G., Hook, M., and Bulau, P. (2016). Rapid characterization of biotherapeutic proteins by size-exclusion chromatography coupled to native mass spectrometry. *mAbs*, **8**(2), 331–339.

35. Schmidt, A. C., Fahlbusch, B., and Otto, M. (2009). Size exclusion chromatography coupled to electrospray ionization mass spectrometry for analysis and quantitative characterization of arsenic interactions with peptides and proteins. *Journal of Mass Spectrometry*, **44**(6), 898–910.

36. Liu, H., Gaza-Bulseco, G., and Chumsae, C. (2009). Analysis of reduced monoclonal antibodies using size exclusion chromatography coupled with mass spectrometry. *Journal of the American Society for Mass Spectrometry*, **20**(12), 2258–2264.

37. Li, Y., Gu, C., Gruenhagen, J., Zhang, K., Yehl, P., Chetwyn, N. P., and Medley, C. D. (2015). A size exclusion-reversed phase two dimensional-liquid chromatography methodology for stability and

small molecule related species in antibody drug conjugates. *Journal of Chromatography A*, **1393**, 81–88.

38. Li, Y., Stella, C., Zheng, L., Bechtel, C., Gruenhagen, J., Jacobson, F., and Medley, C. D. (2016). Investigation of low recovery in the free drug assay for antibody drug conjugates by size exclusion—reversed phase two dimensional-liquid chromatography. *Journal of Chromatography B*, **1032**, 112–118.
39. He, Y., Friese, O. V., Schlittler, M. R., Wang, Q., Yang, X., Bass, L. A., and Jones, M. T. (2012). On-line coupling of size exclusion chromatography with mixed-mode liquid chromatography for comprehensive profiling of biopharmaceutical drug product. *Journal of Chromatography A*, **1262**, 122–129.
40. Gazal, E. (2014). Can size exclusion chromatography (SEC) be done on sub-3 um particles. *In 17th annual meeting of the Israel Analytical Chemistry Society*, Tel Aviv, Israel.
41. Rea, J. C., Moreno, G. T., Vampola, L., Lou, Y., van Haan, B., Tremintin, G., and Farnan, D. (2012). Capillary size exclusion chromatography with picogram sensitivity for analysis of monoclonal antibodies purified from harvested cell culture fluid. *Journal of Chromatography A*, **1219**, 140–146.
42. Farnan, D., Moreno, G. T., Stults, J., Becker, A., Tremintin, G., and van Gils, M. (2009). Interlaced size exclusion liquid chromatography of monoclonal antibodies. *Journal of Chromatography A*, **1216**(51), 8904–8909.

Chapter 17

Countercurrent Chromatography Using Biphasic Molecular Liquids and Ionic Liquids: Is This the Future for Laboratory and Process Purifications?

Leslie Brown and Martyn J. Earle

AECS-QuikPrep Ltd, Unit 5, Higher Trevonick Business Park, Winnards Perch, St Columb Major, Cornwall, England TR9 6DH, United Kingdom

aecs@gmx.com, martynearle@sky.com

17.1 Introduction

Countercurrent chromatography is a separation technique for the isolation of a mixture using a liquid stationary phase held in place by centrifugal force. Many texts are available that describe countercurrent chromatography and related instrumentation [1–6]. In addition, several papers by Berthod et al. discuss the use of ionic liquids in centrifugal partition chromatography (CPC)

Chromatographic Method Development
Edited by Gregory K. Webster and Laila Kott

ISBN 978-981-4800-53-2 (Hardcover), 978-0-429-20172-1 (eBook)
www.jennystanford.com

and CCC [7–12]. It has been suggested, that the high viscosity of ionic liquids would be applicable in hydrodynamic (planetary centrifuges) or hydrostatic (sun centrifuges), countercurrent chromatography [8]. Berthod et al. reported that the use of ionic liquids in CCC was found to have a number of drawbacks such as *"UV absorbance limiting the use of the convenient UV detector,"* *"non-volatility precluding the use of the evaporative light-scattering detector for continuous detection"* and *"high viscosity producing pressure build-up"* [8]. The authors conclude that CCC was unsuitable at that time, as the instrumentation available *"was not able to cope with the high back-pressures generated by the large viscosities of many ionic liquids"* [8]. This view of the use of ionic liquids in CCC was correct with the instrumentation available at the time. However, with careful experimental design and the use of wider coils and pipework, it has been possible to use ionic liquids in CCC. Several papers have been published that show the successful use of ionic liquids and related deep eutectic solvents (DES) in CCC [13–19].

17.1.1 Countercurrent Chromatography/Centrifugal Partition Chromatography Instrumentation for Ionic Liquid Applications

Research into the fundamental understanding of differences and similarities between 2D and 3D coil winding in CCC is ongoing. This coil arrangement is unique in enabling multi-layer 2D coil interactions between the phases of a biphasic solvent system to be visualized and recorded. This 2D coil system can be interchanged with 3D multi-layer coils of various beta vales (ratio of sun to planet radii). It has been confirmed that ionic liquid solvent systems behave differently than biphasic solvent systems created from molecular solvents.

17.1.2 CCC/CPC Separations Using Ionic Liquids

Earlier publications stating that ionic liquids could not be easily utilized with CCC/CPC [8] may have held back the development of ionic liquid solvent systems in CCC/CPC. However, recently biphasic IL's were successfully employed in the direct separation of a mixture of $CoCl_2$, $NiCl_2$, and $CuCl_2$, the separation of two monosaccharides from a disaccharide, and the extraction of cumene from hexane [15]. By the use of hexane/ionic liquid solvent system, the separation of hydrocarbons from oxygenated hydrocarbons in vetiver oil was achieved [20]. In addition, ionic liquid solvent systems have been studied in the countercurrent chromatographic separation nonpolar lipids [16]. Cao et al. have used dilute solutions of ionic liquids in CCC, in the determination of *Alternaria* mycotoxins, and the determination of chlorophenols in red wine [13, 14]. Also, four patents have been published on the use of ionic liquid in CCC solvent systems in general [21], in bio-organic [22], organic [23], and inorganic separations [24].

Other projects have carried out research into the use of chiral ionic liquids [25, 26] as the stationary phase in chiral separations. This research has been based on earlier chiral separations using liquid–liquid chromatography in the literature. Two examples of chiral separations by CCC are in the use of solutions of chiral salts [27] (similar to an ionic liquid), derived from cinchona alkaloid, in the separation of N-(3,5-dinitrobenzoyl)-(+/-)-leucine. Similarly, cyclodextrin and a chiral copper complex have been used in the enantioseparation of naringenin by CCC [28].

17.1.3 Aqueous Biphasic Solvent Systems Using Ionic Liquids

Aqueous biphasic solvent (ABS) systems [29] in CCC are used for the separation of polar, or strongly hydrogen bonding solutes [30]. They are usually made from mixtures of concentrated inorganic salts combined hydrophilic polymers, or alcohols. The two phases generated in ABS systems both have water as a major component in each phase. The first use of ionic liquid-based ABS systems has been described by Berthod et al. [10] and more recently by Earle et al. [15, 22]. The solvent system in these three publications used was based on a mixture of 1-butyl-3-methylimidazolium chloride ([C_4mim]Cl) and 1.5–3.0M $K_2[HPO_4]_{(aq)}$ which forms two relatively viscous liquid phases [10, 31]. The distribution ratios of solutes can be adjusted by varying the concentration of $K_2[HPO_4]$ present in the mobile phase. This solvent system has been used in the separation of glucose or fructose from sucrose [23] and in the purification of the anti-cancer drug lentinan [22]. With a 2.1 mm internal diameter (ID) coil, high backpressures can be encountered at high flow rates and ambient temperatures [15]. In order for the coils to withstand the 70 Bar backpressure generated at >25 cm^3 min^{-1} flow rate and temperatures at <40°C, the coils must be constructed out of stainless steel, rather than the more conventional PTFE tubing. The PTFE flying leads, PTFE tubing and PEEK connectors must also be able to withstand this pressure, and were tested by prolonged running (48 h), at maximum rotation speed of the instrument while maintaining 70 Bar continuous backpressure.

Lentinan, found in shiitake mushrooms (*Lentinus edodes*), has been used in ancient Chinese medicine for the treatment of cancer and is used as an adjunct to therapy in combination with chemotherapeutic drugs such as fluorouracil in order to modulate the body's immune system activity [32]. Lentinan derived from shiitake mushrooms is a β-1,3-1,6-glucan polymer with β-1,6 branching on a β-1,3-glucan backbone. Lentinan naturally exists in a triple-helix form in water and salt solutions [33], but is easily denatured. In one clinical trial, it was used to cure 30% of the patients suffering from terminal gastric cancer, and approximately doubled the life expectancy of 70% of the patients who did not survive [32, 34]. The purification of lentinan is normally an extremely difficult process involving up to 10 steps [35]. The isolation not only requires resolution from crude mushroom extract but also needs the maintenance of the polysaccharide's three-dimensional, triple-helix configuration to maintain its bioactivity. Lentinan is denatured to a single helix form, which is not as biologically or medicinally, active by heating at temperatures above 80°C. This denaturing of lentinan has been observed in DMSO-based solvent systems at 25°C [36]. Lentinan has been purified by HSCCC using an aqueous biphasic solvent system based on polyethylene glycol (PEG) and aqueous mixtures of $K[H_2PO_4]$ and $K_2[HPO_4]$ [37]. The main drawbacks with this process are that the lentinan produced is contaminated with PEG and phosphate salts. Also, the scale of this separation was less than 100 mg per run, which is too low for commercial production. The $[C_4mim]Cl$/2.5M $K_2[HPO_4]_{(aq)}$ (1:1) solvent system [10] has the remarkable property that it does not damage or denature complex biological molecules when they are dissolved, and this property has been used in the purification of lentinan. The ionic liquid-ABS process uses the $K_2[HPO_4]_{(aq)}$ as the stationary phase and $[C_4mim]Cl$ as the mobile phase. The lentinan was separated from the $[C_4mim]Cl$ ionic liquid phase simply by the addition of ethanol to the ionic liquid solution, which then precipitates the triple-helix form of lentinan. In biological testing against cancer cells, the lentinan produced by the ionic liquid process methodology has the highest biological activity available [38].

17.2 Optimization of Ionic Liquid and Molecular Biphasic Solvent Systems for CCC/CPC

17.2.1 Introduction

Ionic liquid research came about as a result of three discoveries: (1) the adoption of water stable neutral ionic liquids such as $[C_4mim][PF_6]$ [39–41] and $[C_4mim][BF_4]$ [42, 43], (2) the discovery that a wide range of chemical reactions can be carried out in ionic liquids [44–46], and (3) the realization that the number and range, and diversity of potential ionic liquids that could be synthesized is enormous [47].

Ionic liquids have unique properties that can help overcome some of the problems associated with the use of conventional organic solvents [44, 47]. Ionic liquids are materials composed entirely of ions, with low melting points (usually <100°C) and can dissolve a huge range of organic and inorganic compounds, polymers, bio-molecules and other salts [48]. They have no measurable vapor pressure at ambient temperatures and pressures [49], resulting in their non-flammability and fire resistance [50]. These properties and their general stability make many ionic liquids inherently safe and environmentally friendly solvents [47]. Their use in synthesis [44], catalysis [51] and separations [52] has expanded significantly in recent years and they are now being applied in a number of industrial processes [53]. Moreover, the properties of ionic liquids can be tuned or adjusted by the choice of the appropriate anion and cation allowing them to be optimized for a particular application [47].

17.2.2 Ionic Liquids in Biphasic Solvent Selection for CCC

Ionic liquids phase behavior can be varied and used to produce two, three [54] or four-phase solvent systems in combination with conventional molecular solvents [55]. The most common classification of ionic liquids is whether they are either hydrophobic or hydrophilic. The term hydrophilic means that the ionic liquid is miscible with water in all proportions and the term hydrophobic corresponds to an ionic liquid that can form a biphasic mixture with water. It should be noted that hydrophobic ionic liquids, when mixed with water, often contain substantial amounts of water in the ionic phase [56]. Another factor to note is that many solvents which form biphasic mixtures with ionic liquids, have very low solubility of the ionic liquid in the molecular solvent phase. This is despite the ionic phase containing appreciable amounts of the molecular solvent. The difference in mutual solubility of ionic liquids and molecular solvents was the miscibility of pentan-1-ol and [C_6mim][BF_4] at different temperatures [57]. It has been observed that ionic liquids are not necessarily miscible with each other, and biphasic ionic liquid mixtures can be produced [55]. This generally occurs when there is a considerable difference in the size and shape of both the cation and anion. The mutual miscibility of two ionic liquids (1-ethyl-3-methylimidazolium bis(trifluoromethanesulfonyl)amide) [C_2mim][NTf_2] and (trihexyltetradecylphosphonium bis(trifluor omethanesulfonyl)amide) are very different, with [C_2mim][NTf_2] being 15 times more soluble in [$P_{6\,6\,6\,14}$][NTf_2] than [$P_{6\,6\,6\,14}$][NTf_2] [C_2mim][NTf_2] [55].

The phenomena of the very low solubility of hydrophobic ionic liquids in water and non-polar solvents like hexane [52] are very important in ionic liquid–liquid chromatography. As a result, the mobile water or organic solvent containing phase is not or minimally contaminated with ionic liquid. This makes product isolation very straightforward, and organic solvent recovery simply involves distilling off the volatile solvent, which can then be reused. The effect of dissolving a molecular solvent in an ionic liquid dramatically reduces the ionic phase's viscosity [58] to the point where it can be effectively pumped by an HPLC. This has enabled viscous hydrophobic ionic liquids such as [$P_{6\,6\,6\,14}$]Cl to be used in CCC (viscosity = 1450 or 2729 cP at 25°C, depending on the source of the measurement) [15, 24, 59]. An example of this (was measured by Stark et al.) is seen in the change in viscosity and density of the ionic liquid $[C_4mim][BF_4]$ as water was added [58]. The result of this was that the viscosity of the ionic was dramatically reduced while the density showed a linear relationship with the change in composition of the mixture. The addition of only 2 wt.% of water to $[C_4mim][BF_4]$ halved the viscosity of the ionic liquid phase. In addition, the reduction in viscosity of the $[C_4mim][BF_4]$ phase is largely independent of the structure of the molecular solvent. The viscosity of the ionic phase can therefore be predicted and modeled by the equation $\eta = \eta_s \exp(-x_{cs}/\alpha)$.

The biphasic solvent systems that employ ionic liquids for use in CCC and CPC generally use mixtures with one or two molecular solvents, rather than three to five components that are commonly found in conventional CCC solvent systems. The approach to solvent selection is to design or select an ionic liquid with a chemical structure which is suited to a particular separation. For polar, strongly hydrogen bonding solutes, ionic liquids with small anions and cations, and for low polarity solutes ionic liquids with long hydrocarbon chains are used. Examples of ionic liquid solvent systems tested in separations are summarized in Table 17.1 and the structure of the ionic liquids is shown in Fig. 17.1.

Table 17.1 Phase combinations used for ILLC separations

Ionic liquid phase	Ionic liquid phase density and pump direction	Mobile phase	Separation
$[C_{12}mim][NTf_2]$	MDP T to H	Hexane	Alkene/oxygenated terpenes in vetiver oil
$[C_{10}mim][OTf]$	MDP T to H	Hexane	Alkane/aromatics
$[C_4mim][OTf]$	MDP T to H	Toluene/hexane	Cyclohexanol from cyclohexanone
$[C_2mim][OTf]$	MDP T to H	Toluene/hexane	Polycyclic aromatic hydrocarbons
$[C_4mim]Cl$	LDP T to H	3.0 M $K_2[HPO_4]$	Glucose/sucrose

(*Continued*)

Table 17.1 (*Continued*)

Ionic liquid phase	Ionic liquid phase density and pump direction	Mobile phase	Separation
$[C_4mim]Cl$	LDP H to T	2.5 M $K_2[HPO_4]$	Lentinan
$[P_{6\,6\,6\,14}]Cl$/ ethyl ethanoate	LDP H to T	Water	Aqueous Co(II), Ni(II), Cu(II) chlorides

Note: Light gray = stationary phase; dark gray = mobile phase; MDP = most dense phase; LDP = least dense phase; T to H = tail to head of coil pump direction; H to T = head to tail of coil pump direction.

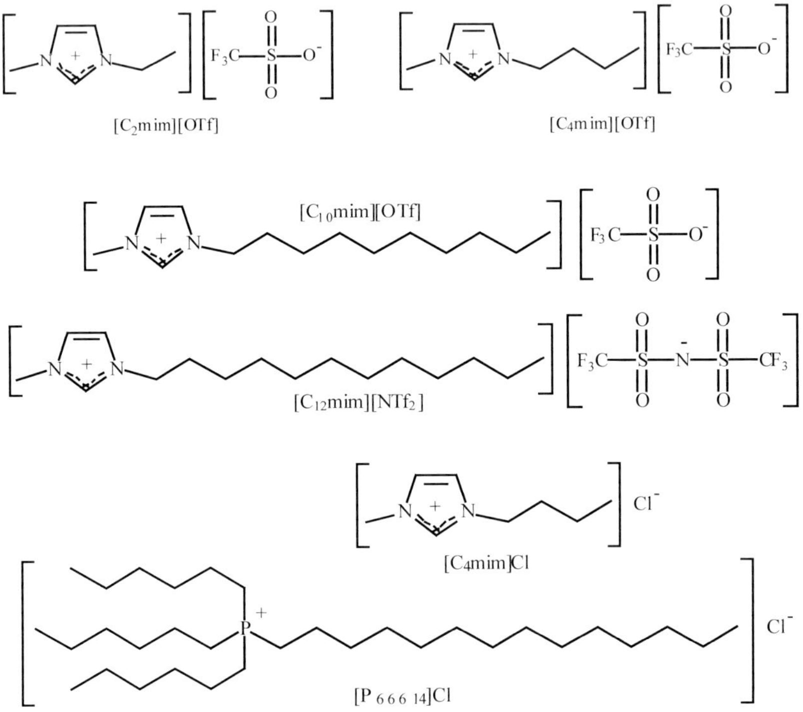

Figure 17.1 The structure of the ionic liquids used in Table 17.1.

17.2.3 CCC/CPC Method Development

CCC/CPC users should apply their fundamental chemical and chromatographic knowledge to solvent systems and separations, and not simply base their research on well-known biphasic mixtures. The latter approach is still too common in CCC/CPC [60, 61]. There are very early research publications on the generic solvent systems such as the *n*-hexane:ethyl acetate:methanol:distilled water solvent systems, which, for a decade or more, was referred to as the "Arizona system" [62] but was eventually renamed to become the "HEMWat system" [63]. By varying the ratios of *n*-hexane: ethyl acetate:methanol:water, these solvents can be used to separate compounds with a large range of polarities. A ratio of 1:0:1:0 (*n*-hexane: methanol), would be a suitable biphasic solvent system for non-polar compounds. A ratio of 0:1:X:1 is a good starting point for the separation of polar compounds, and a ratio of 1:1:1:1 is a good starting position for a generic screen of solute distribution ratios [64–66]. Changing methanol to ethanenitrile is an option since this can improve settling times of these biphasic solvent systems.

This use of complex mixtures approach to solvent selection would be to assess first the 1:1:1:1 solvent mixture and then use a sequence of pre-set, stated ratios of solvents to separate compounds of different polarities, until a partition coefficient (K) of between 0.5 and 3.0 is obtained [64–66]. A value of 1.0 for K is often seen as optimal for CCC separations. $K = 1$ will in CCC/CPC separations be equivalent to one coil or rotor volume of solvent. Usually a K value of ≤ 0.5 is considered to give rise to excessively early analyte elution and $K \geq 3$ too excessively late values of elution times for a solute or analyte. If no suitable K values or solvencies are found, addition solvents, again with pre-set ratios would be sequentially tried [64–66].

CCC/CPC separations need a more thought through and less trial-and-error–based long-winded approach [65]. The principle of LIKE/LIKES/LIKE [67] is fundamental to all chromatographic method development. It is recommended that chromatographers should always consider the chemistry and aim to choose functionality, affinity, solvency (the extent to which a solute dissolves in a given solvent (in units of g cm^{-3})) [68], and polarity of solvent, to match the key functionality of target compound(s) in the initial stages of the biphasic solvent choice. Tests should be carried out (as would normally be carried out for recrystallization of products), for not just solvency in single solvent, but rather for solvency in a biphasic solvent system. These can be quite different from single phase or single constituent solvents.

In preparative chromatography, solvency [68] is a key factor, as higher solvency will result in a greater mass of crude product per unit volume of the instrument coils or rotors that can be reliably injected and separated. An example of this is in the purification of lentinan [22], where the lentinan is much more soluble in the ionic liquid-based solvent system 2.5 M $K_2[HPO_4]/[C_4mim]Cl$, than the more conventional aqueous $K_2[HPO_4]/K[H_2PO_4]$/PEG-1000-based solvent system [37]. The result is that much higher sample concentrations can be used, and the space time yield [69] of the lentinan produced is increased by a factor estimated to be 5 to 10 times [22, 37].

Familiarity with the Snyder Polarity Triangle [70] and various classifications of solvent polarity [71], and its implications can be used in achieving orthogonal selectivity [72, 73]. The octanol: water partition coefficients (OWPCs) [74] are available in many publications, and websites, and an approximate value for these OWPCs can be calculated from a compounds empirical formula and other data [75–80]. If the functionality of a solute is known, and a simple solvency value is obtained, plus, if the crude determination of the OWPC's is known, then the biphasic solvent system should become obvious to chromatographers, without the need for large-scale trial-and-error studies. Biphasic solvent choice tests are very simple and described in Table 17.2.

17.2.4 The "Attraction-Repulsion Principle" of Biphasic Eluent Optimization

The original attraction-repulsion principle (ARP) has been recently updated to accommodate the different behavior of ionic liquid containing biphasic eluents. ARP methodology is based on fundamental chemical principles, rather than the utilization of set ratios of pre-chosen solvents that make up biphasic solvent systems of fixed composition, as is often utilized in CCC and CPC [81].

Table 17.2 The procedure for developing CCC/CPC solvent systems

Step	Procedure
1	A range of solvents of varying structure and polarity include *n*-hexane, *n*-heptane, limonene, liquid carbon dioxide,* sub-critical fluids such as R134a,* as relatively non-polar or low polarity choices; ethyl ethanoate, MTBE, *n*-butanol, chloroform, dichloromethane as mid polarity options; and ethanenitrile, ethanol, methanol and distilled water as high polarity options. For unknown compounds or analytes, screen the above solvents as a single liquid to determine the solvency. Choose those solvents which form a rapid settling biphasic mixture, and then test the solvency and partition coefficient in this biphasic mixture.
2	In a 10 ml stoppered measuring cylinder, add a few milligrams of the analyte to 2 ml of solvent to be used separately and note their solvency.
3	Mix pairs of solvents known to form biphasic solvent systems when combined (2 ml of each). For non-polar analytes, *n*-heptane or *n*-hexane, etc., mixed with ethanol/methanol/ethanenitrile/distilled water gives some preferred solvent system options. For mid to high polarity analytes, combinations such as *n*-butanol, ethyl acetate, chloroform, dichloromethane, PEG, or dextran with distilled water are options. Next, if appropriate, measure the partition coefficient (K) by determining concentration of the target or analyte in both the upper and lower phase, where $K = [\text{analyte}]_{SP}/[\text{analyte}]_{MP}$. Also note the settling time of the biphasic solvent systems created. This is accomplished by stoppering the measuring cylinder and vigorously shaking for 60 s then recording the settling times and volumes of upper and lower phase.
4	In over 80% of the cases, this simple procedure will accurately predict the elution position of targets in CCC or CPC. Where $K = 1$, the analyte will elute in one coil volume, where $K = 0.5$ is half a coil volume, and for $K = 2$, 2 coil volumes, etc.
5	For cases where the above prediction fails, this is most often caused by emulsion formation in the CCC or CPC instrument, and experiments utilizing the instruments, in different modes and using different injected masses may be needed to optimize the method.

(*Continued*)

Table 17.2 (*Continued*)

Step	Procedure
6	For biphasic solvent systems, where inappropriate or unacceptable settling times and/or partition coefficients are found, the ARP is used by the addition of a modifier solvent, which is miscible with the least favored solvent phase, or a solvent which is miscible with the most favored solvent can be added to adjust the partition coefficient to preferred range.
7	For laboratory-only methods, simple optimization as in (1) to (6) can be used with 1, 2, or even 3 additional solvents to rapidly achieve the required partition coefficients, without the need to test a pre-set number of set solvent ratios.
8	For process applications and recycling of solvents, the least number of solvents necessary to obtain the desired distribution ratios is preferred and often necessary. This can be achieved using ionic liquid solvent systems that usually have only two or at most three components. Here the ionic liquid is designed to suit the separation, rather than using complex mixtures to give optimal separations. This is where the use of ionic liquids can pay major dividends.
9	For CCC with molecular solvents, it is very important that the biphasic solvent system chosen has a settling time of less than 60 s and ideally less than 30 s, to achieve good stationary phase retention. For CPC, slower settling times for the solvent system can be used successfully, but in our experience, this often achieves a lower stationary phase retention than is normally found in CCC for rapid settling biphasic eluents.

*Requires autoclave or high-pressure apparatus.

Abbreviations: MP, mobile phase; SP, stationary phase; MTBE, methyl *t*-butyl ether or 1-methoxy-1,1-dimethylethane; R134a, 1,1,1,2-Tetrafluoroethane.

When utilizing the ARP with molecular solvents and ionic liquid-based biphasic eluents, consider using two immiscible solvents as potential biphasic eluents. These are chosen based on the known (or sometimes unknown) experimentally determined solubility of the solute or analyte in a particular solvent (similar to experimentation in recrystallization optimizations). With reference to the settling time and partition coefficient of solutes (if the target is known) (or use the polarities or bioactivity if the target compound is not known), consider the addition of a partition coefficient modifier (PCM) solvent or solvents. Based on the procedure in Table 17.2, it will be found that certain solvents will have a high affinity for compounds of interest, whilst other solvents will display immiscibility or low solubility.

- For solvents with high solubility for the analytes, consider these solvents as "Attractors."
- For solvents with low solubility for the analytes, consider these solvents as "Repulsors."

For example, mixing *n*-hexane, *n*-heptane, cyclohexane, pentane, toluene, chloroform, dichloromethane, ethyl acetate, MTBE, *n*-butanol, etc., with distilled water will all create excellent biphasic solvents. If methanol or ethanenitrile are added to these biphasic mixtures, both will predominately distribute them into the water phase (there are few exceptions, such as with ethanenitrile preferentially dissolving in the organic phase of a water-butanenitrile biphasic mixture). If both methanol and ethyl acetate are added to a biphasic water-organic solvent mixture, with which they are mutually soluble, then each will draw the other into its preferred phase. Adding too much ethyl acetate and/ or methanol to a biphasic solvent system will result in the settling time increasing, and eventually, at critical point, the solution will become mono-phasic.

The solvency, *K* value, and settling times are all easily obtained by adding crude mixture to be separated to the biphasic solvent system (this only needs to be a few milligrams), and measuring the concentration of each component of the mixture contained in each phase. This can be achieved using the following methodologies: GC (volatile compounds, but not ionic liquid containing samples), HPLC, ion chromatography (for salts and ionic liquids), ^{1}H or ^{13}C NMR (for organic compounds and ionic liquids), UV-Vis spectrophotometry (for UV active or colored compounds), polarimetry (for chiral compounds), and AA/ICP (for metal compounds and salts). These measurements can also be run on the individual components of the mixture providing that they are available.

For the partitioning of the analyte into the starting biphasic solvent system, which has been chosen to have high solubility of the analyte, the ideal partition coefficient should be between

0.75 and 3.0. The partitioning of the analyte can be modified in two ways. The addition of an "Attractor" solvent to the phase in which the analyte is least soluble (in the biphasic mixture), and which has high solvent to solvent affinity for the least favored phase, to help draw the analyte into the least favored phase, to better equate the partition coefficient of the analyte. For example, if an analyte in a water/hexane solvent system is too soluble in the water phase then the addition of ethyl ethanoate (which is miscible with hexane and immiscible with water) will help draw the analyte into the organic hexane/ethyl ethanoate phase.

Equally using the same logic, a "Repulsor" can be used to drive the target compound from the more favored phase into the least favored phase. An example with a moderately polar solute dissolved in a water/ethyl ethanoate biphasic solvent system that is preferentially soluble in the ethyl acetate phase. The solute can be forced into the water phase by the addition of hexane to the ethyl acetate phase. Both options can be utilized together as required. The choice "attractor" or "repulsor" solvents or additives will be driven by chemical structural knowledge of target (if known), the solubility of the analyte in the solvents involved (defined by very simple solubility tests), and by biphasic settling tests (discussed below). The greater the stationary phase retention value for biphasic solvent systems, the greater will be the resolution, the amount of solute which can be separated, and the greater flow rate/linear flow rate that can be achieved, which reduces separation times.

To summarize, the appropriate biphasic solvent system can be derived from solvency alone, since solvency is a key indicator of a solutes polarity. As is the case with solid–liquid chromatography used for ionic compounds, or with the use of a pH gradients in separations, and in frontal chromatography, similar techniques in CCC and CPC can be used, and have been found to be very effective [82–84]. This was sometimes referred to as "pH Zone Refining" for CCC or CPC, and this terminology has been generally accepted by CCC/CPC chromatographers [85]. In both solid liquid and liquid–liquid chromatography, as the pH changes, compounds will, at their pKas, change their mobility and move from the stationary phase to the mobile phase. If a compound (ideally your target) has a high enough relative loading, it will act as a buffer for the eluting phase. Having buffered the eluent, initially all compounds with that pKa will elute. If one solute (ideally your target) is in relative excess, it will move only, at its buffered pH, giving a square wave shape signal in the detector. The retention will be determined by its solubility in the eluent, and will be the only component to elute. Once this compound has been completely eluted, the pH gradient will be reestablished and will not depend on other compounds that can elute at later pHs. The use of this methodology can often allow an increase in sample loading by 2 to 4 times, over and above the already high loading of standard CCC/CPC separations [86]. A list of simple solvent systems for use in CCC is shown in Table 17.3. The solvent systems that are suitable depend largely on the polarity of the compounds being separated.

One important factor, particularly with Flash and HPLC chromatography, is how much mass of analyte can be retained when complex mixtures are put onto silica or end-capped modern reverse phase silica chromatography columns. This problem of sample loss does not occur with CCC/CPC, when the elution/extrusion separations are carried out. The full mass balance is nearly always obtained with CCC/CPC type separations. Silica and reversed-phase silica stationary phases, including fully end-capped reverse phases, can change the nature of eluting compounds, with respect to those injected. This can be due to on-column degradation, and in particular, degradation of acid labile compounds [87]. Generally, CCC/CPC techniques do not decompose compounds, assuming that the compounds are stable in the solvent system used. For bio-active samples, if samples are run on a CCC/CPC instrument, they generally remain bioactive. However, if samples are run in solid-phase columns, such as Flash or HPLC columns, the bioactive complex can be damaged, and the bioactivity lost. CCC/CPC techniques and solvent systems can allow the retention of the nature or structure of these molecules in the form that they naturally occur, thereby maintaining the bioactivity. As many natural bioactive compounds occur as complexes chromatographers using solid–liquid chromatography denature the complexes found in their natural state.

Table 17.3 The choice of standard biphasic solvent systems for molecular solvents and ionic liquids (in gray background)

Polarity of analyte or solute	Conventional solvent system
	Ionic liquid solvent system
Very high polarity	Aqueous biphasic solvents systems (ABS), such as PEG or dextran/phosphate salt/water
Very high polarity	$[C_4mim]Cl$/1.5 M–3.0 M $K_2[HPO_4]$ [choline][leucine]/25 wt.% $K_3[PO_4]$ [88]
High polarity (not readily recovered from above biphasic mixtures)	*n*-Butanol/methanol/water *n*-Butanenitrile/ethanenitrile /water Ethyl acetate/*n*-butanol/water DMSO/ethanol: salt solution Consider additives such as MTBE, DCM, chloroform, etc., based on target functionality if known.
High polarity	Water/[cholinium]$[NTf_2]$, or Water/methanol/$[C_2mim][NTf_2]$ $[C_4mim][OTf]$/0.5 M–2.0 M $K_2[HPO_4]$
Medium Polarity	Arizona (HEMWat) solvent system [63, 66]. Consider additives such as MTBE, DCM, Chloroform, etc., based on target functionality if known.
Medium polarity	$[C_2mim][OTf]$/toluene $[C_2mim][OMs]$/ethyl acetate
Low polarity	*n*-Hexane or heptane/methanol and/ethanenitrile. Consider additives such as MTBE, DCM, Chloroform, etc., based on target functionality if known.
Low polarity	$[C_{10}mim][OTf]$/hexane $[C_{12}mim][NTf_2]$/heptane
Very low polarity	Not Available
Very low polarity	$[P_{6\,6\,6\,14}][NTf_2]$/hexane

As can be seen, CCC/CPC with standard solvents has considerable potential in laboratory for both pilot and process purifications. However, as the scale of separations increases, the cost of recycling the solvents can become a key factor. Often, in CCC/CPC, 3 or 4 (or more) solvents are used to form biphasic solvent systems, which makes solvent recovery and recycling of the solvents difficult and costly. An alternative to the use of molecular solvents, supercritical fluids have been researched for their use in CCC/CPC [89, 90]. Research with liquid carbon dioxide and potentially other sub-critical fluids is ongoing, as a means of separating mixtures without the use of volatile organic solvents. Another approach to CCC separations is to simplify the biphasic solvent system used to preferably two components [90]. This requires custom engineered, green, low risk, non-flammable, low volatility, easily recyclable solvents, such as ionic liquids, combined with a single molecular solvent or supercritical solvent such as CO_2 or subcritical solvents such as R134a (1,1,1,2-tetrafluoroethane).

17.2.5 Ionic Liquid Solvent System Selection

For ionic liquid biphasic solvent systems, knowledge of analyte chemistry and likely interactions with ionic liquids can be used to choose readily available ionic liquids for solvent systems, or this information can be used to custom synthesize the most appropriate ionic liquid in terms of functionality, solvency and the properties of the ionic liquid. For ionic liquids solvent system development, a similar procedure to that in Table 17.2 is used. For molecular solvents, it would be unusual to use only two solvents, but this is quite common with ionic liquid solvent systems. This is a large advantage particularly when considering process applications since this and makes solvent recycling much easier. Usually, the ionic liquid is selected or designed to have the correct range of properties for a given separation. The term used to describe the design of ionic liquids to suit a particular end use is called solvent engineering.

For ionic liquid-based biphasic solvent systems in both CCC and CPC, their successful use is very much less dependent on rapid settling times. This is an advantageous situation for ionic liquid-based methods, since many ionic liquids are surface active and would be expected to form emulsions. Research is ongoing as to why ionic liquid solvent systems are so much less sensitive to rapid settling times that are required for molecular solvents in CCC. This difference is thought to be due in part, to ionic liquid solvent systems having a generally greater density contrast between the two phases of the biphasic solvent system, which helps the two phases separate under high "g" loadings in CCC/CPC rotors.

Ionic liquids are said to be immiscible with other solvents when there is a visible phase boundary between two phases where at least one phase contains an ionic liquid. Immiscibility is not however, the same as insolubility. CCC separations require biphasic solvent systems with the phases chosen being either ionic liquid-organic solvent, ionic liquid–water, or mixtures of two or more immiscible ionic liquids. All the components of these biphasic solvent systems will have a degree of solubility in each other, but this mutual solubility can vary considerably. In most cases, biphasic ionic liquid solvent systems have very low ionic liquid solubility in the molecular solvent phase, and as a result, molecular solvents are usually used as the mobile phase and the ionic liquid containing phase as the stationary phase as in Table 17.2. Solvent recovery often involves distillation of the volatile molecular solvent from the mobile phase containing the separated product, which allows the solvent to be reused [91]. Fewer volatile components are used in ionic liquid solvent systems than the Arizona or HEMWat solvent systems [62, 92, 93], makes solvent recovery much more straightforward due to the lower number of volatile components that require separation [81]. Some solutes can remain dissolved in an ionic liquid phase, after the volatile solvent recovery stage. These can be separated from the ionic liquid by a number of different methods, such as solvent extraction, precipitation with an antisolvent, or filtering a solution of the ionic liquid through silica or charcoal, or a combination of these methods [94].

The procedure for the determination of the suitability of a biphasic ionic liquid solvent system for a given separation is as follows: (1) add each solvent of the biphasic solvent system to the sample to be separated, (2) measure and record the solvency in each phase, (3) combine the two phases containing the sample and vigorously shake for 60 s, then record settling time, (4) measure the percentage of each compound dissolved in both the upper & lower phase. With this information, the distribution ratio $K_{D(X)}$ of each component (X) of the sample can be measured. $K_{D(X)}$ = [Solute X]$_{UP}$/[Solute X]$_{LP}$ where X = component of sample, UP = upper phase and LP = lower phase. Conventionally, in CCC separations, one of the phases (usually the ionic phase) is used as the stationary phase. Therefore, $K_{D(X)}$ = [Solute X]$_{SP}$/[Solute X]$_{MP,}$ where $K_{D(X)}$ = distribution coefficient of Solute X, SP = stationary phase and MP = mobile phase. In most cases, these simple tests can be used to predict CCC/CPC elution times and the elution order relative to the elution volume/coil volume. In some cases, the CCC/CPC chromatogram will need to be run, before an optimization of the solvent system composition can be carried out.

For a given biphasic solvent system, the value of K_D can be adjusted by employing the Attraction Repulsion Principle described in Section 17.2.4 using an antisolvent to cause a given solute to move into the opposite phase. As an example, the distribution ratio of lentinan and saccharides in the aqueous $K_2[HPO_4]/[C_4mim]Cl$ solvent system (Table 17.1) can be modified by the addition of more $K_2[HPO_4]_{(s)}$ to the water phase. The $K_2[HPO_4]$ salt is very water soluble and insoluble in the $[C_4mim]Cl$ phase. This causes the lentinan or saccharides to become more soluble in the ionic liquid phase; hence, $K_2[HPO_4]$ can be said to be a "repulsor" compound for the aqueous phase. These solute attraction and repulsion effects can be utilized to fully and predictably adjust K values. This approach, based on adjusting the solvent system by adding to or modifying a solvent system [95], contrasts with the systematic measurement of distribution ratios from a large number of solvent systems described in the literature [64, 81], or by using a computational approach [96, 97].

The performance of CCC/CPC separations depends on differences in the distribution ratios (K_D) of compounds distributed between two liquid phases. Also, mass transfer of solutes between these two phases is a factor that affects separation quality. With ionic liquid solvent systems, the viscosity of the two liquid phases is one of the most important factors. Viscous phases in CCC/CPC can result in excessive bandspreading [98] and high backpressures which require low mobile phase flow rates. Since commonly used ionic liquids have viscosities in the range of 35–2000 cP [99], compared with molecular solvents with viscosities in the range of 0.3 to 10 cP, the use ionic liquids in CCC can be problematic. The high viscosities encountered with ionic liquid solvent systems can be minimized by running CCC/CPC separations at higher temperatures than are conventionally used [100–102]. The properties ionic liquids, including viscosity, density, solvent behavior, and interfacial tensions are available in the literature [103–106] and in the IST Standard Reference Database [107]. This data should be treated with caution, because when an ionic liquid is mixed with a molecular solvent, the properties of the subsequent mixture will be different and with the viscosity of the ionic phase being greatly reduced [59, 108]. This can be estimated using Eyring's absolute rate theory [108].

For the separations in Table 17.1, the biphasic combinations of ionic liquids and molecular solvents were chosen based on previous research into ionic liquid phase behavior and solvent extraction studies [21, 23, 24, 52, 53, 55, 109–114]. Factors affecting the miscibility of solutes in an ionic liquid include the size of the ions, ability of the ions to form hydrogen bonds, its hydrophilicity/hydrophobicity, and the basicity of the anion. Large mostly aliphatic cations (e.g., $[P_{6\ 6\ 6\ 14}]^+$) with very weakly basic anions (e.g., $[NTf_2]^-$ or $[OTf]^-$) produce a phase which is highly immiscible with polar organic compounds (e.g., water or water/methanol mixtures), whereas small polar or strongly hydrogen bonding cations (e.g., cholinium) with small or basic anions (e.g., chloride, ethanoate or hydroxide) are very miscible with polar organic solvents solutes (e.g., water and methanol). The hydrophobicity/hydrophilicity of ionic liquids can also be calculated using software such as COSMO-RS [103].

For effective solvent selection, the distribution ratios of the solutes between the two phases of the solvent system should be in the range of 0.5 to 3.0. Distribution ratios (K_D) can be measured by several means, including electronic absorption, HPLC, NMR spectroscopy of the solute dissolved in each phase. Gas chromatography must not be used for ionic liquid containing phases, because the ionic liquid will not be eluted from the column, and the properties of the column will be altered. For simple biphasic ionic liquid/molecular solvent mixtures, the addition of a co-solvent (usually a second molecular solvent) will significantly affect the distribution ratios of the solutes, as well as significantly reducing the viscosity of the ionic phase [8]. This is a very effective way of fine-tuning the solute distribution ratios to optimize separations [8].

The combination of ionic liquid and molecular solvent(s) chosen must have a large enough density difference between these two phases (>0.05 g cm^{-3}). The hydrophobic ionic liquid [C_{10}mim][OTf] (1.17 g cm^{-3} when pure) [115] cannot be used in combination with water, because the density of the two phases (water and water saturated ionic liquid) have similar densities are slow to separate from each other. However, [C_6mim][OTf] (1.22 g cm^{-3} when pure), and [C_8mim][OTf] (1.19 g cm^{-3} when pure) are slightly more dense than water and readily separate [115]. Some combinations of ionic liquid and molecular solvent should be avoided due to the formation of inseparable emulsions. Although some ionic liquids can be used to break (demulsify) emulsions [116], surface active ionic liquids like [C_{12}mim]Cl [117], when combined with halogenated solvents (CH_2Cl_2 or $CHCl_3$), can produce emulsions which will not separate during CCC operation.

17.2.6 Summary of the CCC Use of Ionic Liquid–Liquid Chromatography

Ionic liquid solvent system advantages

- Can be designed for a specific separation
- Greater range of possible solvent systems
- Greater solute capacity
- Process intensification
- Greater versatility in separations and operating conditions
- Greater range of solutes that can be separated
- Greater range of biphasic solvent systems available
- Allows separations of previously insoluble materials
- Compatible with biological compounds
- Very polar and non-polar compound separations
- Direct metal separations
- Simple solvent systems
- Easy solvent recycling
- Ionic liquids can be recycled
- Best for process scale

Ionic liquid solvent system disadvantages

- Requires solvent engineering to design produce ionic liquids

- Lack of research and development of ionic liquid solvent systems
- Higher viscosities and backpressures
- Bandspreading from higher viscosity of solvent system

Ionic liquid general properties

- Have no vapor pressure at atmospheric pressures [49]
- Millions of potential ionic liquids [47]
- Can form triphasic and tetraphasic solvent systems [40, 55]
- Can be designed for specific end use or phase behavior [105]
- High solute capacity allowing improved space time yields [106]

17.3 Conclusion

The use of molecular solvents as biphasic eluents has been shown to be excellent for a huge variety of applications in laboratory scale preparations of up to kilo or multi-kilo of target. But the complexity of molecular solvent biphasic eluents with three, four or on occasions five solvents, becomes less favored in larger-scale processes. For these larger-scale process applications, of tens of kilos to many tons per annum, the unique properties of ionic liquids can provide many advantages, which we have discussed in the chapter. This can make the additional effort of developing ionic liquid eluents well worthwhile. The use of ionic liquids in countercurrent chromatography instruments was initially found to be problematic due to high backpressures and consequent low mobile phase flow rates. With careful design of the fluid flow paths and by eliminating pressure bottlenecks, this problem has now been solved. Currently there are very few papers in the literature describing the use of solvent systems containing a high percentage of ionic liquids in CCC or LLC. Control over the structure and design of ionic liquids and ionic liquid containing solvent systems (solvent engineering) has enabled us to select or alter solute distribution ratios of solutes that are to be separated.

Peptides, proteins, monoclonal antibodies (mAbs), enzymes, bioactive compounds, precious metals, actinides and lanthanides, are all ideal candidates for ionic liquid CCC/CPC separation processes, as all are high value, but can be potentially difficult to purify cost-effectively from their start crude matrices. Ionic liquid-CCC/CPC/CCE/CPE science and research are likely going to be driven by commercial applications rather than in the academic arenas. Research groups are already in place in the USA and Europe to develop the processes necessary for very large-scale process applications.

Acknowledgements

We would like to express our gratitude to Invest NI for the Collaborative Grants for R&D: RD0210506 and RD1111867, the EPSRC, and the QUILL Research Centre for funding this work. We would also like to thank Professor Paul Davey and Givaudan for useful discussions.

References

1. Snyder, L. R., Kirkland, J. J. (1979). *Introduction to Modern Liquid Chromatography*, 2nd ed., John Wiley & Sons Inc., New York.
2. Conway, W. D., Petroski, R. J. (1995). *Modern Countercurrent Chromatography,* American Chemical Society, Washington, DC.
3. Degenhardt, A., Winterhalter, P. (2001). HSCCC–A powerful tool for the preparative isolation of bioactive compounds, Pfannhauser W, Fenwick GR, Khokhar S in *Biologically-Active Phytochemicals in Food*, eds. Pfannhauser W, Fenwick GR, Khokhar S, pp. 143–146.
4. Berthod, A. (2009). Countercurrent Chromatography: From the Milligram to the Kilogram, Grushka E, Grinberg N in *Advances in Chromatography, Vol 47*, eds. Grushka E, Grinberg N, CRC Press-Taylor & Francis Group, Boca Raton, pp. 323–352.
5. Cazes, J. (2009). *Encyclopedia of Chromatography*, 3rd ed., Taylor & Francis, Boca Raton.
6. Sutherland, I. A. (2010). Industrial applications of CCC, Cazes J in *Encyclopedia of Chromatography*, eds. Cazes J, 3 ed., Taylor & Francis, Boca Raton, pp. 2116–2119.
7. Berthod, A., Carda-Broch, S. (2003). A new class of solvents for CCC: Room temperature ionic liquids, *J. Liq. Chromatogr. Relat. Technol.*, **26**, pp. 1493–1508.
8. Berthod, A., Carda-Broch, S. (2004). Use of the ionic liquid 1-butyl-3-methylimidazolium hexafluorophosphate in countercurrent chromatography, *Anal. Bioanal. Chem.*, **380**, pp. 168–177.

9. Berthod, A., Ruiz-Ángel, M. J., Huguet, S. (2005). Nonmolecular solvents in separation methods: Dual nature of room temperature ionic liquids, *Anal. Chem.*, **77**, pp. 4071–4080.
10. Ruiz-Ángel, M. J., Pino, V., Carda-Broch, S., Berthod, A. (2007). Solvent systems for countercurrent chromatography: An aqueous two phase liquid system based on a room temperature ionic liquid, *J. Chromatogr. A*, **1151**, pp. 65–73.
11. Berthod, A., Ruiz-Ángel, M., Carda-Broch, S. (2008). Ionic liquids in separation techniques, *J. Chromatogr. A*, **1184**, pp. 6–18.
12. Berthod, A., Ruiz-Ángel, M.-J., Carda-Broch, S. (2009). Ionic Liquids as Stationary Phases in Countercurrent Chromatography, Koel M in *Ionic Liquids in Chemical Analysis*, eds. Koel M, CRC Press, Boca Raton, pp. 211–227.
13. Fan, C., Li, N., Cao, X. L. (2015). Determination of chlorophenols in red wine using ionic liquid countercurrent chromatography as a new pretreatment method followed by high-performance liquid chromatography, *J. Sep. Sci.*, **38**, pp. 2109–2116.
14. Fan, C., Cao, X. L., Liu, M., Wang, W. (2016). Determination of Alternaria mycotoxins in wine and juice using ionic liquid modified countercurrent chromatography as a pretreatment method followed by high-performance liquid chromatography, *J. Chromatogr. A*, **1436**, pp. 133–140.
15. Brown, L., Earle, M. J., Gîlea, M. A., Plechkova, N. V., Seddon, K. R. (2017). Ionic liquid-liquid separations using countercurrent chromatography: A new general purpose separation methodology, *Aust. J. Chem.*, **70**, pp. 923–932.
16. Müller, M., Englert, M., Earle, M. J., Vetter, W. (2017). Development of solvent systems with room temperature ionic liquids for the countercurrent chromatographic separation of very nonpolar lipid compounds, *J. Chromatogr. A*, **1488**, pp. 68–76.
17. Roehrer, S., Bezold, F., Garcia, E. M., Minceva, M. (2016). Deep eutectic solvents in countercurrent and centrifugal partition chromatography, *J. Chromatogr. A*, **1434**, pp. 102–110.
18. Ma, S. F., Hu, L. M., Ma, C. Y., Lv, W. P., Wang, H. X. (2014). Application and recovery of ionic liquids in the preparative separation of four flavonoids from Rhodiola rosea by on-line three-dimensional liquid chromatography, *J. Sep. Sci.*, **37**, pp. 2314–2321.
19. Cai, M., Su, J., Jin, M., Song, Z. (2015). Extraction of triptolide and tripdiolide from Tripterygium wilfordii using ionic liquid. Patent CN104710502A.

20. Brown, L., Earle, M. J., Gîlea, M. A., Plechkova, N. V., Seddon, K. R. (2017). Ionic liquid-liquid chromatography: A new general purpose separation methodology, *Top. Curr. Chem.*, **74**, pp. 1–41.

21. Earle, M. J., Seddon, K. R., Self, R., Brown, L. (2013). Ionic Liquid Separations. Patent WO2013121218A1.

22. Earle, M. J., Gilea, M. A. (2013). Lentinan extraction process from mushrooms using ionic liquid. Patent WO2013140185A1.

23. Earle, M., Seddon, K. (2013). Ionic liquid separations. Patent WO2013121219A1.

24. Earle, M. J., Seddon, K. R. (2013). Ionic Liquid Separations. Patent WO2013121220A1.

25. Foreiter, M. B., Gunaratne, H. Q. N., Nockemann, P., Seddon, K. R., Srinivasan, G. (2014). Novel chiral ionic liquids: physicochemical properties and investigation of the internal rotameric behaviour in the neat system, *Phys. Chem. Chem. Phys.*, **16**, pp. 1208–1226.

26. Payagala, T., Armstrong, D. W. (2012). Chiral ionic liquids: A compendium of syntheses and applications (2005–2012), *Chirality*, **24**, pp. 17–53.

27. Franco, P., Blanc, J., Oberleitner, W. R., Maier, N. M., Lindner, W., Minguillon, C. (2002). Enantiomer separation by countercurrent chromatography using cinchona alkaloid derivatives as chiral selectors, *Anal. Chem.*, **74**, pp. 4175–4183.

28. Wang, S. S., Han, C., Wang, S. S., Bai, L. J., Li, S. S., Luo, J. G., Kong, L. Y. (2016). Development of a high speed counter-current chromatography system with Cu(II)-chiral ionic liquid complexes and hydroxypropyl-beta-cyclodextrin as dual chiral selectors for enantioseparation of naringenin, *J. Chromatogr. A*, **1471**, pp. 155–163.

29. Willauer, H. D., Huddleston, J. G., Rogers, R. D. (2002). Solvent properties of aqueous biphasic systems composed of polyethylene glycol and salt characterized by the free energy of transfer of a methylene group between the phases and by a linear solvation energy relationship, *Ind. Eng. Chem. Res.*, **41**, pp. 2591–2601.

30. Ishii, K., Tanaka, Y., Hata, K., Goto, M., Saitoh, K., Minamisawa, H., Shibukawa, M. (2004). Separation of inorganic ions by high-speed countercurrent chromatography with an aqueous biphasic system, *Bunseki Kagaku*, **53**, pp. 911–917.

31. Freire, M. G., Claudio, A. F. M., Araujo, J. M. M., Coutinho, J. A. P., Marrucho, I. M., Lopes, J. N. C., Rebelo, L. P. N. (2012). Aqueous biphasic systems: a boost brought about by using ionic liquids, *Chem. Soc. Rev.*, **41**, pp. 4966–4995.

32. Ina, K., Furuta, R., Kataoka, T., Kayukawa, S., Yoshida, T., Miwa, T., Yamamura, Y., Takeuchi, Y. (2011). Lentinan prolonged survival in patients with gastric cancer receiving S-1-based chemotherapy, *World Journal of Clinical Oncology*, **2**, pp. 339–343.

33. Xu, X. J., Wang, X. H., Cai, F., Zhang, L. N. (2010). Renaturation of triple helical polysaccharide lentinan in water-diluted dimethylsulfoxide solution, *Carbohydr. Res.*, **345**, pp. 419–424.

34. Ina, K., Kataoka, T., Ando, T. (2013). The Use of Lentinan for Treating Gastric Cancer, *Anti-Cancer Agents in Medicinal Chemistry*, **13**, pp. 681–688.

35. Chihara, G., Hamuro, J., Maeda, Y. Y., Arai, Y., Fukuoka, F. (1970). Fractionation and purification of the polysaccharides with marked antitumor activity, especially lentinan, from Lentinus edodes (Berk.) Sing.(an edible mushroom), *Cancer research*, **30**, pp. 2776–2781.

36. Xu, X. J., Zhang, X. F., Zhang, L. N., Wu, C. (2004). Collapse and association of denatured lentinan in water/dimethlysulfoxide solutions, *Biomacromolecules*, **5**, pp. 1893–1898.

37. Jiang, Z. G., Du, Q. Z., Sheng, L. Y. (2009). Separation and purification of lentinan by preparative high speed counter current chromatography, *Chin. J. Anal. Chem.*, **37**, pp. 412–416.

38. Hughes, K., Paisley, N. (2015). Hughes Mushrooms Ltd.

39. Earle, M. J., McCormac, P. B., Seddon, K. R. (1998). Regioselective alkylation in ionic liquids, *Chem. Commun.*, pp. 2245–2246.

40. Carmichael, A. J., Earle, M. J., Holbrey, J. D., McCormac, P. B., Seddon, K. R. (1999). The Heck reaction in ionic liquids: A multiphasic catalyst system, *Org. Lett.*, **1**, pp. 997–1000.

41. Earle, M. J., McCormac, P. B., Seddon, K. R. (2000). The first high yield green route to a pharmaceutical in a room temperature ionic liquid, *Green Chem.*, **2**, pp. 261–262.

42. Holbrey, J. D., Seddon, K. R. (1999). The phase behaviour of 1-alkyl-3-methylimidazolium tetrafluoroborates, ionic liquids and ionic liquid crystals, *J. Chem. Soc.-Dalton Trans.*, pp. 2133–2139.

43. Seddon, K. R., Stark, A., Torres, M. J. (2002). Viscosity and density of 1-alkyl-3-methylimidazolium ionic liquids, Abraham MA, Moens L in *Clean Solvents: Alternative Media for Chemical Reactions and Processing*, eds. Abraham MA, Moens L, Amer. Chem. Soc., Washington, pp. 34–49.

44. Earle, M. J., Seddon, K. R. (2000). Ionic liquids. Green solvents for the future, *Pure Appl. Chem.*, **72**, pp. 1391–1398.

45. Earle, M. J. (2002). Ionic liquids: Solvents for the twenty-first century, Rogers RD, Seddon KR in *Ionic Liquids: Industrial Applications for Green Chemistry*, eds. Rogers RD, Seddon KR, Amer. Chem. Soc., Washington, pp. 90–105.
46. Earle, M. J., Seddon, K. R. (2002). Clean synthesis in ionic liquids, Trulove PC, DeLong HC, Mantz RA, Stafford GR, Matsunaga M in *Molten Salts Xiii*, eds. Trulove PC, DeLong HC, Mantz RA, Stafford GR, Matsunaga M, Electrochemical Soc Inc, Pennington, pp. 177–189.
47. Freemantle, M. (2010). *An Introduction to Ionic Liquids* Royal Society of Chemistry, Cambridge, UK.
48. Wasserscheid, P., Welton, T. (2008). *Ionic Liquids in Synthesis*, 2nd ed., Wiley-VCH, Weinheim, Germany.
49. Earle, M. J., Esperança, J. M. S. S., Gîlea, M. A., Lopes, J. N. C., Rebelo, L. P. N., Magee, J. W., Seddon, K. R., Widegren, J. A. (2006). The distillation and volatility of ionic liquids, *Nature*, **439**, pp. 831–834.
50. Kim, H.-T., Kang, J., Mun, J., Oh, S. M., Yim, T., Kim, Y. G. (2016). Pyrrolinium-based ionic liquid as a flame retardant for binary electrolytes of lithium ion batteries, *ACS Sustain. Chem. Eng.*, **4**, pp. 497–505.
51. Earle, M. J., Wasserscheid, P., Schulz, P. C., Olivier-Bourbigou, H., Favre, F., Vaultier, M., Kirschning, A., Singh, V., Riisager, A., Fehrmann, R., Kuhlmann, S. (2008). Organic Synthesis, Welton T, Wasserscheid P in *Ionic Liquids in Synthesis*, eds. Welton T, Wasserscheid P, 2nd ed., Wiley-VCH, Weinheim, pp. 265–568.
52. Arce, A., Earle, M. J., Rodriguez, H., Seddon, K. R. (2007). Separation of aromatic hydrocarbons from alkanes using the ionic liquid 1-ethyl-3-methylimidazolium bis{(trifluoromethyl) sulfonyl} amide, *Green Chem.*, **9**, pp. 70-74.
53. Plechkova, N. V., Seddon, K. R. (2008). Applications of ionic liquids in the chemical industry, *Chem. Soc. Rev.*, **37**, pp. 123–150.
54. Earle, M. J., Katdare, S. P., Seddon, K. R. (2004). Paradigm confirmed: The first use of ionic liquids to dramatically influence the outcome of chemical reactions, *Org. Lett.*, **6**, pp. 707–710.
55. Arce, A., Earle, M. J., Katdare, S. P., Rodriguez, H., Seddon, K. R. (2006). Mutually immiscible ionic liquids, *Chem. Commun.*, pp. 2548–2550.
56. Martins, V. L., Nicolau, B. G., Urahata, S. M., Ribeiro, M. C. C., Torresi, R. M. (2013). Influence of the Water Content on the Structure and Physicochemical Properties of an Ionic Liquid and Its Li+ Mixture, *J. Phys. Chem. B*, **117**, pp. 8782–8792.
57. Wagner, M., Stanga, I., Schroer, W. (2004). Critical viscosity near the liquid-liquid phase transition in the solution of the ionic liquid 1-

methyl-3-hexylimidazolium tetrafluoroborate in 1-pentanol, *Phys. Chem. Chem. Phys.*, **6**, pp. 1750–1757.

58. Seddon, K. R., Stark, A., Torres, M. J. (2000). Influence of chloride, water, and organic solvents on the physical properties of ionic liquids, *Pure Appl. Chem.*, **72**, pp. 2275–2287.
59. Neves, C., Carvalho, P. J., Freire, M. G., Coutinho, J. A. P. (2011). Thermophysical properties of pure and water-saturated tetradecyltrihexylphosphonium-based ionic liquids, *J. Chem. Thermodyn.*, **43**, pp. 948–957.
60. CamachoFrias, E., Foucault, A. (1996). Solvent systems in centrifugal partition chromatography, *Analusis*, **24**, pp. 159–167.
61. Sutherland, I. A. (2007). Review of centrifugal liquid-liquid chromatography using aqueous two-phase solvent systems: Its scale-up and prospects for the future production of high-value biologicals, *Curr. Opin. Drug Discov. Dev.*, **10**, pp. 540–549.
62. Berthod, A., Hassoun, M., Ruiz-Angel, M. J. (2005). Alkane effect in the Arizona liquid systems used in countercurrent chromatography, *Anal. Bioanal. Chem.*, **383**, pp. 327–340.
63. Chadwick, L. R., Fong, H. H. S., Farnsworth, N. R., Pauli, G. F. (2005). CCC sample cutting for isolation of prenylated phenolics from hops, *J. Liq. Chromatogr. Relat. Technol.*, **28**, pp. 1959–1969.
64. Dubant, S., Mathews, B., Higginson, P., Crook, R., Snowden, M., Mitchell, J. (2008). Practical solvent system selection for counter-current separation of pharmaceutical compounds, *J. Chromatogr. A*, **1207**, pp. 190–192.
65. Brown, L., Luu, T. A. (2010). An introduction to discussions on liquid-liquid/counter current/centrifugal partition chromatography biphasic solvent selection methodologies and instrumentation for target compound preparation from complex matrices, *J. Sep. Sci.*, **33**, pp. 999–1003.
66. Liang, J. L., Meng, J., Wu, D. F., Guo, M. Z., Wu, S. H. (2015). A novel 9 x 9 map-based solvent selection strategy for targeted counter-current chromatography isolation of natural products, *J. Chromatogr. A*, **1400**, pp. 27–39.
67. Carraher Jr., C. E. (2017). *Introduction to Polymer Chemistry*, 4th ed., CRC press, Boca Raton, FL.
68. Fleer, G. J., Skvortsov, A. M. (2005). Theory for concentration and solvency effects in size-exclusion chromatography of polymers, *Macromolecules*, **38**, pp. 2492–2505.

69. Kohlmann, C., Leuchs, S., Greiner, L., Leitner, W. (2011). Continuous biocatalytic synthesis of (R)-2-octanol with integrated product separation, *Green Chem.*, **13**, pp. 1430–1436.

70. Dolan, J. W. (2010). Selectivity in reversed-phase LC separations, Part I: Solvent-type selectivity, *LCGC North America*, **28**, pp. 1022–1027.

71. Lesellier, E. (2015). Sigma pider diagram: A universal and versatile approach for system comparison and classification Application to solvent properties, *J. Chromatogr. A*, **1389**, pp. 49–64.

72. Tan, X., Tan, D. M., Brown, L., Yang, D. P., Zhu, L. P., Chen, Y. L., Chang, D. H. T., Wang, D. M. (2013). Isolation and purification of four bioactive constituents from *alpinia officinarum hance* utilizing high-speed counter-current chromatography, *J. Liq. Chromatogr. Relat. Technol.*, **36**, pp. 1355–1365.

73. Pellett, J., Lukulay, P., Mao, Y., Bowen, W., Reed, R., Ma, M., Munger, R. C., Dolan, J. W., Wrisley, L., Medwid, K., Toltl, N. P., Chan, C. C., Skibic, M., Biswas, K., Wells, K. A., Snyder, L. R. (2006). "Orthogonal" separations for reversed-phase liquid chromatography, *J. Chromatogr. A*, **1101**, pp. 122–135.

74. Kamath, G., Bhatnagar, N., Baker, G. A., Baker, S. N., Potoff, J. J. (2012). Computational prediction of ionic liquid 1-octanol/water partition coefficients, *Phys. Chem. Chem. Phys.*, **14**, pp. 4339–4342.

75. Soskic, M., Plavsic, D. (2005). Modeling the octanol-water partition coefficients by an optimized molecular connectivity index, *J. Chem Inf. Model.*, **45**, pp. 930–938.

76. Sedykh, A. Y., Klopman, G. (2006). A structural analogue approach to the prediction of the octanol-water partition coefficient, *J. Chem Inf. Model.*, **46**, pp. 1598–1603.

77. Xie, Y. J., Liu, H., Liu, H. X., Zhai, Z. C., Wang, Z. Y. (2008). Determination of solubilities and n-Octanol/Water partition coefficients and QSPR study for substituted phenols, *Bull. Environ. Contam. Toxicol.*, **80**, pp. 319–323.

78. Hsieh, C. M., Lin, S. T. (2009). Prediction of 1-octanol-water partition coefficient and infinite dilution activity coefficient in water from the PR plus COSMOSAC model, *Fluid Phase Equilib.*, **285**, pp. 8–14.

79. Souza, E. S., Zaramello, L., Kuhnen, C. A., Junkes, B. D., Yunes, R. A., Heinzen, V. E. F. (2011). Estimating the octanol/water partition coefficient for aliphatic organic compounds using semi-empirical electrotopological index, *Int. J. Mol. Sci.*, **12**, pp. 7250–7264.

80. Bhatnagar, N., Kamath, G., Chelst, I., Potoff, J. J. (2012). Direct calculation of 1-octanol-water partition coefficients from adaptive biasing force molecular dynamics simulations, *J. Chem. Phys.*, **137**, pp. 014502.
81. Garrard, I. J. (2005). Simple approach to the development of a CCC solvent selection protocol suitable for automation, *J. Liq. Chromatogr. Relat. Technol.*, **28**, pp. 1923–1935.
82. Kukula-Koch, W., Koch, W., Angelis, A., Halabalaki, M., Aligiannis, N. (2016). Application of pH-zone refining hydrostatic countercurrent chromatography (hCCC) for the recovery of antioxidant phenolics and the isolation of alkaloids from Siberian barberry herb, *Food Chem.*, **203**, pp. 394–401.
83. Kotland, A., Chollet, S., Autret, J. M., Diard, C., Marchal, L., Renault, J. H. (2015). Modeling pH-zone refining countercurrent chromatography: A dynamic approach, *J. Chromatogr. A*, **1391**, pp. 80–87.
84. Vieira, M. N., Leitao, S. G., Porto, P. C. C., Oliveira, D. R., Pinto, S. C., Braz Filho, R., Leitao, G. G. (2013). Application of pH-zone-refining countercurrent chromatography for the separation of indole alkaloids from Aspidosperma rigidum Rusby, *J. Chromatogr. A*, **1319**, pp. 166–171.
85. Wang, T., Jiang, X., Yang, L., Wu, S. (2008). pH-gradient counter-current chromatography isolation of natural antioxidant chlorogenic acid from Lonicera japonica Thumb. using an upright coil planet centrifuge with three multi-layer coils connected in series, *J. Chromatogr. A*, **1180**, pp. 53–58.
86. Ito, Y. (2013). pH-zone-refining counter-current chromatography: Origin, mechanism, procedure and applications, *J. Chromatogr. A*, **1271**, pp. 71–85.
87. Connolly, B. A. (1987). The synthesis of oligonucleotides containing a primary amino group at the 5’-terminus, *Nucleic acids research*, **15**, pp. 3131–3139.
88. Wang, R., Chang, Y., Tan, Z., Li, F. (2016). Applications of choline amino acid ionic liquid in extraction and separation of flavonoids and pectin from ponkan peels, *Sep. Sci. Technol.*, **51**, pp. 1093–1102.
89. Yeh, C. S., Yu, T., Berthod, A. (1999). Separation of steroids by countercurrent chromatography using supercritical fluid carbon dioxide, *J. Liq. Chromatogr. Relat. Technol.*, **22**, pp. 345–356.
90. Yu, T., Chang, J. Y. (2001). Measurements of partition coefficients in water-1,1,1,2-tetrafluoroethane by countercurrent chromatography, *J. Liq. Chromatogr. Relat. Technol.*, **24**, pp. 1607–1615.

91. Mai, N. L., Ahn, K., Koo, Y. M. (2014). Methods for recovery of ionic liquids-A review, *Process Biochem.*, **49**, pp. 872–881.

92. Faure, K., Bouju, E., Doby, J., Berthod, A. (2014). Limonene in Arizona liquid systems used in countercurrent chromatography. II Polarity and stationary-phase retention, *Anal. Bioanal. Chem.*, **406**, pp. 5919–5926.

93. Friesen, J. B., Pauli, G. F. (2005). GUESS - A generally useful estimate of solvent systems in CCC, *J. Liq. Chromatogr. Relat. Technol.*, **28**, pp. 2777–2806.

94. Earle, M. J., Gordon, C. M., Plechkova, N. V., Seddon, K. R., Welton, T. (2007). Decolorization of ionic liquids for spectroscopy, *Anal. Chem.*, **79**, pp. 758–764.

95. Costa, F. D., Leitao, G. G. (2010). Strategies of solvent system selection for the isolation of flavonoids by countercurrent chromatography, *J. Sep. Sci.*, **33**, pp. 336–347.

96. Hopmann, E., Arlt, W., Minceva, M. (2011). Solvent system selection in counter-current chromatography using conductor-like screening model for real solvents, *J. Chromatogr. A*, **1218**, pp. 242–250.

97. Yin, L. H., Xu, L. N., Wang, X. N., Lu, B. N., Liu, Y. T., Peng, J. Y. (2010). An economical method for isolation of dioscin from dioscorea nipponica makino by HSCCC coupled with ELSD, and a computer-aided UNIFAC mathematical model, *Chromatographia*, **71**, pp. 15–23.

98. Berthod, A., Friesen, J. B., Inui, T., Pauli, G. F. (2007). Elution-extrusion countercurrent chromatography: Theory and concepts in metabolic analysis, *Anal. Chem.*, **79**, pp. 3371–3382.

99. Cocalia, V. A., Visser, A. E., Rogers, R. D., Holbrey, J. D. (2008). Physicochemical Properties, Wasserscheid P, Welton T in *Ionic Liquids in Synthesis*, eds. Wasserscheid P, Welton T, 2nd ed., Wiley-VCH, Weinheim, pp. 89–102.

100. Domanska, U., Wlazlo, M. (2016). Thermodynamics and limiting activity coefficients measurements for organic solutes and water in the ionic liquid 1-dodecyl-3-methylimidzolium bis(trifluoromethylsu lfonyl) imide, *J. Chem. Thermodyn.*, **103**, pp. 76–85.

101. Clough, M. T., Crick, C. R., Grasvik, J., Hunt, P. A., Niedermeyer, H., Welton, T., Whitaker, O. P. (2015). A physicochemical investigation of ionic liquid mixtures, *Chem. Sci.*, **6**, pp. 1101–1114.

102. Tariq, M., Shimizu, K., Esperança, J. M. S. S., Lopes, J. N. C., Rebelo, L. P. N. (2015). Viscosity minima in binary mixtures of ionic liquids plus molecular solvents, *Phys. Chem. Chem. Phys.*, **17**, pp. 13480–13494.

103. Zhou, T., Chen, L., Ye, Y., Chen, L., Qi, Z., Freund, H., Sundmacher, K. (2012). An overview of mutual solubility of ionic liquids and water predicted by COSMO-RS, *Ind. Eng. Chem. Res.*, **51**, pp. 6256–6264.

104. Liu, W., Cheng, L., Zhang, Y., Wang, H., Yu, M. (2008). The physical properties of aqueous solution of room-temperature ionic liquids based on imidazolium: Database and evaluation, *J. Mol. Liq.*, **140**, pp. 68–72.

105. Zhang, S. J., Sun, N., He, X. Z., Lu, X. M., Zhang, X. P. (2006). Physical properties of ionic liquids: Database and evaluation, *J. Phys. Chem. Ref. Data*, **35**, pp. 1475–1517.

106. Pereiro, A. B., Araujo, J. M. M., Teixeira, F. S., Marrucho, I. M., Pineiro, M. M., Rebelo, L. P. N. (2015). Aggregation behavior and total miscibility of fluorinated ionic liquids in water, *Langmuir*, **31**, pp. 1283–1295.

107. NIST Standard Reference Database #147, (2016) http://ilthermo.boulder.nist.gov/index.html (verified 19-12-2016).

108. Atashrouz, S., Zarghampour, M., Abdolrahimi, S., Pazuki, G., Nasernejad, B. (2014). Estimation of the viscosity of ionic liquids containing binary mixtures based on the Eyring's theory and a modified Gibbs energy model, *J. Chem. Eng. Data*, **59**, pp. 3691–3704.

109. Liu, W. W., Cheng, L. Y., Zhang, Y. M., Wang, H. P., Yu, M. F. (2008). The physical properties of aqueous solution of room-temperature ionic liquids based on imidazolium: Database and evaluation, *J. Mol. Liq.*, **140**, pp. 68–72.

110. Arce, A., Earle, M. J., Rodriguez, H., Seddon, K. R., Soto, A. (2010). Isomer effect in the separation of octane and xylenes using the ionic liquid 1-ethyl-3-methylimidazolium bis{(trifluoromethyl)sulfonyl} amide, *Fluid Phase Equilib.*, **294**, pp. 180–186.

111. Arce, A.; Earle, M. J., Rodriguez, H., Seddon, K. R., Soto, A. (2009). Bis{(trifluoromethyl)sulfonyl}amide ionic liquids as solvents for the extraction of aromatic hydrocarbons from their mixtures with alkanes: effect of the nature of the cation, *Green Chem.*, **11**, pp. 365–372.

112. Arce, A., Earle, M. J., Rodriguez, H., Seddon, K. R., Soto, A. (2008). 1-Ethyl-3-methylimidazolium bis{(trifluoromethyl)sulfonyl}amide as solvent for the separation of aromatic and aliphatic hydrocarbons by liquid extraction–extension to C-7- and C-8-fractions, *Green Chem.*, **10**, pp. 1294–1300.

113. Arce, A., Earle, M. J., Rodriguez, H., Seddon, K. R. (2007). Separation of benzene and hexane by solvent extraction with 1-alkyl-3-

methylimidazolium bis{(trifluoromethyl)sulfonyl}amide ionic liquids: Effect of the alkyl-substituent length, *J. Phys. Chem. B*, **111**, pp. 4732–4736.

114. Arce, A., Earle, M. J., Katdare, S. P., Rodriguez, H., Seddon, K. R. (2007). Phase equilibria of mixtures of mutually immiscible ionic liquids, *Fluid Phase Equilib.*, **261**, pp. 427–433.

115. Mantz, R. A., Trulove, P. C. (2008). Physicochemical Properties, Wasserscheid P, Welton T in *Ionic Liquids in Synthesis*, eds. Wasserscheid P, Welton T, 2nd ed., Wiley-VCH, Weinheim, pp. 72–88.

116. Abullah, M. M. S., Al-Lohedan, H. A., Attah, A. M. (2016). Synthesis and application of amphiphilic ionic liquid based on acrylate copolymers as demulsifier and oil spill dispersant, *J. Mol. Liq.*, **219**, pp. 54–62.

117. Hezave, A. Z., Dorostkar, S., Ayatollahi, S., Nabipour, M., Hemmateenejad, B. (2013). Investigating the effect of ionic liquid 1-dodecyl-3-methylimidazolium chloride ([C_{12}mim]Cl) on the water/oil interfacial tension as a novel surfactant, *Colloid Surf. A-Physicochem. Eng. Asp.*, **421**, pp. 63–71.

Index

For Product Safety Concerns and Information please contact our EU representative GPSR@taylorandfrancis.com Taylor & Francis Verlag GmbH, Kaufingerstraße 24, 80331 München, Germany

Batch number: 10777142

Printed by Printforce, the Netherlands